METHODS IN ENZYMOLOGY

Editors-in-Chief

ANNA MARIE PYLE
Departments of Molecular, Cellular and Developmental Biology and Department of Chemistry
Investigator, Howard Hughes Medical Institute
Yale University

DAVID W. CHRISTIANSON
Roy and Diana Vagelos Laboratories
Department of Chemistry
University of Pennsylvania
Philadelphia, PA

Founding Editors

SIDNEY P. COLOWICK and NATHAN O. KAPLAN

VOLUME SIX HUNDRED AND THIRTY ONE

METHODS IN ENZYMOLOGY

Tumor Immunology and Immunotherapy — Cellular Methods Part A

VOLUME SIX HUNDRED AND THIRTY ONE

METHODS IN ENZYMOLOGY

Tumor Immunology and Immunotherapy — Cellular Methods Part A

Edited by

LORENZO GALLUZZI

Department of Radiation Oncology, Weill Cornell Medical College; Sandra and Edward Meyer Cancer Center, Caryl and Israel Englander Institute for Precision Medicine New York, NY; Department of Dermatology, Yale University School of Medicine, New Haven, CT, United States; Université de Paris, Paris, France

NILS-PETTER RUDQVIST

Department of Radiation Oncology, Weill Cornell Medical College, New York, NY, United States

ELSEVIER

ACADEMIC PRESS

An imprint of Elsevier

Academic Press is an imprint of Elsevier
50 Hampshire Street, 5th Floor, Cambridge, MA 02139, United States
525 B Street, Suite 1650, San Diego, CA 92101, United States
The Boulevard, Langford Lane, Kidlington, Oxford OX5 1GB, United Kingdom
125 London Wall, London, EC2Y 5AS, United Kingdom

First edition 2020

ISBN: 978-0-12-818673-2
ISSN: 0076-6879

For information on all Academic Press publications
visit our website at https://www.elsevier.com/books-and-journals

Publisher: Zoe Kruze
Acquisition Editor: Ashlie M. Jackman
Editorial Project Manager: Shellie Bryant
Production Project Manager: James Selvam
Cover Designer: Greg Harris

Typeset by SPi Global, India

Contents

Contributors

Mohsen Abu-Akeel
Ludwig Collaborative and Swim Across America Laboratory, Memorial Sloan Kettering Cancer Center, New York, NY, United States

Abdulaziz Alamri
Department of Immunology, University of Manitoba, Winnipeg, MB, Canada

Mahmoud S. Alghamri
Department of Neurosurgery; Department of Cell and Developmental Biology, University of Michigan Medical School, Ann Arbor, MI, United States

David Altshuler
Department of Neurosurgery; Department of Cell and Developmental Biology, University of Michigan Medical School, Ann Arbor, MI, United States

Maite Alvarez
Program of Immunology and Immunotherapy, CIMA Universidad de Navarra; Navarra Institute for Health Research (IDISNA), Pamplona; Centro de Investigación Biomédica en Red de Cáncer (CIBERONC), Madrid, Spain

Alexandre Bayard
De Duve Institute, Université catholique de Louvain, Brussels, Belgium

Pedro Berraondo
Program of Immunology and Immunotherapy, CIMA Universidad de Navarra; Navarra Institute for Health Research (IDISNA), Pamplona; Centro de Investigación Biomédica en Red de Cáncer (CIBERONC), Madrid, Spain

Jay A. Berzofsky
Vaccine Branch, Center for Cancer Research, National Cancer Institute, National Institutes of Health, Bethesda, MD, United States

Nicole Bidmon
Translational Oncology at the University Medical Center of the Johannes-Gutenberg University Mainz (TRON gGmbH), Mainz, Germany

Francisco Borrego
Biocruces Bizkaia Health Research Institute, Immunopathology Group, Barakaldo; Ikerbasque, Basque Foundation for Science, Bilbao, Spain

Annika M. Bruger
De Duve Institute, Université catholique de Louvain, Brussels, Belgium

Sadna Budhu
Ludwig Collaborative and Swim Across America Laboratory, Memorial Sloan Kettering Cancer Center, New York, NY, United States

Kyle Burrows
Department of Immunology, University of Toronto, Toronto, ON, Canada

Anne Caignard
INSERM, U1160, Institut de Recherche Saint-Louis, Hôpital Saint-Louis, Paris, France

Giulia Campo
Unit of Tumor Immunology and Immunotherapy, Department of Research, Advanced Diagnostics and Technological Innovation, IRCCS Regina Elena National Cancer Institute, Rome, Italy

Jordan J. Cardenas
Department of Immunobiology, Yale School of Medicine, New Haven, CT, United States

Stephen Carney
Department of Neurosurgery; Department of Cell and Developmental Biology, University of Michigan Medical School, Ann Arbor, MI, United States

Maria G. Castro
Department of Neurosurgery; Department of Cell and Developmental Biology, University of Michigan Medical School, Ann Arbor, MI, United States

Maria Teresa Catanese
Nouscom S.R.L., Rome, Italy

Pailin Chiaranunt
Department of Immunology, University of Toronto, Toronto, ON, Canada

Didier Colau
De Duve Institute, Université catholique de Louvain, Brussels, Belgium

George Coukos
Ludwig Institute for Cancer Research, University of Lausanne; Department of Oncology, University Hospital of Lausanne, Lausanne, Switzerland

Isabelle Cremer
Team Inflammation, Complement and Cancer, INSERM, U1138, Centre de Recherche des Cordeliers; Sorbonne Universite de Paris, Paris, France

Ralf Dressel
Institute of Cellular and Molecular Immunology, University Medical Center Göttingen; DZHK (German Center for Cardiovascular Research), Partner Site Göttingen, Göttingen, Germany

Leslie Elsner
Institute of Cellular and Molecular Immunology, University Medical Center Göttingen, Göttingen, Germany

Leisha A. Emens
Department of Medicine, UPMC Hillman Cancer Center, and UPMC Magee Women's Hospital, University of Pittsburgh, Pittsburgh, PA, United States

Jeremy B. Foote
Department of Microbiology, University of Alabama at Birmingham, Birmingham, AL, United States

Alexandra Frazao
INSERM, U1160, Institut de Recherche Saint-Louis, Hôpital Saint-Louis, Paris, France

Jitka Fucikova
Department of Immunology, Charles University, 2nd Faculty of Medicine and University Hospital Motol; Sotio, Prague, Czech Republic

Silvia A. Fuertes Marraco
Department of Oncology, University of Lausanne, Lausanne, Switzerland

Claudia Galassi
Istituto di Patologia Generale, Università Cattolica del Sacro Cuore, Rome, Italy

Alejandra Gomez-Cadena
Department of Oncology UNIL CHUV; Ludwig Institute for Cancer Research, University of Lausanne, Lausanne, Switzerland

Segundo González
Departamento de Biología Funcional, Área de Inmunología, Universidad de Oviedo; Instituto Universitario de Oncología del Principado de Asturias (IUOPA); Instituto de Investigación Sanitaria del Principado de Asturias (ISPA), Oviedo, Spain

Monica Gordon-Alonso
De Duve Institute, Université catholique de Louvain, Brussels, Belgium; F-star Biotechnology, Babraham Research Campus, Cambridge, United Kingdom

Cécile Gouttefangeas
Department of Immunology, Interfaculty Institute for Cell Biology, University of Tuebingen, Tuebingen; German Cancer Consortium (DKTK), German Cancer Research Center (DKFZ) Partner Site Tübingen, Tübingen, Germany

Philippe Guillaume
Ludwig Institute for Cancer Research, University of Lausanne; Department of Oncology, University Hospital of Lausanne, Lausanne, Switzerland

Alexandre Harari
Ludwig Institute for Cancer Research, University of Lausanne; Department of Oncology, University Hospital of Lausanne, Lausanne, Switzerland

Jolly Hipolito
Department of Physics and Astronomy; Department of Immunology, University of Manitoba, Winnipeg, MB, Canada

Peter Holicek
Department of Immunology, Charles University, 2nd Faculty of Medicine and University Hospital Motol; Sotio, Prague, Czech Republic

Monika Holubova
Biomedical Center, Medical Faculty in Pilsen, Charles University, Pilsen, Czech Republic

Camilla Jandus
Department of Oncology UNIL CHUV; Ludwig Institute for Cancer Research, University of Lausanne, Lausanne, Switzerland

Dieter Kabelitz
Institute of Immunology, University Hospital Schleswig-Holstein and Christian-Albrechts University Kiel, Kiel, Germany

Neha Kamran
Department of Neurosurgery; Department of Cell and Developmental Biology, University of Michigan Medical School, Ann Arbor, MI, United States

Lenka Kasikova
Department of Immunology, Charles University, 2nd Faculty of Medicine and University Hospital Motol; Sotio, Prague, Czech Republic

Léonce Kouakanou
Institute of Immunology, Christian-Albrechts-University Kiel, Kiel, Germany

Sam K.P. Kung
Department of Immunology, University of Manitoba, Winnipeg, MB, Canada

Rosa Ana Lacalle
Department of Immunology and Oncology, Centro Nacional de Biotecnología (CNB/CSIC), Madrid, Spain

Francis Lin
Department of Physics and Astronomy; Department of Biosystems Engineering; Department of Immunology, University of Manitoba, Winnipeg, MB, Canada

Alejandro López-Soto
Departamento de Biología Funcional, Área de Inmunología, Universidad de Oviedo; Instituto Universitario de Oncología del Principado de Asturias (IUOPA); Instituto de Investigación Sanitaria del Principado de Asturias (ISPA), Oviedo, Spain

Seila Lorenzo-Herrero
Departamento de Biología Funcional, Área de Inmunología, Universidad de Oviedo; Instituto Universitario de Oncología del Principado de Asturias (IUOPA); Instituto de Investigación Sanitaria del Principado de Asturias (ISPA), Oviedo, Spain

Pedro R. Lowenstein
Department of Neurosurgery; Department of Cell and Developmental Biology, University of Michigan Medical School, Ann Arbor, MI, United States

Francesca Lucibello
INSERM U932, PSL University, Institut Curie, Paris, France

Enrico Lugli
Laboratory of Translational Immunology; Flow Cytometry Core, Humanitas Clinical and Research Center, Rozzano, Milan, Italy

Daniel Lysak
Department of Hematology and Oncology, Hospital in Pilsen, Pilsen, Czech Republic

Morgane Magnin
Ludwig Institute for Cancer Research, University of Lausanne; Department of Oncology, University Hospital of Lausanne, Lausanne, Switzerland

Santos Mañes
Department of Immunology and Oncology, Centro Nacional de Biotecnología (CNB/CSIC), Madrid, Spain

Gwenola Manic
Unit of Tumor Immunology and Immunotherapy, Department of Research, Advanced Diagnostics, and Technological Innovation, IRCCS Regina Elena National Cancer Institute, Rome, Italy

Christophe Martignier
Department of Oncology, University of Lausanne, Lausanne, Switzerland

Ignacio Melero
Program of Immunology and Immunotherapy, CIMA Universidad de Navarra; Navarra Institute for Health Research (IDISNA); Department of Immunology and Immunotherapy, Clinica Universidad de Navarra, Pamplona; Centro de Investigación Biomédica en Red de Cáncer (CIBERONC), Madrid, Spain

Silvia Menegatti
INSERM U932, PSL University, Institut Curie, Paris, France

Laurie Menger
INSERM U932, PSL University, Institut Curie, Paris, France

Taha Merghoub
Ludwig Collaborative and Swim Across America Laboratory; Parker Institute for Cancer Immunotherapy; Department of Medicine, Memorial Sloan Kettering Cancer Center; Weill Cornell Medicine, New York, NY, United States

Luna Minute
Program of Immunology and Immunotherapy, CIMA Universidad de Navarra; Navarra Institute for Health Research (IDISNA), Pamplona, Spain

Arthur Mortha
Department of Immunology, University of Toronto, Toronto, ON, Canada

Martina Musella
Unit of Tumor Immunology and Immunotherapy, Department of Research, Advanced Diagnostics, and Technological Innovation, IRCCS Regina Elena National Cancer Institute; Department of Molecular Medicine, University "La Sapienza", Rome, Italy

Natalie J. Neubert
Department of Oncology, University of Lausanne, Lausanne, Switzerland

Louis Ngai
Department of Immunology, University of Toronto, Toronto, ON, Canada

Paola Nisticò
Unit of Tumor Immunology and Immunotherapy, Department of Research, Advanced Diagnostics and Technological Innovation, IRCCS Regina Elena National Cancer Institute, Rome, Italy

Felipe J. Núñez
Department of Neurosurgery; Department of Cell and Developmental Biology, University of Michigan Medical School, Ann Arbor, MI, United States

Hans-Heinrich Oberg
Institute of Immunology, University Hospital Schleswig-Holstein and Christian-Albrechts University Kiel, Kiel, Germany

Maria C. Ochoa
Program of Immunology and Immunotherapy, CIMA Universidad de Navarra; Navarra Institute for Health Research (IDISNA); Department of Immunology and Immunotherapy, Clinica Universidad de Navarra, Pamplona; Centro de Investigación Biomédica en Red de Cáncer (CIBERONC), Madrid, Spain

Jesús Ogando
Department of Immunology and Oncology, Centro Nacional de Biotecnología (CNB/CSIC), Madrid, Spain

Purevdorj B. Olkhanud
Vaccine Branch, Center for Cancer Research, National Cancer Institute, National Institutes of Health, Bethesda, MD, United States

Ane Orrantia
Biocruces Bizkaia Health Research Institute, Immunopathology Group, Barakaldo, Spain

Belinda Palermo
Unit of Tumor Immunology and Immunotherapy, Department of Research, Advanced Diagnostics and Technological Innovation, IRCCS Regina Elena National Cancer Institute, Rome, Italy

Mariangela Panetta
Unit of Tumor Immunology and Immunotherapy, Department of Research, Advanced Diagnostics and Technological Innovation, IRCCS Regina Elena National Cancer Institute, Rome, Italy

Christian Peters
Institute of Immunology, University Hospital Schleswig-Holstein and Christian-Albrechts University Kiel, Kiel, Germany

Karolina Pilipow
Laboratory of Translational Immunology, Humanitas Clinical and Research Center, Rozzano, Milan, Italy

Katerina Politi
Department of Pathology; Department of Medicine (Section of Medical Oncology), Yale School of Medicine; Yale Cancer Center, Yale University School of Medicine, New Haven, CT, United States

Jana Rakova
Sotio, Prague, Czech Republic

Xiaoou Ren
Department of Physics and Astronomy; Department of Biosystems Engineering, University of Manitoba, Winnipeg, MB, Canada

Louise Rethacker
INSERM, U1160, Institut de Recherche Saint-Louis, Hôpital Saint-Louis, Paris, France

Camila Robles-Oteiza
Department of Immunobiology, Yale School of Medicine, New Haven, CT, United States

Pedro Romero
Translational Tumor Immunology Group, Ludwig Cancer Research; Department of Oncology UNIL CHUV, University of Lausanne, Lausanne, Switzerland

Cyril Salek
Institute of Hematology and Blood Transfusion; Institute of Clinical and Experimental Hematology, 1st Faculty of Medicine, Charles University, Prague, Czech Republic

Javier Santos
Department of Immunology and Oncology, Centro Nacional de Biotecnología (CNB/CSIC), Madrid, Spain

Sujith Sarvesh
Department of Microbiology, University of Alabama at Birmingham, Birmingham, AL, United States

Eloise Scamardella
Laboratory of Translational Immunology, Humanitas Clinical and Research Center, Rozzano, Milan, Italy

Julien Schmidt
Ludwig Institute for Cancer Research, University of Lausanne; Department of Oncology, University Hospital of Lausanne, Lausanne, Switzerland

Antonella Sistigu
Istituto di Patologia Generale, Università Cattolica del Sacro Cuore; Unit of Tumor Immunology and Immunotherapy, Department of Research, Advanced Diagnostics, and Technological Innovation, IRCCS Regina Elena National Cancer Institute, Rome, Italy

Christian Sordo-Bahamonde
Departamento de Biología Funcional, Área de Inmunología, Universidad de Oviedo; Instituto Universitario de Oncología del Principado de Asturias (IUOPA); Instituto de Investigación Sanitaria del Principado de Asturias (ISPA), Oviedo, Spain

Daniel E. Speiser
Department of Oncology, University of Lausanne, Lausanne, Switzerland

Radek Spisek
Department of Immunology, Charles University, 2nd Faculty of Medicine and University Hospital Motol; Sotio, Prague, Czech Republic

Masaki Terabe
Neuro-Oncology Branch, Center for Cancer Research, National Cancer Institute, National Institutes of Health, Bethesda, MD, United States

Iñigo Terrén
Biocruces Bizkaia Health Research Institute, Immunopathology Group, Barakaldo, Spain

Laure Tillé
Department of Oncology, University of Lausanne, Lausanne, Switzerland

Sara Trabanelli
Department of Oncology UNIL CHUV; Ludwig Institute for Cancer Research, University of Lausanne, Lausanne, Switzerland

Iva Truxova
Sotio, Prague, Czech Republic

Pierre van der Bruggen
De Duve Institute, Université catholique de Louvain; WELBIO, Brussels, Belgium

Sjoerd H. van der Burg
Department of Medical Oncology, Leiden University Medical Center, Leiden, The Netherlands

Christophe Vanhaver
De Duve Institute, Université catholique de Louvain, Brussels, Belgium

Ilio Vitale
Italian Institute for Genomic Medicine (IIGM), Turin; Candiolo Cancer Institute—FPO, IRCCS, Candiolo, Italy

Joana Vitallé
Biocruces Bizkaia Health Research Institute, Immunopathology Group, Barakaldo, Spain

Sarka Vosahlikova
Sotio, Prague, Czech Republic

Mathias Wenes
Translational Tumor Immunology Group, Ludwig Cancer Research; Department of Oncology UNIL CHUV, University of Lausanne, Lausanne, Switzerland

Daniela Wesch
Institute of Immunology, University Hospital Schleswig-Holstein and Christian-Albrechts University Kiel, Kiel, Germany

Roberta Zappasodi
Ludwig Collaborative and Swim Across America Laboratory; Parker Institute for Cancer Immunotherapy, Memorial Sloan Kettering Cancer Center, New York, NY, United States

Olatz Zenarruzabeitia
Biocruces Bizkaia Health Research Institute, Immunopathology Group, Barakaldo, Spain

Lianjun Zhang
Translational Tumor Immunology Group, Ludwig Cancer Research; Department of Oncology UNIL CHUV, University of Lausanne, Lausanne, Switzerland; Center for Systems Medicine, Institute of Basic Medical Sciences, Chinese Academy of Medical Sciences and Peking Union Medical College, Beijing; Suzhou Institute of Systems Medicine, Suzhou, Jiangsu, China

Preface: More than two decades of modern tumor immunology

Lorenzo Galluzzi[a,b,c,d,e,*], **Nils-Petter Rudqvist**[a,*]
[a]Department of Radiation Oncology, Weill Cornell Medical College, New York, NY, United States
[b]Sandra and Edward Meyer Cancer Center, New York, NY, United States
[c]Caryl and Israel Englander Institute for Precision Medicine
[d]Department of Dermatology, Yale University School of Medicine, New Haven, CT, United States
[e]Université de Paris, Paris, France
*Corresponding authors: e-mail address: deadoc80@gmail.com; nils.rudqvist@gmail.com

The roots of modern tumor immunology extend back to the pioneering work of William B. Coley, an American surgeon who—at the end of the 19th century—was the first to successfully use a partially inactivated bacterial preparation (the so-called "Coley toxin") for treating soft tissue sarcoma (Coley, 1893), and the revolutionary proposals of Paul Ehlrich, a German-Jewish physician who—at the beginning of the 20th century—introduced for the first time in history the concept of cancer immunosurveillance (Ehrlich, 1909; Rao, Gharib, & Han, 2019). However, with the exception of punctual episodes including the establishment of the Cancer Research Institute of New York (in 1957), whose specific purpose was the development of immunological approaches to treat cancer (Decker et al., 2017), and the publication of the first cancer vaccine study (in 1959) (Graham & Graham, 1959), the field remained virtually unborn until the 1990s. Indeed, the 1990s witnessed not only the discovery that cytotoxic T lymphocyte associated protein 4 (CTLA4) mediates robust immunosuppressive functions, hence constituting a potential target for cancer therapy (Krummel & Allison, 1995; Leach, Krummel, & Allison, 1996), but also the realization that oncogenesis occurs in the context of a bidirectional crosstalk between malignant cells and the immune system of the host (Dighe, Richards, Old, & Schreiber, 1994; Kaplan et al., 1998), laying the foundation for the modern cancer immunosurveillance and immunoediting theories (Dunn, Bruce, Ikeda, Old, & Schreiber, 2002; Dunn, Old, & Schreiber, 2004). In the 2000s, the field gathered additional momentum from preclinical and clinical findings elucidating multiple mechanisms that underlie the ability of the immune system to control developing tumors and the capacity of progressing neoplasms to evade immunosurveillance (Bargou et al., 2008; Dudley et al., 2002; Galon et al., 2006;

Hildner et al., 2008; Jaiswal et al., 2009; Koebel et al., 2007; Majeti et al., 2009; Morgan et al., 2006; Savage et al., 2008; Shankaran et al., 2001), as well as from the discovery that the efficacy of multiple conventional anticancer agents relies (at least in part) on the (re-)activation of tumor-targeting immune responses as a consequence of immunogenic cell death (ICD) induction (Bezu et al., 2018; Casares et al., 2005; Galluzzi, Bravo-San Pedro, Demaria, Formenti, & Kroemer, 2017; Galluzzi, Buque, Kepp, Zitvogel, & Kroemer, 2015, 2017; Galluzzi, Kepp, Chan, & Kroemer, 2017; Galluzzi, Vitale, et al., 2018; Galluzzi, Zitvogel, & Kroemer, 2016; Obeid et al., 2007; Pol et al., 2015). These data paved the way to the cancer immunotherapy revolution of the 2010s (Kelly, 2018).

The 2010s represent indeed the decade in which modern tumor immunology revolutionized the clinical management of multiple tumor types upon the approval of no less than 12 different immunotherapeutic agents for use in cancer patients by the US Food and Drug Administration (FDA) and various equivalent regulatory agencies worldwide (Galluzzi, Chan, Kroemer, Wolchok, & Lopez-Soto, 2018). In January 2019, when this Preface was being redacted, these agents included: (1) sipuleucel T, a myeloid cell-derived product originally approved for the treatment of asymptomatic or minimally symptomatic metastatic castration resistant (hormone refractory) prostate cancer in April 2010 (Hagihara et al., 2019; Kantoff et al., 2010); (2) ipilimumab, a CTLA4-targeting immune checkpoint inhibitor (ICI) (Aranda et al., 2014) initially licensed for use in patients with unresectable or metastatic melanoma who received at least one prior systemic treatment in March 2011 (Hodi et al., 2010; McDermott, Haanen, Chen, Lorigan, & O'Day, 2013; Robert, Schadendorf, Messina, Hodi, & O'Day, 2013), and currently approved for no less than two additional indications (Coens et al., 2017; Eggermont et al., 2015; Eggermont et al., 2016; Geoerger et al., 2017; Merchant et al., 2016); (3) nivolumab, an ICI targeting programmed cell death 1 (PDCD1, best known as PD-1) first licensed for the management of patients with unresectable or metastatic melanoma and disease progression following ipilimumab and, if $BRAF^{V600}$ mutation positive, a BRAF inhibitor in December 2014 (Weber et al., 2015), and currently available for no less than 10 additional indications (Ansell et al., 2015; Borghaei et al., 2015; Cella et al., 2016; El-Khoueiry et al., 2017; Escudier, Motzer, et al., 2017; Escudier, Sharma, et al., 2017; Ferris et al., 2016; Harrington et al., 2017; Horn et al., 2017; Kazandjian et al., 2016; Kiyota et al., 2017; Motzer et al., 2015; Overman et al., 2017; Sharma et al., 2017; Vokes et al., 2018; Weber et al., 2017; Xu et al., 2017; Younes et al., 2016);

(4) pembrolizumab, a PD-1-targeting ICI originally licensed for the treatment of advanced or unresectable melanoma patients who are no longer responding to ipilimumab and BRAF inhibitors (when indicated) in September 2014 (Hodi et al., 2016; Joseph et al., 2018; Patnaik et al., 2015; Ribas et al., 2016; Robert et al., 2018, 2014), and currently approved for no less than 10 additional indications (Alley et al., 2017; Balar, Castellano, et al., 2017; Brahmer et al., 2017; Brogden et al., 2018; Chatterjee et al., 2016; Chen et al., 2017; Chow et al., 2016; Doi et al., 2018; Frenel et al., 2017; Fuchs et al., 2018; Gandhi et al., 2018; Garon et al., 2015; Hsu et al., 2017; Hui et al., 2017; Le et al., 2017, 2015; Muro et al., 2016; Nanda et al., 2016; O'Neil et al., 2017; Ott, Elez, et al., 2017; Ott, Piha-Paul, et al., 2017; Patnaik et al., 2015; Petrella et al., 2017; Plimack et al., 2017; Reck et al., 2016; Robert et al., 2015; Schachter et al., 2017; Seiwert et al., 2016; Shaverdian et al., 2017; Tahara et al., 2018; Zinzani et al., 2017); (5) cemiplimab, yet another ICI specific for PD-1 originally approved for the treatment of metastatic or locally advanced cutaneous squamous cell carcinoma in September 2018 (Migden et al., 2018); (6) durvalumab, an ICI specific for CD274 (best known as PD-L1) originally licensed for the treatment of locally advanced or metastatic urothelial carcinoma progressing during or following platinum-containing chemotherapy in May 2017 (Levy, Massard, Soria, & Deutsch, 2016; Powles et al., 2017), and currently also available for the management of patients with unresectable nonsmall cell lung carcinoma (Antonia et al., 2017); (7) avelumab, a PD-L1-targeting ICI originally approved for the treatment of adult and pediatric patients with metastatic Merkel cell carcinoma in March 2017 (D'Angelo et al., 2018; Kaufman et al., 2016, 2018), and currently licensed also for use in patients with locally advanced or metastatic urothelial carcinoma progressing during or following platinum-containing chemotherapy (Apolo et al., 2017; Heery et al., 2017); (8) atezolizumab, yet another PD-L1 inhibitor originally approved for the treatment of locally advanced or metastatic urothelial carcinoma progressing during or following platinum-containing chemotherapy in May 2016 (Balar, Galsky, et al., 2017; Necchi et al., 2017; Perez-Gracia et al., 2018; Rosenberg et al., 2016), and currently approved for no less than two additional indications (Fehrenbacher et al., 2016; Hida et al., 2018; Rittmeyer et al., 2017); (9) PEGylated interferon alpha 2B, a recombinant cytokine (Garcia-Martinez et al., 2018) originally approved for the treatment of stage III melanoma patients after surgery in March 2011 (Bottomley et al., 2009; Bouwhuis et al., 2010; Di Trolio et al., 2012; Eggermont et al., 2008; Eggermont et al., 2010; Fusi et al., 2009);

(10) talimogene laherparepvec, an oncolytic virus originally approved for the treatment of metastatic, unresectable melanoma patients in October 2015 (Andtbacka et al., 2015; Pol et al., 2018; Pol, Kroemer, & Galluzzi, 2016; Vacchelli et al., 2013), (11) tisagenlecleucel, a preparation of autologous T cells engineered to express a chimeric antigen receptor (CAR) (Borrie & Maleki Vareki, 2018; Raikar et al., 2018) that recognizes CD19, originally approved for the treatment of pediatric and young adults with acute lymphocytic leukemia who relapsed or are refractory to recommended prior therapy in August 2017 (Maude et al., 2018), and now also licensed for the therapy of subjects with various B-cell lymphomas relapsing on or refractory to recommended prior therapy (Schuster et al., 2019); and (12) axicabtagene ciloleucel, another variant of CAR-expressing T cells specific for CD19 (Kellner, Peipp, Gramatzki, Schrappe, & Schewe, 2018), originally approved for the treatment of several variants of refractory or relapsing B-cell lymphoma in October 2017 (Neelapu et al., 2017).

Alongside, considerable translational advances have been made in the 2010s, including (but not limited to): (1) the realization of the major impact of the immunological tumor microenvironment on disease outcome among cancer patients treated with a variety of therapeutic regimens beyond immunotherapy (Atherton et al., 2018; Bantug, Galluzzi, Kroemer, & Hess, 2018; Chevrier et al., 2017; De Henau et al., 2016; Herbst et al., 2014; Humbert & Hugues, 2019; Lavin et al., 2017; Li et al., 2019; Mariathasan et al., 2018; Riaz et al., 2017; Spranger, Bao, & Gajewski, 2015; Straussman et al., 2012; Tauriello et al., 2018; Zheng et al., 2017); (2) the identification of the gut microbiota as a major determinant of the sensitivity of cancer patients to multiple treatments including ICIs (Bhutiani et al., 2018; Gopalakrishnan et al., 2018; Iida et al., 2013; Routy et al., 2018; Vetizou et al., 2015; Viaud et al., 2013); (3) the discovery that novel antigens expressed by cancer cells as a consequence of nonsynonymous mutations play a key role in tumor progression and response to treatment (Gubin et al., 2014; Janikovits et al., 2018; Ozcan, Janikovits, von Knebel Doeberitz, & Kloor, 2018; Rizvi et al., 2015; Schumacher & Schreiber, 2015; Vitale et al., 2019); (4) the elucidation of innate immune mechanisms that provide a major contribution to both natural and therapy-elicited immunosurveillance (Andre et al., 2018; Dou et al., 2017; Ferrari de Andrade et al., 2018; Galluzzi, Vanpouille-Box, Bakhoum, & Demaria, 2018; Harding et al., 2017; Lopez-Soto, Gonzalez, Smyth, & Galluzzi, 2017; Mackenzie et al., 2017; Oyer, Gitto, Altomare, & Copik, 2018; Ruscetti et al., 2018;

van Montfoort et al., 2018; Vanpouille-Box et al., 2017; Vanpouille-Box, Demaria, Formenti, & Galluzzi, 2018); and (5) the identification of exosomes as a prominent vehicle for malignant cells to condition distant anatomical sites and interface with the host immune system (Chen et al., 2018; Diamond et al., 2018; Gabrusiewicz et al., 2018; Hoshino et al., 2015; Mirzaei et al., 2018; Yaddanapudi et al., 2019; Zhang et al., 2015). Finally, major technological advances have been achieved in the 2010s, including (but not limited to): (1) the development of reliable and cost-effective platforms for single cell DNA or RNA sequencing (Alemany, Florescu, Baron, Peterson-Maduro, & van Oudenaarden, 2018; Mulqueen et al., 2018; Zhang et al., 2018); (2) the design of robust algorithms enabling not only the analysis of large single cell omic datasets (Becht et al., 2019; Grobner et al., 2018; Wu et al., 2014), but also various other *in silico* functions that are widely employed in modern tumor immunology, such as neoantigen and TCR motif prediction (Balachandran et al., 2017; Brown & Holt, 2019; Dash et al., 2017; Formenti et al., 2018; Glanville et al., 2017; Hundal et al., 2019; Karasaki et al., 2017); (3) the refinement of the CRISPR/Cas system for gene editing (Halperin et al., 2018; Hille et al., 2018; Kalhor et al., 2018; Nishimasu et al., 2018; Shen et al., 2018; Wang et al., 2018; Yan et al., 2019); and (4) the introduction of mass cytometry (Chevrier et al., 2017; Galluzzi, Yamazaki, & Demaria, 2017; Lavin et al., 2017; Spitzer & Nolan, 2016; Wierz, Janji, Berchem, Moussay, & Paggetti, 2018). Beyond any reasonable doubt, the preclinical, clinical and technological progress we witnessed in the 2010s will pave the way to an exciting decade of achievements in the fields of tumor immunology and immunotherapy as the 2020s are on their way.

In *Tumor Immunology and Immunotherapy* (a group of volumes of the *Methods in Enzymology* series), world leaders in the field gathered to provide the first comprehensive compendium of methods for the study of cancer immunology in both its fundamental and translational components. Each of the chapters of *Tumor Immunology and Immunotherapy* provides a detailed description of one or a few protocols that are suitable for the characterization of (components of) the tumor–host interaction, *in vitro, in vivo* or *in silico*. The book consists of five volumes: "Molecular Methods," "Cellular Methods—Part A," "Cellular Methods—Part B," "Integrated Methods—Part A," and "Integrated Methods—Part B," and is equally addressed to expert researchers who may want to extend their technical portfolio, and to beginners in this exciting and rapidly growing field of investigation.

Acknowledgments

L.G. is supported by a Breakthrough Level 2 grant from the US Department of Defense (DoD), Breast Cancer Research Program (BRCP) [#BC180476P1], by a startup grant from the Department of Radiation Oncology at Weill Cornell Medicine (New York, USA), by industrial collaborations with Lytix (Oslo, Norway) and Phosplatin (New York, USA), and by donations from Phosplatin (New York, USA), the Luke Heller TECPR2 Foundation (Boston, USA) and Sotio a.s. (Prague, Czech Republic).

Disclosures

L.G. provides remunerated consulting to OmniSEQ (Buffalo, NY, USA), Astra Zeneca (Gaithersburg, MD, USA), Boehringer Ingelheim (Vienna, Austria), VL47 (New York, NY, USA), and the Luke Heller TECPR2 Foundation (Boston, MA, USA), and is member of the Scientific Advisory Committee of OmniSEQ (Buffalo, NY, USA).

References

Alemany, A., Florescu, M., Baron, C. S., Peterson-Maduro, J., & van Oudenaarden, A. (2018). Whole-organism clone tracing using single-cell sequencing. *Nature, 556*(7699), 108–112. https://doi.org/10.1038/nature25969.

Alley, E. W., Lopez, J., Santoro, A., Morosky, A., Saraf, S., Piperdi, B., et al. (2017). Clinical safety and activity of pembrolizumab in patients with malignant pleural mesothelioma (KEYNOTE-028): Preliminary results from a non-randomised, open-label, phase 1b trial. *The Lancet Oncology, 18*(5), 623–630. https://doi.org/10.1016/s1470-2045(17)30169-9.

Andre, P., Denis, C., Soulas, C., Bourbon-Caillet, C., Lopez, J., Arnoux, T., et al. (2018). Anti-NKG2A mAb is a checkpoint inhibitor that promotes anti-tumor immunity by unleashing both T and NK cells. *Cell, 175*(7). 1731–1743.e1713. https://doi.org/10.1016/j.cell.2018.10.014.

Andtbacka, R. H., Kaufman, H. L., Collichio, F., Amatruda, T., Senzer, N., Chesney, J., et al. (2015). Talimogene laherparepvec improves durable response rate in patients with advanced melanoma. *Journal of Clinical Oncology, 33*(25), 2780–2788. https://doi.org/10.1200/jco.2014.58.3377.

Ansell, S. M., Lesokhin, A. M., Borrello, I., Halwani, A., Scott, E. C., Gutierrez, M., et al. (2015). PD-1 blockade with nivolumab in relapsed or refractory Hodgkin's lymphoma. *The New England Journal of Medicine, 372*(4), 311–319. https://doi.org/10.1056/NEJMoa1411087.

Antonia, S. J., Villegas, A., Daniel, D., Vicente, D., Murakami, S., Hui, R., et al. (2017). Durvalumab after chemoradiotherapy in stage III non-small-cell lung cancer. *The New England Journal of Medicine, 377*(20), 1919–1929. https://doi.org/10.1056/NEJMoa1709937.

Apolo, A. B., Infante, J. R., Balmanoukian, A., Patel, M. R., Wang, D., Kelly, K., et al. (2017). Avelumab, an anti-programmed death-ligand 1 antibody, in patients with refractory metastatic urothelial carcinoma: Results from a multicenter, phase Ib study. *Journal of Clinical Oncology, 35*(19), 2117–2124. https://doi.org/10.1200/jco.2016.71.6795.

Aranda, F., Vacchelli, E., Eggermont, A., Galon, J., Fridman, W. H., Zitvogel, L., et al. (2014). Trial watch: Immunostimulatory monoclonal antibodies in cancer therapy. *Oncoimmunology, 3*(1), e27297. https://doi.org/10.4161/onci.27297.

Atherton, M. J., Stephenson, K. B., Tzelepis, F., Bakhshinyan, D., Nikota, J. K., Son, H. H., et al. (2018). Transforming the prostatic tumor microenvironment with oncolytic virotherapy. *Oncoimmunology*, 7(7), e1445459. https://doi.org/10.1080/2162402X.2018.1445459.

Balachandran, V. P., Luksza, M., Zhao, J. N., Makarov, V., Moral, J. A., Remark, R., et al. (2017). Identification of unique neoantigen qualities in long-term survivors of pancreatic cancer. *Nature*, *551*(7681), 512–516. https://doi.org/10.1038/nature24462.

Balar, A. V., Castellano, D., O'Donnell, P. H., Grivas, P., Vuky, J., Powles, T., et al. (2017). First-line pembrolizumab in cisplatin-ineligible patients with locally advanced and unresectable or metastatic urothelial cancer (KEYNOTE-052): A multicentre, single-arm, phase 2 study. *The Lancet Oncology*, *18*(11), 1483–1492. https://doi.org/10.1016/s1470-2045(17)30616-2.

Balar, A. V., Galsky, M. D., Rosenberg, J. E., Powles, T., Petrylak, D. P., Bellmunt, J., et al. (2017). Atezolizumab as first-line treatment in cisplatin-ineligible patients with locally advanced and metastatic urothelial carcinoma: A single-arm, multicentre, phase 2 trial. *Lancet*, *389*(10064), 67–76. https://doi.org/10.1016/s0140-6736(16)32455-2.

Bantug, G. R., Galluzzi, L., Kroemer, G., & Hess, C. (2018). The spectrum of T cell metabolism in health and disease. *Nature Reviews. Immunology*, *18*(1), 19–34. https://doi.org/10.1038/nri.2017.99.

Bargou, R., Leo, E., Zugmaier, G., Klinger, M., Goebeler, M., Knop, S., et al. (2008). Tumor regression in cancer patients by very low doses of a T cell-engaging antibody. *Science*, *321*(5891), 974–977. https://doi.org/10.1126/science.1158545.

Becht, E., McInnes, L., Healy, J., Dutertre, C. A., Kwok, I. W. H., Ng, L. G., et al. (2019). Dimensionality reduction for visualizing single-cell data using UMAP. *Nature Biotechnology*, *37*, 38–44. https://doi.org/10.1038/nbt.4314.

Bezu, L., Sauvat, A., Humeau, J., Leduc, M., Kepp, O., & Kroemer, G. (2018). eIF2alpha phosphorylation: A hallmark of immunogenic cell death. *Oncoimmunology*, 7(6). e1431089https://doi.org/10.1080/2162402X.2018.1431089.

Bhutiani, N., Li, Q., Anderson, C. D., Gallagher, H. C., De Jesus, M., Singh, R., et al. (2018). Enhanced gut barrier integrity sensitizes colon cancer to immune therapy. *Oncoimmunology*, 7(11), e1498438. https://doi.org/10.1080/2162402X.2018.1498438.

Borghaei, H., Paz-Ares, L., Horn, L., Spigel, D. R., Steins, M., Ready, N. E., et al. (2015). Nivolumab versus docetaxel in advanced nonsquamous non-small-cell lung cancer. *The New England Journal of Medicine*, *373*(17), 1627–1639. https://doi.org/10.1056/NEJMoa1507643.

Borrie, A. E., & Maleki Vareki, S. (2018). T lymphocyte-based cancer immunotherapeutics. *International Review of Cell and Molecular Biology*, *341*, 201–276. https://doi.org/10.1016/bs.ircmb.2018.05.010.

Bottomley, A., Coens, C., Suciu, S., Santinami, M., Kruit, W., Testori, A., et al. (2009). Adjuvant therapy with pegylated interferon alfa-2b versus observation in resected stage III melanoma: A phase III randomized controlled trial of health-related quality of life and symptoms by the European Organisation for Research and Treatment of Cancer Melanoma Group. *Journal of Clinical Oncology*, *27*(18), 2916–2923. https://doi.org/10.1200/jco.2008.20.2069.

Bouwhuis, M. G., Suciu, S., Testori, A., Kruit, W. H., Sales, F., Patel, P., et al. (2010). Phase III trial comparing adjuvant treatment with pegylated interferon Alfa-2b versus observation: Prognostic significance of autoantibodies—EORTC 18991. *Journal of Clinical Oncology*, *28*(14), 2460–2466. https://doi.org/10.1200/jco.2009.24.6264.

Brahmer, J. R., Rodriguez-Abreu, D., Robinson, A. G., Hui, R., Csoszi, T., Fulop, A., et al. (2017). Health-related quality-of-life results for pembrolizumab versus chemotherapy in advanced, PD-L1-positive NSCLC (KEYNOTE-024): A multicentre, international, randomised, open-label phase 3 trial. *The Lancet Oncology*, *18*(12), 1600–1609. https://doi.org/10.1016/s1470-2045(17)30690-3.

Brogden, K. A., Parashar, D., Hallier, A. R., Braun, T., Qian, F., Rizvi, N. A., et al. (2018). Genomics of NSCLC patients both affirm PD-L1 expression and predict their clinical responses to anti-PD-1 immunotherapy. *BMC Cancer, 18*(1), 225. https://doi.org/10.1186/s12885-018-4134-y.

Brown, S. D., & Holt, R. A. (2019). Neoantigen characteristics in the context of the complete predicted MHC class I self-immunopeptidome. *Oncoimmunology, 8*(3), 1556080. https://doi.org/10.1080/2162402X.2018.1556080.

Casares, N., Pequignot, M. O., Tesniere, A., Ghiringhelli, F., Roux, S., Chaput, N., et al. (2005). Caspase-dependent immunogenicity of doxorubicin-induced tumor cell death. *The Journal of Experimental Medicine, 202*(12), 1691–1701. https://doi.org/10.1084/jem.20050915.

Cella, D., Grunwald, V., Nathan, P., Doan, J., Dastani, H., Taylor, F., et al. (2016). Quality of life in patients with advanced renal cell carcinoma given nivolumab versus everolimus in CheckMate 025: A randomised, open-label, phase 3 trial. *The Lancet Oncology, 17*(7), 994–1003. https://doi.org/10.1016/s1470-2045(16)30125-5.

Chatterjee, M., Turner, D. C., Felip, E., Lena, H., Cappuzzo, F., Horn, L., et al. (2016). Systematic evaluation of pembrolizumab dosing in patients with advanced non-small-cell lung cancer. *Annals of Oncology, 27*(7), 1291–1298. https://doi.org/10.1093/annonc/mdw174.

Chen, G., Huang, A. C., Zhang, W., Zhang, G., Wu, M., Xu, W., et al. (2018). Exosomal PD-L1 contributes to immunosuppression and is associated with anti-PD-1 response. *Nature, 560*(7718), 382–386. https://doi.org/10.1038/s41586-018-0392-8.

Chen, R., Zinzani, P. L., Fanale, M. A., Armand, P., Johnson, N. A., Brice, P., et al. (2017). Phase II study of the efficacy and safety of pembrolizumab for relapsed/refractory classic Hodgkin lymphoma. *Journal of Clinical Oncology, 35*(19), 2125–2132. https://doi.org/10.1200/jco.2016.72.1316.

Chevrier, S., Levine, J. H., Zanotelli, V. R. T., Silina, K., Schulz, D., Bacac, M., et al. (2017). An immune atlas of clear cell renal cell carcinoma. *Cell, 169*(4), 736–749.e718. https://doi.org/10.1016/j.cell.2017.04.016.

Chow, L. Q. M., Haddad, R., Gupta, S., Mahipal, A., Mehra, R., Tahara, M., et al. (2016). Antitumor activity of pembrolizumab in biomarker-unselected patients with recurrent and/or metastatic head and neck squamous cell carcinoma: Results from the Phase Ib KEYNOTE-012 expansion cohort. *Journal of Clinical Oncology, 34*(32), 3838–3845. https://doi.org/10.1200/jco.2016.68.1478.

Coens, C., Suciu, S., Chiarion-Sileni, V., Grob, J. J., Dummer, R., Wolchok, J. D., et al. (2017). Health-related quality of life with adjuvant ipilimumab versus placebo after complete resection of high-risk stage III melanoma (EORTC 18071): Secondary outcomes of a multinational, randomised, double-blind, phase 3 trial. *The Lancet Oncology, 18*(3), 393–403. https://doi.org/10.1016/s1470-2045(17)30015-3.

Coley, W. B. (1893). The treatment of malignant tumors by repeated inoculations of erysipelas. With a report of ten original cases. *The American Journal of the Medical Sciences, 105*, 487–511.

D'Angelo, S. P., Russell, J., Lebbe, C., Chmielowski, B., Gambichler, T., Grob, J. J., et al. (2018). Efficacy and safety of first-line avelumab treatment in patients with Stage IV metastatic Merkel cell carcinoma: A preplanned interim analysis of a clinical trial. *JAMA Oncology, 4*(9). e180077. https://doi.org/10.1001/jamaoncol.2018.0077.

Dash, P., Fiore-Gartland, A. J., Hertz, T., Wang, G. C., Sharma, S., Souquette, A., et al. (2017). Quantifiable predictive features define epitope-specific T cell receptor repertoires. *Nature, 547*(7661), 89–93. https://doi.org/10.1038/nature22383.

De Henau, O., Rausch, M., Winkler, D., Campesato, L. F., Liu, C., Cymerman, D. H., et al. (2016). Overcoming resistance to checkpoint blockade therapy by targeting PI3Kgamma in myeloid cells. *Nature, 539*(7629), 443–447. https://doi.org/10.1038/nature20554.

Decker, W. K., da Silva, R. F., Sanabria, M. H., Angelo, L. S., Guimaraes, F., Burt, B. M., et al. (2017). Cancer immunotherapy: Historical perspective of a clinical revolution and emerging preclinical animal models. *Frontiers in Immunology*, *8*, 829. https://doi.org/10.3389/fimmu.2017.00829.

Di Trolio, R., Simeone, E., Di Lorenzo, G., Grimaldi, A. M., Romano, A., Ayala, F., et al. (2012). Update on PEG-interferon alpha-2b as adjuvant therapy in melanoma. *Anticancer Research*, *32*(9), 3901–3909.

Diamond, J. M., Vanpouille-Box, C., Spada, S., Rudqvist, N. P., Chapman, J. R., Ueberheide, B. M., et al. (2018). Exosomes shuttle TREX1-sensitive IFN-stimulatory dsDNA from irradiated cancer cells to DCs. *Cancer Immunology Research*, *6*(8), 910–920. https://doi.org/10.1158/2326-6066.cir-17-0581.

Dighe, A. S., Richards, E., Old, L. J., & Schreiber, R. D. (1994). Enhanced in vivo growth and resistance to rejection of tumor cells expressing dominant negative IFN gamma receptors. *Immunity*, *1*(6), 447–456.

Doi, T., Piha-Paul, S. A., Jalal, S. I., Saraf, S., Lunceford, J., Koshiji, M., et al. (2018). Safety and antitumor activity of the anti-programmed death-1 antibody pembrolizumab in patients with advanced esophageal carcinoma. *Journal of Clinical Oncology*, *36*(1), 61–67. https://doi.org/10.1200/jco.2017.74.9846.

Dou, Z., Ghosh, K., Vizioli, M. G., Zhu, J., Sen, P., Wangensteen, K. J., et al. (2017). Cytoplasmic chromatin triggers inflammation in senescence and cancer. *Nature*, *550*(7676), 402–406. https://doi.org/10.1038/nature24050.

Dudley, M. E., Wunderlich, J. R., Robbins, P. F., Yang, J. C., Hwu, P., Schwartzentruber, D. J., et al. (2002). Cancer regression and autoimmunity in patients after clonal repopulation with antitumor lymphocytes. *Science*, *298*(5594), 850–854. https://doi.org/10.1126/science.1076514.

Dunn, G. P., Bruce, A. T., Ikeda, H., Old, L. J., & Schreiber, R. D. (2002). Cancer immunoediting: From immunosurveillance to tumor escape. *Nature Immunology*, *3*(11), 991–998. https://doi.org/10.1038/ni1102-991.

Dunn, G. P., Old, L. J., & Schreiber, R. D. (2004). The three Es of cancer immunoediting. *Annual Review of Immunology*, *22*, 329–360. https://doi.org/10.1146/annurev.immunol.22.012703.104803.

Eggermont, A. M., Bouwhuis, M. G., Kruit, W. H., Testori, A., ten Hagen, T., Yver, A., et al. (2010). Serum concentrations of pegylated interferon alpha-2b in patients with resected stage III melanoma receiving adjuvant pegylated interferon alpha-2b in a randomized phase III trial (EORTC 18991). *Cancer Chemotherapy and Pharmacology*, *65*(4), 671–677. https://doi.org/10.1007/s00280-009-1072-z.

Eggermont, A. M., Chiarion-Sileni, V., Grob, J. J., Dummer, R., Wolchok, J. D., Schmidt, H., et al. (2015). Adjuvant ipilimumab versus placebo after complete resection of high-risk stage III melanoma (EORTC 18071): A randomised, double-blind, phase 3 trial. *The Lancet Oncology*, *16*(5), 522–530. https://doi.org/10.1016/s1470-2045(15)70122-1.

Eggermont, A. M., Chiarion-Sileni, V., Grob, J. J., Dummer, R., Wolchok, J. D., Schmidt, H., et al. (2016). Prolonged survival in stage III melanoma with ipilimumab adjuvant therapy. *The New England Journal of Medicine*, *375*(19), 1845–1855. https://doi.org/10.1056/NEJMoa1611299.

Eggermont, A. M., Suciu, S., Santinami, M., Testori, A., Kruit, W. H., Marsden, J., et al. (2008). Adjuvant therapy with pegylated interferon alfa-2b versus observation alone in resected stage III melanoma: Final results of EORTC 18991, a randomised phase III trial. *Lancet*, *372*(9633), 117–126. https://doi.org/10.1016/s0140-6736(08)61033-8.

Ehrlich, P. (1909). Über den jetzigen stand der karzinomforschung. *Nederlands Tijdschrift voor Geneeskunde*, *5*, 273–290.

El-Khoueiry, A. B., Sangro, B., Yau, T., Crocenzi, T. S., Kudo, M., Hsu, C., et al. (2017). Nivolumab in patients with advanced hepatocellular carcinoma (CheckMate 040): An open-label, non-comparative, phase 1/2 dose escalation and expansion trial. *Lancet, 389*(10088), 2492–2502. https://doi.org/10.1016/s0140-6736(17)31046-2.

Escudier, B., Motzer, R. J., Sharma, P., Wagstaff, J., Plimack, E. R., Hammers, H. J., et al. (2017). Treatment beyond progression in patients with advanced renal cell carcinoma treated with nivolumab in CheckMate 025. *European Urology, 72*(3), 368–376. https://doi.org/10.1016/j.eururo.2017.03.037.

Escudier, B., Sharma, P., McDermott, D. F., George, S., Hammers, H. J., Srinivas, S., et al. (2017). CheckMate 025 randomized phase 3 study: Outcomes by key baseline factors and prior therapy for nivolumab versus everolimus in advanced renal cell carcinoma. *European Urology, 72*(6), 962–971. https://doi.org/10.1016/j.eururo.2017.02.010.

Fehrenbacher, L., Spira, A., Ballinger, M., Kowanetz, M., Vansteenkiste, J., Mazieres, J., et al. (2016). Atezolizumab versus docetaxel for patients with previously treated non-small-cell lung cancer (POPLAR): A multicentre, open-label, phase 2 randomised controlled trial. *Lancet, 387*(10030), 1837–1846. https://doi.org/10.1016/s0140-6736(16)00587-0.

Ferrari de Andrade, L., Tay, R. E., Pan, D., Luoma, A. M., Ito, Y., Badrinath, S., et al. (2018). Antibody-mediated inhibition of MICA and MICB shedding promotes NK cell-driven tumor immunity. *Science, 359*(6383), 1537–1542. https://doi.org/10.1126/science.aao0505.

Ferris, R. L., Blumenschein, G., Jr., Fayette, J., Guigay, J., Colevas, A. D., Licitra, L., et al. (2016). Nivolumab for recurrent squamous-cell carcinoma of the head and neck. *The New England Journal of Medicine, 375*(19), 1856–1867. https://doi.org/10.1056/NEJMoa1602252.

Formenti, S. C., Rudqvist, N. P., Golden, E., Cooper, B., Wennerberg, E., Lhuillier, C., et al. (2018). Radiotherapy induces responses of lung cancer to CTLA-4 blockade. *Nature Medicine, 24*(12), 1845–1851. https://doi.org/10.1038/s41591-018-0232-2.

Frenel, J. S., Le Tourneau, C., O'Neil, B., Ott, P. A., Piha-Paul, S. A., Gomez-Roca, C., et al. (2017). Safety and efficacy of pembrolizumab in advanced, programmed death ligand 1-positive cervical cancer: Results from the phase Ib KEYNOTE-028 trial. *Journal of Clinical Oncology, 35*(36), 4035–4041. https://doi.org/10.1200/jco.2017.74.5471.

Fuchs, C. S., Doi, T., Jang, R. W., Muro, K., Satoh, T., Machado, M., et al. (2018). Safety and efficacy of pembrolizumab monotherapy in patients with previously treated advanced gastric and gastroesophageal junction cancer: Phase 2 clinical KEYNOTE-059 trial. *JAMA Oncology, 4*(5). e180013. https://doi.org/10.1001/jamaoncol.2018.0013.

Fusi, A., Collette, S., Busse, A., Suciu, S., Rietz, A., Santinami, M., et al. (2009). Circulating melanoma cells and distant metastasis-free survival in stage III melanoma patients with or without adjuvant interferon treatment (EORTC 18991 side study). *European Journal of Cancer, 45*(18), 3189–3197. https://doi.org/10.1016/j.ejca.2009.09.004.

Gabrusiewicz, K., Li, X., Wei, J., Hashimoto, Y., Marisetty, A. L., Ott, M., et al. (2018). Glioblastoma stem cell-derived exosomes induce M2 macrophages and PD-L1 expression on human monocytes. *Oncoimmunology, 7*(4), e1412909. https://doi.org/10.1080/2162402X.2017.1412909.

Galluzzi, L., Bravo-San Pedro, J. M., Demaria, S., Formenti, S. C., & Kroemer, G. (2017). Activating autophagy to potentiate immunogenic chemotherapy and radiation therapy. *Nature Reviews. Clinical Oncology, 14*(4), 247–258. https://doi.org/10.1038/nrclinonc.2016.183.

Galluzzi, L., Buque, A., Kepp, O., Zitvogel, L., & Kroemer, G. (2015). Immunological effects of conventional chemotherapy and targeted anticancer agents. *Cancer Cell, 28*(6), 690–714. https://doi.org/10.1016/j.ccell.2015.10.012.

Galluzzi, L., Buque, A., Kepp, O., Zitvogel, L., & Kroemer, G. (2017). Immunogenic cell death in cancer and infectious disease. *Nature Reviews. Immunology, 17*(2), 97–111. https://doi.org/10.1038/nri.2016.107.

Galluzzi, L., Chan, T. A., Kroemer, G., Wolchok, J. D., & Lopez-Soto, A. (2018). The hallmarks of successful anticancer immunotherapy. *Science Translational Medicine*, *10*(459), eaat7807. https://doi.org/10.1126/scitranslmed.aat7807.

Galluzzi, L., Kepp, O., Chan, F. K., & Kroemer, G. (2017). Necroptosis: Mechanisms and relevance to disease. *Annual Review of Pathology*, *12*, 103–130. https://doi.org/10.1146/annurev-pathol-052016-100247.

Galluzzi, L., Vanpouille-Box, C., Bakhoum, S. F., & Demaria, S. (2018). SnapShot: CGAS-STING signaling. *Cell*, *173*(1), 276–276.e271. https://doi.org/10.1016/j.cell.2018.03.015. .

Galluzzi, L., Vitale, I., Aaronson, S. A., Abrams, J. M., Adam, D., Agostinis, P., et al. (2018). Molecular mechanisms of cell death: Recommendations of the Nomenclature Committee on Cell Death 2018. *Cell Death and Differentiation*, *25*(3), 486–541. https://doi.org/10.1038/s41418-017-0012-4.

Galluzzi, L., Yamazaki, T., & Demaria, S. (2017). Heavy metal to rock the immune infiltrate. *Trends in Immunology*, *38*(8), 539–541. https://doi.org/10.1016/j.it.2017.05.007.

Galluzzi, L., Zitvogel, L., & Kroemer, G. (2016). Immunological mechanisms underneath the efficacy of cancer therapy. *Cancer Immunology Research*, *4*(11), 895–902. https://doi.org/10.1158/2326-6066.CIR-16-0197.

Galon, J., Costes, A., Sanchez-Cabo, F., Kirilovsky, A., Mlecnik, B., Lagorce-Pages, C., et al. (2006). Type, density, and location of immune cells within human colorectal tumors predict clinical outcome. *Science*, *313*(5795), 1960–1964. https://doi.org/10.1126/science.1129139.

Gandhi, L., Rodriguez-Abreu, D., Gadgeel, S., Esteban, E., Felip, E., De Angelis, F., et al. (2018). Pembrolizumab plus chemotherapy in metastatic non-small-cell lung cancer. *The New England Journal of Medicine*, *378*(22), 2078–2092. https://doi.org/10.1056/NEJMoa1801005.

Garcia-Martinez, E., Smith, M., Buque, A., Aranda, F., de la Pena, F. A., Ivars, A., et al. (2018). Trial watch: Immunostimulation with recombinant cytokines for cancer therapy. *Oncoimmunology*, 7(6), e1433982. https://doi.org/10.1080/2162402X.2018.1433982.

Garon, E. B., Rizvi, N. A., Hui, R., Leighl, N., Balmanoukian, A. S., Eder, J. P., et al. (2015). Pembrolizumab for the treatment of non-small-cell lung cancer. *The New England Journal of Medicine*, *372*(21), 2018–2028. https://doi.org/10.1056/NEJMoa1501824.

Geoerger, B., Bergeron, C., Gore, L., Sender, L., Dunkel, I. J., Herzog, C., et al. (2017). Phase II study of ipilimumab in adolescents with unresectable stage III or IV malignant melanoma. *European Journal of Cancer*, *86*, 358–363. https://doi.org/10.1016/j.ejca.2017.09.032.

Glanville, J., Huang, H., Nau, A., Hatton, O., Wagar, L. E., Rubelt, F., et al. (2017). Identifying specificity groups in the T cell receptor repertoire. *Nature*, *547*(7661), 94–98. https://doi.org/10.1038/nature22976.

Gopalakrishnan, V., Spencer, C. N., Nezi, L., Reuben, A., Andrews, M. C., Karpinets, T. V., et al. (2018). Gut microbiome modulates response to anti-PD-1 immunotherapy in melanoma patients. *Science*, *359*(6371), 97–103. https://doi.org/10.1126/science.aan4236.

Graham, J. B., & Graham, R. M. (1959). The effect of vaccine on cancer patients. *Surgery, Gynecology & Obstetrics*, *109*(2), 131–138.

Grobner, S. N., Worst, B. C., Weischenfeldt, J., Buchhalter, I., Kleinheinz, K., Rudneva, V. A., et al. (2018). The landscape of genomic alterations across childhood cancers. *Nature*, *555*(7696), 321–327. https://doi.org/10.1038/nature25480.

Gubin, M. M., Zhang, X., Schuster, H., Caron, E., Ward, J. P., Noguchi, T., et al. (2014). Checkpoint blockade cancer immunotherapy targets tumour-specific mutant antigens. *Nature*, *515*(7528), 577–581. https://doi.org/10.1038/nature13988.

Hagihara, K., Chan, S., Zhang, L., Oh, D. Y., Wei, X. X., Simko, J., et al. (2019). Neoadjuvant sipuleucel-T induces both Th1 activation and immune regulation in localized prostate cancer. *Oncoimmunology*, *8*(1), e1486953. https://doi.org/10.1080/2162402X.2018.1486953.

Halperin, S. O., Tou, C. J., Wong, E. B., Modavi, C., Schaffer, D. V., & Dueber, J. E. (2018). CRISPR-guided DNA polymerases enable diversification of all nucleotides in a tunable window. *Nature*, *560*(7717), 248–252. https://doi.org/10.1038/s41586-018-0384-8.

Harding, S. M., Benci, J. L., Irianto, J., Discher, D. E., Minn, A. J., & Greenberg, R. A. (2017). Mitotic progression following DNA damage enables pattern recognition within micronuclei. *Nature*, *548*(7668), 466–470. https://doi.org/10.1038/nature23470.

Harrington, K. J., Ferris, R. L., Blumenschein, G., Jr., Colevas, A. D., Fayette, J., Licitra, L., et al. (2017). Nivolumab versus standard, single-agent therapy of investigator's choice in recurrent or metastatic squamous cell carcinoma of the head and neck (CheckMate 141): Health-related quality-of-life results from a randomised, phase 3 trial. *The Lancet Oncology*, *18*(8), 1104–1115. https://doi.org/10.1016/s1470-2045(17)30421-7.

Heery, C. R., O'Sullivan-Coyne, G., Madan, R. A., Cordes, L., Rajan, A., Rauckhorst, M., et al. (2017). Avelumab for metastatic or locally advanced previously treated solid tumours (JAVELIN Solid Tumor): A phase 1a, multicohort, dose-escalation trial. *The Lancet Oncology*, *18*(5), 587–598. https://doi.org/10.1016/s1470-2045(17)30239-5.

Herbst, R. S., Soria, J. C., Kowanetz, M., Fine, G. D., Hamid, O., Gordon, M. S., et al. (2014). Predictive correlates of response to the anti-PD-L1 antibody MPDL3280A in cancer patients. *Nature*, *515*(7528), 563–567. https://doi.org/10.1038/nature14011.

Hida, T., Kaji, R., Satouchi, M., Ikeda, N., Horiike, A., Nokihara, H., et al. (2018). Atezolizumab in Japanese patients with previously treated advanced non-small-cell lung cancer: A subgroup analysis of the phase 3 OAK study. *Clinical Lung Cancer*, *19*(4), e405–e415. https://doi.org/10.1016/j.cllc.2018.01.004.

Hildner, K., Edelson, B. T., Purtha, W. E., Diamond, M., Matsushita, H., Kohyama, M., et al. (2008). Batf3 deficiency reveals a critical role for CD8alpha+ dendritic cells in cytotoxic T cell immunity. *Science*, *322*(5904), 1097–1100. https://doi.org/10.1126/science.1164206.

Hille, F., Richter, H., Wong, S. P., Bratovic, M., Ressel, S., & Charpentier, E. (2018). The biology of CRISPR-Cas: Backward and forward. *Cell*, *172*(6), 1239–1259. https://doi.org/10.1016/j.cell.2017.11.032.

Hodi, F. S., Hwu, W. J., Kefford, R., Weber, J. S., Daud, A., Hamid, O., et al. (2016). Evaluation of immune-related response criteria and RECIST v1.1 in patients with advanced melanoma treated with pembrolizumab. *Journal of Clinical Oncology*, *34*(13), 1510–1517. https://doi.org/10.1200/jco.2015.64.0391.

Hodi, F. S., O'Day, S. J., McDermott, D. F., Weber, R. W., Sosman, J. A., Haanen, J. B., et al. (2010). Improved survival with ipilimumab in patients with metastatic melanoma. *The New England Journal of Medicine*, *363*(8), 711–723. https://doi.org/10.1056/NEJMoa1003466.

Horn, L., Spigel, D. R., Vokes, E. E., Holgado, E., Ready, N., Steins, M., et al. (2017). Nivolumab versus docetaxel in previously treated patients with advanced non-small-cell lung cancer: Two-year outcomes from two randomized, open-label, phase III trials (CheckMate 017 and CheckMate 057). *Journal of Clinical Oncology*, *35*(35), 3924–3933. https://doi.org/10.1200/jco.2017.74.3062.

Hoshino, A., Costa-Silva, B., Shen, T. L., Rodrigues, G., Hashimoto, A., Tesic Mark, M., et al. (2015). Tumour exosome integrins determine organotropic metastasis. *Nature*, *527*(7578), 329–335. https://doi.org/10.1038/nature15756.

Hsu, C., Lee, S. H., Ejadi, S., Even, C., Cohen, R. B., Le Tourneau, C., et al. (2017). Safety and antitumor activity of pembrolizumab in patients with programmed death-ligand 1-positive nasopharyngeal carcinoma: results of the KEYNOTE-028 study. *Journal of Clinical Oncology*, *35*(36), 4050–4056. https://doi.org/10.1200/jco.2017.73.3675.

Hui, R., Garon, E. B., Goldman, J. W., Leighl, N. B., Hellmann, M. D., Patnaik, A., et al. (2017). Pembrolizumab as first-line therapy for patients with PD-L1-positive advanced non-small cell lung cancer: A phase 1 trial. *Annals of Oncology*, *28*(4), 874–881. https://doi.org/10.1093/annonc/mdx008.

Humbert, M., & Hugues, S. (2019). Warming up the tumor microenvironment in order to enhance immunogenicity. *Oncoimmunology*, *8*(1), e1510710. https://doi.org/10.1080/2162402X.2018.1510710.

Hundal, J., Kiwala, S., Feng, Y. Y., Liu, C. J., Govindan, R., Chapman, W. C., et al. (2019). Accounting for proximal variants improves neoantigen prediction. *Nature Genetics*, *51*(1), 175–179. https://doi.org/10.1038/s41588-018-0283-9.

Iida, N., Dzutsev, A., Stewart, C. A., Smith, L., Bouladoux, N., Weingarten, R. A., et al. (2013). Commensal bacteria control cancer response to therapy by modulating the tumor microenvironment. *Science*, *342*(6161), 967–970. https://doi.org/10.1126/science.1240527.

Jaiswal, S., Jamieson, C. H., Pang, W. W., Park, C. Y., Chao, M. P., Majeti, R., et al. (2009). CD47 is upregulated on circulating hematopoietic stem cells and leukemia cells to avoid phagocytosis. *Cell*, *138*(2), 271–285. https://doi.org/10.1016/j.cell.2009.05.046.

Janikovits, J., Muller, M., Krzykalla, J., Korner, S., Echterdiek, F., Lahrmann, B., et al. (2018). High numbers of PDCD1 (PD-1)-positive T cells and B2M mutations in microsatellite-unstable colorectal cancer. *Oncoimmunology*, 7(2), e1390640. https://doi.org/10.1080/2162402X.2017.1390640.

Joseph, R. W., Elassaiss-Schaap, J., Kefford, R., Hwu, W. J., Wolchok, J. D., Joshua, A. M., et al. (2018). Baseline tumor size is an independent prognostic factor for overall survival in patients with melanoma treated with pembrolizumab. *Clinical Cancer Research*, *24*(20), 4960–4967. https://doi.org/10.1158/1078-0432.ccr-17-2386.

Kalhor, R., Kalhor, K., Mejia, L., Leeper, K., Graveline, A., Mali, P., et al. (2018). Developmental barcoding of whole mouse via homing CRISPR. *Science*, *361*(6405), eaat9804. https://doi.org/10.1126/science.aat9804.

Kantoff, P. W., Higano, C. S., Shore, N. D., Berger, E. R., Small, E. J., Penson, D. F., et al. (2010). Sipuleucel-T immunotherapy for castration-resistant prostate cancer. *The New England Journal of Medicine*, *363*(5), 411–422. https://doi.org/10.1056/NEJMoa1001294.

Kaplan, D. H., Shankaran, V., Dighe, A. S., Stockert, E., Aguet, M., Old, L. J., et al. (1998). Demonstration of an interferon gamma-dependent tumor surveillance system in immunocompetent mice. *Proceedings of the National Academy of Sciences of the United States of America*, *95*(13), 7556–7561.

Karasaki, T., Nagayama, K., Kuwano, H., Nitadori, J. I., Sato, M., Anraku, M., et al. (2017). Prediction and prioritization of neoantigens: Integration of RNA sequencing data with whole-exome sequencing. *Cancer Science*, *108*(2), 170–177. https://doi.org/10.1111/cas.13131.

Kaufman, H. L., Russell, J., Hamid, O., Bhatia, S., Terheyden, P., D'Angelo, S. P., et al. (2016). Avelumab in patients with chemotherapy-refractory metastatic Merkel cell carcinoma: A multicentre, single-group, open-label, phase 2 trial. *The Lancet Oncology*, *17*(10), 1374–1385. https://doi.org/10.1016/s1470-2045(16)30364-3.

Kaufman, H. L., Russell, J. S., Hamid, O., Bhatia, S., Terheyden, P., D'Angelo, S. P., et al. (2018). Updated efficacy of avelumab in patients with previously treated metastatic Merkel cell carcinoma after >/=1 year of follow-up: JAVELIN Merkel 200, a phase 2 clinical trial. *Journal for Immunotherapy of Cancer*, *6*(1), 7. https://doi.org/10.1186/s40425-017-0310-x.

Kazandjian, D., Suzman, D. L., Blumenthal, G., Mushti, S., He, K., Libeg, M., et al. (2016). FDA approval summary: Nivolumab for the treatment of metastatic non-small cell lung cancer with progression on or after platinum-based chemotherapy. *The Oncologist, 21*(5), 634–642. https://doi.org/10.1634/theoncologist.2015-0507.

Kellner, C., Peipp, M., Gramatzki, M., Schrappe, M., & Schewe, D. M. (2018). Perspectives of Fc engineered antibodies in CD19 targeting immunotherapies in pediatric B-cell precursor acute lymphoblastic leukemia. *Oncoimmunology*, 7(8), e1448331. https://doi.org/10.1080/2162402X.2018.1448331.

Kelly, P. N. (2018). The cancer immunotherapy revolution. *Science, 359*(6382), 1344–1345. https://doi.org/10.1126/science.359.6382.1344.

Kiyota, N., Hasegawa, Y., Takahashi, S., Yokota, T., Yen, C. J., Iwae, S., et al. (2017). A randomized, open-label, phase III clinical trial of nivolumab vs. therapy of investigator's choice in recurrent squamous cell carcinoma of the head and neck: A subanalysis of Asian patients versus the global population in checkmate 141. *Oral Oncology, 73*, 138–146. https://doi.org/10.1016/j.oraloncology.2017.07.023.

Koebel, C. M., Vermi, W., Swann, J. B., Zerafa, N., Rodig, S. J., Old, L. J., et al. (2007). Adaptive immunity maintains occult cancer in an equilibrium state. *Nature, 450*(7171), 903–907. https://doi.org/10.1038/nature06309.

Krummel, M. F., & Allison, J. P. (1995). CD28 and CTLA-4 have opposing effects on the response of T cells to stimulation. *The Journal of Experimental Medicine, 182*(2), 459–465.

Lavin, Y., Kobayashi, S., Leader, A., Amir, E. D., Elefant, N., Bigenwald, C., et al. (2017). Innate immune landscape in early lung adenocarcinoma by paired single-cell analyses. *Cell, 169*(4), 750–765.e717. https://doi.org/10.1016/j.cell.2017.04.014.

Le, D. T., Durham, J. N., Smith, K. N., Wang, H., Bartlett, B. R., Aulakh, L. K., et al. (2017). Mismatch repair deficiency predicts response of solid tumors to PD-1 blockade. *Science, 357*(6349), 409–413. https://doi.org/10.1126/science.aan6733.

Le, D. T., Uram, J. N., Wang, H., Bartlett, B. R., Kemberling, H., Eyring, A. D., et al. (2015). PD-1 blockade in tumors with mismatch-repair deficiency. *The New England Journal of Medicine, 372*(26), 2509–2520. https://doi.org/10.1056/NEJMoa1500596.

Leach, D. R., Krummel, M. F., & Allison, J. P. (1996). Enhancement of antitumor immunity by CTLA-4 blockade. *Science, 271*(5256), 1734–1736.

Levy, A., Massard, C., Soria, J. C., & Deutsch, E. (2016). Concurrent irradiation with the anti-programmed cell death ligand-1 immune checkpoint blocker durvalumab: Single centre subset analysis from a phase 1/2 trial. *European Journal of Cancer, 68*, 156–162. https://doi.org/10.1016/j.ejca.2016.09.013.

Li, H., van der Leun, A. M., Yofe, I., Lubling, Y., Gelbard-Solodkin, D., van Akkooi, A. C. J., … Amit, I. (2019). Dysfunctional CD8 T cells form a proliferative, dynamically regulated compartment within human melanoma. Cell, 176(4), 775-789 e718. doi:https://doi.org/10.1016/j.cell.2018.11.043

Lopez-Soto, A., Gonzalez, S., Smyth, M. J., & Galluzzi, L. (2017). Control of metastasis by NK cells. *Cancer Cell, 32*(2), 135–154. https://doi.org/10.1016/j.ccell.2017.06.009.

Mackenzie, K. J., Carroll, P., Martin, C. A., Murina, O., Fluteau, A., Simpson, D. J., et al. (2017). cGAS surveillance of micronuclei links genome instability to innate immunity. *Nature, 548*(7668), 461–465. https://doi.org/10.1038/nature23449.

Majeti, R., Chao, M. P., Alizadeh, A. A., Pang, W. W., Jaiswal, S., Gibbs, K. D., Jr., et al. (2009). CD47 is an adverse prognostic factor and therapeutic antibody target on human acute myeloid leukemia stem cells. *Cell, 138*(2), 286–299. https://doi.org/10.1016/j.cell.2009.05.045.

Mariathasan, S., Turley, S. J., Nickles, D., Castiglioni, A., Yuen, K., Wang, Y., et al. (2018). TGFbeta attenuates tumour response to PD-L1 blockade by contributing to exclusion of T cells. *Nature, 554*(7693), 544–548. https://doi.org/10.1038/nature25501.

Maude, S. L., Laetsch, T. W., Buechner, J., Rives, S., Boyer, M., Bittencourt, H., et al. (2018). Tisagenlecleucel in children and young adults with B-cell lymphoblastic leukemia. *The New England Journal of Medicine*, *378*(5), 439–448. https://doi.org/10.1056/NEJMoa1709866.

McDermott, D., Haanen, J., Chen, T. T., Lorigan, P., & O'Day, S. (2013). Efficacy and safety of ipilimumab in metastatic melanoma patients surviving more than 2 years following treatment in a phase III trial (MDX010-20). *Annals of Oncology*, *24*(10), 2694–2698. https://doi.org/10.1093/annonc/mdt291.

Merchant, M. S., Wright, M., Baird, K., Wexler, L. H., Rodriguez-Galindo, C., Bernstein, D., et al. (2016). Phase I clinical trial of ipilimumab in pediatric patients with advanced solid tumors. *Clinical Cancer Research*, *22*(6), 1364–1370. https://doi.org/10.1158/1078-0432.ccr-15-0491.

Migden, M. R., Rischin, D., Schmults, C. D., Guminski, A., Hauschild, A., Lewis, K. D., et al. (2018). PD-1 blockade with cemiplimab in advanced cutaneous squamous-cell carcinoma. *The New England Journal of Medicine*, *379*(4), 341–351. https://doi.org/10.1056/NEJMoa1805131.

Mirzaei, R., Sarkar, S., Dzikowski, L., Rawji, K. S., Khan, L., Faissner, A., et al. (2018). Brain tumor-initiating cells export tenascin-C associated with exosomes to suppress T cell activity. *Oncoimmunology*, 7(10), e1478647. https://doi.org/10.1080/2162402X.2018.1478647.

Morgan, R. A., Dudley, M. E., Wunderlich, J. R., Hughes, M. S., Yang, J. C., Sherry, R. M., et al. (2006). Cancer regression in patients after transfer of genetically engineered lymphocytes. *Science*, *314*(5796), 126–129. https://doi.org/10.1126/science.1129003.

Motzer, R. J., Escudier, B., McDermott, D. F., George, S., Hammers, H. J., Srinivas, S., et al. (2015). Nivolumab versus everolimus in advanced renal-cell carcinoma. *The New England Journal of Medicine*, *373*(19), 1803–1813. https://doi.org/10.1056/NEJMoa1510665.

Mulqueen, R. M., Pokholok, D., Norberg, S. J., Torkenczy, K. A., Fields, A. J., Sun, D., et al. (2018). Highly scalable generation of DNA methylation profiles in single cells. *Nature Biotechnology*, *36*(5), 428–431. https://doi.org/10.1038/nbt.4112.

Muro, K., Chung, H. C., Shankaran, V., Geva, R., Catenacci, D., Gupta, S., et al. (2016). Pembrolizumab for patients with PD-L1-positive advanced gastric cancer (KEYNOTE-012): A multicentre, open-label, phase 1b trial. *The Lancet Oncology*, *17*(6), 717–726. https://doi.org/10.1016/s1470-2045(16)00175-3.

Nanda, R., Chow, L. Q., Dees, E. C., Berger, R., Gupta, S., Geva, R., et al. (2016). Pembrolizumab in patients with advanced triple-negative breast cancer: Phase Ib KEYNOTE-012 study. *Journal of Clinical Oncology*, *34*(21), 2460–2467. https://doi.org/10.1200/jco.2015.64.8931.

Necchi, A., Joseph, R. W., Loriot, Y., Hoffman-Censits, J., Perez-Gracia, J. L., Petrylak, D. P., et al. (2017). Atezolizumab in platinum-treated locally advanced or metastatic urothelial carcinoma: Post-progression outcomes from the phase II IMvigor210 study. *Annals of Oncology*, *28*(12), 3044–3050. https://doi.org/10.1093/annonc/mdx518.

Neelapu, S. S., Locke, F. L., Bartlett, N. L., Lekakis, L. J., Miklos, D. B., Jacobson, C. A., et al. (2017). Axicabtagene ciloleucel CAR T-cell therapy in refractory large B-cell lymphoma. *The New England Journal of Medicine*, *377*(26), 2531–2544. https://doi.org/10.1056/NEJMoa1707447.

Nishimasu, H., Shi, X., Ishiguro, S., Gao, L., Hirano, S., Okazaki, S., et al. (2018). Engineered CRISPR-Cas9 nuclease with expanded targeting space. *Science*, *361*(6408), 1259–1262. https://doi.org/10.1126/science.aas9129.

Obeid, M., Tesniere, A., Ghiringhelli, F., Fimia, G. M., Apetoh, L., Perfettini, J. L., et al. (2007). Calreticulin exposure dictates the immunogenicity of cancer cell death. *Nature Medicine*, *13*(1), 54–61. https://doi.org/10.1038/nm1523.

O'Neil, B. H., Wallmark, J. M., Lorente, D., Elez, E., Raimbourg, J., Gomez-Roca, C., et al. (2017). Safety and antitumor activity of the anti-PD-1 antibody pembrolizumab in patients with advanced colorectal carcinoma. *PLoS One*, *12*(12), e0189848. https://doi.org/10.1371/journal.pone.0189848.

Ott, P. A., Elez, E., Hiret, S., Kim, D. W., Morosky, A., Saraf, S., et al. (2017). Pembrolizumab in patients with extensive-stage small-cell lung cancer: Results from the phase Ib KEYNOTE-028 study. *Journal of Clinical Oncology*, *35*(34), 3823–3829. https://doi.org/10.1200/jco.2017.72.5069.

Ott, P. A., Piha-Paul, S. A., Munster, P., Pishvaian, M. J., van Brummelen, E. M. J., Cohen, R. B., et al. (2017). Safety and antitumor activity of the anti-PD-1 antibody pembrolizumab in patients with recurrent carcinoma of the anal canal. *Annals of Oncology*, *28*(5), 1036–1041. https://doi.org/10.1093/annonc/mdx029.

Overman, M. J., McDermott, R., Leach, J. L., Lonardi, S., Lenz, H. J., Morse, M. A., et al. (2017). Nivolumab in patients with metastatic DNA mismatch repair-deficient or microsatellite instability-high colorectal cancer (CheckMate 142): An open-label, multicentre, phase 2 study. *The Lancet Oncology*, *18*(9), 1182–1191. https://doi.org/10.1016/s1470-2045(17)30422-9.

Oyer, J. L., Gitto, S. B., Altomare, D. A., & Copik, A. J. (2018). PD-L1 blockade enhances anti-tumor efficacy of NK cells. *Oncoimmunology*, 7(11), e1509819. https://doi.org/10.1080/2162402X.2018.1509819.

Ozcan, M., Janikovits, J., von Knebel Doeberitz, M., & Kloor, M. (2018). Complex pattern of immune evasion in MSI colorectal cancer. *Oncoimmunology*, 7(7), e1445453. https://doi.org/10.1080/2162402X.2018.1445453.

Patnaik, A., Kang, S. P., Rasco, D., Papadopoulos, K. P., Elassaiss-Schaap, J., Beeram, M., et al. (2015). Phase I study of pembrolizumab (MK-3475; Anti-PD-1 monoclonal antibody) in patients with advanced solid tumors. *Clinical Cancer Research*, *21*(19), 4286–4293. https://doi.org/10.1158/1078-0432.ccr-14-2607.

Perez-Gracia, J. L., Loriot, Y., Rosenberg, J. E., Powles, T., Necchi, A., Hussain, S. A., et al. (2018). Atezolizumab in platinum-treated locally advanced or metastatic urothelial carcinoma: Outcomes by prior number of regimens. *European Urology*, *73*, 462–468. https://doi.org/10.1016/j.eururo.2017.11.023.

Petrella, T. M., Robert, C., Richtig, E., Miller, W. H., Jr., Masucci, G. V., Walpole, E., et al. (2017). Patient-reported outcomes in KEYNOTE-006, a randomised study of pembrolizumab versus ipilimumab in patients with advanced melanoma. *European Journal of Cancer*, *86*, 115–124. https://doi.org/10.1016/j.ejca.2017.08.032.

Plimack, E. R., Bellmunt, J., Gupta, S., Berger, R., Chow, L. Q., Juco, J., et al. (2017). Safety and activity of pembrolizumab in patients with locally advanced or metastatic urothelial cancer (KEYNOTE-012): A non-randomised, open-label, phase 1b study. *The Lancet Oncology*, *18*(2), 212–220. https://doi.org/10.1016/S1470-2045(17)30007-4.

Pol, J., Kroemer, G., & Galluzzi, L. (2016). First oncolytic virus approved for melanoma immunotherapy. *Oncoimmunology*, *5*(1), e1115641. https://doi.org/10.1080/2162402X.2015.1115641.

Pol, J., Levesque, S., Workenhe, S. T., Gujar, S., Le Boeuf, F., D, R. C., et al. (2018). Trial watch: Oncolytic viro-immunotherapy of hematologic and solid tumors. *Oncoimmunology*, 7(12), e1503032. https://doi.org/10.1080/2162402X.2018.1503032.

Pol, J., Vacchelli, E., Aranda, F., Castoldi, F., Eggermont, A., Cremer, I., et al. (2015). Trial watch: Immunogenic cell death inducers for anticancer chemotherapy. *Oncoimmunology*, *4*(4). e1008866https://doi.org/10.1080/2162402X.2015.1008866.

Powles, T., O'Donnell, P. H., Massard, C., Arkenau, H. T., Friedlander, T. W., Hoimes, C. J., et al. (2017). Efficacy and safety of durvalumab in locally advanced or metastatic urothelial carcinoma: Updated results from a Phase 1/2 open-label study. *JAMA Oncology*, *3*(9), e172411. https://doi.org/10.1001/jamaoncol.2017.2411.

Raikar, S. S., Fleischer, L. C., Moot, R., Fedanov, A., Paik, N. Y., Knight, K. A., et al. (2018). Development of chimeric antigen receptors targeting T-cell malignancies using two structurally different anti-CD5 antigen binding domains in NK and CRISPR-edited T cell lines. *Oncoimmunology*, 7(3), e1407898. https://doi.org/10.1080/2162402X.2017.1407898.

Rao, S., Gharib, K., & Han, A. (2019). Cancer immunosurveillance by T cells. *International Review of Cell and Molecular Biology*, *342*, 149–173. https://doi.org/10.1016/bs.ircmb.2018.08.001.

Reck, M., Rodriguez-Abreu, D., Robinson, A. G., Hui, R., Csoszi, T., Fulop, A., et al. (2016). Pembrolizumab versus chemotherapy for PD-L1-positive non-small-cell lung cancer. *The New England Journal of Medicine*, *375*(19), 1823–1833. https://doi.org/10.1056/NEJMoa1606774.

Riaz, N., Havel, J. J., Makarov, V., Desrichard, A., Urba, W. J., Sims, J. S., et al. (2017). Tumor and microenvironment evolution during immunotherapy with nivolumab. *Cell*, *171*(4), 934–949.e916. https://doi.org/10.1016/j.cell.2017.09.028.

Ribas, A., Hamid, O., Daud, A., Hodi, F. S., Wolchok, J. D., Kefford, R., et al. (2016). Association of pembrolizumab with tumor response and survival among patients with advanced melanoma. *JAMA*, *315*(15), 1600–1609. https://doi.org/10.1001/jama.2016.4059.

Rittmeyer, A., Barlesi, F., Waterkamp, D., Park, K., Ciardiello, F., von Pawel, J., et al. (2017). Atezolizumab versus docetaxel in patients with previously treated non-small-cell lung cancer (OAK): A phase 3, open-label, multicentre randomised controlled trial. *Lancet*, *389*(10066), 255–265. https://doi.org/10.1016/s0140-6736(16)32517-x.

Rizvi, N. A., Hellmann, M. D., Snyder, A., Kvistborg, P., Makarov, V., Havel, J. J., et al. (2015). Cancer immunology. Mutational landscape determines sensitivity to PD-1 blockade in non-small cell lung cancer. *Science*, *348*(6230), 124–128. https://doi.org/10.1126/science.aaa1348.

Robert, C., Ribas, A., Hamid, O., Daud, A., Wolchok, J. D., Joshua, A. M., et al. (2018). Durable complete response after discontinuation of pembrolizumab in patients with metastatic melanoma. *Journal of Clinical Oncology*, *36*(17), 1668–1674. https://doi.org/10.1200/jco.2017.75.6270.

Robert, C., Ribas, A., Wolchok, J. D., Hodi, F. S., Hamid, O., Kefford, R., et al. (2014). Anti-programmed-death-receptor-1 treatment with pembrolizumab in ipilimumab-refractory advanced melanoma: A randomised dose-comparison cohort of a phase 1 trial. *Lancet*, *384*(9948), 1109–1117. https://doi.org/10.1016/s0140-6736(14)60958-2.

Robert, C., Schachter, J., Long, G. V., Arance, A., Grob, J. J., Mortier, L., et al. (2015). Pembrolizumab versus ipilimumab in advanced melanoma. *The New England Journal of Medicine*, *372*(26), 2521–2532. https://doi.org/10.1056/NEJMoa1503093.

Robert, C., Schadendorf, D., Messina, M., Hodi, F. S., & O'Day, S. (2013). Efficacy and safety of retreatment with ipilimumab in patients with pretreated advanced melanoma who progressed after initially achieving disease control. *Clinical Cancer Research*, *19*(8), 2232–2239. https://doi.org/10.1158/1078-0432.ccr-12-3080.

Rosenberg, J. E., Hoffman-Censits, J., Powles, T., van der Heijden, M. S., Balar, A. V., Necchi, A., et al. (2016). Atezolizumab in patients with locally advanced and metastatic urothelial carcinoma who have progressed following treatment with platinum-based chemotherapy: A single-arm, multicentre, phase 2 trial. *Lancet*, *387*(10031), 1909–1920. https://doi.org/10.1016/s0140-6736(16)00561-4.

Routy, B., Le Chatelier, E., Derosa, L., Duong, C. P. M., Alou, M. T., Daillere, R., et al. (2018). Gut microbiome influences efficacy of PD-1-based immunotherapy against epithelial tumors. *Science*, *359*(6371), 91–97. https://doi.org/10.1126/science.aan3706.

Ruscetti, M., Leibold, J., Bott, M. J., Fennell, M., Kulick, A., Salgado, N. R., et al. (2018). NK cell-mediated cytotoxicity contributes to tumor control by a cytostatic drug combination. *Science*, *362*(6421), 1416–1422. https://doi.org/10.1126/science.aas9090.

Savage, P. A., Vosseller, K., Kang, C., Larimore, K., Riedel, E., Wojnoonski, K., et al. (2008). Recognition of a ubiquitous self antigen by prostate cancer-infiltrating CD8+ T lymphocytes. *Science*, *319*(5860), 215–220. https://doi.org/10.1126/science.1148886.

Schachter, J., Ribas, A., Long, G. V., Arance, A., Grob, J. J., Mortier, L., et al. (2017). Pembrolizumab versus ipilimumab for advanced melanoma: final overall survival results of a multicentre, randomised, open-label phase 3 study (KEYNOTE-006). *Lancet*, *390*(10105), 1853–1862. https://doi.org/10.1016/s0140-6736(17)31601-x.

Schumacher, T. N., & Schreiber, R. D. (2015). Neoantigens in cancer immunotherapy. *Science*, *348*(6230), 69–74. https://doi.org/10.1126/science.aaa4971.

Schuster, S. J., Bishop, M. R., Tam, C. S., Waller, E. K., Borchmann, P., McGuirk, J. P., et al. (2019). Tisagenlecleucel in adult relapsed or refractory diffuse large B-cell lymphoma. *The New England Journal of Medicine*, *380*(1), 45–56. https://doi.org/10.1056/NEJMoa1804980.

Seiwert, T. Y., Burtness, B., Mehra, R., Weiss, J., Berger, R., Eder, J. P., et al. (2016). Safety and clinical activity of pembrolizumab for treatment of recurrent or metastatic squamous cell carcinoma of the head and neck (KEYNOTE-012): An open-label, multicentre, phase 1b trial. *The Lancet Oncology*, *17*(7), 956–965. https://doi.org/10.1016/s1470-2045(16)30066-3.

Shankaran, V., Ikeda, H., Bruce, A. T., White, J. M., Swanson, P. E., Old, L. J., et al. (2001). IFNgamma and lymphocytes prevent primary tumour development and shape tumour immunogenicity. *Nature*, *410*(6832), 1107–1111. https://doi.org/10.1038/35074122.

Sharma, P., Retz, M., Siefker-Radtke, A., Baron, A., Necchi, A., Bedke, J., et al. (2017). Nivolumab in metastatic urothelial carcinoma after platinum therapy (CheckMate 275): A multicentre, single-arm, phase 2 trial. *The Lancet Oncology*, *18*(3), 312–322. https://doi.org/10.1016/s1470-2045(17)30065-7.

Shaverdian, N., Lisberg, A. E., Bornazyan, K., Veruttipong, D., Goldman, J. W., Formenti, S. C., et al. (2017). Previous radiotherapy and the clinical activity and toxicity of pembrolizumab in the treatment of non-small-cell lung cancer: A secondary analysis of the KEYNOTE-001 phase 1 trial. *The Lancet Oncology*, *18*(7), 895–903. https://doi.org/10.1016/s1470-2045(17)30380-7.

Shen, M. W., Arbab, M., Hsu, J. Y., Worstell, D., Culbertson, S. J., Krabbe, O., et al. (2018). Predictable and precise template-free CRISPR editing of pathogenic variants. *Nature*, *563*(7733), 646–651. https://doi.org/10.1038/s41586-018-0686-x.

Spitzer, M. H., & Nolan, G. P. (2016). Mass cytometry: Single cells, many features. *Cell*, *165*(4), 780–791. https://doi.org/10.1016/j.cell.2016.04.019.

Spranger, S., Bao, R., & Gajewski, T. F. (2015). Melanoma-intrinsic beta-catenin signalling prevents anti-tumour immunity. *Nature*, *523*(7559), 231–235. https://doi.org/10.1038/nature14404.

Straussman, R., Morikawa, T., Shee, K., Barzily-Rokni, M., Qian, Z. R., Du, J., et al. (2012). Tumour micro-environment elicits innate resistance to RAF inhibitors through HGF secretion. *Nature*, *487*(7408), 500–504. https://doi.org/10.1038/nature11183.

Tahara, M., Muro, K., Hasegawa, Y., Chung, H. C., Lin, C. C., Keam, B., et al. (2018). Pembrolizumab in Asia-Pacific patients with advanced head and neck squamous cell carcinoma: Analyses from KEYNOTE-012. *Cancer Science*, *109*(3), 771–776. https://doi.org/10.1111/cas.13480.

Tauriello, D. V. F., Palomo-Ponce, S., Stork, D., Berenguer-Llergo, A., Badia-Ramentol, J., Iglesias, M., et al. (2018). TGFbeta drives immune evasion in genetically reconstituted colon cancer metastasis. *Nature, 554*(7693), 538–543. https://doi.org/10.1038/nature25492.
Vacchelli, E., Eggermont, A., Sautes-Fridman, C., Galon, J., Zitvogel, L., Kroemer, G., et al. (2013). Trial watch: Oncolytic viruses for cancer therapy. *Oncoimmunology, 2*(6). e24612https://doi.org/10.4161/onci.24612.
van Montfoort, N., Borst, L., Korrer, M. J., Sluijter, M., Marijt, K. A., Santegoets, S. J., et al. (2018). NKG2A blockade potentiates CD8 T cell immunity induced by cancer vaccines. *Cell, 175*(7), 1744–1755.e1715. https://doi.org/10.1016/j.cell.2018.10.028.
Vanpouille-Box, C., Alard, A., Aryankalayil, M. J., Sarfraz, Y., Diamond, J. M., Schneider, R. J., et al. (2017). DNA exonuclease Trex1 regulates radiotherapy-induced tumour immunogenicity. *Nature Communications, 8*, 15618. https://doi.org/10.1038/ncomms15618.
Vanpouille-Box, C., Demaria, S., Formenti, S. C., & Galluzzi, L. (2018). Cytosolic DNA sensing in organismal tumor control. *Cancer Cell, 34*(3), 361–378. https://doi.org/10.1016/j.ccell.2018.05.013.
Vetizou, M., Pitt, J. M., Daillere, R., Lepage, P., Waldschmitt, N., Flament, C., et al. (2015). Anticancer immunotherapy by CTLA-4 blockade relies on the gut microbiota. *Science, 350*(6264), 1079–1084. https://doi.org/10.1126/science.aad1329.
Viaud, S., Saccheri, F., Mignot, G., Yamazaki, T., Daillere, R., Hannani, D., et al. (2013). The intestinal microbiota modulates the anticancer immune effects of cyclophosphamide. *Science, 342*(6161), 971–976. https://doi.org/10.1126/science.1240537.
Vitale, I., Sistigu, A., Manic, G., Rudqvist, N. P., Trajanoski, Z., & Galluzzi, L. (2019). Mutational and antigenic landscape in tumor progression and cancer immunotherapy. *Trends in Cell Biology, 29*, 395–416. https://doi.org/10.1016/j.tcb.2019.01.003.
Vokes, E. E., Ready, N., Felip, E., Horn, L., Burgio, M. A., Antonia, S. J., et al. (2018). Nivolumab versus docetaxel in previously treated advanced non-small-cell lung cancer (CheckMate 017 and CheckMate 057): 3-year update and outcomes in patients with liver metastases. *Annals of Oncology, 29*(4), 959–965. https://doi.org/10.1093/annonc/mdy041.
Wang, H., Xu, X., Nguyen, C. M., Liu, Y., Gao, Y., Lin, X., et al. (2018). CRISPR-mediated programmable 3D genome positioning and nuclear organization. *Cell, 175*(5), 1405–1417.e1414. https://doi.org/10.1016/j.cell.2018.09.013.
Weber, J. S., D'Angelo, S. P., Minor, D., Hodi, F. S., Gutzmer, R., Neyns, B., et al. (2015). Nivolumab versus chemotherapy in patients with advanced melanoma who progressed after anti-CTLA-4 treatment (CheckMate 037): A randomised, controlled, open-label, phase 3 trial. *The Lancet Oncology, 16*(4), 375–384. https://doi.org/10.1016/s1470-2045(15)70076-8.
Weber, J., Mandala, M., Del Vecchio, M., Gogas, H. J., Arance, A. M., Cowey, C. L., et al. (2017). Adjuvant nivolumab versus ipilimumab in resected stage III or IV melanoma. *The New England Journal of Medicine, 377*(19), 1824–1835. https://doi.org/10.1056/NEJMoa1709030.
Wierz, M., Janji, B., Berchem, G., Moussay, E., & Paggetti, J. (2018). High-dimensional mass cytometry analysis revealed microenvironment complexity in chronic lymphocytic leukemia. *Oncoimmunology*, 7(8), e1465167. https://doi.org/10.1080/2162402X.2018.1465167.
Wu, Y., Williams, E. G., Dubuis, S., Mottis, A., Jovaisaite, V., Houten, S. M., et al. (2014). Multilayered genetic and omics dissection of mitochondrial activity in a mouse reference population. *Cell, 158*(6), 1415–1430. https://doi.org/10.1016/j.cell.2014.07.039.
Xu, J. X., Maher, V. E., Zhang, L., Tang, S., Sridhara, R., Ibrahim, A., et al. (2017). FDA approval summary: nivolumab in advanced renal cell carcinoma after anti-angiogenic therapy and exploratory predictive biomarker analysis. *The Oncologist, 22*(3), 311–317. https://doi.org/10.1634/theoncologist.2016-0476.

Yaddanapudi, K., Meng, S., Whitt, A. G., Al Rayyan, N., Richie, J., Tu, A., et al. (2019). Exosomes from GM-CSF expressing embryonic stem cells are an effective prophylactic vaccine for cancer prevention. *Oncoimmunology, 8*(3), 1561119. https://doi.org/10.1080/2162402X.2018.1561119.

Yan, W. X., Hunnewell, P., Alfonse, L. E., Carte, J. M., Keston-Smith, E., Sothiselvam, S., et al. (2019). Functionally diverse type V CRISPR-Cas systems. *Science, 363*(6422), 88–91. https://doi.org/10.1126/science.aav7271.

Younes, A., Santoro, A., Shipp, M., Zinzani, P. L., Timmerman, J. M., Ansell, S., et al. (2016). Nivolumab for classical Hodgkin's lymphoma after failure of both autologous stem-cell transplantation and brentuximab vedotin: A multicentre, multicohort, single-arm phase 2 trial. *The Lancet Oncology, 17*(9), 1283–1294. https://doi.org/10.1016/s1470-2045(16)30167-x.

Zhang, S. Q., Ma, K. Y., Schonnesen, A. A., Zhang, M., He, C., Sun, E., et al. (2018). High-throughput determination of the antigen specificities of T cell receptors in single cells. *Nature Biotechnology*. https://doi.org/10.1038/nbt.4282.

Zhang, L., Zhang, S., Yao, J., Lowery, F. J., Zhang, Q., Huang, W. C., et al. (2015). Microenvironment-induced PTEN loss by exosomal microRNA primes brain metastasis outgrowth. *Nature, 527*(7576), 100–104. https://doi.org/10.1038/nature15376.

Zheng, C., Zheng, L., Yoo, J. K., Guo, H., Zhang, Y., Guo, X., et al. (2017). Landscape of infiltrating T cells in liver cancer revealed by single-cell sequencing. *Cell, 169*(7), 1342–1356.e1316. https://doi.org/10.1016/j.cell.2017.05.035.

Zinzani, P. L., Ribrag, V., Moskowitz, C. H., Michot, J. M., Kuruvilla, J., Balakumaran, A., et al. (2017). Safety and tolerability of pembrolizumab in patients with relapsed/refractory primary mediastinal large B-cell lymphoma. *Blood, 130*(3), 267–270. https://doi.org/10.1182/blood-2016-12-758383.

CHAPTER ONE

Cytokine profiling of tumor-infiltrating T lymphocytes by flow cytometry

Jeremy B. Foote[a,]*, Sujith Sarvesh[a], Leisha A. Emens[b,]*

[a]Department of Microbiology, University of Alabama at Birmingham, Birmingham, AL, United States
[b]Department of Medicine, UPMC Hillman Cancer Center, and UPMC Magee Women's Hospital, University of Pittsburgh, Pittsburgh, PA, United States
*Corresponding authors: e-mail address: jbf130@uab.edu; emensla@upmc.edu

Contents

Abstract

Intracellular cytokine staining (ICS) utilizing fluorescently labeled, cytokine-specific antibodies is a powerful technique utilized to evaluate cytokine expression that provides resolution at the single cell level. Combined with multi-parameter flow cytometry, ICS can provide detailed information on complex cytokine profiles and cellular phenotypes within the tumor microenvironment, particularly for the $CD4^+$ T helper and $CD8^+$ cytotoxic T cell compartments. This technique provides critical information concerning tumor-infiltrating T cell function that is essential for evaluating existing or therapeutically-induced tumor antigen-specific T cell responses in both preclinical models and cancer patients. In this chapter we will discuss in detail the critical steps necessary for a successful ICS assay and outline common protocols for the evaluation of cytokine production from T cell subsets present within the tumor microenvironment.

Methods in Enzymology, Volume 631
ISSN 0076-6879
https://doi.org/10.1016/bs.mie.2019.08.015

1. Introduction

The presence of tumor-infiltrating T lymphocytes (TILs) is a positive prognostic marker for both response to treatment and survival in multiple tumor types (Fridman et al., 2013). However, in the absence of successful therapeutic interventions, endogenous responses mediated by tumor-infiltrating $CD4^+$ and $CD8^+$ T lymphocytes are frequently insufficient for tumor regression. Failure of T cell-mediated control of tumor progression is a consequence of the evolution of a variety of immune suppressive cellular and molecular networks within the tumor microenvironment, resulting in progressive $CD4^+$ and $CD8^+$ T cell dysfunction (Thommen & Schumacher, 2018). Understanding how immunotherapies impact tumor antigen-specific T cell function in individual patients will provide a better understanding of the mechanisms underlying the success or failure of a particular therapeutic approach. Furthermore, increased understanding of the mechanistic bases of T cell dysfunction is essential for selection of the appropriate immunotherapy agents for each individual patient. Therefore, multiplex assays capable of characterizing tumor-infiltrating T cell functional status are a critical analytic tool for advancing our understanding of T cell dysfunction. The use of multi-parameter flow cytometry provides a highly informative technique for resolving cytokine expression patterns in T cells at the single cell level, and can be adapted using selective stimulation methods to profile either antigen-specific responses or general cytokine profiles in response to mitogen stimulation. Furthermore, combining ICS with antibodies specific for markers of cellular phenotype, immune checkpoint receptor expression, intranuclear transcription factors, cellular proliferation, cell activation, and cell death provide comprehensive details about the functional features of tumor-infiltrating T cells. This important information defining the T cell phenotype can be coupled with cytokine expression in order to better examine types and functional status of tumor-infiltrating T cells in patients with cancer.

In this chapter, we will outline widely accepted protocols (Foster, Prussin, Liu, & Whitton, 2007; Lovelace & Maecker, 2018) that we adapted for multi-parameter intracellular cytokine staining, focusing on IL-2, TNFα, and IFNγ as a measure of tumor-infiltrating $CD4^+$ and $CD8^+$ T cell function. There are several important considerations when performing intracellular cytokine staining that are critical to successful evaluation of cytokine expression profiles in tumor-infiltrating T cells at the single cell level. These include: (i) effective tumor dissociation and recovery of viable tumor-infiltrating T cells; (ii) *in vitro* stimulation conditions; (iii) cell surface staining protocols; and (iv) intracellular cytokine profiling methodologies. Understanding the essential details for these methods is critical for the successful profiling of cytokine expression within tumor-infiltrating T cells. The goal of this chapter is to review accepted methods for intracellular cytokine staining in tumor-infiltrating T cells using murine and human pancreatic cancer as a model tumor. Pancreatic cancer was chosen due to the challenges it presents in obtaining sufficient numbers of viable immune cells to characterize due to the extensive extracellular matrix and the relatively low frequency of immune cells present in this tumor type. Summaries of commonly used protocols (Foster et al., 2007; Lovelace & Maecker, 2018) for *in vitro* stimulation, cell surface and intracellular cytokine staining will be detailed in this chapter.

2. Tumor dissociation and T cell isolation

The tumor microenvironment is composed of various cell types including infiltrating immune cells, stromal cells, endothelial cells and tumor cells (Gajewski, Schreiber, & Fu, 2013). *Ex vivo* studies of tumor-infiltrating lymphocytes from fresh human and murine tumors are complicated by the small tumor samples obtained, the low frequency of immune cells typically present, and the potential for loss of immune cell viability during extraction (Garaud et al., 2014). Because immune cells infiltrating tumors are usually present as passengers rather than permanent residents they are generally amenable to release from the tissue matrix using enzymatic (Grange et al., 2011; Gros et al., 2012; Zirakzadeh, Marits, Sherif, & Winqvist, 2013) and/or mechanical dissociation (Garaud et al., 2014). Prior to beginning the dissociation process, it is important to identify and remove potential contaminating sources of T cells present outside the tumor in adjacent tissues, as they can skew the phenotype and functional characteristics of T cells present physically within the TME. Such contamination would result in inaccurate resolution of tumor-infiltrating T cell phenotypic and functional states.

2.1 Tumor T and B lymphocyte contamination from regional tissues

The presence of contaminating leukocytes from adjacent tissues such as adipose tissue (subcutaneous fat, omentum, and mesentery), blood, and draining lymph nodes can result in an inaccurate phenotypic and functional profile of lymphocytes within the tumor microenvironment. If included in the gross tumor specimen, any of these sources would alter the proportions of lymphocyte and myeloid cells interpreted to be within the tumor immune infiltrate. In some instances, normal tissues may be included with tumor biopsies due to locoregional or metastatic spread, both of which are common features of advanced, malignant tumors. Pancreatic cancer, for example, exhibits aggressive locoregional and distant metastatic invasion in both patients and mouse models of advanced disease (Hingorani et al., 2005; Majumder et al., 2016; Peixotoa et al., 2015; Yachida & Iacobuzio-Donahue, 2009). Dissemination of pancreatic tumors through the omentum, mesentery, and regional lymph nodes is common behavior of this tumor type, making gross dissection away from these adjacent tissues a difficult challenge in obtaining transcriptional, phenotypic, and functional information on cells present exclusively within the tumor (Heller et al., 2015).

When profiling the phenotype and function of infiltrating T cells localized within human or murine tumors such as pancreatic cancer, it is important to consider the potential impact of these factors on the outcome of the intracellular cytokine assay. Therefore, if characterization of the phenotype and function of the T cell infiltrate within the tumor microenvironment is the goal, it is important to develop rigorous practices to identify and then separate regional tissues (omentum, adiopose tissue, and regional draining lymph nodes) and blood contamination from the tumor (Fig. 1A–C). When conducting flow cytometric evaluation of tumor-infiltrating T cell function, using spontaneous pancreatic tumors from the pdx1-Cre; LSL-Kras^{G12D}; LSL-TP53$^{FL/+}$ as an example, omentum, regional lymph nodes, and the blood supply are the most likely sources of T and B lymphocytes (Fig. 1D and E). Depending on the source of T cells, differences in the response to mitogenic stimulation may result in different cytokine profiles.

2.2 Tumor dissociation

The vast majority of protocols utilize enzymatic digestion with a combination of collagenases, particularly those specific for type I, III and IV collagens. DNase type 1 and hyaluronidase are also frequently employed to digest non-cellular components of the extracellular matrix (ECM) and liberate cells into suspension. Depending on the cell type and the degree of association with the ECM, both the digestion time and enzymatic components may require adjustment to optimize cell isolation. The ECM plays a critical role in tumor development and progression. Aggressive tumors often display increased deposition of altered extracellular matrix

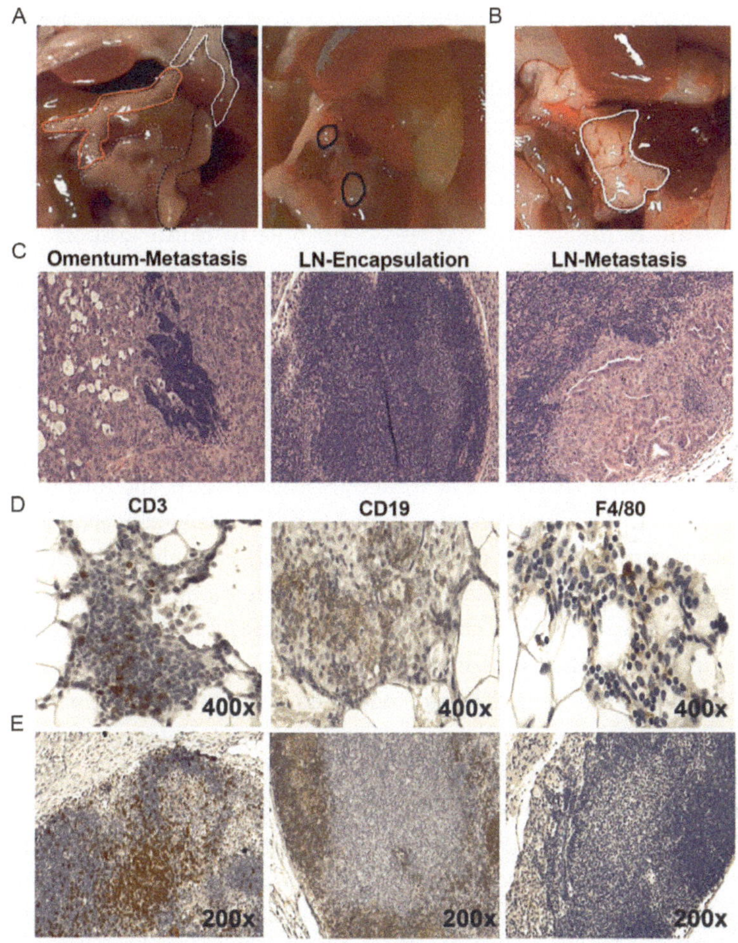

Fig. 1 Potential sources of T cell contamination of the pancreatic tumor immune infiltrate. (A) Gross dissection of healthy C57BL/6 mouse reveals omentum (white line), tail (dashed black line), body (dashed gray line), and head (dashed red line) of the pancreas. Draining lymph nodes (black line), which typically are located ventral to the body of the pancreas and adjacent to the greater curvature of the stomach, are visible after retraction of the pancreas. (B) Locally infiltrative pancreatic ductal adenocarcinoma (white line) in a Pdx1-Cre;LSL-KRas^{G12D};LSL-TP53$^{FL/+}$ mouse. Identification of the omentum and lymph nodes can be difficult due to the invasive nature of the tumor. (C) Invasion of omentum and draining lymph nodes are commonly noted in spontaneous PDAC in the Pdx1-Cre;LSL-KRas^{G12D};LSL-TP53$^{FL/+}$ mouse model. Both the omentum (D) and draining lymph nodes (E) contain significant numbers of CD3^{+} T cells and CD19^{+} B cells, but not F4/80^{+} macrophages.

components characterized by increased amounts of distinct collagen types. In cancers, type IV collagen is a major component of both the basement membrane (Paulsson, 1992) and the extracellular matrix, particularly those that exhibit desmoplastic responses such as pancreatic cancer. Furthermore, the tumor stroma contains fibrillar type I, III, and V collagens (Bachem et al., 2005; Ohlund, Franklin, Lundenberg, Lundin, & Sund, 2013; Weniger, Honselmann, & Liss, 2018) that also contribute to tumor-associated ECM deposition.

Similar to the normal distribution of various collagens in specific organs (Fang, Yuan, Peng, & Li, 2014), distinct tumors types can exhibit differences in the collagen composition within the tumor microenvironment. Therefore, dissociation protocols involving enzyme-based digestion strategies need to be adapted to the particular tumor type of interest, and optimized for cell yield, viability, the retention of cell surface epitopes for detection, and the proportions of different immune subsets present. Type I and IV collagenases are widely used in tumor digestion protocols to liberate both tumor epithelial cells and infiltrating leukocytes (http://www.worthington-biochem.com/tissueDissociation/Tumor.html) across murine and human tumor types. DNAse type I is typically used to supplement proteases for tissue dissociation as it helps to reduce viscosity resulting from DNA released from damaged cells during harvesting (Foster et al., 2007). Furthermore, dissociation media can be supplemented with hyaluronidase due to its abilities to catalyze the random hydrolysis of 1,4-β-D-glycosidic linkages between *N*-acetyl-galactosamine or *N*-acetylgalactosamine sulfate on proteoglycan components such as glucuronic acid in hyaluronic acid, chondroitin, chondroitin 4- and 6-sulfates, and dermatan (Stern & Jedrzejas, 2006).

Typically, tumors are gradually minced from larger fragments into smaller 2 × 2 mm pieces to provide sufficient surface area for liberation of infiltrating lymphocytes using digestive enzymes or mechanical measures. With enzyme-based digestion protocols, tumor fragments are incubated at 37 °C in media containing a digestive enzyme mixture to facilitate enzymatic digestion of the ECM. Incubation time is critical and must be optimized for each tumor type as excessive incubation times in dissociative enzymes can decrease both cell viability and destroy cell surface epitope expression (Autengruber, Gereke, Hansen, Hennig, & Bruder, 2012; Quatromoni et al., 2015). Alternative non-enzymatic mechanical methods may also be used (Crossey et al., 2018; Garaud et al., 2014). Furthermore, commercially available combinations of mechanical and enzyme-based extraction (Miltenyi GentleMACS) can provide potentially better yield and viability using protocols and proprietary reagents adapted to various human and murine tumor types. It is up to the investigator to determine the optimal combination of enzymatic and/or mechanical disruption and the processing times necessary to achieve the best yield, viability, and retention of cell surface epitopes.

After enzymatic and/or mechanical dissociation of tumor stroma, cell suspensions are filtered to separate cells from undigested stroma and then either further separated using density centrifugation with Ficoll or Percoll gradients (Li and Kurlander, 2010). Isolated cell suspensions may then be used as is, or further separated using magnetic enrichment and/or cell sorting. CD45- or Thy1.1-specific magnetic microbeads may be used to enrich tumor-infiltrating leukocytes from human and mouse tumors (Salvagno & de Visser, 2016). Notably, isolating T cells from TIL is not critical for either mitogenic stimulants (PMA/ionomycin) or the direct stimulation of TCR using plate-bound anti-CD3 and soluble anti-CD28 for co-stimulation (Li and Kurlander, 2010). For this protocol, enrichment was not utilized as the total numbers of tumor-infiltrating lymphocytes from spontaneous models are limited. We will outline in detail commonly accepted protocols for both PMA-ionomycin and CD3-CD28 stimulation, as either can be utilized for the mitogenic stimulation of mouse and human tumor-infiltrating T cell isolates.

3. *In vitro* T cell stimulation

3.1 PMA-ionomycin

Analysis of cytokine expression at the single cell level requires, *in vitro* activation. There are various methods of stimulation, some of which involve MHC-TCR interactions and some of which do not. One of the most common *in vitro* stimulation methods bypass TCR stimulation through direct activation of protein kinase C using phorbol myristate acetate (PMA) together with the calcium ionophore ionomycin. These mitogenic stimuli act together as to induce cytokine expression, in particular IL-2, IFNγ, and TNFα. This method can be used to stimulate human (Muul et al., 2011) or murine lymphocytes (Kruisbeek, Shevach, & Thorton, 2004) derived from peripheral blood or other sources such as spleen, lymph node, or tumor. The use of an inhibitor such as monensin or brefeldin A is required to block cytokine secretion, thereby increasing intracellular cytokine retention and enabling the detection of cytokines within the cell by flow cytometry. Both monensin (an ionophore) and brefeldin A (a Golgi transport inhibitor) inhibit cytokine export; however, they have variable efficacy depending on the cytokine in question. In the protocol described here, brefeldin A alone was chosen due to its specific activity in inhibiting vesicle transport. In contrast, monensin inhibits recycling of labeled antibodies in the lysosome, and is thus required for the flow cytometric evaluation of IL-10 in B cells (Yanaba et al., 2008), CD107a/b and CD154. If these markers are part of the profile, both monensin and brefeldin A are used (Lovelace & Maecker, 2011). The following protocol can be adapted for the *in vitro* stimulation of either human or murine lymphocytes from multiple sources and can be performed with or without *in vitro* T cell expansion. In this example, we use single cell suspensions derived from spontaneous pancreatic tumors obtained from Pdx1-Cre;LSL-Kras^{G12D}; LSL-TP53$^{FL/+}$ (KPC$_{null}$) mice, and stimulate with PMA-ionomycin for 6h using an established protocol (Foster et al., 2007) to evaluate CD4$^+$ and CD8$^+$ T cell production of IL-2, TNFα, and IFNγ as indicators of tumor-infiltrating T cell functional status.

Materials

- Lymphocyte sources such as peripheral blood mononuclear cells, splenocytes, or single cell suspensions from dissociated tumor (TIL)
- Complete RPMI containing 10% Fetal Bovine Serum, 1 mM glutamate, 1 mM Penicillin and Streptomycin
- 10 mM ionomycin (Ca^{2+} salt) in DMSO
- 200 μg/mL phorbol myristate acetate (PMA) in 100% ethanol
- 10 mg/mL brefeldin A in DMSO
- 3 mg/mL DNase 1 (Worthington-Biochem, bovine DNase 1)
- Live/Dead Aqua Dye (Invitrogen) or equivalent product (optimize based on fluorochrome combinations)
- Phosphate-buffered saline (PBS)
- 96-Well round bottom plate (Costar)
- Disposable pipette tips (p1000 and p200)
- Multi-channel pipettor
- Table-top centrifuge

Protocol

1. Resuspend the isolated cells (TILs, PBMC's, splenocytes) in pre-warmed, complete RPMI at a concentration of 4×10^6 cells/mL ($10 \times$ cell concentration for stimulation using 96-well round bottom plates).

   Notes: (i) This number can be scaled up to 40×10^6 cells/6-well plate or 4×10^6 cells in a 12-well plate. (ii) For TIL the number of cells will contain a variable number of $CD45^+$ leukocytes and variable numbers of $CD3^+$ T cells, which might affect the results. Therefore, either normalization with magnetic enrichment of $CD45^+$ leukocytes or a modification of the initial cell numbers may be necessary.

2. Prepare 2× concentration of PMA/ionomycin/brefeldin A mixture using 5 mL of complete RPMI medium +1 μL of PMA (stock concentration = 200 μg/mL), ionomycin (stock concentration = 10 mM), and brefeldin A (stock concentration = 10 mg/mL).
3. Aliquot 100 μL of cells (4×10^5) into 96-well roundbottom plate and add 100 μL of RPMI complete medium containing PMA (final concentration = 20 ng/mL), ionomycin (final concentration = 1 μM), and brefeldin A (final concentration = 10 μg/mL).
4. Incubate for 6 h in a 37 °C, 5% CO_2 incubator. A 6-h stimulation is typically sufficient for detection of IL-2, IFNγ, and TNFα. However, each investigator should determine optimal conditions for different stimulation conditions and also for additional cytokines of interest.

   Notes: (i) PMA can induce downregulation of both CD3 and CD4 through recycling of the TCR (Kemp and Bruunsgaard, 2001). Therefore, intracellular staining in addition to surface staining of these molecules may aid in visualization of T cell subsets. (ii) After the 6-h incubation prior to cell surface labeling and intracellular staining, DNase 1 can be utilized to digest extracellular DNA generated by activation-induced cell death and/or debris from mesenchymal and epithelial cell contaminants from the single cell tumor lysate (Foster et al., 2007). This may reduce cell clumping and viscosity that can lead additional cell loss during centrifugation due to cell shearing.

3.2 CD3- and CD28-mediated T cell stimulation

Mature T cells recognize and respond to the antigen/MHC complex through their antigen-specific TCR-CD3 complex. Direct stimulation of the TCR signaling through plate-bound CD3 ligation and CD28 co-stimulation mimic antigen presentation by providing appropriate signals for both TCR stimulation and the necessary co-stimulation vital for T cell proliferation and acquisition of effector function. In the following protocol we will utilize plate-bound CD3 (which mimics antigen presentation and TCR ligation) and soluble anti-CD28 (which provides the necessary co-stimulation to reduce activation-induced cell death) to more effectively mimic the antigen-specific induction of effector T cell cytokine responses represented by IL-2, TNFα, and IFNγ production than PMA and ionomycin. The purpose of the following protocol is to elicit an immediate cytokine response from effector CD4 and CD8 T cells present within the tumor microenvironment. The 6-h stimulus is sufficient to assay for IL-2, TNFα, and IFNγ production from single cell isolates containing tumor-infiltrating T cells from the pancreatic tumor microenvironment (Nomura, Walker, & Maecker, 2000).

Materials

- Lymphocyte sources such as peripheral blood mononuclear cells, splenocytes, or single cell suspensions from dissociated tumor (TIL)
- Complete RPMI containing 10% Fetal Bovine Serum, 1 mM glutamate, 1 mM Penicillin and Streptomycin
- Anti-mouse CD3 (clone 145-2C11, no azide/low endotoxin) or anti-human CD3 (clone UCHT1, NA/LE)
- Anti-mouse CD28 (clone 37.51, no azide/low endotoxin) or anti-human CD28 (clone CD28.2, NA/LE)
- 10 mg/mL brefeldin A in DMSO
- 3 mg/mL DNase 1 (Worthington-Biochem)
- Live/Dead Aqua Dye (Invitrogen) or equivalent product (can optimize based on fluorochrome combinations)
- Phosphate-buffered saline (PBS)
- 96-Well round bottom plate (Costar)
- Disposable pipette tips (p1000 and p200)
- Multi-channel pipettor
- Refrigerated centrifuge

Protocol

1. Prepare dilutions of anti-mouse CD3 (Clone 145-2C11) ranging from 1.0 to 10 μg/mL and coat 96-well round bottom plates using 30 μL of anti-CD3/well. For controls include 30uL of PBS. Cover plates and incubate either 2 h at 37 °C or overnight at 4 °C.

   Note: A range of 1–10 μg/mL will typically define the most effective concentration for *in vitro* stimulation.
2. After the end of the incubation period, wash plates with sterile PBS by adding 200 μL of PBS 3 × to remove excess, unbound antibody. Plates are now ready to use.
3. Add 4×10^5 cells per well in 100 μL of RPMI complete media. For anti-CD28 co-stimulation add 100 μL containing 2 μg/mL of anti-CD28 and 20 μg/mL of brefeldin A into RPMI complete medium to obtain final concentrations of 1 μg/mL and 10 μg/mL, respectively. Pipette to mix adequately with T cells.

Notes: (i) Titration of the cell number, particularly for single cell suspensions containing tumor-infiltrating lymphocytes, is critical to obtaining sufficient numbers of T cells for intracellular cytokine studies as the number of T cells within isolated cell suspensions can vary based on the tumor type and the model system. (ii) Magnetic enrichment of leukocytes using anti-CD45 microbeads (Miltenyi Biotech, Auburn CA) can potentially reduce variability due to insufficient numbers of T cells (unless the TIL fraction is devoid of or contains minimal T cells). It is important to keep in mind that the use of magnetic enrichment will decrease the overall cell yields, leading to reduced sample sizes for analysis. This can be quite problematic for tumors with very low frequencies of T cells. (iii) Cross linking CD28 mimics co-stimulation activated by the binding of CD28 with its natural ligands CD80 (B7.1) and CD86 (B7.2) on antigen presenting cells. CD28 co-stimulation is essential for optimal activation of both human and murine T cells. (iv) Anti-human CD49 can be used as an additional co-stimulus for human T cell activation (Foster et al., 2007). CD49d is the integrin subunit α4, which associates with β1 or β7 to form the integrin subunits α4β1 and α4β7. Cross-linking CD49d mimics interactions with fibronectin, eliciting enhanced $CD4^+$ and $CD8^+$ T cell proliferation (Davis, Oppenheimer-Marks, Bednarczyk, McIntyre, & Lipsky, 1990; Shimizu, Van Seventer, Horgan, & Shaw, 1990) and optimizing the *in vitro* culture of human T cells.

4. Incubation for 6 h at 37 °C is typically sufficient for the production and detection of IL-2, IFNγ, and TNFα. However, similar to determining the best cell numbers to use for a particular application, incubation times should also be optimized per investigator.

   Note: CD3 stimulation can induce a modest downregulation of both CD3 and CD4 through recycling of the TCR (Foster et al., 2007). Therefore, intracellular staining may aid in visualizing CD3, CD4, and CD8 on T cell subsets post-stimulation.

4. Combining cell surface and intracellular cytokine staining

Intracellular cytokine staining is an essential technique in the evaluation of cytokine expression levels at the single cell level. Combining this technique with multi-parameter flow cytometry of cell surface markers allows increasing cellular resolution through the use of a panel of fluorochrome conjugates. Advances in laser-mediated excitation and absorption spectral capabilities may provide an opportunity for greater resolution of subpopulations of T cells within the $CD4^+$ and $CD8^+$ subsets. Several critical issues can limit the diagnostic sensitivity of multi-parameter flow cytometric profiling of T cell phenotype and cytokine expression. These include (i) autofluorescence; (ii) increased non-specific binding of antibodies by cytosolic and nuclear proteins; (iii) the degree of spectral overlap of fluorochrome conjugates; (iv) the choice of fluorochrome-antibody conjugates; (v) tandem fluorochrome conjugate degradation; and (vi) instrument set-up, capabilities, and calibration (Cossarizza et al., 2017; Maecker, Frey, Nomura, & Trotter, 2004).

Tumor dissociation creates an abundance of cellular debris as discussed previously in this chapter. The use of viability dyes capable of dead cell/debris discrimination is critical to remove autofluorescence emissions by cellular components such as DNA, cellular enzymes, and small molecules such NADPH. Each of these components can contribute to false positives when evaluating cytokine expression at the single cell level due to their capacity to emit different light spectra when excited by specific wavelengths. The use of "fluorescence staining minus-1," where cells are stained with all antibodies with exception of the antibody-fluorochrome conjugate thought to be causing a false fluorescence signal allows the investigator to determine potential issues with autoflourescence that may contribute to false positives. In most instances, the removal of dead cells enhances the sensitivity and resolution of single cell cytokine profiling in tumor-infiltrating T cells. Importantly, cell clumping and debris can also contribute to increased false positives. Using single cell discrimination parameters (FSCA *vs.* FSCH) can overcome this obstacle. Cellular permeabilization for intracellular staining promotes non-specific binding to the detection antibodies, and the addition of a protein like BSA or Fc-block into Perm/Wash reagents can help reduce both Fc-receptor (cell surface) and intracytoplasmic/nuclear non-specific binding and retention of cytokine-specific antibodies. To evaluate the extent of non-specific binding it is useful to include samples where cytokine-specific antibodies are substituted with isotype-matched-control antibodies at the same concentration and with the same fluorochrome conjugation. Quantifying the degree of non-specific binding will reduce incidence of false positive results as the % of isotype control staining can be subtracted from the % of cytokine-specific staining.

In addition to cellular artifacts, technical issues involving the fluorochrome choice can influence the sensitivity of the ICS assay. As a general rule, the use of intense fluorochromes (PE, PECy5, PECy7, APC, AlexaFluor 647, BV421, BV650, and BV721) is recommended to increase the sensitivity of the ICS assay for lower abundance cytokine antigens, whereas the use of fluorochromes whose emission spectra are intermediate to weak (FITC, AlexaFluor 488, PerCP, PerCPCy5.5, APC-Cy7, AlexaFluor 700) should be utilized for high density antigens such as CD3, CD4, CD8, CD45 (Lovelace & Maecker, 2011; Maecker et al., 2004; Maecker, 2009). When using the photo-unstable tandem dye conjugates APC-Cy7 or PE-Cy7, lower expression target antigens should be chosen if possible. Otherwise, photo-induced cleavage may create spill-over that cannot be compensated into the parent fluorochrome channel (PE and APC), thus increasing false positives and/or rendering data for a particular cytokine non-diagnostic (Maecker, 2009; Maecker et al., 2004). Lastly, the flow cytometer instrument setup is critical to the quality of the analysis. Whether or not the flow cytometer has the optimal configuration of photomultipliers and filters capable of detecting the fluorochrome-antibody conjugates you intend to use, and whether it is proper working order with regard to laser function, alignment, and adequate intra-experimental compensation are critical considerations. There is both an increasing number of fluorochrome options and an increasing ability to resolve their particular emission spectra. However, given the overall limitations of the optical spectrum, the addition of novel fluorochrome-antibody conjugates to a multi-color panel may create compensation issues resulting from increased spill-over into detectors. This is the result of overlap between emission spectra of fluorochromes excited by ultraviolet, violet, red, and green lasers (Cossarizza et al., 2017; Maecker et al., 2004). The ability to provide ideal compensation to reduce this spill-over becomes increasingly challenging with expansion of the fluorochrome-antibody panel. For simplicity of this chapter we will present an optimized 8 fluorochrome-antibody panel used to evaluate IL-2, TNFα, and IFNγ production in pancreatic tumor-infiltrating $CD4^+$ and $CD8^+$ T cell subsets.

Materials

Antibodies (BioLegend): 8 Color Setup

- Anti-mouse TNFα (Clone MP6-XT22; BioLegend catalog # 506313)
- Anti-mouse CD8 PE (Clone 53-6.7; BioLegend catalog # 100708)
- Anti-mouse CD45 PerCP (Clone 30-F11; BioLegend catalog # 103130)
- Anti-mouse CD3 PECy7 (Clone 17A2; BioLegend catalog # 100220)
- Anti-mouse CD4 APC (Clone GK1.5; BioLegend catalog # 100412)
- Anti-mouse IFNγ APC Cy7 (Clone XMG1.2; BioLegend catalog # 505850)

- Anti-mouse IL-2 BV 421 (Clone JES6-5H4; BioLegend catalog # 503826)
- Live/Dead Aqua (ThermoFisher catalog # L34966; CD19 BV 510 (Clone 6D5; BioLegend catalog # 115546), and CD11b BV510 (Clone M1/70; BioLegend catalog # 101263) (dump/dead channel))

Other reagents

- FACS buffer-0.5% BSA in PBS
- Cell fixation reagent, BD Fixation/Permeabilization Solution cat # 554714 (BD Biosciences, San Jose, CA)
- Cell permeabilization reagent, BD Perm/Wash 10× Buffer cat # 554714 (BD Biosciences, San Jose, CA)
- 96-Well round bottom polypropylene plates with lids (BD Falcon)
- Plate holder for centrifuge (Sorvall Instruments, Newton, CT)

Equipment

- BD LSR II equipped with Cytometer Setup and Tracking beads and software (BD Biosciences)
- Table-top centrifuge
- Multi-channel pipettor

Protocol

1. Pipette 1×10^6 cells/well from a single cell suspension of dissociated tumor, lymph node, or spleen (10^7/mL starting concentration) into 96-well plate to use as single color control samples.
2. Centrifuge at $500 \times g$ for 5 min to pellet cells.
3. Stain cells with 100 μL containing 1 μg/mL of Live/Dead Fixable Aqua or Yellow Dead Cell Stain (ThermoFisher Scientific) for 30 min on ice or 15 min at room temperature.

   Note: The use of Fc block (BD Biosciences) at 5 μg/mL is recommended to block non-specific Fc receptor-mediated interactions with antibodies in the master mix.
4. Wash with 200 μL of FACS buffer and centrifuge at $500 \times g$ for 5 min.
5. Prepare a master mix of antibodies for cell surface staining in FACS buffer. Add 50 μL of the master mix to each sample, and incubate for 30 min on ice. Cover the plate in aluminum foil during incubation to reduce light exposure, as light exposure not only bleaches the fluorochromes but also catalyzes breakdown of tandem dye conjugates APC-Cy7 and PE-Cy7 into APC/PE and Cy7 (Le Roy, Varin-Blank, Ajchenbaum-Cymbalista, & Letestu, 2009; Maecker et al., 2004).

   Notes: (i) A typical concentration range for antibodies is 0.5–5 μg/mL. Start with an antibody concentration based on manufacturer recommendations, and then optimize it to reach the maximal signal:noise;

(ii) CD3, CD4, and CD8 can either be stained during the surface staining steps or during the intracellular cytokine staining steps due to the stability of these antigens to formaldehyde-based fixatives; (iii) tandem dye conjugates, in addition to being photo-unstable, also are sensitive to formaldehyde-based-fixatives (Maecker et al., 2004) and free radicals (Le Roy et al., 2009).

6. For single color compensation controls, we recommend utilizing fluorochrome-bead conjugates (BD CompBeads) for consistency, or alternatively each antibody used in the panel can be utilized for generation of single color controls.
7. After cell surface staining, wash cells with 200 μL of FACS buffer, and centrifuge at 500 × *g*.
8. Resuspend cells in 100 μL of BD Cytofix/Cytoperm buffer for 30 min on ice at 4 °C, or use another appropriate fixative such as 4% paraformaldehyde.
9. Add 200 μL of 1 × Perm/Wash (or a comparable saponin-based fixative) to the fixed cells, resuspend, and centrifuge at 600 × *g* for 5 min.

   Notes: (i) Centrifugation at higher force than 500 × *g* is recommended to increase cell recovery as permeabilized cells will take on greater buoyancy post-permeabilization (Lovelace & Maecker, 2011).
10. Make a master mix of cytokine-specific antibodies and isotype-matched controls in 1 × perm/wash buffer. Resuspend cell pellets in 50 μL/sample of master mix and incubate for 30 min at 4 °C in the dark.
11. Add 200 μL of perm/wash buffer and centrifuge at 600 × *g* for 5 min to remove excess antibody. The cells can now be resuspended in FACS buffer at the desired volume for acquisition.

    Notes: (i) Keep cells covered and away from light to prevent tandem fluorochrome conjugate breakdown. Cells may be evaluated immediately or stored overnight in the dark at 4 °C for analysis the next day. Longer term storage at −80 °C is possible (Foster et al., 2007) if paraformaldehyde is utilized.

5. Data acquisition and analysis

Prior to beginning data acquisition, verify the flow cytometer calibration and photomultiplier setup necessary to detect the emission spectra of the fluorochromes employed in the antibody-conjugate panel. This is essential to the success of ICS assay. For detailed information and protocols necessary for optimal flow cytometer function, refer to protocols by Perfetto and colleagues (2012). Prior to beginning sample acquisition, it is necessary to first determine the optimal voltages in the photomultiplier tube (PMT) settings on the flow cytometer using both unstained and single color controls to allow for optimal adjustment. To generate the sample compensation matrix necessary for resolving fluorochrome spill-over, the use of appropriate single color controls and the use the software's automated algorithms are necessary to calculate compensation. The choice of reagents to use in compensation can include antibody-fluorochrome conjugates used in the panel, a common antigen of intermediate density conjugated with every fluorochrome used in the panel (e.g., CD19, CD4, CD8, etc.), or beads conjugated with the fluorochromes of interest. The use of stained cells with a defined phenotype most replicates the unknown samples stained with the antibody-fluorochrome. However, the epitope density or the amount of biologic sample may preclude the successful use of the exact fluorochrome panel for compensation. The use of a common, intermediate density antigen such as CD19, CD4, or CD8 and a source of abundant immune cells such as the spleen can facilitate suitable compensation for most samples (Cossarizza et al., 2017; Maecker et al., 2004; Maecker, 2009). Alternatively, conjugated beads may provide a bright and homogeneous population of events and eliminate the need for biologic materials. Anti-immunoglobulin-coated beads will allow the user to use the exact antibodies in the panel without the need for sufficient expression on cells within the biologic samples. One disadvantage of beads is that the use of beads with a higher intensity of epitope expression than the epitope expressed on cells within the tumor microenvironment may result in overcompensation (Cossarizza et al., 2017; Maecker, 2009). Compensation matrices can be evaluated post-compensation, and adjusted in the event of inadequate compensation. After successful compensation, create a template for gating lymphoid cells, excluding doublets and dead cells, and gating $CD3^+$ $CD4^+$ and $CD3^+$ $CD8^+$ T cells (Fig. 2). Collecting a minimum of 250,000 events should allow for adequate visualization of $CD4^+$ and $CD8^+$, however, this can be scaled up depending on the initial cellular input (4×10^5 were utilized in these experiments). Data derived from these protocols was analyzed using FlowJo (TreeStar). Other software options include FACS DIVA (BD), and Kaluza (Beckman-Coulter).

6. Concluding remarks

Intracellular cytokine analysis combined with multi-parameter flow cytometry is a valuable technique for profiling single cell functional phenotypes of tumor-infiltrating T cells. It is important to validate each step of the

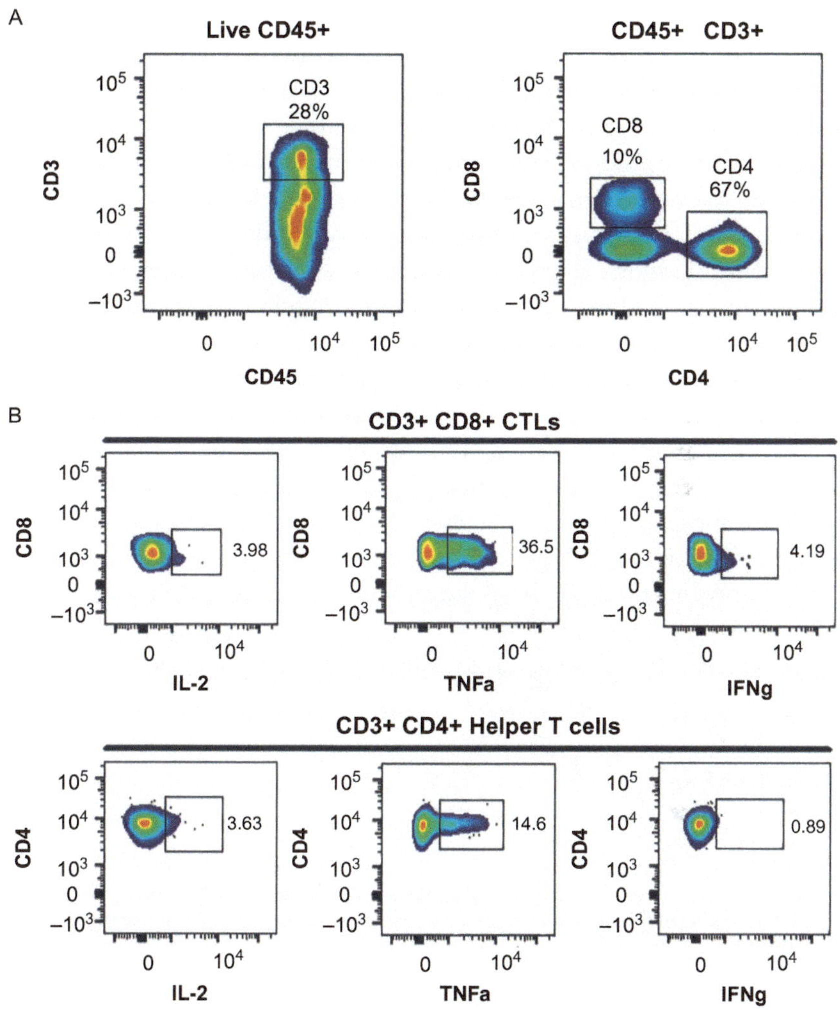

Fig. 2 Evaluation of TNFα, IL-2, and IFNγ production 6 h post αCD3/αCD28 stimulation *in vitro*. (A) Tumor-infiltrating lymphocytes were isolated from a pancreatic tumor derived from a 12 week old, pdx1-Cre;LSL-Kras^{G12D};LSL-TP53$^{FL/+}$ mouse, and immune cells were stimulated for 6 h with plate-bound CD3 (5 μg/mL) and soluble CD28 (1 μg/mL). Following stimulation, cells were removed from 96-well plates using 0.1% EDTA in PBS, then surface-stained for CD3, CD4, CD8, and CD45, followed by intracellular staining for TNFα, IL-2, and IFNγ. T lymphocytes were identified on the basis of FSC and SSC properties as well as expression of CD45 and CD3, then doublets and dead cells were excluded as shown. (B) CD3$^+$ CD4$^+$ T helper and CD3$^+$ CD8$^+$ cytotoxic lymphocytes were evaluated for expression of TNFα, IL-2, and IFNγ.

ICS assay in order to obtain meaningful data sets from preclinical mouse models and clinical patient samples. Protocols outlined in this book chapter were adapted for rigorously validated and accepted protocols used for single cell profiling of T cells from peripheral blood. Analysis of intracellular cytokine staining profiles from tumor-infiltrating T cells provided additional challenges to companion assays based on peripheral blood due the difficulty in immune cell isolation from solid tumors, variability of immune composition within tumors, degree of invasiveness of tumor into surrounding tissues, and state of function of intratumoral T cells. Therefore, it is highly recommended that evaluating other sources of T cells such as peripheral blood or spleen be utilized in addition to sample specimens from otherwise healthy animals and humans to determine the optimal performance of these assays prior to evaluating IL-2, TNFα, and IFNγ production by tumor-infiltrating T cells. For more in depth information on all aspects of these protocols, please reference Foster et al. (2007), Cossarizza et al. (2017), and Perfetto et al. (2012), as they cover all aspects of intracellular cytokine staining assays in extensive detail. Lastly and importantly, the use of complementary assays such as ELISpot or single cell RNA sequence analysis will help to validate ICS findings from tumor-infiltrating T cells.

References

Autengruber, A., Gereke, M., Hansen, G., Hennig, C., & Bruder, D. (2012). Impact of enzymatic tissue disintegration on the level of surface molecule expression and immune cell function. *European Journal of Microbiology and Immunology*, *2*, 112–120.

Bachem, M. G., et al. (2005). Pancreatic carcinoma cells induce fibrosis by stimulating proliferation and matrix synthesis of stellate cells. *Gastroenterology*, *128*, 907–921.

Cossarizza, A., et al. (2017). Guidelines for the use of flow cytometry and cell sorting in immunological studies. *European Journal of Immunology*, *47*, 1584–1797.

Crossey, F., et al. (2018). Robust method for isolation of tumor infiltrating lymphocytes with a high vital cell yield from small samples of renal cell carcinomas by a new collagenase-free mechanical procedure. *Urologic Oncology*, *36*, 402e1–402e10.

Davis, L. S., Oppenheimer-Marks, N., Bednarczyk, J. L., McIntyre, B. W., & Lipsky, P. E. (1990). Fibronectin promotes proliferation of naive and memory T cells by signaling through both the VLA-4 and VLA-5 integrin molecules. *Journal of Immunology*, *145*, 785–793.

Fang, M., Yuan, J., Peng, C., & Li, Y. (2014). Collagen as a double-edged sword in tumor progression. *Tumour Biology*, *35*, 2781–2882.

Foster, B., Prussin, C., Liu, F., & Whitton, J. L. (2007). Detection of intracellular cytokines by flow cytometry. *Current Protocols in Immunology*, *6*, 6.24.1–6.24.21.

Fridman, W. H., Dieu-Nosjean, M. C., Pages, F., Cremer, I., Damotte, D., Sautes-Fridman, C., et al. (2013). The immune microenvironment of human tumors: General significance and clinical impact. *Cancer Microenvironment*, *6*, 117–122.

Gajewski, T. F., Schreiber, H., & Fu, Y. X. (2013). Innate and adaptive immune cells in the tumor microenvironment. *Nature Immunology*, *14*, 1014–1022.

Garaud, S., et al. (2014). A simple and rapid protocol to non-enzymatically dissociate fresh human tissues for the analysis of infiltrating lymphocytes. *Journal of Visualized Experiments*, (94), e52392.

Grange, C., et al. (2011). Phenotypic characterization and functional analysis of human tumor infiltration after mechanical and enzymatic disaggregation. *Journal of Immunological Methods*, *372*, 119–126.

Gros, A., Turcotte, S., Wunderlich, J. R., Ahmadzadeh, M., Dudley, M. E., & Rosenberg, S. A. (2012). Myeloid cells obtained from blood but not tumor can suppress T-cell proliferation in patients with melanoma. *Clinical Cancer Research*, *18*, 5212–5223.

Heller, A., et al. (2015). Stratification of pancreatic tissue samples for molecular studies: RNA-based cellular annotation procedure. *Pancreatology*, *15*, 423–431.

Hingorani, S. R., et al. (2005). Trp53R172H and KrasG12D cooperate to promote chromosomal instability and widely metastatic ductal adenocarcinoma in mice. *Cancer Cell*, 7, 469–483.

Kemp, K., & Bruunsgaard, H. (2001). Identification of IFN-gamma-producing CD4+ T cells following PMA stimulation. *Journal of Interferon and Cytokine Research*, *21*, 503–506.

Kruisbeek, A. M., Shevach, E., & Thorton, A. M. (2004). Proliferative assays for T cell function. *Current Protocols in Immunology*, *3*, 3.12.1–3.12.20.

Le Roy, C., Varin-Blank, N., Ajchenbaum-Cymbalista, F., & Letestu, R. (2009). Flow cytometry APC-tandem dyes are degraded through a cell-dependent mechanism. *Cytometry Part A*, *75A*, 882–809.

Li, Y., & Kurlander, R. J. (2010). Comparison of anti-CD3 and anti_CD28 coated beads with soluble anti-CD3 for expanding human T cells: Differing impact of CD8 T cell phenotype and responsiveness to restimulation. *Journal of Translational Medicine*, *8*, 104.

Lovelace, P., & Maecker, H. T. (2011). Multiparameter intracellular cytokine staining. *Methods in Molecular Biology*, *699*, 165–178.

Lovelace, P., & Maecker, H. T. (2018). Multiparameter intracellular cytokine staining. *Methods in Molecular Biology*, *1678*, 151–166.

Maecker, H. T. (2009). Multiparameter flow cytometry monitoring of T cell responses. *Methods in Molecular Biology*, *485*, 375–391.

Maecker, H. T., Frey, T., Nomura, L. E., & Trotter, J. (2004). Selecting fluorochrome conjugates for maximum sensitivity. *Cytometry. Part A*, *62*, 169–173.

Majumder, K., et al. (2016). A novel immune competent mouse model of pancreatic cancer with robust stroma: A valuable tool for preclinical evaluation of new therapies. *Journal of Gastrointestinal Surgery*, *20*, 53–65.

Muul, L. M., et al. (2011). Measurement of proliferative responses of cultured lymphocytes. *Current Protocols in Immunology*, 7, 7.10.1–7.10.26.

Nomura, L. E., Walker, J. M., & Maecker, H. T. (2000). Optimization of whole blood antigen-specific cytokine assays for CD4(+) T cells. *Cytometry*, *40*, 60–68.

Ohlund, D., Franklin, O., Lundenberg, E., Lundin, C., & Sund, M. (2013). Type IV collagen stimulates pancreatic cancer cell proliferation, migration, and inhibits apoptosis through an autocrine loop. *BMC Cancer*, *13*, 154.

Paulsson, M. (1992). Basement membrane proteins: Structure, assembly, and cellular interactions. *Critical Reviews in Biochemistry and Molecular Biology*, *27*, 93–127.

Peixotoa, R. D. A., Speers, C., McGahan, C. E., Renouf, D. J., Schaeffer, D. F., & Kennecke, H. F. (2015). Prognostic factors and sites of metastasis in unresectable locally advanced pancreatic cancer. *Cancer Medicine*, *4*, 1171–1177.

Perfetto, S. P., Ambrozak, D., Nguyen, R., Chattopadhyay, P. K., & Roederer, M. (2012). Quality assurance for polychromatic flow cytometry using a suite of calibration beads. *Nature Protocols*, 7, 2067–2079.

Quatromoni, J. G., Singhal, S., Bhojnagarwala, P., Hancock, W. W., Albelda, S. M., & Eruslanov, E. (2015). An optimized disaggregation method for human lung tumors that preserves the phenotype and function of immune cells. *Journal of Leukocyte Biology*, *97*, 201–209.

Salvagno, C., & de Visser, K. E. (2016). Purification of immune cell populations from freshly isolated murine tumors and organs by consecutive magnetic cell sorting and multi-parameter flow cytometry-based sorting. *Methods in Molecular Biology*, *1458*, 125–135.
Shimizu, Y., Van Seventer, G. A., Horgan, K. J., & Shaw, S. (1990). Costimulation of proliferative responses of resting CD4+ T cells by the interaction of VLA-4 and VLA-5 with fibronectin or VLA-6 with laminin. *Journal of Immunology*, *145*, 59–67.
Stern, R., & Jedrzejas, M. J. (2006). The hyaluronidases: Their genomics, structures, and mechanisms of action. *Chemical Reviews*, *106*, 818–839.
Thommen, D. S., & Schumacher, T. N. (2018). T cell dysfunction in Cancer. *Cancer Cell*, *33*, 547–562.
Weniger, M., Honselmann, K. C., & Liss, A. S. (2018). The extracellular matrix and pancreatic cancer: A complex relationship. *Cancers (Basel)*, *10*, E316.
Yachida, S., & Iacobuzio-Donahue, C. A. (2009). The pathology and genetics of metastatic pancreatic cancer. *Archives of Pathology & Laboratory Medicine*, *133*, 413–422.
Yanaba, K., Bouaziz, J. D., Haas, K. M., Poe, J. C., Fujimoto, M., & Tedder, T. F. (2008). A regulatory B cell subset with a unique $CD1d^{hi}$ $CD5^{+}$ phenotype controls T cell-dependent inflammatory responses. *Immunity*, *28*, 630–650.
Zirakzadeh, A. A., Marits, P., Sherif, A., & Winqvist, O. (2013). Multiplex B cell characterization in blood, lymph nodes, and tumors from patients with malignancies. *Journal of Immunology*, *190*, 5847–5855.

CHAPTER TWO

High-throughput identification of human antigen-specific CD8$^+$ and CD4$^+$ T cells using soluble pMHC multimers

Morgane Magnin[a,b], Philippe Guillaume[a,b], George Coukos[a,b,*], Alexandre Harari[a,b], Julien Schmidt[a,b]

[a]Ludwig Institute for Cancer Research, University of Lausanne, Lausanne, Switzerland
[b]Department of Oncology, University Hospital of Lausanne, Lausanne, Switzerland
*Corresponding author: e-mail address: george.coukos@chuv.ch

Contents

Abstract

Peptide major histocompatibility complex (pMHC) multimers have been used since decades to identify, isolate and analyze antigen-specific T cells by flow (and more recently mass) cytometry. Yet well established as a standard technology, improvements are still required to face the growing needs of personalized immune monitoring. Here we review the latest developments about (i) the quality of pMHC class I and II

Methods in Enzymology, Volume 631
ISSN 0076-6879
https://doi.org/10.1016/bs.mie.2019.05.019

monomers, (ii) the importance of the multimeric scaffold, (iii) the staining conditions and (iv) the high-throughput synthesis of pMHC monomers. Finally, innovative multiplexed, combinatorial strategies for parallel detection of antigen-specific T cells in a single sample are discussed.

1. Introduction

$CD8^+$ and $CD4^+$ T lymphocytes are main actors of adaptive cellular immunity and play a central role in protecting individuals against pathogen infections and neoplastic transformations (Gorska & Alam, 2003). While $CD8^+$ T cells mainly differentiate in cytotoxic T lymphocytes with killing capacities toward target cells displaying pMHC class I complexes, $CD4^+$ T cells differentiate in T helper cells (e.g., Th1, Th2) or regulatory T cells (Treg) and recognize cell surface pMHC class II complexes (Fooksman et al., 2010).

Detection, isolation and analysis of antigen-specific $CD8^+$ and $CD4^+$ T cells is of major importance to monitor disease progression as well as for the development of vaccines and personalized immunotherapies (Bobisse et al., 2018; Tanyi et al., 2018).

While initially determined by limiting dilution and assessment of frequencies of T cells mediating functional responses, antigen-specific $CD8^+$ and $CD4^+$ T cells are now routinely determined by using soluble pMHC multimers, labeled either with fluorescent dyes or isotopes, by flow and mass cytometry, respectively (Brett, Kingston, & Colston, 1987; Schmidt, Dojcinovic, Guillaume, & Luescher, 2013; Sharrock, Kaminski, & Man, 1990). These methods have the advantage of being highly sensitive and not dependent on any functional capacity of T cells. However, besides requiring a detailed knowledge of HLA typing and antigen restriction, pMHC multimers can also failed to detect cells bearing very low-affinity TCRs. In 1992, Garboczi et al. published the first refolded peptide-HLA-A0201 complex (Garboczi, Hung, & Wiley, 1992) and in 1996, Altman and colleagues reported the first soluble pMHC tetramer for direct identification and enumeration of antigen-specific $CD8^+$ T cells (Altman et al., 1996).

Peptide-MHC class I monomers can be obtained by a simple, cost efficient procedure that consists of refolding denatured heavy and light (beta-2-microglobulin, β2m) chains in the presence of an 8–11 residue long, synthetic peptide of interest. On the other hand, pMHC class II monomers are more difficult to obtain in reasonable yields by refolding or by mammalian expressions systems. Indeed, folding conditions of pMHC class II have to be optimized individually for each allotype (Vollers & Stern, 2008). The most commonly adopted strategy is the expression of "empty" MHC class II molecules expressed by insect cells (such as Drosophila S2 or SF9 cells) and further loaded with synthetic peptides of 10–16 residues (Frayser, Sato, Xu, & Stern, 1999; Fremont, Hendrickson, Marrack, & Kappler, 1996). This method relies on the lack of antigen processing and loading machinery in insect cells and the recovery of stable pure MHC class II proteins without any peptide.

Due to the low affinity of pMHC class I/II monomers for T-cell receptors (TCR), multimeric structures are required to enable efficient staining of antigen-specific T-cell populations. Enzymatic biotinylation of pMHC monomers allows multimerization thanks to fluorescent streptavidin conjugates, increasing their overall avidity for TCRs. Performances of pMHC class II multimers are worse than their class I counterparts and can be accounted for (i) the lower affinity TCR expressed by $CD4^+$ T cells, (ii) the lack of contribution of the CD4 coreceptor in stabilizing the binding to pMHC class II, (iii) the high conformational diversity of pMHC complexes and (iv) the poor quality of pMHC class II reagents. Over the last decade, many advances have been made to improve the quality of pMHC class I/II multimers, the staining protocols and the developments of high dimensional detection strategies.

2. Monitoring of antigen-specific $CD8^+$ and $CD4^+$ T cells with soluble pMHC I & II

Optimal detection of antigen-specific $CD8^+$ and $CD4^+$ T cell is mainly driven by the binding strength between multimeric pMHC and their cognate TCRs (Rius et al., 2018). Several parameters can be optimized to increase the avidity of pMHC multimers.

2.1 Improved pMHC monomer quality

Following the refolding of peptides with heavy and light chains, pMHC class I monomers are purified by size exclusion chromatography, eliminating unfolded proteins. Production yields as well as stability of pMHC class I complexes are correlated with the peptide binding affinity for a given HLA molecule. While some pMHC class I complexes have half-lives of several hours at 37 °C, others are highly unstable (Schmidt et al., 2017).

Great care should be taken when using low stability pMHC multimers. Lowering working temperature as well as addition of free β2m and synthetic peptides can help preserving pMHC integrity (Wooldridge et al., 2009).

Peptide binding affinity to MHC class II also correlates with peptide loading efficiency of "empty" MHC class II molecules (Frayser et al., 1999). The fraction of peptide loaded complexes cannot be separated from "empty" monomers by means of size exclusion chromatography. Resulting pMHC class II multimers contain a fraction of active monomers, thus precluding efficient staining of low affinity antigen-specific $CD4^+$ T cells (Ayyoub, Pignon, et al., 2010; Gebe et al., 2003). Purification of peptide-containing complexes can be achieved by careful ion exchange chromatography but remains very challenging. Alternatively, Luescher et al. have reported the use of molecularly defined pMHC class II multimers restoring staining of NY-ESO-I-specific $CD4^+$ T cells that were not detected with conventional multimers (Ayyoub, Dojcinovic, et al., 2010). Molecularly defined pMHC class II monomers are produced with long peptides carrying a specific tag (i.e., Histag) enabling the tag-purification of peptide-containing monomers and subsequent assembly of pMHC multimers composed of fully loaded monomers. As it has been reported that flanking residues can lead to unwanted side effects (Carson, Vignali, Woodland, & Vignali, 1997; Sant'Angelo, Robinson, Janeway, & Denzin, 2002), introduction of a photocleavable residue allows efficient tag removal. If the gain in staining efficiency remains marginal for high avidity $CD4^+$ T cells, it becomes dramatic for low affinity ones, typically tumor-specific T cells (Carson et al., 1997; Sant'Angelo et al., 2002).

Another hurdle in the preparation of pMHC class I/II monomers is the biotinylation step needed for further multimerization using streptavidin-dyes conjugates. Routinely performed using the BirA enzyme, with overnight exposure at room temperature in the presence of biotin and ATP, this step can lead to degradation of unstable complexes. To circumvent this, it is possible to perform biotinylation directly in bacteria by co-transformation with the BirA enzyme and a given HLA leading directly to biotinylated inclusion bodies readily usable for refolding (Leisner et al., 2008). Working with highly pure, molecularly defined pMHC class I/II monomers is a prerequisite for further building of improved quality multimers.

2.2 Choosing the best multimeric scaffold

The dissociation constant (K_D) of pMHC-TCR is in the range of 1–1000 μM (Boniface et al., 1998), precluding staining of T cells with soluble monomeric complexes. Increasing pMHC valency by multimerization with streptavidin (SA) conjugates leads to increased pMHC-TCR avidity and allows the effective detection of antigen-specific T cells by flow cytometry (Altman et al., 1996). However, it has been shown that, in addition to the number of pMHC monomers displayed by multimeric scaffolds, their spatial arrangement is also an important parameter.

Elegant studies using synthetic linkers of different length and flexibility shown that maximal binding avidity is obtained with very short linkers (<20 Å) and gradually decreases with the spacer length. Similarly, binding avidity is also increasing with valency, octamers being better binders than tetramers and than dimers (Angelov et al., 2006; Cebecauer, Guillaume, Mark, et al., 2005).

So-called pMHC tetramers are usually made by reacting biotinylated pMHC class I/II monomers with heterogenous streptavidin-PE (or APC) and should therefore be considered as multimers. A PE (or APC)-labeled multimer carries in average 8–12 pMHC monomers making allowing highly efficient binding to antigen-specific T cells. Conversely, true tetrameric pMHC complexes can be obtained using a tetra-grade PE (with equimolar ratio PE/SA) (Hugues, Malherbe, Filippi, & Glaichenhaus, 2002). The latter exhibit much lower binding avidities compared to conventional multimers. This is also true when using SA labeled with small molecular weight fluorochromes (i.e., FITC, Alexa Fluor or Cyanine derivatives). The choice of the multimeric scaffold is therefore important to avoid inefficient staining of low affinity T cells.

Conventional pMHC multimers made of SA-PE (or -APC) are available from different companies or academic core facilities but other multimeric scaffolds have emerged and are extensively used nowadays (Fig. 1). Pro-Immune is commercializing MHC class I and II "Pentamers" made of a coiled coil protein carrying five PE and five pMHC monomers (Leisner et al., 2008). "Dextramers" are available from Immudex and are built on a dextran polymer carrying multiple fluorochromes and up to 12 pMHC monomers (Batard et al., 2006). More recently, "Dodecamers" were reported (Huang et al., 2016). They are molecularly well defined and display 12 pMHC monomers but unfortunately are not commercially available. "Dodecamers," "Dextramers" and conventional multimers confer the highest and comparable staining efficiencies, with a slight advantage for "dodecamers." They outperform "Pentamers" and true tetramers, which is accounted for lower valency of these latter (Dolton et al., 2014; Schmidt et al., 2011).

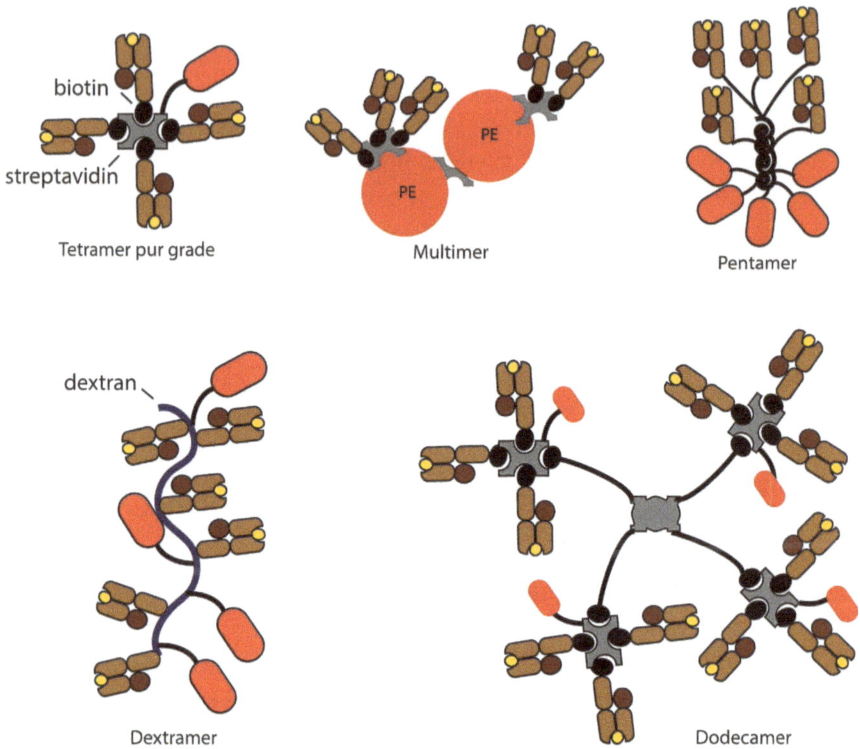

Fig. 1 Representation of the most commonly used soluble pMHC-based reagents. Cartoons of pMHC oligomers and their multimeric scaffolds. Red circles represent PE fluorochromes.

2.3 An improved staining protocol

Despite all efforts to improve the quality of pMHC multimers, they can still fail to detect relevant functional T-cell clonotypes. Several reports highlighted discrepancies between frequencies of antigen-specific $CD8^+$ and $CD4^+$ T cells determined by functional assays (ELISpot and ICS) and by multimer staining (Rius et al., 2018; Rubio-Godoy et al., 2001). This is mainly due to the difficulty to stain T cells bearing TCRs of low affinity for cognate antigen as pMHC-TCR binding strength needed for efficient staining is superior to the one needed for T-cell activation (Laugel et al., 2007).

Several optimizations in staining protocols have led to improved methodology for detection of very low affinity T cells such as $CD4^+$ T cells or those directed against self-antigens (i.e., anticancer and autoimmune). First, the use of the protein kinase inhibitor Dasatinib improves the physical detection of $CD8^+$ T cells by inhibiting TCR downregulation pathways preventing TCR internalization, therefore increasing cell surface TCR presentation (Lissina et al., 2009; Weichsel et al., 2008). However, it cannot be used for functional profiling assays as it inhibits cellular activation. Second, the use of an anti-PE monoclonal antibody targeting cell surface bound pMHC multimers helps stabilizing their binding. This anti-PE can directly be coupled to magnetic beads allowing at the same time antigen-specific $CD8^+$ T cells enrichment (Dolton et al., 2018).

Staining of antigen-specific $CD4^+$ T cells can be improved, first by using molecularly defined pMHC class II multimers lowering the affinity threshold of detectable clones and second by using glycoside hydrolase enzymes, like neuraminidase (i.e., from *Vibrio cholerae*), as staining efficiency critically depends on $CD4^+$ T cells surface glycosylation (Massilamany, Gangaplara, Chapman, Rose, & Reddy, 2011; Reddy et al., 2003).

It is important to point out that effect of Dasatinib and/or neuraminidase on T cell viability, proliferation or functional analysis has never been clearly investigated and even they provide a significant gain in staining efficiency, care should be considered for T cell sorting and cloning.

Temperature is also an important parameter influencing staining efficiencies. It has been reported that the binding of pMHC class I multimers to antigen-specific $CD8^+$ T cells can strongly induce cell activation, sometimes leading to apoptosis. This TCR-induced cell death is directly correlated to and increases with TCR affinity. This negative effect can be circumvented by the use of dasatinib and manipulation of cells at 4 °C (Dolton et al., 2018). On the other side, staining of antigen-specific $CD4^+$ T cells is better when performed at 37 °C and it has never been reported any TCR-induced cell death of these latter with soluble pMHC class II multimers.

The choice of anti-CD8 and/or anti-CD4 antibodies can also affect multimer staining. Generally, these anti-coreceptor antibodies can be disruptive and should therefore be used in a second step after cells were stained with multimers (Wooldridge et al., 2006).

Finally, it is important to highlight that both pMHC class I and II multimers can be aliquoted and frozen. A comparative study showed they can be cryopreserved for at least 6 months, which is advantageous for low stability pMHC complexes and better reliability (Hadrup et al., 2015).

3. Staining of antigen-specific $CD8^+$ and $CD4^+$ T cells: Materials and methods

The protocol described below is intended for the detection of a single antigen-specific $CD8^+$ T cell and a single antigen-specific $CD4^+$ T cell population in a given sample of PBMCs. Detection of multiple pMHC specificities in one sample is discussed later in this chapter. This protocol can also be adapted for detection of only $CD8^+$ or $CD4^+$ antigen-specific T cells.

General recommendations before starting are (i) work fast with freshly-prepared reagents and solutions and (ii) avoid light-exposure for all reagents containing fluorochromes. Indicated volumes are intended for 10^7 T cells but can be adapted for other cell numbers.

3.1 Materials

3.1.1 Chemicals and reagents

1. Pan human T cell isolation kit, ref# 130-096-535, Miltenyi
2. Alexa Fluor 488 anti-human CD8 antibody, clone SK1, ref# 344716, Biolegend
3. PE/Cy7 anti-human CD4 antibody, clone A161A1, ref# 357409, Biolegend
4. PE-labeled pMHC class I multimer
5. APC-labeled pMHC class II multimer
6. Mouse anti-PE primary unconjugated monoclonal antibody, clone PE001, ref# 408101, BioLegend
7. Mouse anti-APC primary unconjugated monoclonal antibody, clone APC003, ref# 408001, BioLegend
8. Live/Dead cell marker (4′,6-diamidino-2-phenylindole, DAPI), ref# D1306, ThermoFisher Scientific
9. Dasatinib, ref# 1392, Axon Medchem
10. Neuraminidase, ref# 11585886001, Sigma-Aldrich

3.1.2 Instruments and software

1. Flow cytometer LSR II, Becton Dickinson
2. FlowJo software, Tristar
3. Centrifuge 5810R, Eppendorf

3.1.3 Solutions

1. FACS buffer: PBS pH 7.2, 0.5% BSA, 2 mM EDTA, 0.02% NaN_3.
2. Dasatinib: 1 mM single use DMSO stocks at −80 °C. Final experimental concentration is 50 nM in PBS.
3. Neuraminidase: final experimental concentration is 0.2 U/mL in PBS.
4. pMHC class I/II multimers: stock solution at 200 nM. Final experimental concentration is usually 4 nM in FACS buffer but should be determined by titration if possible.
5. Mouse anti-PE (or APC) antibody: final experimental concentration is 10 μg/mL in FACS buffer.
6. Anti-CD8 and anti-CD4 antibodies: final experimental concentration should be determined by titration.
7. DAPI: 1 mg/mL H_2O stocks at 4 °C. Final experimental concentration is 1 μg/mL FACS buffer.

3.1.4 Methods

1. T cells isolation from PBMCs using Pan T cell isolation kit
   - ➢ Count PBMCs with trypan blue.
   - ➢ Centrifuge PBMCs 5 min at 400*g*. Discard supernatant and resuspend pellet in 40 μL of FACS buffer per 10^7 cells.
   - ➢ Add 10 μL of Pan T cell biotin-antibody cocktail per 10^7 cells.
   - ➢ Mix well and incubate for 5 min at 4 °C in the refrigerator.
   - ➢ Add 30 μL of FACS buffer per 10^7 cells.
   - ➢ Add 20 μL of Pan T cell microbead cocktail per 10^7 cells.
   - ➢ Mix well and incubate for 10 min at 4 °C in the refrigerator.
   - ➢ Proceed to subsequent magnetic cell separation. A minimum volume of 500 μL is required. If needed add FACS buffer to the cell suspension.
   - ➢ Place LS column in the magnetic field of a suitable MACS Separator.
   - ➢ Wash column three times with 3 mL of FACS buffer.
   - ➢ Apply cell suspension and collect flow-through containing unlabeled cells, representing the enriched T cells in a 15 mL Falcon tube.
   - ➢ Rinse column with 3 mL of FACS buffer and keep collecting in the same Falcon tube.
   - ➢ Count T cells with trypan blue.
2. Staining of antigen-specific $CD8^+$ and $CD4^+$ T cells
   - ➢ Centrifuge T cells 5 min at 400*g*, discard supernatant and resuspend pellet in 1 mL of dasatinib/neuraminidase solution per 10^7 cells.
   - ➢ Incubate 30 min at 37 °C.

- Add 5 mL of FACS buffer and centrifuge 5 min at 400*g*.
- Discard supernatant and resuspend pellet in 100 μL of pMHC class I/II multimer solution per 10^7 cells.
- Mix well and incubate 40 min at 4 °C in the refrigerator.
- Add 5 mL of FACS buffer and centrifuge 5 min at 400*g*.
- Discard supernatant and resuspend pellet in 100 μL of anti-PE and anti-APC antibodies.
- Mix well and incubate 20 min at 4 °C in the refrigerator.
- Add 5 mL of FACS buffer and centrifuge 5 min at 400*g*.
- Discard supernatant and resuspend pellet in 100 μL of antibody cocktail solution per 10^7 cells.
- Mix well and incubate 20 min at 4 °C in the refrigerator.
- Add 5 mL of FACS buffer and centrifuge 5 min at 400*g*.
- Add 5 mL of FACS buffer and centrifuge 5 min at 400*g*.
- Discard supernatant and resuspend pellet in 200 μL of DAPI solution.
- Keep sample on ice and proceed immediately with FACS analysis.

3. Flow cytometer acquisition and analysis
- Calibrate the flow cytometer using reference beads.
- Using unstained control cells and single-color stained samples adjust the location of viable lymphocyte gate (FSC vs SSC) as well as single color signals (by changing voltages).
- Automatically calculate compensations.
- Run negative control samples using irrelevant cells (alternatively irrelevant pMHC multimers) stained identically.
- Acquire data by collecting events of each sample tube and negative control tube.
- Gating strategy and analysis
  - Import FCS files and open them with FlowJo. Start analyzing a negative control sample.
  - Plot FSC-A vs SSC-A and draw a gate around live lymphocytes.
  - Plot FSC-A vs FSC-H and gate on single cells.
  - Plot SSC-H vs SSC-A and gate on single cells.
  - Plot FSC-A vs DAPI and gate on live, DAPI negative cells.
  - Plot CD8 vs CD4 and draw gates around $CD8^+$ and $CD4^+$ populations.
  - Plot separately CD8 vs pMHC class I multimer and CD4 vs pMHC class II multimer and set quadrant boundaries according to multimer backgrounds determined with the negative irrelevant sample tubes.
  - Analyze all samples without changing gates.

4. High dimensional T-cell staining

Developments of personalized immunotherapies call for a growing need for high-throughput, multiple screening of T-cell specificities in limited samples. Recently, innovative combinatorial staining strategies using pMHC class I multimers were made possible to screen for >1000 specificities in a single sample, better mapping immune responses at a genome-wide level (Bentzen & Hadrup, 2017).

4.1 High-throughput pMHC monomer synthesis

Generation of pMHC class I monomers is tedious and time-consuming. Because empty MHC class I molecules are unstable, each new pMHC complex has to be individually refolded and purified before multimerization, limiting their use to only few selected epitopes at one time. This prompted the development of new methods for multiplexed parallel synthesis of pMHC class I monomers. Strategies are based on the large-scale production of a single pMHC class I complex used for subsequent peptide exchange with peptides of interest. Rodenko et al. reported an elegant way of high-throughput synthesis using a pMHC class I complex refolded with a photocleavable peptide that is cleaved upon UV irradiation (Rodenko et al., 2006). Addition of a "rescue" peptide in the solution, if of sufficient affinity, can restore the pMHC complex by displacing cleaved fragments from the binding groove. In situ-produced pMHC monomers can then be multimerized and directly used for staining purposes. This UV-conditional approach allows as well the quantification of peptide binding to a given MHC molecule thanks to the use of a conformational antibody and has facilitated the fast discovery of multiple epitopes. Such UV-conditional peptides have now been designed for a large number of MHC class I alleles, of both human and murine origin. However, this approach requires the use of a specific photocleavable peptide for each MHC allele. UV exposure can also damage MHC molecules and "rescue" peptides, create nitroso reactive species and lead to sample evaporation due to overheating. A similar approach used a chemically-cleavable peptide to generate pMHC that this time is cleaved by addition of periodate or dithionite and replaced by a peptide of interest (Rodenko et al., 2009). Again, the need to find an efficient conditional peptide for each MHC allele and the possible chemical side effects have limited its widespread use. Dipeptides can also act as catalysts and facilitate the folding reaction and peptide exchange by binding to the F pocket of MHC class I molecules but have never been applied in high-throughput generation of pMHC multimers (Saini et al., 2015). More recently the thermo-labile production of pMHC class I monomers was introduced (Luimstra et al., 2018). A pMHC class I complex is refolded with a peptide having a low off-rate at 4 °C. Because complexes undergo conformational changes at a certain temperature leading to full peptide dissociation, increasing temperature will promote exchange with higher affinity peptides creating new stable pMHC complexes. This strategy does not require the use of specific, non-natural amino acids nor deleterious triggers and can be applied directly after pMHC multimer formation. Even if the design of thermo-labile peptides is not trivial and only reported so far for H-2K^b and HLA-A0201 in monitoring viral responses, there is no doubt that this principle will be expanded. Thanks to these different methods the synthesis of pMHC class I monomers is no longer a time limiting factor.

4.2 Fluorochrome-based combinatorial staining

Parallel identification of multiple antigen-specific CD8$^+$ and CD4$^+$ T cells with conventional pMHC multimers is limited by the low number of streptavidin-conjugated fluorochromes available. In 2009, two innovative strategies using fluorescent multidimensional encoding of pMHC class I multimers were reported.

Assuming that a single T cell will only bind one pMHC complex, distinct color-coding can be assigned to each specific pMHC multimer. Hadrup et al. reported a dual-color coding, where each antigen-specific population is stained using a unique combination of two specific pMHC multimers, providing a detection limit of 0.002% in terms of frequency (Hadrup et al., 2009). In the meantime, Newell et al. reported the use of multivalent color coding, using combinations of one to four colors to stain individual antigen-specific populations (Newell, Klein, Yu, & Davis, 2009). If the multivalent color-coding strategy theoretically provides a higher number of color combinations, it has the disadvantage of reducing staining intensity as the number of colors is increasing. The dual-color coding, thus, provides a lower number of unique combinations but a better staining intensity and a lower background as any T cell not matching the color combination is gated out and recorded as a background event. This approach has also been successfully tested with pMHC class II multimers directly ex vivo (Uchtenhagen et al., 2016).

4.3 Heavy-metal ions-based combinatorial staining

With the advent of mass cytometry, using heavy metal ions, the number of parallel labels detectable increased dramatically (Chen & Weng, 2012). Newell et al. reported the screening of 109 different antigen-specificities in a single sample combined with phenotype and functional analysis (Newell, Sigal, Bendall, Nolan, & Davis, 2012). Combinatorial staining by mass cytometry offers higher order of combinations and simplifies the analysis as no/little compensations are required, each metal ion being detectable in a single channel. Yet this technology has several drawbacks including low resolution and processing speed, high instrument and reagents costs and the lack of cell recovery as cells are nebulized during the process.

4.4 DNA-barcoded pMHC multimers

More recently, Bentzen et al. have reported the use of DNA-barcoded pMHC multimers (Bentzen et al., 2016) (Fig. 2). These reagents are made of a polysaccharide dextran backbone (like Dextramers) carrying a high number of SA and PE. Biotinylated pMHC as well as biotinylated DNA barcodes can be coupled to free SA binding sites generating high avidity reagents with a unique combination of theoretically up to 10^{10} single DNA barcodes. After combinatorial staining of a single sample (with >1000 pMHC specificities), antigen-specific $CD8^+$ T cells can be sorted by means of the PE signal and subsequently, DNA barcodes amplified and identified by high-throughput sequencing. The number of reads for a specific DNA barcode assigned to a given pMHC correlates with the frequency of antigen-specific T cells within the sorted population. Even if TCR expression and affinity might introduce a bias to be fully quantitative, comparison between fluorochrome-based and DNA barcode combinatorial determination of T-cell frequencies showed high correlation ($r^2 > 0.9$ across 10 healthy donors and 11 melanoma patients looking at 32 and 15 T-cell responses, respectively). Moreover, even cells labeled with low levels of DNA-barcodes will be detected and counted as positives after DNA amplification where fluorescent labeled would not have provided sufficient separation of multimer positive and negative cells. This is of particular interest for low affinity clones like $CD4^+$ T cells.

The invaluable advantage of DNA barcoded pMHC multimers is the possibility to screen an unprecedented number of antigens in a single, often limited sample. Due to higher sensitivity and lower detection limits compared to other previous methodologies, it enables earlier screening of

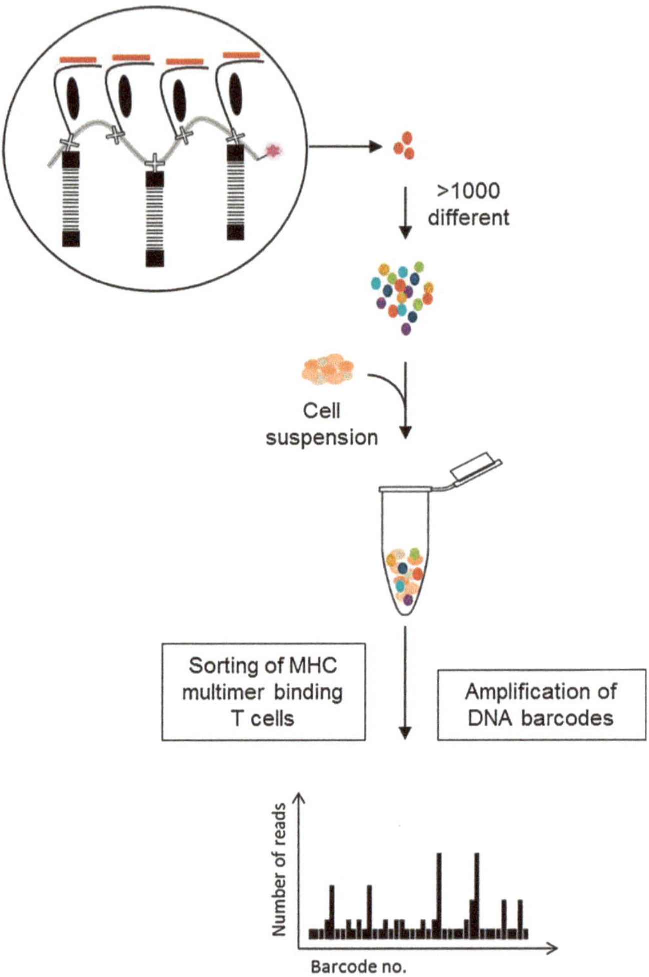

Fig. 2 Detection of antigen-responsive T cells using the DNA barcode-labeled MHC multimer methodology. A PE-labeled dextran backbone carrying a number of SA binding sites (illustrated as an X) is applied to co-attach biotinylated molecules; DNA barcodes and pMHCs. Thus, >1000 pMHC multimers can be generated, each carrying a different DNA barcode. All MHC multimer binding T cells are sorted based on the common fluorescent label and the associated DNA barcodes are amplified and sequenced. The relative numbers of DNA barcode reads are used to determine the composition of antigen-responsive T cells in the sample. *Reproduced from Bentzen, A. K., & Hadrup, S. R. (2017). Evolution of MHC-based technologies used for detection of antigen-responsive T cells.* Cancer Immunology, Immunotherapy, *66(5), 657–666. doi:10.1007/s00262-017-1971-5, according to the terms of the Creative Commons Attribution 4.0 International License (http://creativecommons.org/licenses/by/4.0/).*

PBMCs or TILS that could be less expanded and then less prone to cell culture bias. The possibility for high dimensional screening in a sample allows unbiased interpretation of antigen-specific T-cell frequencies as all precursor T cells will be in presence of the entire pMHC library. The DNA-barcoded methodology has been applied in both peripheral blood and tissue biopsies (Bentzen et al., 2016). However, even if in theory the combination of DNA barcodes would be sufficient to cover antigens at a genome-wide level, many limitations are still to be addressed. First, this strategy does not allow, as for the mass cytometry approach, recovery of viable cells since cells are lyzed during processing. Second, the manufacturing of such large libraries is limited by the inherent production costs and turnover times of synthetic peptides that are incompressible. Also, as one pMHC can be recognized by several TCRs, cross-reactivity cannot be evaluated by this approach as the number of reads cannot be associated with any visual assessment. Very recently, it has been applied at the single cell level enabling integration of transcriptome, cell-surface protein, immune repertoire and TCR-antigen specificity measurements. Combining DNA-barcoded pMHC dextramers (Immudex) with single cell immune profiling (10 × Genomics) allows direct correlation between TCR sequences and their antigen specificity at unprecedented resolution. This high-throughput, highly sensitive and unbiased methodology will rapidly increase our understanding of T-cell recognition. Finally, the group of Hadrup showed the feasibility of applying the DNA-barcoded multimers strategy to determine TCR cross-recognition patterns, which they termed the TCR fingerprinting (Bentzen et al., 2018). The authors investigated the recognition pattern of two distinct TCRs recognizing each a 10mer epitope derived from Merkel cell polyomavirus. They mutated each position of the two epitopes with all possible amino acid available and used these large peptides variants libraries for high-throughput generation of DNA-barcoded multimers. By measuring the relative affinity of TCRs to libraries of barcoded pMHC variants it is possible to determine the TCR fingerprinting and to fully characterize TCR cross-recognition. Even if the study was only extended to a limited number of TCRs, the power of this approach is invaluable in evaluating potential clinical risks of cross-reactivity in TCR-based therapies.

5. Reversible pMHC multimer for comprehensive analysis of antigen-specific T cells

Reversible pMHC multimers are reagents in which pMHC monomers can be disrupted from the multimeric scaffold upon addition of a specific stimulus. Because of the weak affinity of pMHC monomers for TCRs, staining with reversible multimers allows the removal of all pMHC monomers from the cell surface.

5.1 Isolation of "untouched" antigen-specific CD8 T cells

It is known that staining of antigen-specific $CD8^+$ T cells with pMHC multimers can lead to TCR-mediated activation and even in some cases to TCR-induced cell death (Cebecauer, Guillaume, Hozak, et al., 2005). In 2002, Knabel et al. reported the use of *Streptamers*, reversible pMHC class I multimers commercially available from IBA Life Science (Knabel et al., 2002). *Streptamers* are built on an optimized streptactin-streptag scaffold mimicking the streptavidin-biotin interaction. Because streptactin binds more avidly biotin than streptag, addition of free biotin allows the dissociation of pMHC monomers from the multimeric scaffold. More recently, reversible multimers built on Ni^{2+}-NTA-Histag chelates complexes that can be disrupted upon addition of imidazole (so-called NTAmers) were described (Schmidt et al., 2011). If NTAmers appear more stable and provide better staining efficiencies than *Streptamers*, both reversible reagents showed similar advantages for isolation of "untouched" antigen-specific $CD8^+$ T cells. Sorting of antigen-specific $CD8^+$ T cells with NTAmers or *Streptamers* leads to better cloning efficiencies by preventing the loss of high affinity T cell clones that can fully undergo TCR-induced apoptosis when conventional multimers are used (Neudorfer et al., 2007; Schmidt et al., 2011). This approach then allows the analysis of complete TCR repertoires of antigen-specific $CD8^+$ T cells. Even applicable to antigen-specific $CD4^+$ T cells, reversible sorting of $CD4^+$ T cells has not been yet fully evaluated.

5.2 Measurement of pMHC-TCR monomeric dissociation rates

A major advantage of reversible multimers is the measurement of pMHC-TCR dissociation kinetics. By addition of a fluorescent label on the β2m (i.e., Cy5), it is possible to follow over time the dissociation of pMHC monomers from the engaged TCR after dissociation of the multimeric scaffold. Briefly, antigen-specific $CD8^+$ T cells are stained in the cold to avoid internalization and cell-associated PE (multimeric scaffold) and Cy5 (pMHC monomers) fluorescence are recorded before and after the addition of the disruptive stimulus. Nauerth et al. used *Streptamers* in real-time microscopic-based assay to measure monomeric pMHC-TCR dissociation kinetics on live virus-specific $CD8^+$ T cells and showed a correlation between off-rates and T cell responsiveness in vitro as well as a better protection capacity in vivo (Nauerth et al., 2013). More recently, similar work was performed on self tumor-specific $CD8^+$ T cells using NTAmers, as these reagents, due to faster decay under the monomeric form upon addition of imidazole, allow measurement of lower affinity T cells (Hebeisen et al., 2015). Allard et al. have used the NTAmer technology for a comprehensive analysis of structural and functional avidities of virus and tumor-specific $CD8^+$ T cells (Allard et al., 2017). They showed a strong correlation between pMHC-TCR off-rates and cytokine production, polyfunctionality, cell proliferation, activating/inhibitory receptor expression and in vivo tumor control. Moreover, the pMHC-TCR off-rate is a more stable and more reliable parameter than conventional functional assays that depend on the T cell's shape and activation state.

The use of reversible pMHC multimers has proven its importance in sorting "untouched" $CD8^+$ T cells and characterization of their TCR structural affinities. Despite several drawbacks, like the need to clone and expand specific T cells before affinity measurements or their restriction to $CD8^+$ T cells analysis as $CD4^+$ T cells display too weak affinity TCR precluding their efficient use, they remain a useful tool for rapid screening of functionally relevant $CD8^+$ T cells.

6. Conclusion

Technical advances have been recently made to identify and isolate antigen-specific $CD8^+$ and $CD4^+$ T cells using soluble pMHC class I and II multimers. Detection of low avidity T cells, especially self-reactive and $CD4^+$ T cells, has been dramatically improved using molecularly defined pMHC class II monomers and optimized staining protocols. Developments of high-throughput pMHC class I synthesis as well as combinatorial strategies allow rapid and parallel screening of thousands of antigen specificities in a precious and limited blood sample or tissue biopsies. Finally, developments of reversible pMHC class I multimers allow high-throughput screening of functionally relevant viral or self tumor-specific T cell clones. Application of these methodologies at the single cell level will greatly improve our understanding of mechanisms governing T cell recognition. Future developments will be needed to reduce costs and automate processes.

References

Allard, M., Couturaud, B., Carretero-Iglesia, L., Duong, M. N., Schmidt, J., Monnot, G. C., et al. (2017). TCR-ligand dissociation rate is a robust and stable biomarker of CD8+ T cell potency. *JCI Insight*, *2*(14). https://doi.org/10.1172/jci.insight.92570.

Altman, J. D., Moss, P. A., Goulder, P. J., Barouch, D. H., McHeyzer-Williams, M. G., Bell, J. I., et al. (1996). Phenotypic analysis of antigen-specific T lymphocytes. *Science*, *274*(5284), 94–96.

Angelov, G. S., Guillaume, P., Cebecauer, M., Bosshard, G., Dojcinovic, D., Baumgaertner, P., et al. (2006). Soluble MHC-peptide complexes containing long rigid linkers abolish CTL-mediated cytotoxicity. *Journal of Immunology*, *176*(6), 3356–3365.

Ayyoub, M., Dojcinovic, D., Pignon, P., Raimbaud, I., Schmidt, J., Luescher, I., et al. (2010). Monitoring of NY-ESO-1 specific CD4+ T cells using molecularly defined MHC class II/His-tag-peptide tetramers. *Proceedings of the National Academy of Sciences of the United States of America*, *107*(16), 7437–7442. https://doi.org/10.1073/pnas.1001322107.

Ayyoub, M., Pignon, P., Dojcinovic, D., Raimbaud, I., Old, L. J., Luescher, I., et al. (2010). Assessment of vaccine-induced CD4 T cell responses to the 119-143 immunodominant region of the tumor-specific antigen NY-ESO-1 using DRB1*0101 tetramers. *Clinical Cancer Research*, *16*(18), 4607–4615. https://doi.org/10.1158/1078-0432.CCR-10-1485.

Batard, P., Peterson, D. A., Devevre, E., Guillaume, P., Cerottini, J. C., Rimoldi, D., et al. (2006). Dextramers: New generation of fluorescent MHC class I/peptide multimers for visualization of antigen-specific CD8+ T cells. *Journal of Immunological Methods*, *310*(1–2), 136–148. https://doi.org/10.1016/j.jim.2006.01.006.

Bentzen, A. K., & Hadrup, S. R. (2017). Evolution of MHC-based technologies used for detection of antigen-responsive T cells. *Cancer Immunology, Immunotherapy*, *66*(5), 657–666. https://doi.org/10.1007/s00262-017-1971-5.

Bentzen, A. K., Marquard, A. M., Lyngaa, R., Saini, S. K., Ramskov, S., Donia, M., et al. (2016). Large-scale detection of antigen-specific T cells using peptide-MHC-I multimers labeled with DNA barcodes. *Nature Biotechnology*, *34*(10), 1037–1045. https://doi.org/10.1038/nbt.3662.

Bentzen, A. K., Such, L., Jensen, K. K., Marquard, A. M., Jessen, L. E., Miller, N. J., et al. (2018). T cell receptor fingerprinting enables in-depth characterization of the interactions governing recognition of peptide-MHC complexes. *Nature Biotechnology*, *36*, 1191–1196. https://doi.org/10.1038/nbt.4303.

Bobisse, S., Genolet, R., Roberti, A., Tanyi, J. L., Racle, J., Stevenson, B. J., et al. (2018). Sensitive and frequent identification of high avidity neo-epitope specific CD8 (+) T cells in immunotherapy-naive ovarian cancer. *Nature Communications*, *9*(1), 1092. https://doi.org/10.1038/s41467-018-03301-0.

Boniface, J. J., Rabinowitz, J. D., Wulfing, C., Hampl, J., Reich, Z., Altman, J. D., et al. (1998). Initiation of signal transduction through the T cell receptor requires the multivalent engagement of peptide/MHC ligands [corrected]. *Immunity*, *9*(4), 459–466.

Brett, S. J., Kingston, A. E., & Colston, M. J. (1987). Limiting dilution analysis of the human T cell response to mycobacterial antigens from BCG vaccinated individuals and leprosy patients. *Clinical and Experimental Immunology*, *68*(3), 510–520.

Carson, R. T., Vignali, K. M., Woodland, D. L., & Vignali, D. A. (1997). T cell receptor recognition of MHC class II-bound peptide flanking residues enhances immunogenicity and results in altered TCR V region usage. *Immunity*, 7(3), 387–399.

Cebecauer, M., Guillaume, P., Hozak, P., Mark, S., Everett, H., Schneider, P., et al. (2005). Soluble MHC-peptide complexes induce rapid death of CD8+ CTL. *Journal of Immunology*, *174*(11), 6809–6819.

Cebecauer, M., Guillaume, P., Mark, S., Michielin, O., Boucheron, N., Bezard, M., et al. (2005). CD8+ cytotoxic T lymphocyte activation by soluble major histocompatibility complex-peptide dimers. *The Journal of Biological Chemistry, 280*(25), 23820–23828. https://doi.org/10.1074/jbc.M500654200.

Chen, G., & Weng, N. P. (2012). Analyzing the phenotypic and functional complexity of lymphocytes using CyTOF (cytometry by time-of-flight). *Cellular & Molecular Immunology, 9*(4), 322–323. https://doi.org/10.1038/cmi.2012.16.

Dolton, G., Lissina, A., Skowera, A., Ladell, K., Tungatt, K., Jones, E., et al. (2014). Comparison of peptide-major histocompatibility complex tetramers and dextramers for the identification of antigen-specific T cells. *Clinical and Experimental Immunology, 177*(1), 47–63. https://doi.org/10.1111/cei.12339.

Dolton, G., Zervoudi, E., Rius, C., Wall, A., Thomas, H. L., Fuller, A., et al. (2018). Optimized peptide-MHC multimer protocols for detection and isolation of autoimmune T-Cells. *Frontiers in Immunology, 9*, 1378. https://doi.org/10.3389/fimmu.2018.01378.

Fooksman, D. R., Vardhana, S., Vasiliver-Shamis, G., Liese, J., Blair, D. A., Waite, J., et al. (2010). Functional anatomy of T cell activation and synapse formation. *Annual Review of Immunology, 28*, 79–105. https://doi.org/10.1146/annurev-immunol-030409-101308.

Frayser, M., Sato, A. K., Xu, L., & Stern, L. J. (1999). Empty and peptide-loaded class II major histocompatibility complex proteins produced by expression in *Escherichia coli* and folding in vitro. *Protein Expression and Purification, 15*(1), 105–114. https://doi.org/10.1006/prep.1998.0987.

Fremont, D. H., Hendrickson, W. A., Marrack, P., & Kappler, J. (1996). Structures of an MHC class II molecule with covalently bound single peptides. *Science, 272*(5264), 1001–1004.

Garboczi, D. N., Hung, D. T., & Wiley, D. C. (1992). Hla-A2-peptide complexes—Refolding and crystallization of molecules expressed in *Escherichia coli* and complexed with single antigenic peptides. *Proceedings of the National Academy of Sciences of the United States of America, 89*(8), 3429–3433. https://doi.org/10.1073/pnas.89.8.3429.

Gebe, J. A., Falk, B. A., Rock, K. A., Kochik, S. A., Heninger, A. K., Reijonen, H., et al. (2003). Low-avidity recognition by CD4+ T cells directed to self-antigens. *European Journal of Immunology, 33*(5), 1409–1417. https://doi.org/10.1002/eji.200323871.

Gorska, M. M., & Alam, R. (2003). Signaling molecules as therapeutic targets in allergic diseases. *Journal of Allergy and Clinical Immunology, 112*(2), 241–250. https://doi.org/10.1067/mai.2003.1667.

Hadrup, S. R., Bakker, A. H., Shu, C. J., Andersen, R. S., van Veluw, J., Hombrink, P., et al. (2009). Parallel detection of antigen-specific T-cell responses by multidimensional encoding of MHC multimers. *Nature Methods, 6*(7), 520–526. https://doi.org/10.1038/nmeth.1345.

Hadrup, S. R., Maurer, D., Laske, K., Frosig, T. M., Andersen, S. R., Britten, C. M., et al. (2015). Cryopreservation of MHC multimers: Recommendations for quality assurance in detection of antigen specific T cells. *Cytometry. Part A, 87*(1), 37–48. https://doi.org/10.1002/cyto.a.22575.

Hebeisen, M., Schmidt, J., Guillaume, P., Baumgaertner, P., Speiser, D. E., Luescher, I., et al. (2015). Identification of rare high-avidity, tumor-reactive CD8+ T cells by monomeric TCR-ligand off-rates measurements on living cells. *Cancer Research, 75*(10), 1983–1991. https://doi.org/10.1158/0008-5472.CAN-14-3516.

Huang, J., Zeng, X., Sigal, N., Lund, P. J., Su, L. F., Huang, H., et al. (2016). Detection, phenotyping, and quantification of antigen-specific T cells using a peptide-MHC dodecamer. *Proceedings of the National Academy of Sciences of the United States of America, 113*(13), E1890–E1897. https://doi.org/10.1073/pnas.1602488113.

Hugues, S., Malherbe, L., Filippi, C., & Glaichenhaus, N. (2002). Generation and use of alternative multimers of peptide/MHC complexes. *Journal of Immunological Methods*, *268*(1), 83–92.

Knabel, M., Franz, T. J., Schiemann, M., Wulf, A., Villmow, B., Schmidt, B., et al. (2002). Reversible MHC multimer staining for functional isolation of T-cell populations and effective adoptive transfer. *Nature Medicine*, *8*(6), 631–637. https://doi.org/10.1038/nm0602-631.

Laugel, B., van den Berg, H. A., Gostick, E., Cole, D. K., Wooldridge, L., Boulter, J., et al. (2007). Different T cell receptor affinity thresholds and CD8 coreceptor dependence govern cytotoxic T lymphocyte activation and tetramer binding properties. *The Journal of Biological Chemistry*, *282*(33), 23799–23810. https://doi.org/10.1074/jbc.M700976200.

Leisner, C., Loeth, N., Lamberth, K., Justesen, S., Sylvester-Hvid, C., Schmidt, E. G., et al. (2008). One-pot, mix-and-read peptide-MHC tetramers. *PLoS One*, *3*(2), e1678. https://doi.org/10.1371/journal.pone.0001678.

Lissina, A., Ladell, K., Skowera, A., Clement, M., Edwards, E., Seggewiss, R., et al. (2009). Protein kinase inhibitors substantially improve the physical detection of T-cells with peptide-MHC tetramers. *Journal of Immunological Methods*, *340*(1), 11–24. https://doi.org/10.1016/j.jim.2008.09.014.

Luimstra, J. J., Garstka, M. A., Roex, M. C. J., Redeker, A., Janssen, G. M. C., van Veelen, P. A., et al. (2018). A flexible MHC class I multimer loading system for large-scale detection of antigen-specific T cells. *The Journal of Experimental Medicine*, *215*(5), 1493–1504. https://doi.org/10.1084/jem.20180156.

Massilamany, C., Gangaplara, A., Chapman, N., Rose, N., & Reddy, J. (2011). Detection of cardiac myosin heavy chain-alpha-specific CD4 cells by using MHC class II/IA (k) tetramers in A/J mice. *Journal of Immunological Methods*, *372*(1–2), 107–118. https://doi.org/10.1016/j.jim.2011.07.004.

Nauerth, M., Weissbrich, B., Knall, R., Franz, T., Dossinger, G., Bet, J., et al. (2013). TCR-ligand koff rate correlates with the protective capacity of antigen-specific CD8+ T cells for adoptive transfer. *Science Translational Medicine*, *5*(192), 192ra187. https://doi.org/10.1126/scitranslmed.3005958.

Neudorfer, J., Schmidt, B., Huster, K. M., Anderl, F., Schiemann, M., Holzapfel, G., et al. (2007). Reversible HLA multimers (streptamers) for the isolation of human cytotoxic T lymphocytes functionally active against tumor- and virus-derived antigens. *Journal of Immunological Methods*, *320*(1–2), 119–131. https://doi.org/10.1016/j.jim.2007.01.001.

Newell, E. W., Klein, L. O., Yu, W., & Davis, M. M. (2009). Simultaneous detection of many T-cell specificities using combinatorial tetramer staining. *Nature Methods*, *6*(7), 497–499. https://doi.org/10.1038/nmeth.1344.

Newell, E. W., Sigal, N., Bendall, S. C., Nolan, G. P., & Davis, M. M. (2012). Cytometry by time-of-flight shows combinatorial cytokine expression and virus-specific cell niches within a continuum of CD8+ T cell phenotypes. *Immunity*, *36*(1), 142–152. https://doi.org/10.1016/j.immuni.2012.01.002.

Reddy, J., Bettelli, E., Nicholson, L., Waldner, H., Jang, M. H., Wucherpfennig, K. W., et al. (2003). Detection of autoreactive myelin proteolipid protein 139-151-specific T cells by using MHC II (IAs) tetramers. *Journal of Immunology*, *170*(2), 870–877.

Rius, C., Attaf, M., Tungatt, K., Bianchi, V., Legut, M., Bovay, A., et al. (2018). Peptide-MHC class I tetramers can fail to detect relevant functional T cell clonotypes and underestimate antigen-reactive T cell populations. *Journal of Immunology*, *200*(7), 2263–2279. https://doi.org/10.4049/jimmunol.1700242.

Rodenko, B., Toebes, M., Celie, P. H., Perrakis, A., Schumacher, T. N., & Ovaa, H. (2009). Class I major histocompatibility complexes loaded by a periodate trigger. *Journal of the American Chemical Society*, *131*(34), 12305–12313. https://doi.org/10.1021/ja9037565.

Rodenko, B., Toebes, M., Hadrup, S. R., van Esch, W. J., Molenaar, A. M., Schumacher, T. N., et al. (2006). Generation of peptide-MHC class I complexes through UV-mediated ligand exchange. *Nature Protocols*, *1*(3), 1120–1132. https://doi.org/10.1038/nprot.2006.121.

Rubio-Godoy, V., Dutoit, V., Rimoldi, D., Lienard, D., Lejeune, F., Speiser, D., et al. (2001). Discrepancy between ELISPOT IFN-gamma secretion and binding of A2/peptide multimers to TCR reveals interclonal dissociation of CTL effector function from TCR-peptide/MHC complexes half-life. *Proceedings of the National Academy of Sciences of the United States of America*, *98*(18), 10302–10307. https://doi.org/10.1073/pnas.181348898.

Saini, S. K., Schuster, H., Ramnarayan, V. R., Rammensee, H. G., Stevanovic, S., & Springer, S. (2015). Dipeptides catalyze rapid peptide exchange on MHC class I molecules. *Proceedings of the National Academy of Sciences of the United States of America*, *112*(1), 202–207. https://doi.org/10.1073/pnas.1418690112.

Sant'Angelo, D. B., Robinson, E., Janeway, C. A., Jr., & Denzin, L. K. (2002). Recognition of core and flanking amino acids of MHC class II-bound peptides by the T cell receptor. *European Journal of Immunology*, *32*(9), 2510–2520. https://doi.org/10.1002/1521-4141(200209)32:9<2510::AID-IMMU2510>3.0.CO;2-Q.

Schmidt, J., Dojcinovic, D., Guillaume, P., & Luescher, I. (2013). Analysis, isolation, and activation of antigen-specific CD4(+) and CD8(+) T cells by soluble MHC-peptide complexes. *Frontiers in Immunology*, *4*, 218. https://doi.org/10.3389/fimmu.2013.00218.

Schmidt, J., Guillaume, P., Dojcinovic, D., Karbach, J., Coukos, G., & Luescher, I. (2017). In silico and cell-based analyses reveal strong divergence between prediction and observation of T-cell-recognized tumor antigen T-cell epitopes. *The Journal of Biological Chemistry*, *292*(28), 11840–11849. https://doi.org/10.1074/jbc.M117.789511.

Schmidt, J., Guillaume, P., Irving, M., Baumgaertner, P., Speiser, D., & Luescher, I. F. (2011). Reversible major histocompatibility complex I-peptide multimers containing Ni(2+)-nitrilotriacetic acid peptides and histidine tags improve analysis and sorting of CD8(+) T cells. *The Journal of Biological Chemistry*, *286*(48), 41723–41735. https://doi.org/10.1074/jbc.M111.283127.

Sharrock, C. E., Kaminski, E., & Man, S. (1990). Limiting dilution analysis of human T cells: A useful clinical tool. *Immunology Today*, *11*(8), 281–286.

Tanyi, J. L., Bobisse, S., Ophir, E., Tuyaerts, S., Roberti, A., Genolet, R., et al. (2018). Personalized cancer vaccine effectively mobilizes antitumor T cell immunity in ovarian cancer. *Science Translational Medicine*, *10*(436). https://doi.org/10.1126/scitranslmed.aao5931.

Uchtenhagen, H., Rims, C., Blahnik, G., Chow, I. T., Kwok, W. W., Buckner, J. H., et al. (2016). Efficient ex vivo analysis of CD4+ T-cell responses using combinatorial HLA class II tetramer staining. *Nature Communications*, *7*, 12614. https://doi.org/10.1038/ncomms12614.

Vollers, S. S., & Stern, L. J. (2008). Class II major histocompatibility complex tetramer staining: Progress, problems, and prospects. *Immunology*, *123*(3), 305–313. https://doi.org/10.1111/j.1365-2567.2007.02801.x.

Weichsel, R., Dix, C., Wooldridge, L., Clement, M., Fenton-May, A., Sewell, A. K., et al. (2008). Profound inhibition of antigen-specific T-cell effector functions by dasatinib. *Clinical Cancer Research*, *14*(8), 2484–2491. https://doi.org/10.1158/1078-0432.CCR-07-4393.

Wooldridge, L., Lissina, A., Cole, D. K., van den Berg, H. A., Price, D. A., & Sewell, A. K. (2009). Tricks with tetramers: How to get the most from multimeric peptide-MHC. *Immunology*, *126*(2), 147–164. https://doi.org/10.1111/j.1365-2567.2008.02848.x.

Wooldridge, L., Scriba, T. J., Milicic, A., Laugel, B., Gostick, E., Price, D. A., et al. (2006). Anti-coreceptor antibodies profoundly affect staining with peptide-MHC class I and class II tetramers. *European Journal of Immunology*, *36*(7), 1847–1855. https://doi.org/10.1002/eji.200635886.

CHAPTER THREE

In vitro assays for effector T cell functions and activity of immunomodulatory antibodies

Roberta Zappasodi[a,b,*], Sadna Budhu[a], Mohsen Abu-Akeel[a], Taha Merghoub[a,b,c,d]

[a]Ludwig Collaborative and Swim Across America Laboratory, Memorial Sloan Kettering Cancer Center, New York, NY, United States
[b]Parker Institute for Cancer Immunotherapy, Memorial Sloan Kettering Cancer Center, New York, NY, United States
[c]Department of Medicine, Memorial Sloan Kettering Cancer Center, New York, NY, United States
[d]Weill Cornell Medicine, New York, NY, United States
*Corresponding author: e-mail address: zappasor@mskcc.org

Contents

Abstract

The recent clinical success of cancer immunotherapy with checkpoint blockade has led to renewed interest into the development of immune modulatory agents with the capacity to activate anti-tumor T cell responses. Standardization of optimized in vitro

ISSN 0076-6879
https://doi.org/10.1016/bs.mie.2019.08.012

assays for efficient assessment of immune function of such new drugs is thus needed to facilitate clinical development of the optimal drug candidates.

Here, we describe an optimized version of T cell suppression assay designed to test the effect of immunomodulatory agents on T cell function and activation. We apply this assay to investigate the agonist activity of the T cell co-stimulatory molecule glucocorticoid-induced TNFR-related protein (GITR). We detail a protocol for concurrent assessment of multiple levels of T cell functional modulation upon GITR engagement, including T cell priming, activation and effector function, in a single assay. As human GITR agonist antibodies are currently under development, availability of standardized cell-based functional assays of GITR agonism is instrumental to translate anti-GITR therapy into the clinical setting.

1. Introduction

The unprecedented clinical success of immune checkpoint blockade for the treatment of cancer has triggered a substantial interest in developing drugs that can modulate T cell responses to cancer (Khalil et al., 2015). Since the FDA approval of CTLA-4 blockade with ipilimumab in 2011 and PD-1 blockade with pembrolizumab and nivolumab in 2014, immunotherapy is now at the cutting edge of cancer care (Chen & Han, 2015; Wolchok et al., 2013). Despite the success of these drugs leading the durable responses in the clinic, the majority of cancer patients do not benefit from current immunotherapies. With the exception of few particularly sensitive disease types, such as Hodgkin lymphoma, the clinical success rate of approved checkpoint blockade therapies remains relatively low, with durable clinical responses seen in 20–40% of patients treated with single-agent immunotherapy and in up to 60% of patients treated with combination regimens (Zappasodi, Merghoub, & Wolchok, 2018; Zappasodi, Wolchok, & Merghoub, 2018). Thus, there is a need for novel and more effective immunotherapies. Currently, there are several immune modulatory agents at various stages of clinical development that either block alternative inhibitory T cell checkpoints (e.g., LAG-3, TIM-3) or activate T cell co-stimulatory receptors (e.g., GITR, OX40, 4-1BB) (Khalil et al., 2015). Other forms of immunotherapies are also under development, such as adoptive cell transfer, with tumor-infiltrating lymphocytes (TILs), chimeric antigen receptor (CAR) T cells, or T cell receptor (TCR)-transgenic T cells, cytokines and vaccines. While some therapies target the tumor cells directly (e.g., CAR T cells), others act indirectly by enhancing pre-existing tumor immunity (cytokines, immune-modulating antibodies) or inducing de novo T cell responses (vaccines). There has been some success of these agents as monotherapies; however, it is becoming apparent that the combination of these agents together or with conventional therapies may provide greater clinical outcomes (Zappasodi, Merghoub, & Wolchok, 2018). Further improvement of immunotherapies is possible through rational design of pre-clinical and clinical trials with a focus on the optimal means to combine two or more therapies. In addition, optimization of molecular design, regimens and combinations of approved immunotherapies, guided by our current understanding of their mechanisms of action, has the potential to increase the response rates in patients in the clinic.

Prior to moving a new drug into pre-clinical or clinical evaluation, the design, screen and ultimate choice of the compound must undergo rigorous testing in order to maximize the likelihood of anti-tumor efficacy in in vivo animal models, and achievement of clinical benefit in patients. As part of this process, proper in vitro assays are needed to efficiently screen the drug candidates with the desired biologic activity. In order to optimize an assay specifically for a target molecule expressed by T cells, several considerations must be taken into account. First, the target molecule must be expressed by T cells in the culture conditions chosen as part of the selected assay(s). It is also important to determine which cell subsets (e.g., $CD4^+$, $CD8^+$ T cells, regulatory T cells (Tregs)) express the target molecule. Second, knowledge of the expression pattern of the target molecule over time during T cell activation may be needed to develop the proper assay conditions, including the definition of the appropriate length of in vitro incubation with the immunomodulatory agent for maximal effects and readout detectability. Lastly, it is important to consider that in vitro assays, while useful to determine the functional activity of immunomodulatory agents on immune cells, may not always anticipate the level of in vivo efficacy, as observed for example with PD-1 blocking antibodies (Wang et al., 2014).

In this chapter, we describe an optimized version of T cell suppression assay designed to test the capacity of immune co-stimulatory agents to enhance priming and activation of T cells in the presence of Tregs. More specifically, we describe a T cell functional assay optimized to test the activity of agents stimulating the T cell co-stimulatory molecule glucocorticoid-induced TNFR-related protein (GITR). GITR is expressed at high baseline levels on Tregs and upregulated on activated $CD4^+$ and $CD8^+$ effector T cells (Teff) (Nocentini, Ronchetti, Petrillo, & Riccardi, 2012; Schaer, Murphy, & Wolchok, 2012). Thus, engagement of GITR affects both Teff and Tregs. The assays described here were optimized to test GITR stimulation using a recombinant human GITR ligand (rhGITRL) to overcome the suppressive effects of Tregs on CD4$^+$ and CD8$^+$ Teff activation. Concurrently, these assays allow to test direct agonistic effects of GITR simulation on activation and effector function of T cells, in the absence of Tregs. It is important to note that a good commend and knowledge of flow cytometry is necessary in order to perform these assays since the readout for this procedure is analyzed by flow cytometry.

2. Functional assays for T cell co-stimulatory molecules: Principles

T cell priming (de novo activation of T cell responses) requires concurrent activation of two signals in T cells: recognition of peptides bound to MHC molecules through the TCR, and CD28 co-stimulation via CD80/CD86 on antigen presenting cells (APCs). The fate of activated T cells is further modulated by additional interactions between co-signaling (co-stimulatory/co-inhibitory) receptors on T cells and their ligands generally expressed on APCs (Chen & Flies, 2013). It is now clear that co-signaling molecules regulate T cell activation, effector function, survival and memory development. The availability of in vitro assays for the characterization of the effects of T cell co-signaling molecules is thus critical to understand their biology and develop therapeutics mimicking their function.

T cell activation can be measured by assessing (1) proliferation, (2) up-regulation of activation markers (e.g., IL2RA/CD25) and (3) production of effector cytokines (IFN-γ, TNF-α). To test the contribution of co-stimulatory receptor engagement to these effects, T cells are typically activated with agonist anti-CD3 antibodies, for the activation of signal 1, and sub-optimal concentrations of agonist anti-CD28 antibodies, for a signal 2 that can be improved by engaging the co-stimulatory molecules (e.g., GITR) under investigation. Conventional immunosuppressive Tregs can be added for a more stringent version of the assay in which to test the capacity of the co-stimulatory agents in exam to bypass Treg suppression and activate Teff even in their presence. If this is true, it may be important to further dissect the mechanism of action of the co-stimulatory agent under investigation in the Teff:Treg co-culture assay and test whether it acts on Teff, Tregs or both. This can be assessed by pre-incubating Tregs or Teff with the co-stimulatory agent in exam before the use of these cell subsets in the co-culture assay. If selective engagement of the co-stimulatory receptor on one cell type can reproduce the effects obtained in the co-culture system, it means that the co-stimulatory pathway has a unique or predominant cellular target.

Technically, T cell proliferation can be assessed by staining T cells with a fluorescent tracking dye, such as carboxyfluorescein succinimidyl ester (CFSE), before starting the culture and by monitoring dilution of the dye in daughter cells as cells get activated and divide over time. These are typically very bright and stable dyes that stain lymphocytes with minimal variation. With cell division, the amount of dye in each cell is halved and this produces discrete shifts or peaks in fluorescence intensity associated with each cell generation. These discrete peaks in fluorescence intensity are highly amenable to mathematical modeling for quantifying cell proliferation. Importantly, dilution of the fluorescent tracking dyes in proliferating T cells nicely correlates with up-regulation of activation markers and production of effector cytokines. Human T cells generally need to be incubated with anti-CD3 and anti-CD28 antibodies for 3–5 days to detect discrete proliferation peaks. Treg suppression can be measured in these assays by co-culturing CFSE-labeled Teff (target cells) in the presence of Tregs (suppressor cells) and monitoring inhibition of proliferation and activation of target cells (Fig. 1A). In these suppression assays, Tregs can also be labeled with a fluorescent tracking dye—different from the one used to label Teff—such as CellTrace Violet dye (CTV), to confidently exclude these cells from the target Teff gate and monitor concurrent changes in Tregs from the same co-cultures (Fig. 1A). As T cell activation is influenced by cell concentration, it is important that Treg suppression assay cultures are properly controlled for cell number. Therefore, in addition to testing CFSE-labeled target Teff alone to measure baseline proliferation in the absence of Tregs, a condition where Teff are co-cultured with an equal amount of CTV-labeled Teff should be included to control for metabolic/nutrient competition (Fig. 1A). This allows to reliably assess and distinguish actual T-cell suppression from metabolic/nutrient competition, which can also result in inhibition of target Teff proliferation, although through different mechanisms. Teff capacity to undergo proliferation upon anti-CD3 and anti-CD28 stimulation may vary from donor to donor. Therefore, it is advisable to plate extra wells with Teff to check CFSE dilution starting on day 3 after stimulation to understand when to stop the incubation and acquire the samples. Following such incubation, cells are stained with a cocktail of fluorochrome-labeled antibodies specific for the T cell subsets in culture (CD4 and/or CD8) and co-stimulatory molecules (e.g., CD25) and a

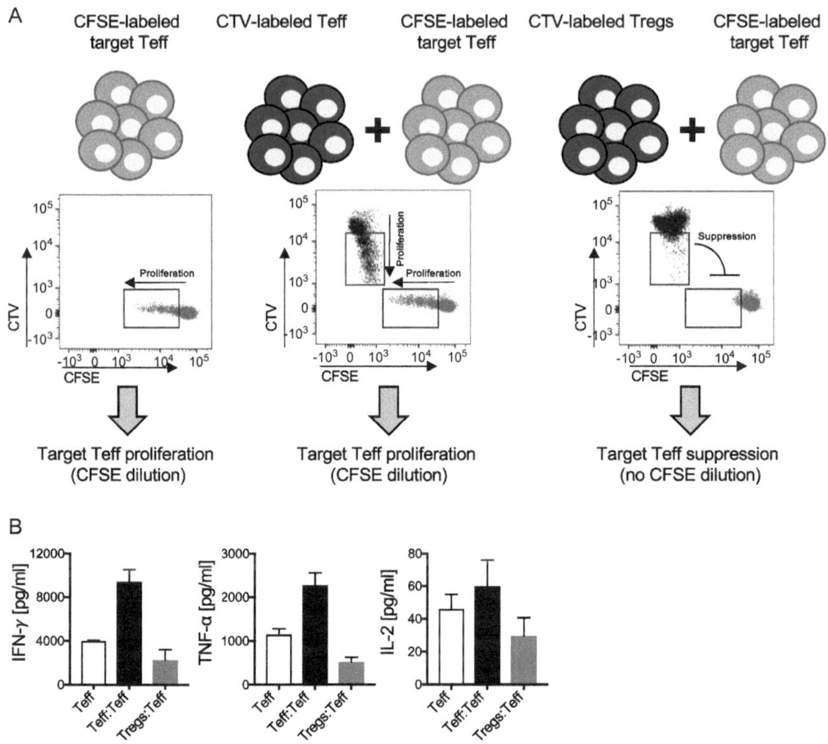

Fig. 1 Principle of Treg suppression assay. CFSE-labeled Teff are cultured alone, or with an equal amount of CTV-labeled Teff or Tregs and stimulated for 4 days with 1 μg/mL plate-bound anti-CD3 (clone SP34) and 0.1 μg/mL soluble anti-CD28 (clone CD28.2). Cultures are then analyzed for proliferation by FACS (A) and the indicated cytokines are quantified in culture supernatants by Luminex-based bead multiplex immunoassay (B). (A) Representative bivariate FACS plots (CFSE (FITC) vs CTV (Pacific Blue)) of live single $CD4^+$ T cells in each culture condition are reported.

viability dye thus allowing for concurrent flow cytometry-based analysis of T cell proliferation and activation. Fluorescent cell tracking dyes are compatible with fixation/permeabilization buffers for detection of intracellular proteins, including cytokines and transcription factors. For a more comprehensive analysis of cytokine production, culture supernatants can be subjected to multiplex ELISA covering cytokine specificities across multiple T cell differentiation lineages (Fig. 1B).

Here we detail a basic protocol for assessing the effects of GITR co-stimulation on proliferation and activation of human Teff cultured alone or in the presence of human Tregs isolated from healthy donor peripheral blood. $CD4^+$ Teff and Tregs are isolated by magnetic bead purification from peripheral blood mononuclear cells (PBMC) obtained by gradient centrifugation of whole peripheral blood. Alternatively, Teff and Treg subsets can be FACS-sorted from PBMC using proper combinations of fluorochrome-conjugated anti-CD4 and anti-CD25 antibodies. If Teff and Tregs are FACS-sorted, it is important that anti-CD4 and anti-CD25 antibodies conjugated with the same fluorochromes are included in the FACS panel for the acquisition of the assay. We have optimized these assays to provide suboptimal co-stimulation through the CD28 signaling pathway, which can be further improved by the addition of GITR stimulation. By promoting Teff function and reducing Treg suppression, GITR agonist antibodies have shown potent preclinical activity in multiple mouse tumor models (Cohen et al., 2010; Ko et al., 2005; Turk et al., 2004). These promising preclinical results have led to the development of human GITR agonist antibodies, which are currently under clinical investigation. Availability of standardized cell-based functional assays of GITR agonism is thus instrumental to properly translate anti-GITR therapy into the clinical setting.

3. Functional assays for T-cell co-stimulatory molecules: Protocol

3.1 Reagents and instruments

Agonist anti-human CD3 antibody (clone SP34, BD Biosciences)
Agonist anti-human CD28 antibody (clone CD8.2, BD Biosciences)
rhGITRL (Genscript)
Lymphocyte Separation Medium (Corning)
$CD4^+CD25^+$ Human Regulatory T Cell Isolation Kit (Miltenyi)
MACS separation columns (LD and LS, Miltenyi)
MidiMACS separator (Miltenyi)
CFSE and CTV labeling dyes (Invitrogen)
RPMI 1640 supplemented with 10% FBS, 4 mM glutamine, 1 mM sodium pyruvate and 50 μM β-mercaptoethanol
Ca^{2+}/Mg^{2+} free PBS
U bottomed 96-well plates
15 and 50 mL falcon tubes
FcR blocking reagent (Miltenyi)
Refrigerated centrifuge

3.2 Day 1: Coating plates with stimulatory antibodies

The day before the assay, add 50 μL of 2 μg/mL anti-CD3 in PBS to the required number of wells of a U-bottomed 96-well plate. Add 50 μL of 20 μg/mL rhGITRL in PBS or 50 μL PBS for GITR stimulation or control, respectively, for a final concentration of 1 μg/mL anti-CD3 and 10 μg/mL rhGITRL. Mix well, seal the plate with parafilm and incubate at 4 °C for at least 18 h.

3.3 Day 0: Isolation of mononuclear cells from human peripheral blood by density gradient centrifugation

a. Dilute one part of peripheral blood with three parts of Ca^{2+}/Mg^{2+} free PBS.

b. Carefully layer 30 mL diluted blood on 15 mL Lymphocyte Separation Medium (Corning) in a 50 mL conical tube; be sure to maintain a clear interphase between the blood and the lymphocyte separation medium. Mixing of the gradient with the blood prior to centrifugation may result in poor recovery.

c. Centrifuge at 400 × *g* for 30 min at room temperature without brake.

d. Aspirate the upper layer as much as possible leaving the mononuclear cell layer undisturbed at the interphase.

e. Transfer the mononuclear cell layer into a new 50 mL tube, fill with Ca^{2+}/Mg^{2+} free PBS to the 50 mL mark, mix, and centrifuge at 300 × *g* for 5 min.

f. Discard the supernatant and resuspend in an appropriate volume of media for counting the mononuclear cells by hemocytometer.

3.4 Day 0: Teff and Treg immunomagnetic purification

Teff and Tregs are purified from total PBMC using the Miltenyi $CD4^{+}CD25^{+}$ Human Regulatory T Cell Isolation Kit following the manufacturer's instructions. This procedure includes a first step for $CD4^{+}$ T cell enrichment by negative selection followed by separation of $CD25^{+}CD4^{+}$ Tregs and $CD25^{-}CD4^{+}$ Teff using immunomagnetic beads conjugated to an anti-CD25 antibody (Fig. 2A). This method allows to enrich in Foxp3-expressing cells within the $CD25^{+}$ Treg fraction (Fig. 2B), which displays inhibitory function (Fig. 1).

3.5 Day 0: Assay set up

a. Determine Teff and Treg cell number.

b. Label target Teff with CFSE: adjust the volume of Teff to 10^{7}/mL in serum-free RPMI 1640 medium and add CFSE to a final concentration of 5–10 μM. After 10 min incubation at 37 °C, quench free CFSE with a double amount of FBS, top up with fresh cold complete RPMI 1640,

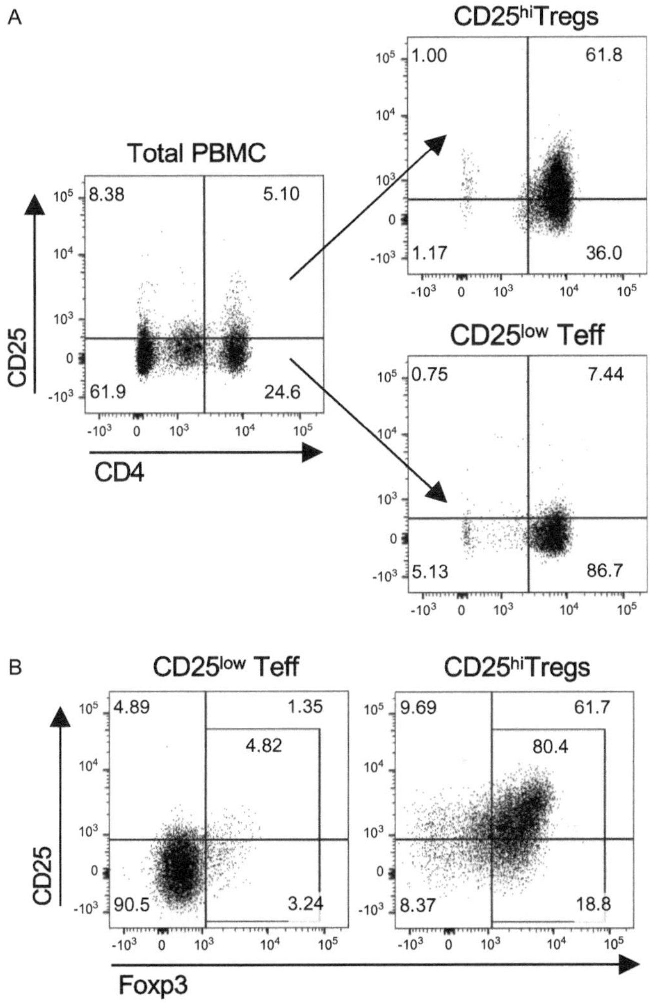

Fig. 2 Immunomagnetic purification of human Tregs. Representative example of Treg and Teff purification from healthy donor derived PBMC using CD4$^+$CD25$^+$ Human Regulatory T Cell Isolation Kit (Miltenyi). (A) FACS plots gated on live single cells showing expression of CD25 and CD4 before (left) and after Tregs (right, top) and Teff (right, bottom) purification. (B) Intracellular staining of Foxp3 in purified Teff and Tregs. FACS plots showing Foxp3 and CD25 expression in CD4$^+$ T cells from Teff and Treg purified fractions.

supplemented with 10% FBS, 4 mM glutamine, 1 mM sodium pyruvate and 50 μM β-mercaptoethanol, and pellet the cells (300 × *g* for 5 min). Resuspend CFSE-labeled cells at the desired concentration in complete RPMI 1640.

c. Label Tregs and another set of Teff with CTV: adjust the cell volume to 5×10^6/mL in serum-free RPMI 1460 medium and add CTV to a final concentration of 5–10 μM. After 20 min incubation at 37 °C, quench free CTV with a double amount of FBS, top up with fresh cold complete RPMI 1640 and pellet the cells ($300 \times g$ for 5 min). Resuspend CTV-labeled cells at the desired concentration in complete RPMI 1640.

d. Plate CTV-labeled cells in 100 μL complete RPMI 1640 per well in the antibody-coated U bottomed 96-well plate. Add 100 μL complete RPMI 1640 per well to the wells where CFSE-labeled target Teff have to be cultured alone. In general, plating Teff and Treg at 1:1 ratio is sufficient to detect suppression. Depending on the recovered quantity of Tregs, determine the number of Tregs to use per well. A number of Teff ranging from 30,000 to 100,000 can be efficiently stimulated in U-bottomed 96-well plate. Therefore, it is recommended to plate a number of Tregs per well within that range. If multiple Treg:Teff ratios need to be tested, perform serial dilution of CTV-labeled Tregs (and CTV-labeled Teff as control) directly in the culture plate.

e. Plate CFSE-labeled target Teff in 50 μL complete RPMI 1640 medium in each well (same number as Tregs for the 1:1 ratio). In case of multiple Treg:Teff ratios to be tested, the number of target Teff has to be fixed.

f. Add 50 μL of 0.4 μg/mL anti-CD28 to all the wells that need to be stimulated for a final concentration of 0.1 μg/mL.

g. Additional control for FACS compensation and set up: plate extra wells with CFSE-labeled or CTV-labeled Teff alone stimulated with anti-CD3 and anti-CD28 for CFSE and CTV single stain controls and to check proliferation between day 3–5 incubation. Plate extra wells with CFSE-labeled and CTV-labeled Teff (1:1 ratio) co-incubated with anti-CD3 and anti-CD28 for isotype control staining. Plate extra wells with CFSE-labeled and CTV-labeled Teff (1:1 ratio) and leave them unstimulated to gate the top CFSE and CTV peak of non-proliferative cells.

h. Incubated plates at 37 °C in a humidified incubator for 3–5 days.

3.6 Day 3–5: FACS analysis

Before proceeding with the following steps, prepare FACS buffer (2 mM EDTA and 1% BSA in Ca^{2+}/Mg^{2+} free PBS), a $2\times$ FcR blocking reagent solution (Miltenyi) and a $2\times$ surface antibody cocktail as described below.

a. Collect 100 µL culture supernatant from each well and store at −80 °C for cytokine analysis by multiplex ELISA.
b. Add 100 µL FACS buffer/well and spin down the plate (2000 × rpm for 3 min at 4 °C).
c. Resuspend cell pellets in 50 µL of 2 × FcR blocking reagent (Miltenyi) in FACS buffer and incubate for 10 min on ice.
d. Add 50 µL of 2 × surface antibody cocktail diluted in FACS buffer, mix and incubate for 30 min on ice in the dark. FACS-based suppression assays have the advantage to allow for concurrent assessment of T-cell proliferation and expression of maturation/activation, as a further measure of functional changes in target cells (Fig. 3). Besides staining for CD4 and with a viability dye, it is thus desirable to include antibodies for such markers, such as CD25. Results shown in Fig. 3B were obtained by using a nine-color FACS panel (Table 1), which can be acquired with a flow cytometry equipped with three lasers (488, 633, and 405 nm).
e. Add 100 µL FACS buffer per well and spin the plate (2000 × rpm for 3 min at 4 °C).
f. Discard the supernatant and resuspend with 200 µL FACS buffer per well for another wash (2000 × rpm for 3 min at 4 °C).
g. Resuspend in 50 µL FACS buffer and transfer to collection tubes for FACS acquisition.

3.7 Multiplex ELISA for cytokine quantification in culture supernatant

As a further readout of T cell co-stimulation, relevant cytokines can be quantified in the culture supernatants collected at the end of the culture incubation. This can be assessed by standard or multiplex ELISA. Fig. 3C shows representative cytokines quantified using the Luminex-based bead multiplex immunoassay Th1/Th2/Th9/Th17 Cytokine 18-Plex Human ProcartaPlex™ Panel (Invitrogen), which is particularly suitable to investigate changes in T cell function (Zappasodi et al., 2018). This type of immunoassays uses the xMAP® technology (multi-analyte profiling beads) for the simultaneous detection and quantification of multiple cytokines in the same sample. All Luminex-based instruments are compatible with this analysis.

4. FACS compensation, set up and data analysis

Flow cytometry is an essential part of an immunologist toolbox. Understanding the principles of flow cytometry is essential for monitoring

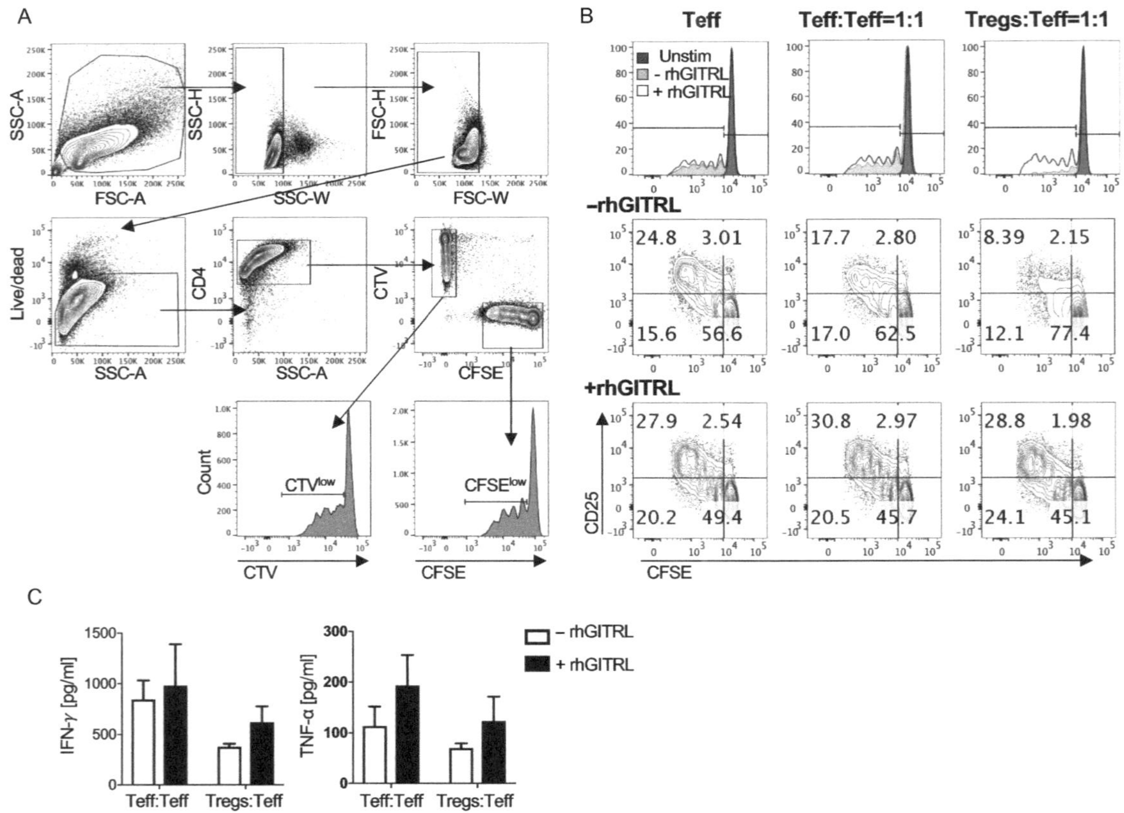

Fig. 3 Effects of GITR agonism in Treg suppression assay. Donor-derived CFSE-labeled Teff were cultured alone, or with an equal amount of CTV-labeled Teff or Tregs and stimulated for 4 days with anti-CD3 and anti-CD28 in the presence or the absence of rhGITRL. (A) Representative gating strategy of proliferating CFSE-labeled and CTV-labeled $CD4^+$ Teff. CFSE-labeled and CTV-labeled cells are gated among live single $CD4^+$ T cells. Proliferating $CFSE^{low}$ and CTV^{low} cells are then gated among the $CFSE^+$ and CTV^+ $CD4^+$ cell subsets in the respective histogram plot. (B) Representative FACS analysis of proliferation (CFSE dilution) and activation (CD25 up-regulation) in target Teff cultured alone or with an equal amount of CTV-labeled Teff or Tregs and stimulated with (+rhGITRL) or without (−rhGITRL) rhGITRL. (C) Quantification of the indicated cytokines in the supernatants from Teff:Teff and Tregs:Teff co-cultures as shown in (B).

Table 1 Example of FACS antibody panel for assay acquisition.

Channel	Target	Dilution
FITC	CFSE	–
PE	**GITR**	**1:50**
PE-TexasRed	**CD45RA**	**1:50**
PerCP-Cy5.5	**PD-1**	**1:50**
PE-Cy7	**CD62L**	**1:50**
Pacific Blue	CTV	–
AmCyan	Viability	1:500
AlexaFluor700	CD4	1:50
APC-Cy7	**CD25**	**1:50**

Highlighted in gray are the non-lineage/non-viability markers that need to be substituted with fluorochrome matched isotype controls in the isotype control samples described in Section 4.

immune responses in preclinical and clinical studies. This section will not cover the principles or basic operations of flow cytometry but instead is tailored to users of intermediate and advance levels. A proficient understanding of multicolor flow cytometry including compensation, knowledge of the FACS Diva and FlowJo softwares is assumed.

4.1 Single stain controls

In general, compensation beads are suitable to properly compensate multicolor flow cytometry antibody panels. However, given the nature of this assay, we recommend using fresh human PBMC for single stain controls and pre-activated T cells to control for CFSE and CTV single staining.

a. Add 100,000–200,000 cells in 100 μL of FACS buffer to each well in a U-bottomed 96-well plate. Plate a well for each fluorochrome included in the FACS panel plus one unstained control. For CTV and CFSE single stain controls, use pre-labeled cells.

b. Add 1 × (use manufacturer recommended amount) of fluorochrome-conjugated antibody to the appropriate well.

c. Mix thoroughly by pipetting several times.

d. Incubate plates for 30 min on ice in the dark.

e. Add 100 μL FACS buffer to each well and spin the plate (2000 × rpm for 3 min at 4 °C).

f. Discard supernatant and resuspend cells in 200 μL FACS buffer and spin the plate again (2000 × rpm for 3 min at 4 °C).
g. Discard supernatant and resuspend cells in 200 μL FACS buffer and transfer to collection tubes for FACS acquisition.
h. Keep tubes in the dark until ready for use.

4.2 Isotype controls

To ensure proper gating of the appropriate T cell populations and their activation markers, it is recommended to prepare additional isotype control samples for each antibody that is not a lineage marker. For example, in the experiment outlined in Section 3 (Fig. 3B), we recommend staining one control sample using pre-activated CFSE:CTV T-cell co-cultures with fluorochrome-matched isotype controls in place of the non-lineage/non-viability markers (Table 1, bold) in addition to the lineage and viability staining included in the acquisition panel.

4.3 Gating strategy and data analysis

Fig. 3A shows the representative gating strategy of live single CFSE-labeled and CTV-labeled $CD4^+$ T cells in co-culture, acquired on an LSRII flow cytometer (BD Biosciences) using BD FACSDiva software (BD Biosciences) and analyzed with the Flow software version 10 (FlowJo, LLC). To gate on the appropriate T cell populations, first gate on the lymphocyte population according to the physical parameters FSC and SSC, then gate on single cells using SSC-H vs SSC-W and FSC-H vs FSC-W (Fig. 3A). Next, gate the live cells using eFluor506 vs SSC-A. The live cells can then be sub-gated into $CD4^+$ T cells (CD4 vs SSC-A), where a clear separation of CFSE-labeled and CTV-labeled T cells can be detected on a bivariate plot (CFSE (FITC) vs CTV (Pacific Blue)). Each T cell population can then be gated and analyzed individually for cell proliferation and activation (Fig. 3). The unstimulated control is used to gate on the $CFSE^{low}$ and CTV^{low} populations, which represent the cells that have undergone one or more rounds of division (Fig. 3B). In addition to proliferation, T cell activation can also be assessed by examining relative expression of the activation markers. Activation markers can be analyzed by gating on the percentage of positive cells using the isotype control or by median fluorescence intensity (MFI).

5. Concluding remarks

The methods describe here are useful to assess Teff and Treg functions. The experiments were designed in a suboptimal setting to examine how the addition of immunomodulatory agents to the assay modifies Teff and Treg behavior. In this protocol, we used GITRL as the immunomodulatory agent. Since GITR stimulation is known to affect both Teff and Tregs, this assay can be used to test two variables: (1) the effect of GITR stimulation on modulating the suppressive function of Tregs and (2) the effect of GITR stimulation on modulating the activation state of effector T cells. In the data shown in Fig. 3B, suppression of Teff proliferation and CD25 expression can be observed when Teff were co-cultured with Tregs at a 1:1 ratio (−rhGITRL). The addition of rhGITRL to this co-culture was able to reverse this suppression. Moreover, the experiment setup described here can be modified to a simpler version where only CFSE or CTV labeled Teff ($CD4^+$ or $CD8^+$) are stimulated. Fig. 3B shows that addition of rhGITRL to CFSE-labeled Teff cultured either alone or in the presence of an equal number of Teff (Teff:Teff) enhances Teff proliferation and activation. The co-stimulatory effects of GITR agonism on Teff proliferation, activation and inhibition of Treg suppression are paralleled by increases in pro-inflammatory cytokine production (Fig. 3C). Thus, the assay described here provides a useful tool in deciphering several levels of T cell and Treg function that can be nicely complemented with the assessment of cytokine production in the supernatants from the same cultures.

In case T-cell antigen specificity is known, this assay can be further modified to use cognate antigen stimulation in place of anti-CD3 polyclonal activation. For example, common viral epitopes (e.g., from Cytomegalovirus, Epstein-Barr and Influenza viruses) can be used to re-stimulate donor-derived PBMC. In this setting, peptide-pulsed irradiated APCs may be used as scaffold for crosslinking the immunomodulatory antibodies in exam, providing even more physiological co-stimulatory signals compared to plastic-bound antibodies.

Upon in vitro selection with these assays, candidate agents with the desired immunomodulatory function can be further studied in proper in vivo model systems to assess therapeutic activity (e.g., in suitable humanized mouse tumor models) and safety (e.g., in primates, in case of cross-reactivity of the drug with the target molecule in these species), before initiating the clinical development.

Acknowledgments

We thank the Ludwig Institute for Cancer Research, Leap Therapeutics, the Swim Across America, the Parker Institute for Cancer Immunotherapy, the Breast Cancer Research Foundation and Hazen Polsky Foundation for supporting this research. This study was supported in part by NIH grants R01CA056821, R01CA215136-02, P01CA33049, and P01CA59350, and MSKCC Cancer Center Core Grant P30CA008748.

Competing interests

R.Z. is inventor on patent applications related to work on GITR, PD-1 and CTLA-4. R.Z. is consultant for Leap Therapeutics. T.M. is consultant for Leap Therapeutics, Immunos Therapeutics and Pfizer, and co-founder of Imvaq therapeutics. T.M. has equity in Imvaq therapeutics. T.M. reports grants from Bristol-Myers Squibb, Surface Oncology, Kyn Therapeutics, Infinity Pharmaceuticals, Peregrine Pharmaceuticals, Adaptive Biotechnologies, Leap Therapeutics, Aprea. T.M. is inventor on patent applications related to work on oncolytic viral therapy, alphavirus-based vaccines, neo-antigen modeling, CD40, GITR, OX40, PD-1 and CTLA-4.

References

Chen, L., & Flies, D. B. (2013). Molecular mechanisms of T cell co-stimulation and co-inhibition. *Nature Reviews. Immunology, 13*, 227–242. https://doi.org/10.1038/nri3405.

Chen, L., & Han, X. (2015). Anti-PD-1/PD-L1 therapy of human cancer: Past, present, and future. *The Journal of Clinical Investigation, 125*, 3384–3391. https://doi.org/10.1172/JCI80011.

Cohen, A. D., et al. (2010). Agonist anti-GITR monoclonal antibody induces melanoma tumor immunity in mice by altering regulatory T cell stability and intra-tumor accumulation. *PLoS One, 5*, e10436. https://doi.org/10.1371/journal.pone.0010436.

Khalil, D. N., et al. (2015). The new era of cancer immunotherapy: Manipulating T-cell activity to overcome malignancy. *Advances in Cancer Research, 128*, 1–68. https://doi.org/10.1016/bs.acr.2015.04.010.

Ko, K., et al. (2005). Treatment of advanced tumors with agonistic anti-GITR mAb and its effects on tumor-infiltrating Foxp3+CD25+CD4+ regulatory T cells. *Journal of Experimental Medicine, 202*, 885–891.

Nocentini, G., Ronchetti, S., Petrillo, M. G., & Riccardi, C. (2012). Pharmacological modulation of GITRL/GITR system: Therapeutic perspectives. *British Journal of Pharmacology, 165*, 2089–2099. https://doi.org/10.1111/j.1476-5381.2011.01753.x.

Schaer, D. A., Murphy, J. T., & Wolchok, J. D. (2012). Modulation of GITR for cancer immunotherapy. *Current Opinion in Immunology, 24*, 217–224. https://doi.org/10.1016/j.coi.2011.12.011.

Turk, M. J., et al. (2004). Concomitant tumor immunity to a poorly immunogenic melanoma is prevented by regulatory T cells. *The Journal of Experimental Medicine, 200*, 771–782. https://doi.org/10.1084/jem.20041130.

Wang, C., et al. (2014). In vitro characterization of the anti-PD-1 antibody nivolumab, BMS-936558, and in vivo toxicology in non-human primates. *Cancer Immunology Research, 2*, 846–856. https://doi.org/10.1158/2326-6066.CIR-14-0040.

Wolchok, J. D., et al. (2013). Development of ipilimumab: A novel immunotherapeutic approach for the treatment of advanced melanoma. *Annals of the New York Academy of Sciences, 1291*, 1–13. https://doi.org/10.1111/nyas.12180.

Zappasodi, R., Merghoub, T., & Wolchok, J. D. (2018). Emerging concepts for immune checkpoint blockade-based combination therapies. *Cancer Cell*, *33*, 581–598. https://doi.org/10.1016/j.ccell.2018.03.005.

Zappasodi, R., Wolchok, J. D., & Merghoub, T. (2018). Strategies for predicting response to checkpoint inhibitors. *Current Hematologic Malignancy Reports*, *13*, 383–395. https://doi.org/10.1007/s11899-018-0471-9.

Zappasodi, R., et al. (2018). Non-conventional inhibitory CD4(+)Foxp3(−)PD-1(hi) T cells as a biomarker of immune checkpoint blockade activity. *Cancer Cell*, *33*, https://doi.org/10.1016/j.ccell.2018.05.009. 1017–1032.e1017.

CHAPTER FOUR

A cytofluorimetric assay to evaluate T cell polyfunctionality

Belinda Palermo*, Mariangela Panetta, Giulia Campo, Paola Nisticò*
Unit of Tumor Immunology and Immunotherapy, Department of Research, Advanced Diagnostics and Technological Innovation, IRCCS Regina Elena National Cancer Institute, Rome, Italy
*Corresponding authors: e-mail address: belinda.palermo@ifo.gov.it; paola.nistico@ifo.gov.it

Contents

Abstract

The magnitude and the quality of T-cell response are critical endpoints in the immune-monitoring of cancer patients. The release of cytokines and lytic factors by T cells operate in each phase of the cancer immunity cycle. The simultaneous cytokine production, defined as T-cell polyfunctionality, is a dynamic state of different T-cell subsets and may represent a surrogate biomarker of response to immune-mediated therapies. To quantify the T-cell immune mediators, different methodologies are available. Herein we describe two flow cytometry-based protocols to detect the simultaneous intracellular cytokine production, by using different methods of T-cell activation. Among the different procedures, multicolor flow cytometry is considered a powerful, quick, and semi-quantitative technique, able to achieve the analysis of both surface and intracellular proteins, expressed by specific T-cell subsets. We report the capability of this technique to study simultaneous cytokine production within a heterogeneous population. As the

Methods in Enzymology, Volume 631
ISSN 0076-6879
https://doi.org/10.1016/bs.mie.2019.07.041

field of novel single-cell high-throughput methodological approaches advance in this exciting era of Immuno-Oncology, we expect to gather even more information on the immunodynamics of polyfunctional T cells and their role in improving disease control.

1. Introduction

T lymphocytes are the major mediators of the anti-tumor immune response. The identification of their traits of diversity is crucial to define subpopulations of T cells linked to disease protection or reflecting effective immune response to clinical interventions.

T cells are highly heterogeneous in both phenotype and functional profile as determined by the distinctive patterns of surface markers and cytokines secreted after the stimulation of the T-cell receptor (TCR). Accordingly, T-cells are commonly classified for both their differentiation stage, gathered on a multitude of surface-expressed proteins, and their functional states based on their capability to produce one or more cytokines, chemokines, inflammatory mediators and lytic factors, which characterize functionally specialized subpopulations (Appay, van Lier, Sallusto, & Roederer, 2008; Masopust, Kaech, Wherry, & Ahmed, 2004). T-cell subset classification, according to cytokine patterns, is a "hot" topic in tumor immunology, with a continuous growth of information gathered by the advance of sophisticated high-throughput technologies. The milieu of cytokines and other immune mediators operate on each phase of the cancer immunity cycle (Chen & Mellman, 2013), by improving the antigen priming and increasing the functionality of effector T cells.

As cancer cells undergo malignant transformation, the immune system eliminates neoplastic cells and sculpts the tumor leading to tumor control or tumor progression, with the generation of variants which escape the immune control (Dunn, Bruce, Ikeda, Old, & Schreiber, 2002). This phase of escape occurs in an immune suppressive tumor microenvironment (TME), regulated by specific soluble factors (*e.g.*, anti-inflammatory cytokines) and regulatory cells along with the expression of immune checkpoint molecules (Mittal, Gubin, Schreiber, & Smyth, 2014).

In the new era of Immuno-Oncology, the immunologists aim to maximize methods to determine both the magnitude and the quality of the T-cell response and identify immune mediators simultaneously expressed on subsets of T cells. These cell subsets are commonly defined as polyfunctional T cells and are widely considered the main actors of an efficacious immune response against pathogens or tumor cells enabling disease control (Almeida et al., 2007; Precopio et al., 2007; Yuan et al., 2008). Polyfunctionality is not a permanent state, and is closely associated with the T-cell differentiation profile (naïve, central and effector memory cells), that also affects the cytokine secretion sequences, adding complexity for identifying effective anti-tumor T cells. Indeed, Han et al. showed that the polyfunctional state is a transient feature of different T cells and that most of these cells start in a monofunctional state and then follow a programmatic pattern of simultaneous cytokine release (Han et al., 2012).

This complex landscape reveals that the polyfunctional definition remains a key objective for defining immunological signatures indicating the efficacy of the Ag-specific T-cell response and their ability to control the disease.

To quantify cytokines, chemokines and lytic molecule production—hereafter defined as T-cell immune mediators (IM)—a plethora of techniques and different protocols are available. The most commonly used being the enzyme-linked immunosorbent assay (ELISA), enzyme-linked immunosorbent spot (ELIspot) assay, bead-based multiplex assay (Luminex technology), immunohistochemistry, single cell PCR, protein microarrays, and intracellular staining (ICS) by flow cytometry. The choice of a selected method depends on the planned experiment and herein we focus on different protocols to detect T-cell IM by flow cytometry technology.

Multicolor flow cytometry is a quick, specific, easy high-throughput technique for the semiquantitative analysis of surface or intracellular proteins that are expressed by specific cell types. While methods such as the traditional ELISA or the more recent Luminex technology measure secreted proteins produced by the entire cell population, the ICS, combined with the staining of surface markers, enables the identification of multiple T-cell IM produced by a single, phenotypically identified cell type. A great advantage of this technique is its ability to determine whether cytokine production results from rare cells producing large amounts of IM or from abundant cells each producing small quantities of IM. Importantly many groups, including ours, have demonstrated a significant correlation between polyfunctionality and the anti-tumor killing capability (Boyd et al., 2015; Franzese et al., 2016; Lin et al., 2009).

Herein, based on our immune-monitoring experience, we describe in detail two different methods for quantifying IM production after T-cell activation, that may be potentially adapted to all cell types with minor variations.

2. Basic principles of intracellular staining

The requirements for the formulation of a standard operating procedure (SOP) of ICS/flow cytometry experiments have to take into consideration three major steps: (1) Immune cell activation; (2) Cytokine transport inhibition; (3) Fixation and permeabilization of cells.

2.1 Immune cell activation

Functional activity of T cells depends on the release of T-cell IM and a beneficial immune response has been linked to the ability of T cells to produce multiple IMs upon activation. Thus, there is a need to standardize methodologies to evaluate T-cell polyfunctionality.

Multiple assays are used to activate T cells and these modalities frequently include a short-term mitogen-stimulation with Concanavalin A, lipopolysaccharide (LSP), Phorbol esters (PMA) plus calcium ionophore or Ionomycin, Phytohaemaglutinin (PHA) or Staphylococcal Enterotoxin B (SEB) being the most used (Ai, Li, Song, Li, & Chen, 2013). Alternatively, a monoclonal antibody directed against subunits of the TCR/CD3 complex, along with or without antibodies directed against costimulatory receptors, such as CD28, can be used as more physiological activation (Andersson & Sander, 1989).

Comparative analysis between different activation methods have been reported with contrasting results. Baran et al. found that the use of PHA may limit *in vitro* artifacts, with respect to PMA/Ionomycin activation, and allows more precise analysis of intracellular cytokine production (Baran, Kowalczyk, Ozóg, & Zembala, 2001). Others showed that the use of PHA alone activates a low number of cytokine producing T cells, while the PHA+PMA activation determine a potent stimulus (Jung, Schauer, Heusser, Neumann, & Rieger, 1993). Comparative analysis between PMA/Ionomycin and anti-CD3/anti-CD28 activation revealed that all these stimuli are able to induce cytokine production with a different amount depending on the time and the dose of the reagent used (Olsen & Sollid, 2013; Sotsios, Blair, Westwick, & Ward, 2000). There is a general consensus on the anti-CD3 stimulation as a preferred choice of T-cell activation.

Finally, antigen presenting cells (APC) pulsed with the cognate peptide or HLA-matched tumor cells expressing known or unknown antigens can be used when searching for antigen-specific polyfunctional T cells. The choice of the T-cell activation assay depends on the effector cells employed, the cytokine/s to be detected and the aim of the experiment conducted.

Since the cell activation is a crucial step in the design of experiments aimed at evaluating an immune response mediated by T-cell polyfunctionality, we will describe in depth some activation protocols in paragraph 3.

2.2 Cytokine transport inhibition

To prevent protein secretion and trap the proteins in the Golgi complex, the intracellular transport process has to be inhibited during the stimulation phase. This results in a detectable accumulation of the cytokines inside the cells. The two most commonly used protein transport inhibitors are carboxylic ionophore monensin and Brefeldin A (O'Neil-Andersen & Lawrence, 2002; Schuerwegh, Stevens, Bridts, & De Clerck, 2001). The time of transport inhibition treatment depends on the nature of the cytokines analyzed, since different cytokines, chemokines and inflammatory mediators have different production peaks, ranging between 4 and 12 h. Prolonged transport inhibition treatment (more than 12–15 h) causes cell toxicity, which represents a limitation of this technique, that does not allow a long period of stimulation.

2.3 Fixation and permeabilization

In 1991, Sander et al. demonstrated that after fixation with the cross-linking reagent paraformaldehyde and permeabilization with the detergent saponin it was possible to detect intracellular cytokines by flow cytometry (Sander, Andersson, & Andersson, 1991).

The protocol requires 3 steps: (1) Fixation; (2) Permeabilization; (3) Intracellular staining.

1. The first step consists of cell fixation by using a procedure that allows the preservation of cell morphology and the retention of antigenicity. Moreover, fixation enables cells to be resistant to the membrane permeabilization, as indicated by loss of integrity and cell lysis when cells are unfixed at the time of permeabilization. Cross-linking agents, such as formaldehyde, paraformaldehyde or glutaraldehyde, are the most common fixation reagents (Oliver & Jamur, 2010).
2. In the second step, cells are permeabilized to provide the access to the intracellular compartment and allow the intracellular epitopes to be exposed to the specific antibody. The permeabilization of cells can be

done with two types of reagents: (i) detergents, among which the most frequently used is saponin, a plant glycoside that interacts with membranes, making holes (~100 Å) by selectively extracting cholesterol. The usage of this detergent has the advantage of forming saponin/cholesterol micelles in which antibodies may be incorporated, facilitating their access into the cells. On the other hand, saponin permeabilization does not act on membranes with reduced cholesterol content. Moreover, Triton X-100 and Tween-20 non-ionic detergents, are not used as often in flow cytometry assays due to their non-selective activity, with false negative results; (ii) organic solvents, such as methanol and acetone, that dissolve membrane lipids. They have the advantage to fix and permeabilize cells simultaneously due to their ability of coagulate membrane proteins (Jamur & Oliver, 2010).

3. The final step consists of cell staining with the fluorescent-conjugated antibodies (a list of the mAbs employed in this paper is reported in Supplementary Table 1 in the online version at https://doi.org/10.1016/bs.mie.2019.07.041).

3. T cells and activation protocols

3.1 *Ex-vivo* lymphocytes isolated from blood or tissue samples

3.1.1 *T cells*

Peripheral blood mononuclear cells (PBMC) are isolated from blood samples using standardized density gradient technique (Ficoll-Paque). Tumor infiltrating lymphocytes (TIL) are isolated from bioptic tissue samples according to different protocols depending on tissue types. Both freshly isolated or thawed T cells can be used (see Note 1). Cells are plated at $1–2 \times 10^6$ cells/mL in 48 or 24 well polystyrene plates, depending on the cell numbers available (see Notes 2–3) and cultured in RPMI-1640, 1% Glutamin, 10% human serum (HS) (see Note 4), 25 U/mL human rIL-2, in standard culture conditions (37 °C, 5% CO_2).

3.1.2 *Activation protocols*

1. To study an antigen-specific T-cell response, add the purified soluble cognate peptide (see Note 5) to the HLA allele haplotype-matched PBMC culture ($3–4 \times 10^5$ cells per well), for ~5 h (see Notes 6–8).

To inhibit the protein secretion add 1 μl/ml monensin (GolgiStop, BD Bioscience) or Brefeldin A (GolgiPlug, BD Bioscience) and mix carefully (see Notes 9–10).

2. To study an unknown antigen T-cell response, an appropriate activator has to be used (see Section 2.1). For a more physiologic activation signal, we suggest the use of monoclonal antibodies directed against subunits of the TCR/CD3 complex, with or without antibodies directed against costimulatory receptors, such as CD28 (see Note 11).

   Anti-CD3 mAbs have to be plate-coated at 2 μg/mL for 1.5 h at 37 °C or O.N. at 4 °C. Supplemental soluble anti-CD28 mAbs may be added at 1 μg/mL. Activate the cells for ~5 h (see Note 7), along with protein secretion inhibition with 1 μl/mL of monensin or Brefeldin A (see Note 9).

3.2 *In vitro* short-term expanded Ag-specific CD8$^+$ T-cell lines and Ag-specific CD8$^+$ T-cell clones

3.2.1 T cells

In vitro short-term expanded Ag-specific CD8$^+$ T-cell lines are obtained as described (Nisticò et al., 2009). Ag-specific T-cell clones are generated from Ag-specific T-cell lines by limiting dilution (Palermo et al., 2001; Palermo et al., 2010). T-cell clones are restimulated biweekly with irradiated allogenic PBMC plus 1 μg/mL PHA, 25 U/mL hrIL-2, and periodically stained with Ag/tetramers to monitor the maintenance of Ag-specificity (Fig. 1A).

3.2.2 Activation protocols

1. To stimulate the Ag-specific cells, pulsed APC have to be prepared. APC may consist of autologous CD8-depleted PBMC (3–4 × 10^5 cells per well) or TAP-deficient T2 cells (when HLA-A*0201 restriction is employed), pulsed with the cognate or the irrelevant peptide (see Note 5) for 1 h at 37 °C in serum-free medium (see Notes 6 and 10).

   Add the Ag-specific T-cell line (E:T ratio 1:1) to the pulsed-APC and co-culture for ~5 h, along with protein secretion inhibition with 1 μL/mL of monensin or Brefeldin A (see Note 9).

2. For Ag-specific T-cell lines expressing a high percentage of Ag-specific population, and for Ag-specific T-cell clones, we use HLA-matched antigen-expressing tumor cell lines to stimulate the cells (see Notes 8 and 12).

   In this activation setting, we seed the tumor cell line in a 48 well plate (1–2 × 10^5 cells per well) and add the Ag-specific T-cell line or the

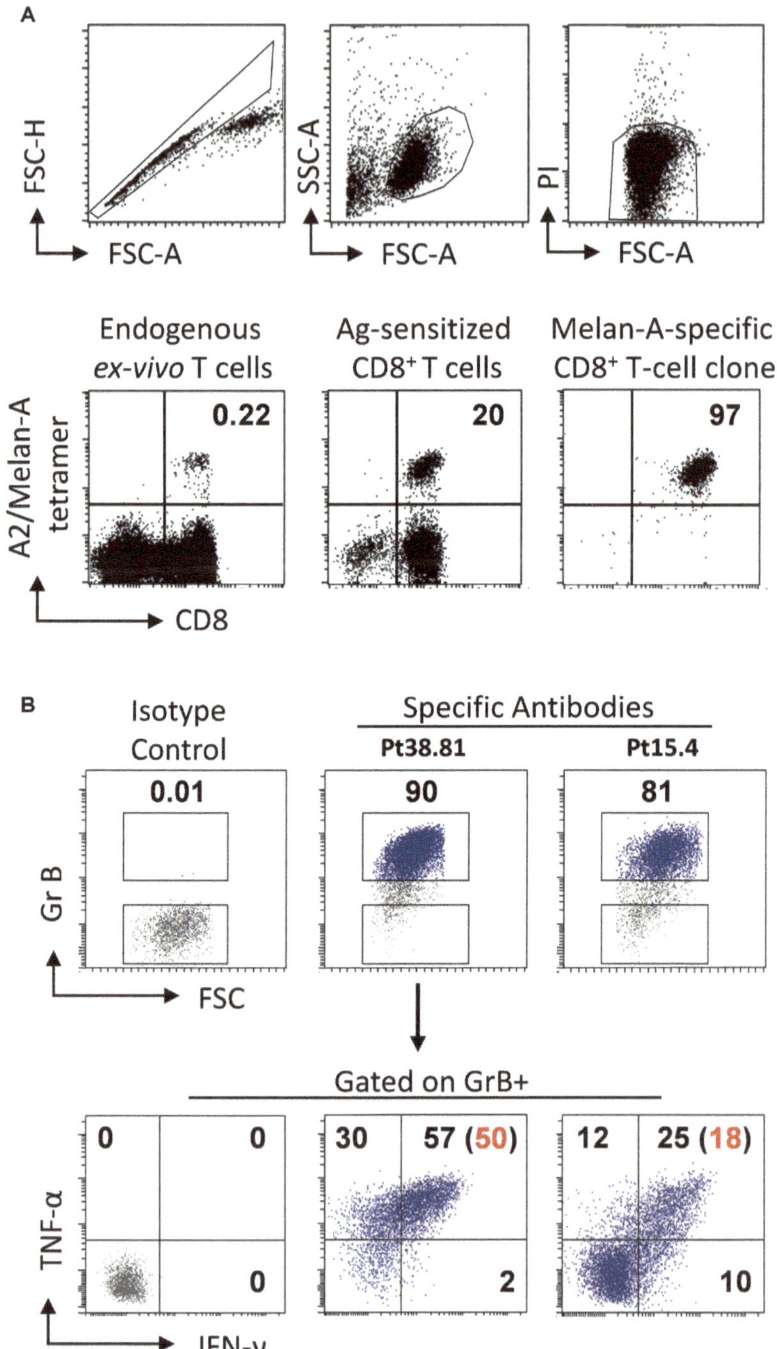

Fig. 1 See legend on opposite page.

Ag-specific T-cell clone after ~18 h (E:T ratio 1:1) and co-culture for ~5 h, along with protein secretion inhibition with 1 μL/mL of monensin or Brefeldin A.

4. ICS protocols

These protocols refer only to cytoplasmic ICS and do not consider the nucleus ICS, which requires different permeabilization conditions.

4.1 In house ICS protocol

This is a low-cost protocol, relatively easy, quick (4–5 h of culture, 2–3 h staining and analysis), and sensitive enough to require only a few cells per test. It is based on an ICS protocol (Sallusto, Lenig, Mackay, & Lanzavecchia, 1998) with minor modification. Due to the negative impact of this fixation/permeabilization procedure on surface molecule staining, we recommend its use only when surface staining is not necessary.

Equipment

1. PBS (without Ca2+ and Mg2+)
2. Fixation buffer: prepare a stock solution of 4% paraformaldehyde in PBS. Warm the solution and use NaOH until dissolved. Bring to room temperature (r.t.) and adjust pH to 7.2 with HCl. Aliquot the solution and store at −20 °C. Thaw just before use. Avoid freeze-thaw cycles.
3. Permeabilization buffer: PBS, 0.5% saponin, 5% FBS. Make it fresh.
4. Washing Buffer: PBS, 0.01% sodium azide, 1.5% FBS

Fig. 1 Characterization of polyfunctionality in Ag-specific T-cell clones. A, Representative example of increasing HLA-A2/Melan-A tetramer staining from *ex-vivo* PBMC (left), to short-term Melan-A-sensitized $CD8^+$ T-cell line (middle) and $CD8^+$ Melan-A-specific T-cell clone (right), obtained from HLA-$A2^+$ melanoma patients (Palermo et al., 2010). Upper panels show the gating strategy for dead cell exclusion. First, debris were excluded in the FSC-A. Then, doublets were removed using a plot with FSC-A/FSC-H and live lymphocytes were gated using Propidium Iodide (PI). B, Representative dot plot gating strategy of multicolor intracellular staining as performed with in house ICS protocol, in two different $gp100^+$ T-cell clones (Palermo et al., 2018). T-cell clones were activated with an HLA-$A2^+$ gp100-expressing melanoma cell line in the presence of GolgiStop (1 μl/ml), for 4 h. Top panels, GrB production. Bottom panels, percentage (black numbers) of IFN-γ and TNF-α expression with respect to the GrB^+ cells. Values in brackets (red numbers) represent the percentage of polyfunctional GrB^+ TNF-α^+ IFN-γ^+ cells among the total lymphocytes.

Protocol

1. Harvest previously activated cells (see Section 3) and transfer in the FACS tubes (see Note 13). Cells from each condition are divided into two FACS tubes, one for unstained cells (or control isotype), and one for the specific antibody staining.
2. Wash the cells with 2mL of PBS. Centrifuge at 350g for 5min to pellet cells.
3. Discard the supernatant and shake vigorously to resuspend the cells.
4. Add 100μL of PBS and after 100μL of cold paraformaldehyde 4%.
   Tip: This sequence is critical to avoid cellular damage.
5. Incubate 15min on ice (see Notes 14).
6. Wash cells with 1–2mL of permeabilization buffer (1mL for $1–2 \times 10^5$ cells). Centrifuge at 350g for 5min to pellet the cells.
7. Discard the supernatant and shake vigorously to resuspend the cells. Add 30μL of permeabilization buffer. After 20min of incubation at r.t., add 20μL of fluorescent antibody cocktail and incubate at r.t. for 30min in the dark. Add 50μL of permeabilization buffer also to the sample tubes designated as autofluorescence controls.
   Tip: Prepare the antibody dilution in permeabilization buffer to warrant the cells remain permeable during the staining process
8. Wash the cells with 1–2mL of permeabilization buffer. Centrifuge at 350g for 5min to pellet the cells.
9. Discard the supernatant and resuspend the cells. Wash cells with 2mL of washing buffer. Centrifuge at 350g for 5min to pellet the cells.
10. Discard the supernatant and resuspend the cells in 200–300μL PBS for the FACS acquisition process. Representative results of this procedure are shown in Fig. 1B.

4.2 ICS assay recommended for simultaneous surface staining

Equipment

1. PBS (without Ca2+ and Mg2+)
2. BD Intrasure kit. The kit components are as follows:
   a. Reagent A (containing 1% formaldehyde and 0.35% methanol)
   b. Reagent B (a proprietary buffered permeabilization solution containing a detergent)
3. Washing Buffer: PBS, 0.01% sodium azide, 1.5% FBS

Protocol

We have partially modified the manufacturer's instructions as follows:

1. Harvest previously activated cells (see Section 3) and transfer in the FACS tubes (see Note 13). Cells from each condition are divided into two FACS tubes, one for unstained cells (or control isotype), and one for the specific antibody staining.
2. Wash cells with 2 mL of cold washing buffer. Centrifuge at 350 g for 5 min to pellet cells. Discard the supernatant and shake vigorously to resuspend the cells.
3. Add 50 μL of fluorescent antibody cocktail and incubate at 4 °C for 30 min in the dark (see Note 15).
4. Wash cells with 1 mL of PBS. Centrifuge at 350 g for 5 min to pellet cells.
5. Discard the supernatant and shake vigorously to resuspend the cells.
6. Add 100 μL of Reagent A to each tube and incubate for 15 min in the dark at r.t .
7. Wash cells with 1 mL of PBS. Centrifuge at 350 g for 5 min to pellet cells.
8. Discard the supernatant and shake vigorously to resuspend the cells.

   Tip: It is recommended to avoid PBS residue by patting tubes on dry absorbent paper.
9. Add 50 μL of the appropriate fluorescent antibody cocktail diluted in Reagent B. Add 50 μL of Reagent B also to sample tubes designated as autofluorescence controls.
10. Incubate at r.t. for 15 min in the dark.
11. Wash cells with 1 mL of washing buffer. Centrifuge at 350 g for 5 min to pellet cells.
12. Discard the supernatant and resuspend the cells in 200–300 μL PBS for the FACS acquisition process. Representative results of this procedure are shown in Fig. 2.

5. Flow cytometric data acquisition and analysis

Set the photo-multiplier detectors and perform the compensation, manually or by the use of commercial compensation beads. Set quadrant markers based on isotype control or unstained cells (see Note 16).

An accurate gating strategy, gathered from the multitude of dot plots generated upon data analysis, have to be performed in order to measure the percentage of cells expressing intracellular cytokines, also along with specific surface cell markers (Fig. 2 for a representative example). So far, a standardized method of gating strategy in ICS experiments is not available, being greatly subjective and operator-dependent. In particular, two steps are critical: (1) the sequence of hierarchy plots to identify the final population of

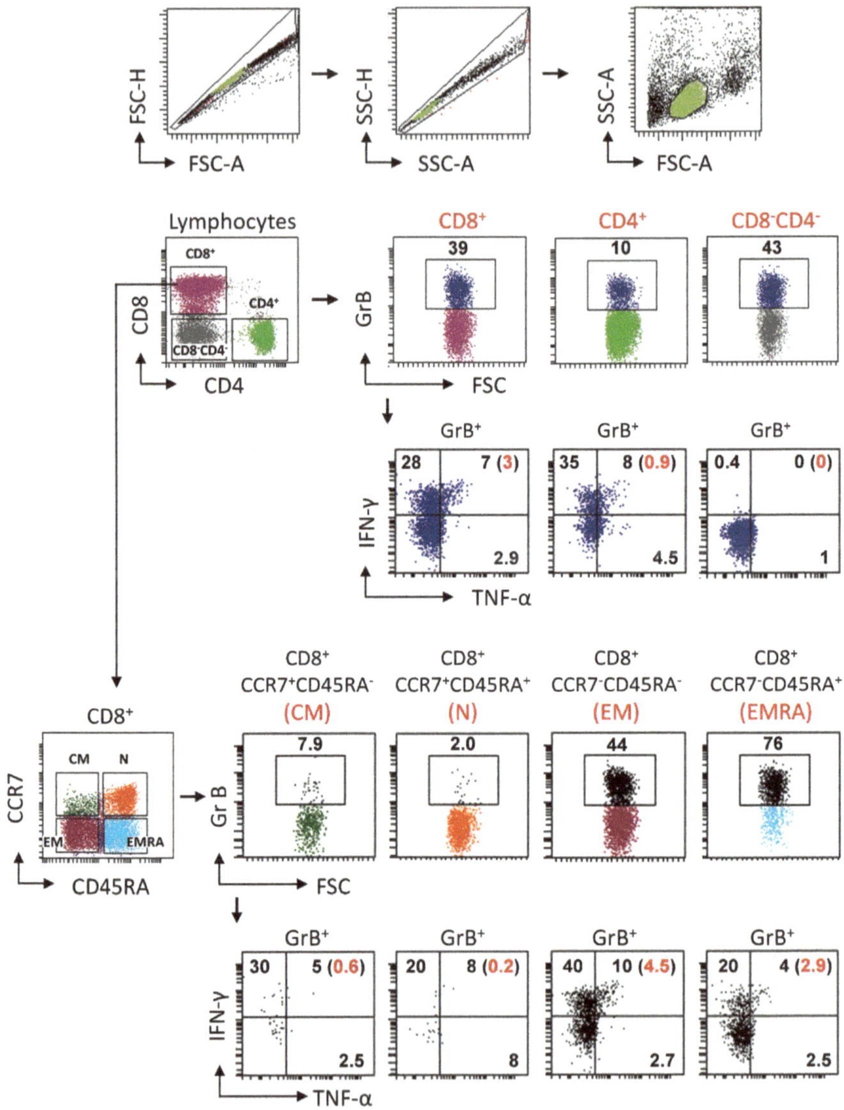

Fig. 2 Representative staining for the evaluation of T-cell polyfunctionality in *ex vivo* PBMC. PBMC from a healthy donor were activated with plate-bound anti-CD3 (2 μg/ml) in the presence of GolgiStop (1 μl/ml), for 5 h. First, debris were excluded in the FSC-A. Then, doublets were removed using two plots with FSC-A/FSC-H and SSC-A/SSC-H, and live lymphocytes were gated (upper panels). Dot plots represent the multiple gating strategies of concomitant multicolor surface and intracellular staining. On the top of each dot-plot diagram, gated T-cell subpopulation is reported. The percentages of GrB, IFN-γ and TNF-α in different T-cell populations ($CD8^+$, $CD4^+$, $CD8^-CD4^-$, upper panels) and in $CD8^+$ T-cell differentiation subsets (CM, N, EM, EMRA, bottom panels) are reported (black numbers). Values in brackets (red numbers) represent the percentage of polyfunctional GrB^+ TNF-α^+ IFN-γ^+ cells among the different subpopulations analyzed.

interest; (2) the precise draw for the cytokine-positive gate, with an adequate proximity from the negative population, to avoid false-positive or -negative results. Several ICS proficiency panels have concluded that the gating strategies have the major impact on the results generated across different institutions (Horton et al., 2007; Jaimes et al., 2011; McNeil et al., 2013; Welters et al., 2012). Thus, to judge and compare published results, it is mandatory to clearly report the employed gating strategy.

A good familiarity with basic principles of flow cytometry is required to perform these experiments. This chapter is not intended to provide guidelines on the procedure of flow cytometry use. For in depth information on flow cytometry, the Shapiro's book is recommended (Shapiro, 2003).

6. Concluding remarks

As the field of Immuno-Oncology continues to grow, with T cells at the center of this galaxy, the identification of T-cell polyfunctionality is at the front-line of novel standardized single-cell based methodological approaches. The immunodynamic endpoints, based on T-cell polyfunctionality, will allow to identify surrogate biomarkers of response to cancer immunotherapy. Of relevance, the longitudinal monitoring during the course of treatment will be crucial to design combined treatments, including vaccines and immune checkpoint blockade therapies.

7. Notes

1. When cells are thawed, we recommend an overnight culture before starting the activation protocol.
2. For more accurate results, T-cell subsets can be purified by the use of specific magnet beads (Miltenyi Biotec), according to the manufacturer's instructions.
3. When few cells are available 96 well plates can be used.
4. We used a noncommercial human serum prepared from a pool of healthy donors.
5. The amount of peptide used for APC pulsing depends on its binding affinity and stability in the HLA molecule complex. We commonly used 1–10 μg/mL.
6. Unpulsed or pulsed cells with an irrelevant peptide provide a good quality control to evaluate background staining.

7. Different cytokines, chemokines and inflammatory mediators, show different production peaks. In order to obtain optimal staining signals, the stimulation/transport inhibition conditions for each analyte need to be optimized.
8. To confirm the MHC-dependent Ag-specific T-cell response, blockade of MHC-I or MHC-II with purified mAbs has to be performed.
9. When longer stimulation condition (up to 5 h) is needed, concentration of the protein transport inhibitor needs to be carefully optimized. For GolgiStop we suggest 0.6 μL/mL for 12–15 h.
10. Performing the tetramer staining along with intracellular cytokine production requires careful attention to the procedure. Indeed, upon T cell activation, the TCR is rapidly down regulated which could potentially affect the evaluation of the tetramer-positive population. To avoid the problem we suggest to perform the tetramer staining before the activation with the TCR-dependent activators.
11. Unstimulated cells provide a good quality control to evaluate background staining.
12. HLA-unmatched and/or Ag-negative tumor cell lines provide a good quality control to evaluate background staining.
13. Staining may be performed both in FACS tubes or in 96-well round plates, which are recommended when few cells are available. It requires a very small volume of wash buffer but it needs more washes (usually 2–3 times with 200 μL).
14. The experiment can be stopped at this point and held in standby for 1–2 days. Wash fixed cells with PBS, centrifuge at 350 g for 5 min, discard the supernatant and resuspend the cells in PBS. Store at 4 °C in the dark.
15. Some surface antibodies are not suitable for fixed cells, thus we perform the surface staining before the fixation/permeabilization procedure.
16. After the fixation/permeabilization procedure, the light scatter profiles of the cells on the flow cytometer may undergo changes. We recommended the use of the fixed/permeabilized sample tubes designated as autofluorescence controls to cytometry setup.

Acknowledgments

This work was supported by the Italian Association for Cancer Research AIRC: 5 × 1000, 12182, and IG 15224.

References

Ai, W., Li, H., Song, N., Li, L., & Chen, H. (2013). Optimal method to stimulate cytokine production and its use in immunotoxicity assessment. *International Journal of Environmental Research and Public Health, 10*, 3834–3842.

Almeida, J. R., Price, D. A., Papagno, L., Arkoub, Z. A., Sauce, D., Bornstein, E., et al. (2007). Superior control of HIV-1 replication by CD8C T cells is reflected by their avidity, polyfunctionality, and clonal turnover. *The Journal of Experimental Medicine, 204*, 2473–2485.

Andersson, U., & Sander, B. (1989). Detection of individual interleukin 2-producing cells after anti-CD3 antibody activation. *Immunology Letters, 20*, 115–120.

Appay, V., van Lier, R. A., Sallusto, F., & Roederer, M. (2008). Phenotype and function of human T lymphocyte subsets: Consensus and issues. *Cytometry, 73*, 975–983 [Review].

Baran, J., Kowalczyk, D., Ozóg, M., & Zembala, M. (2001). Three-color flow cytometry detection of intracellular cytokines in peripheral blood mononuclear cells: Comparative analysis of phorbol myristate acetate-ionomycin and phytohemagglutinin stimulation. *Clinical and Diagnostic Laboratory Immunology, 8*, 303–313.

Boyd, A., Almeida, J. R., Darrah, P. A., Sauce, D., Seder, R. A., Appay, V., et al. (2015). Pathogen-specific T cell Polyfunctionality is a correlate of T cell efficacy and immune protection. *PLoS One, 10,* e0138395.

Chen, D. S., & Mellman, I. (2013). Oncology meets immunology: The cancer-immunity cycle. *Immunity, 39*, 1–10.

Dunn, G. P., Bruce, A. T., Ikeda, H., Old, L. J., & Schreiber, R. D. (2002). Cancer immunoediting: From immunosurveillance to tumor escape. *Nature Immunology, 3*, 991–998 [Review].

Franzese, O., Palermo, B., Di Donna, C., Sperduti, I., Ferraresi, V., Stabile, H., et al. (2016). Polyfunctional Melan-A-specific tumor-reactive CD8(+) T cells elicited by dacarbazine treatment before peptide-vaccination depends on AKT activation sustained by ICOS. *Oncoimmunology, 5,* e1114203.

Han, Q., Bagheri, N., Bradshaw, E. M., Hafler, D. A., Lauffenburger, D. A., & Love, J. C. (2012). Polyfunctional responses by human T cells result from sequential release of cytokines. *Proceedings of the National Academy of Sciences of the United States of America, 109*, 1607–1612.

Horton, H., Thomas, E. P., Stucky, J. A., Frank, I., Moodie, Z., Huang, Y., et al. (2007). Optimization and validation of an 8-color intracellular cytokine staining (ICS) assay to quantify antigen-specific T cells induced by vaccination. *Journal of Immunological Methods, 323*, 39–54.

Jaimes, M. C., Maecker, H. T., Yan, M., Maino, V. C., Hanley, M. B., Greer, A., et al. (2011). Quality assurance of intracellular cytokine staining assays: Analysis of multiple rounds of proficiency testing. *Journal of Immunological Methods, 363*, 143–157.

Jamur, M. C., & Oliver, C. (2010). Permeabilization of cell membranes. *Methods in Molecular Biology, 588*, 63–66.

Jung, T., Schauer, U., Heusser, C., Neumann, C., & Rieger, C. (1993). Detection of intracellular cytokines by flow cytometry. *Journal of Immunological Methods, 159*, 197–207.

Lin, Y., Gallardo, H. F., Ku, G. Y., Li, H., Manukian, G., Rasalan, T. S., et al. (2009). Optimization and validation of a robust human T-cell culture method for monitoring phenotypic and polyfunctional antigen-specific CD4 and CD8 T-cell responses. *Cytotherapy, 11*, 912–922.

Masopust, D., Kaech, S. M., Wherry, E. J., & Ahmed, R. (2004). The role of programming in memory T-cell development. *Current Opinion in Immunology, 16*, 217–225 [Review].

McNeil, L. K., Price, L., Britten, C. M., Jaimes, M., Maecker, H., Odunsi, K., et al. (2013). A harmonized approach to intracellular cytokine staining gating: Results from an international multiconsortia proficiency panel conducted by the Cancer immunotherapy consortium (CIC/CRI). *Cytometry. Part A*, *83*, 728–738.

Mittal, D., Gubin, M. M., Schreiber, R. D., & Smyth, M. J. (2014). New insights into cancer immunoediting and its three component phases–elimination, equilibrium and escape. *Current Opinion in Immunology*, *27*, 16–25.

Nisticò, P., Capone, I., Palermo, B., Del Bello, D., Ferraresi, V., Moschella, F., et al. (2009). Chemotherapy enhances vaccine-induced antitumor immunity in melanoma patients. *International Journal of Cancer*, *124*, 130–139.

Oliver, C., & Jamur, M. C. (2010). Fixation and embedding. *Methods in Molecular Biology*, *588*, 353–362.

Olsen, I., & Sollid, L. M. (2013). Pitfalls in determining the cytokine profile of human T cells. *Journal of Immunological Methods*, *390*, 106–112.

O'Neil-Andersen, N. J., & Lawrence, D. A. (2002). Differential modulation of surface and intracellular protein expression by T cells after stimulation in the presence of monensin or brefeldin a. *Clinical and Diagnostic Laboratory Immunology*, *9*, 243–250.

Palermo, B., Campanelli, R., Mantovani, S., Lantelme, E., Manganoni, A. M., Carella, G., et al. (2001). Diverse expansion potential and heterogeneous avidity in tumor-associated antigen-specific T lymphocytes from primary melanoma patients. *European Journal of Immunology*, *31*, 412–420.

Palermo, B., Del Bello, D., Sottini, A., Serana, F., Ghidini, C., Gualtieri, N., et al. (2010). Dacarbazine treatment before peptide vaccination enlarges T-cell repertoire diversity of melan-a-specific, tumor-reactive CTL in melanoma patients. *Cancer Research*, *70*, 7084–7092.

Palermo, B., Franzese, O., Di Donna, C., Panetta, M., Quintarelli, C., Sperduti, I., et al. (2018). Antigen-specificity and DTIC before peptide-vaccination differently shape immune-checkpoint expression pattern, anti-tumor functionality and TCR repertoire in melanoma patients. *Oncoimmunology*, *7*, e1465163.

Precopio, M. L., Betts, M. R., Parrino, J., Price, D. A., Gostick, E., Ambrozak, D. R., et al. (2007). Immunization with vaccinia virus induces polyfunctional and phenotypically distinctive CD8(+) T cell responses. *The Journal of Experimental Medicine*, *204*, 1405–1416.

Sallusto, F., Lenig, D., Mackay, C. R., & Lanzavecchia, A. (1998). Flexible programs of chemokine receptor expression on human polarized T helper 1 and 2 lymphocytes. *The Journal of Experimental Medicine*, *187*, 875–883.

Sander, B., Andersson, J., & Andersson, U. (1991). Assessment of cytokines by immunofluorescence and the paraformaldehyde-saponin procedure. *Immunological Reviews*, *119*, 65–93.

Schuerwegh, A. J., Stevens, W. J., Bridts, C. H., & De Clerck, L. S. (2001). Evaluation of monensin and brefeldin a for flow cytometric determination of interleukin-1 beta, interleukin-6, and tumor necrosis factor-alpha in monocytes. *Cytometry*, *46*, 172–176.

Shapiro, H. M. (2003). *Pratical flow cytometry* (4th ed.). New York: Wiley-Liss.

Sotsios, Y., Blair, P. J., Westwick, J., & Ward, S. G. (2000). Disparate effects of phorbol esters, CD3 and the costimulatory receptors CD2 and CD28 on RANTES secretion by human T lymphocytes. *Immunology*, *101*, 30–37.

Welters, M. J., Gouttefangeas, C., Ramwadhdoebe, T. H., Letsch, A., Ottensmeier, C. H., Britten, C. M., et al. (2012). Harmonization of the intracellular cytokine staining assay. *Cancer Immunology, Immunotherapy*, *61*, 967–978.

Yuan, J., Gnjatic, S., Li, H., Powel, S., Gallardo, H. F., Ritter, E., et al. (2008). CTLA-4 blockade enhances polyfunctional NY-ESO-1 specific T cell responses in metastatic melanoma patients with clinical benefit. *Proceedings of the National Academy of Sciences of the United States of America*, *105*, 20410–20415.

CHAPTER FIVE

Assessment of memory formation by metabolically engineered antigen-specific CD8 T cells

Mathias Wenes[a,b], Pedro Romero[a,b,*], Lianjun Zhang[a,b,c,d]
[a]Translational Tumor Immunology Group, Ludwig Cancer Research, University of Lausanne, Lausanne, Switzerland
[b]Department of Oncology UNIL CHUV, University of Lausanne, Lausanne, Switzerland
[c]Center for Systems Medicine, Institute of Basic Medical Sciences, Chinese Academy of Medical Sciences and Peking Union Medical College, Beijing, China
[d]Suzhou Institute of Systems Medicine, Suzhou, Jiangsu, China
*Corresponding author: e-mail address: pedro.romero@unil.ch

Contents

Abstract

The formation of memory $CD8^+$ T cells is crucial for host protection upon reencounter of the same pathogen. Moreover, long-lasting anti-tumoral effects of immunomodulatory drugs largely depend on the induction of immunological memory. While the transcriptional processes behind memory T cell differentiation have been described in detail, it is only recently becoming clear that memory T cell formation and maintenance are tightly controlled by specific metabolic pathways. Therefore, metabolic engineering of $CD8^+$ T cells in order to promote their memory formation represents a valuable strategy to improve vaccination efficacy and cancer immunotherapy. Here, we describe several mouse *in vitro* and *in vivo* assays that can be used to evaluate whether a specific metabolic pathway is involved in memory $CD8^+$ T cell differentiation. To this end, a metabolic process can be tested by either genetic deletion or pharmaceutical inhibition/activation of the key enzyme involved. The *in vitro* assays might allow for rapid screening

Methods in Enzymology, Volume 631
ISSN 0076-6879
https://doi.org/10.1016/bs.mie.2019.10.021

of multiple candidate pathways, while the *in vivo* assays are required to reliably determine the quality and functionality of the generated memory $CD8^+$ T cells.

1. Introduction

$CD8^+$ T cells are crucial mediators of the adaptive immune response against cancer cells and pathogenic microorganisms. Upon antigen stimulation, naïve $CD8^+$ T cells undergo extensive clonal expansion and differentiation into effector cells. At the peak of the primary response, a major proportion of these cells are short-lived effector cells (SLEC) that are terminally differentiated and characterized by a potent cytotoxic potential, which mostly undergo apoptosis during the contraction phase. Of note, a small fraction of effector cells are memory precursor effector cells (MPEC), that will further differentiate into long-lived, self-renewing memory $CD8^+$ T cells, able to rapidly generate robust secondary effector cells upon re-challenge (Harty & Badovinac, 2008). This self-renewal and multipotency are capacities that characterize the ideal T cell population, fit for adoptive cell transfer (ACT) immunotherapy against cancer, and able to overcome some of the current issues with ACT (Gattinoni, Klebanoff, & Restifo, 2012). Indeed, despite showing promising results in a fraction of patients, T lymphocytes prepared for ACT are generally terminally differentiated, resulting in inefficient engraftment and cancer recurrence. It has been shown both in mouse and human that the infusion of T cells with a self-renewing, memory phenotype confers a stronger and more sustained anti-tumor response (Fraietta et al., 2018; Klebanoff et al., 2011). Therefore, there is a strong scientific interest in uncovering the molecular and metabolic pathways leading to memory $CD8^+$ T cell differentiation, which can be exploited to improve T cell quality in the context of cancer immunotherapy.

A multitude of molecular pathways and transcription factors has been shown to drive effector versus memory $CD8^+$ T cell differentiation, leading to the differential expression of genes and surface molecules that can be used to discriminate between these subpopulations (Kaech & Cui, 2012). The differentiation of distinct effector or memory $CD8^+$ T cell subpopulations is also regulated at the epigenetic level. Temporal acetylation and methylation processes of histones and DNA may activate or repress the transcription of naïve and memory genes versus effector genes (Gray, Amezquita, Guan, Kleinstein, & Kaech, 2017; Youngblood et al., 2017). A central position in this regulatory network is taken by the mammalian target of rapamycin (mTOR) complex. Constituted by two different complexes (mTORC1 and mTORC2), mTOR converges activating signals from the T cell receptor, co-stimulatory molecules and cytokines together with the metabolic status inside and outside the cell. These upstream signals define the transcriptional and (post-)translational control of transcription factors and genes important for cellular metabolism and effector or memory T cell function (Araki et al., 2009; Chi, 2012; Zhang et al., 2016). Concretely, it has been shown that genetic loss of mTORC2 in T cells leads to decrease kinase activity by AKT, resulting in reduced phosphorylation of FOXO1. This allows for its nuclear accumulation and transcription of important memory genes, such as Eomes and Tcf1. Indeed, mTORC2 deficiency in $CD8^+$ T cells favors their memory differentiation during bacterial infection (Zhang et al., 2016). Thus, by activating mTORC1 and activated by mTORC2, the serine/threonine kinase Akt influences $CD8^+$ T cell fate, and unlike mTOR, might provide a specific and druggable target for the manipulation of $CD8^+$ T cell differentiation. Indeed, small molecule inhibition of Akt skews mouse and human T cells toward a memory phenotype (Crompton et al., 2015; Macintyre et al., 2011). Importantly, *in vitro* Akt inhibition of human tumor-infiltrating lymphocytes or chimeric antigen receptor (CAR)-transduced T cells leads to increased memory features, resulting in improved tumor control when infused into immunodeficient mice bearing a hematologic malignancy (Klebanoff et al., 2017; van der Waart et al., 2014).

It has become clear that those molecular, transcriptional and epigenetic processes driving effector versus memory T cell differentiation are all closely intertwined with the cell's metabolism. Indeed, drastic changes in T cell metabolism are required to support their activation and function. The high energetic and biosynthetic requirements of clonally expanding effector cells are provided by aerobic glycolysis and TCA cycle activity. However, the metabolic machinery driving the differentiation, function and long-term persistence of memory T cells seems to vary depending on subset differentiation (effector memory versus central memory) and the nutritive and other microenvironmental conditions in which these subsets of memory T cells reside (Ecker et al., 2018; O'Sullivan et al., 2014; Pearce et al., 2009; Phan et al., 2016). Of note, the inhibition or stimulation of certain metabolic processes alone is sufficient to induce or inhibit effector versus memory differentiation, with according *in vivo* effects when the cells are re-infused in infected or tumor-bearing mice (Balmer et al., 2016; Phan et al., 2016).

However, due to the complex and extensive nature of metabolic processes, many questions on the involvement of specific metabolic pathways in memory $CD8^+$ T cell differentiation remain open. We will describe here several *in vitro* and *in vivo* experiments that might give a straightforward answer to whether a metabolic pathway under investigation is affecting memory $CD8^+$ T cell differentiation. By making use of OT1 transgenic mice, a large number of antigen-specific $CD8^+$ T cells can be obtained. The metabolic gene of interest can be either investigated by intercrossing KO mice with OT1 transgenic mice, or by simply culturing OT1 cells in the presence of a small molecule inhibitor against the metabolic enzyme. *In vitro* experiments will provide a fast read-out of the induction of memory characteristics and will thus allow for a larger scale screening of multiple targets. Here, we describe several potential phenotypic and metabolic read-outs, indicative of a memory differentiation capacity upon certain pharmacological or genetic manipulation. However, only *in vivo* studies in which the $CD8^+$ T cells are (re-) activated in their physiological environment, consisting of multiple cell types and diverse cytokines, will allow the investigator to determine with certainty whether a specific metabolic pathway is implicated in memory formation. We will describe a bacterial infection model with *Listeria monocytogenes*, generally giving highly reproducible and consistent data on the kinetics and phenotypic subset differentiation of the $CD8^+$ T cells immune response, and a melanoma tumor model, providing a functional read-out of memory T cells with potential therapeutic implications.

2. In vitro assays

2.1 Culture conditions and activation of $CD8^+$ T cells

The spleens are collected under sterile conditions, and smashed on a 70 μm cell strainer. Single cells are collected by centrifugation. Red blood cells are lysed by resuspending the pellet in 2 mL Red Blood Cell Lysis Buffer (Qiagen) and incubated for 5 min at room temperature. Splenocytes are counted and seeded at 1×10^6 cells in 1 mL per well in a 48-well plate in RPMI (completed with 10% fetal bovine serum (FBS), 1% L-glutamine, 1% Penicillin/streptomycin, 1 mM sodium pyruvate, non-essential amino acids, 10 mM HEPES (all GIBCO) and 50 μM β-mercaptoethanol), in the presence of 1 μg/mL ovalbumine-derived SIINFEKL peptide, 100 IU/mL recombinant human IL2, and incubated in a humidified incubator at 37 °C for 72 h. At 72 h, cell viability and expression levels of differentiation markers can be assessed on a fraction of the cells by flow cytometry (see below).

The remaining cells are washed twice with medium, re-seeded in a 48-well plate at 1×10^6 cells/mL and cultured for another 96 h in the presence of 100 U/mL rhIL-2 and 100 U/mL rhIL-7 (Fig. 1A). At day 7 of culture, cell viability and expression levels of differentiation markers is assessed as described below.

Culturing a mixture of OT1 splenocytes with the SIINFEKL peptide allows for an efficient and budget-friendly way of activating $CD8^+$ T cells by making use of the antigen-presenting cells present in the spleen. If pure $CD8^+$ T cells cultures are required, naïve $CD8^+$ T cells can be isolated by using the Naïve $CD8^+$ T Cell Negative Selection Kit (Stem Cell Technologies) and activated by seeding 5×10^5 cells per well in 48-well plates coated with anti-CD3 (1 μg/mL) and anti-CD28 (2 μg/mL); or by culturing with anti-CD3/CD28 coated Dynabeads (1–2 beads per cell, Thermo Fisher Scientific).

2.2 Flow cytometry analysis

At 72 h, cell viability (Fixable Viability Dye, Thermo Fisher Scientific) and surface marker expression (CD25, CD44, CD69 and CD71 as activation/effector markers and CD62L as naïve/memory marker) can be assessed on a fraction of the cells by flow cytometry. At this point, the T cells are efficiently activated and show high expression levels of CD25, CD44, CD69 and CD71, while CD62L expression is downregulated as compared to naïve cells. At day 7 of culture, a clear distinction in phenotype can be made by cells expressing either low or high levels of CD62L (Fig. 1B). Besides CD62L, expression levels of transcription factors TCF1, T-BET and EOMES can also be analyzed. Staining for transcription factors can be successfully executed by using the Foxp3/transcription factor staining buffer set (Thermo Fisher Scientific). TCF1 is an important transcription factor driving memory gene expression (Utzschneider et al., 2016). An increased ratio of EOMES/T-BET expression is also indicative of a memory phenotype (Kaech & Cui, 2012).

2.3 Metabolic flux analysis

Since mitochondrial metabolism plays a major role during T cell differentiation, differences in oxygen consumption or glycolysis might be indicative of a memory versus effector T cell phenotype. Both parameters can be evaluated in real time by using the Seahorse extracellular flux analyzer (Agilent Technologies). Apart from measuring basal oxygen consumption, cells can

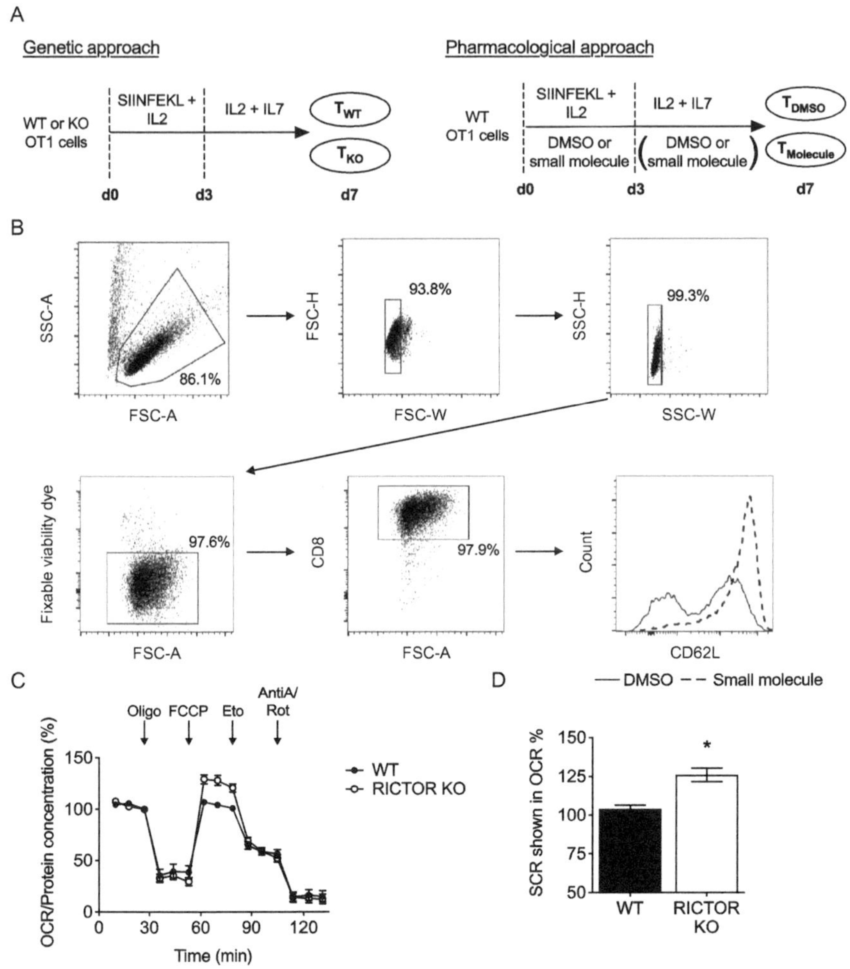

Fig. 1 *In vitro* assays to analyze the induction of memory characteristics. (A) Schematic representation of the *in vitro* culture system. OT1 splenocytes are cultured and activated for 3 days with SIINFEKL peptide and IL2. The peptide is then washed out and the cells are further cultured in IL2 and IL7 for 4 days. The metabolic target of interest can be genetically inhibited by using splenocytes from WT versus KO mice on a OT1 background (left), or by pharmacological intervention with a small molecule either only during activation or during the entire culture period (right). (B) Typical gating strategy on day 7 splenocytes. The final histogram shows a typical pattern with cells having either low or high CD62L expression (full line). The dashed line shows the strong increase in CD62L expression upon culturing the cells in the presence of a small molecule inhibitor (target not disclosed). (C) Oxygen consumption rate (OCR) measured by Seahorse in WT and RICTOR KO cells, treated with oligomycin (Oligo), FCCP, etomoxir (Eto) and antimycin A and rotenone (AntiA/Rot). (D) The spare respiratory capacity (SRC, difference between basal OCR and maximal respiration obtained by uncoupling mitochondrial electron transport chain with FCCP) was calculated. All graphs show mean ± standard error of mean (SEM). $*P < 0.05$.

be stressed by injecting drugs inhibiting key metabolic processes. This allows the evaluation of the maximal metabolic capacity and flexibility in substrate use upon environmental stress. However, since metabolic inhibitors might not be specific, the exact contribution of a substrate to a metabolic process should be evaluated by 13-C metabolic flux analysis.

2.3.1 Seahorse assay

Concretely, OT-1 cells are stimulated *in vitro* with SIINFEKL peptide and 100U/mL rhIL-2 for 3 days (as described above) and continued to culture in the presence of IL2. The oxygen consumption rate (OCR, in pmol/min) and extracellular acidification rate (ECAR, in mpH/min) are measured using the Seahorse XF-96 metabolic extracellular flux analyzer (Seahorse Bioscience). Seahorse cell plates are coated overnight at 4°C with Cell-Tak (BD Bioscience) to allow the attachment of T cells. Day 7 $CD8^+$ T cells are washed and resuspended in unbuffered Seahorse XF base medium supplemented with 25mM Glucose, 2mM Glutamine and 2% FBS at a pH 7.4. 3×10^5 cells are seeded per well in 180 μL medium. The Mito Stress Test as defined by Agilent can be achieved on $CD8^+$ T cells by addition of oligomycin (0.5 μM), FCCP (1.5 μM), rotenone (100 nM) and antimycin A (1 μM; all from Sigma-Aldrich). The following assay conditions should be programmed: 2 min mixture; 2 min wait; and 3 min measurement. Before injecting rotenone and antimycin A, the cells could be also treated with etomoxir (Carnitine Palmitoyltransferase 1A inhibitor) to determine the contribution of fatty acid oxidation. Importantly, and as recently shown, the concentration used should not be too high (we advise 50 μM), because from 100 μM onward, the inhibitor has been shown to have off-target effects (Raud et al., 2018). Finally, the obtained data should be normalized for the cell number. Although the input cell number is controlled, the final cell number should be measured at the end of the experiment as well, to account for potential cell loss due to detachment or loss of viability. This can be determined either by viable cell counting (automated imaging on, e.g., a BioTek Cytation or after cell dissociation), protein content measurement or genomic DNA quantitation. Counting viable cells is by far the preferred method, since different metabolic interventions might influence total cellular protein content as compared to the control conditions (Divakaruni, Paradyse, Ferrick, Murphy, & Jastroch, 2014).

As described by Zhang et al., *Rictor* deletion in $CD8^+$ T cells increases the spare respiratory capacity of these cells, (calculated by subtracting the basal

OCR from the OCR after FCCP addition) possibly due to increased fatty acid oxidation (Fig. 1C and D) (Zhang et al., 2016).

2.3.2 13-C metabolic flux analysis

A Seahorse analysis is a fast and user-friendly way to identify global changes in glycolytic versus oxidative metabolism, and the use of several inhibitors during the assay can give an indication which substrates contribute to the observed metabolic phenotype. In order to more precisely map which nutrient is involved in specific metabolic fluxes and to which extent it gets incorporated in certain cellular metabolites, a 13-C metabolic flux analysis should be executed. Culturing T cells in the presence of a 13-C-labeled metabolite, such as 13-C-glucose or 13-C-glutamine, allows for the mass spectrometry-based measurement of its degree of incorporation in cellular metabolites. One can thus follow the fate of a certain nutrient and determine the predominant metabolic pathway.

Practically, the T cells are activated and cultured as described before. At specific time points of the culture, a labeling medium is prepared. This medium should be ideally devoid of the 12-C isotope of the metabolite under investigation. Therefore, we advise to use RPMI medium without D-glucose and L-glutamine (Biological Industries), supplemented as usual with antibiotics, HEPES and β-mercaptoethanol, and with 10% dialyzed serum (Gibco). The 13-C metabolite then needs to be precisely weighed and added at the same molarity as the normal growth medium. If this 13-C metabolite is not glucose or glutamine, the medium could be supplemented with 11 mM glucose and 4 mM glutamine, which are the concentrations found in the normal growth medium. The T cells are first collected, centrifuged and washed once in RPMI medium without D-glucose and L-glutamine. The pellet is then resuspended in the 13-C-containing labeling medium and cultured at 37 °C for several hours depending the 13-C substrate and on which metabolic flux is under investigation. Finally, the T cells are then collected, centrifuged and washed once with ice-cold saline. The cell pellet is then quickly placed on dry ice and metabolites are extracted according to a protocol allowing the detection of the metabolites of interest. Designing these protocols and detecting the metabolites by mass spectrometry requires expert knowledge and is thus generally performed in collaboration with metabolomics experts.

2.4 Proliferation and cytokine production upon restimulation

As shown above, at day 7 of culture, a fraction of the T cells upregulated CD62L, implying that they are reaching a more quiescent state. At this stage, the T cells could be restimulated in order to mimic a memory response *in vitro*. Memory T cells could be characterized by an increased proliferation rate and IL-2 production in response to the re-challenge. To analyze proliferation, day 7 T cells are collected and washed in PBS. 1×10^6 T cells are then stained in 1 mL of pre-warmed PBS containing 200 nM CSFE (Life Technologies) for 8 min at 37 °C. The excess CSFE is immediately removed by washing the cells 3 times in completed RPMI. The cells are counted again and 2×10^5 cells are seeded in 200 μL per well in a 96-well plate containing anti-CD3/CD28 beads to reactivate the cells. The dilution of CSFE, which is representative of cell division, can be determined by flow cytometry at day 5 of restimulation.

To analyze cytokine production upon restimulation, 2×10^5 cells are seeded in 200 μL per well in a 96-well plate containing anti-CD3/CD28 beads. After 30 min incubation at 37 °C, golgistop and golgiplug are added (1:1000 final dilution, BD Biosciences). The cells are then further incubated at 37 °C for another 4 h before they are collected to analyze the production and intracellular accumulation of IL2 and eventually other cytokines. The cells should be first stained with a fixable viability dye and eventual surface antibodies, before being fixed (Fixation Buffer, Biolegend) and washed and stained for IL2 (Intracellular Staining Permeabilization Wash Buffer, Biolegend). IL2 accumulation is then measured by flow cytometry.

3. In vivo assays

The above described *in vitro* assays can give an indication of memory formation in T cells but are limited by the artificial nature of the culture conditions. Indeed, the memory capacity of T cells can only reliably be determined by an *in vivo* immune response. On one hand, memory T cells formation can by followed during a bacterial infection (Listeria model). On the other hand, in a tumor model, the anti-tumoral activity of T cells can reflect the functionality of the memory T cells.

3.1 Listeria model

By using a genetically modified *Listeria monocytogenes* strain expressing the SIINFEKL peptide, the antigen-specific immune response and memory differentiation can be followed by using OT1 mice on a CD45.1 background. Naïve OT-1 $CD8^+$ cells are first purified from the spleen (Naïve CD8 Isolation Kit, Stem Cell Technologies). 10^4 naïve OT1 cells are then intravenously injected in healthy $CD45.2^+$ mice. Within 24 h after cell transfer, the mice are infected with a sub-lethal dose of SIINFEKL-expressing

Listeria monocytogenes (2000 CFU i.v.). Weekly blood analysis of the mice might reveal differences in the frequency and absolute quantity of the transferred cells during the peak of the immune response, or during the contraction phase (Fig. 2A). At the different stages of the immune reaction, the phenotype of the transferred cells ($CD45.1^+$ cells) in the blood is analyzed by flow cytometry ($CD127^{low}$ and $KLRG1^{high}$ for short-lived effector cells (SLEC), $CD127^{high}$ and $KLRG1^{low}$ for memory precursors effector cells (MPEC) and CD44-CD62L-double positive cells as central memory T cells). Concretely, about five drops of blood are collected in heparin-coated tubes. Red blood cells are lysed by the addition of 1 mL Red Blood Cell Lysis Buffer (Sigma-Aldrich) and incubated for 15 min at 37 ° C. White blood cells are then pelleted by centrifugation and resuspended in FACS buffer (PBS, 2% FBS, 2 mM EDTA) followed by antibody staining.

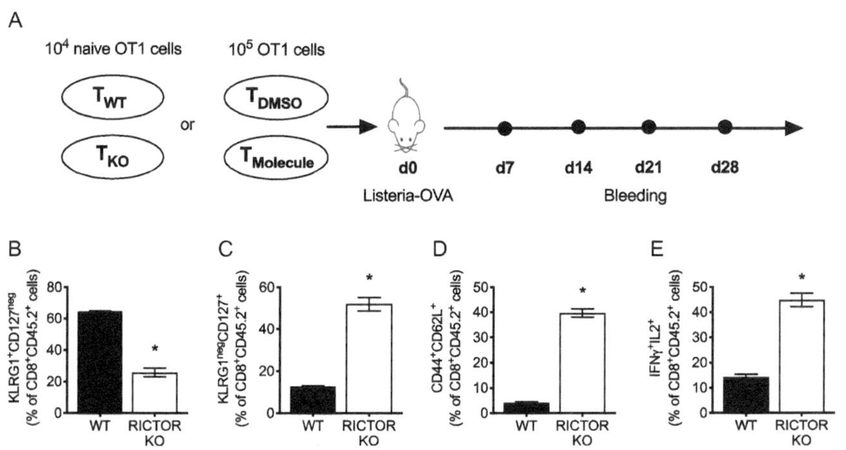

Fig. 2 A Listeria infection model enables a fast determination of memory $CD8^+$ T cell induction *in vivo*. (A) Schematic representation of the SIINFEKL-expressing Listeria infection model. Either 10^4 naïve WT or KO $CD8^+$ T cells or 10^5 day 7 inhibitor-treated $CD8^+$ T cells are injected in the tail vein of healthy mice. Within 24 h, the mice are infected by injecting 2000 CFU Listeria in the tail vein. Blood is taken every week to follow-up the immune reaction. (B–D) The percentage of short-lived effector cells (KLRG1-positive, CD127-negative), memory precursor effector cells (KLRG1-negative, CD127-positive) and central memory T cells (CD44 and CD62L double positive) out of the transferred cell population in the spleen, determined by flow cytometry 7 days after Listeria infection in mice that received an adoptive transfer of 10^4 naïve WT or RICTOR KO $CD8^+$ T cells. (E) At day 7 post-infection, splenocytes were restimulated *ex vivo* with SIINFEKL peptide. The percentage of cells expressing both IL2 and IFNγ out of the transferred cell population was determined by flow cytometry. All graphs show mean ± standard error of mean (SEM). $*P < 0.05$.

Additionally, the transferred cells can be restimulated *ex vivo* to evaluate cytokine production. Whole white blood cells are cultured for 30 min with SIINFEKL at 37 °C, followed by the addition of golgistop and golgiplug. After 4 h incubation, the cells are collected and stained for surface markers, followed by fixation and cytokine staining. The cytokine production is then analyzed by flow cytometry. Upon sacrifice of the mice, the incorporation of the transferred cells in different lymphoid and non-lymphoid organs and their phenotype can be evaluated.

As described by Zhang et al., *Rictor* deficiency in T cells indeed leads to an increased differentiation toward MPECs and central memory T cells. Furthermore, *Rictor* KO cells produce more IL2 upon restimulation (Fig. 2B–E).

If small molecule inhibitors are used to induce a memory phenotype *in vitro*, the stability and re-challenge capacity of the cells can be assessed upon transfer in this Listeria model. OT1 splenocytes are cultured with the small molecules and activated with SIINFEKL *in vitro* as described above. At day 7, 10^5 T cells are then adoptively transferred in CD45.2$^+$ mice which are subsequently infected with Listeria (Figs. 1A and 2A). The immune response of the transferred cells is followed as described above.

3.2 B16 melanoma model

Finally, the anti-tumoral potential of metabolically engineered CD8$^+$ T can be tested in a mouse model of melanoma. Briefly, 10^5 ovalbumine-expressing B16 melanoma cells are injected subcutaneous in 6-week-old CD5.2 mice. At day 6 post-engraftment, when a palpable tumor is present, mice are preferably irradiated with 5Gy, in order to improve the efficiency of T cell engraftment. The next day 10^5 naïve genetically engineered or *in vitro* small molecule-treated OT1 cells are intravenously injected, followed by a subcutaneous vaccination of 50 μg CpG and 10 μg SIINFEKL. Tumor growth is measured 3×/week and weekly blood analysis by flow cytometry allows to follow the immune response and the phenotype of the transferred cells. At the endstage, the tumor, spleen and tumor-draining lymph node is dissected and a single cell suspension is prepared. For the spleen and lymph node, a single cell suspension is prepared as described above. The tumor is first mechanically minced with the gentleMACS Dissociator (program "Imptumor_2," Miltenyi Biotec), followed by enzymatic digestion using the Tumor Dissociation Kit (according to the manufacturer's instructions, Miltenyi Biotec) at 37 °C with gentle agitation. A second mechanical dissociation (program "Imptumor_3") precedes single cells filtration using a 70 μm filter. This suspension is further purified by resuspending the cells in 40% Percoll diluted in completed RPMI and centrifigution on an equal volume of 70% Percoll (20 min at 2000 rpm, room temperature, no brake). The interface is collected and washed with medium and the single cells suspension is ready to be quantified and phenotyped by flow cytometry (besides marker described above in the Listeria model, also the expression of inhibitory/exhaustion markers such as PD1, Lag3, Tim3 should be analyzed). Tumor single cell suspensions are restimulated *ex vivo* with the OVA peptide SIINFEKL, to evaluate cytokine production by the TILs. As an alternative tumor model, ovalbumine-expressing MCA38 colon cancer cells can be injected subcutaneously.

4. Concluding remarks

The methods described here could be used to investigate whether a specific metabolic process is important for the generation of memory $CD8^+$ T cells. We mentioned the possibility of using either mice that have a T cell-specific deletion of the metabolic gene or by culturing T cells in the presence of a small molecule inhibitor. There are several important differences to be taken in consideration when comparing both approaches. By using the genetic approach, the researcher is certain about the specificity of the observed effects. However, it might be that knocking out a certain metabolic gene promotes memory differentiation, but at the same time inhibits effector function, an effect that might become evident in the melanoma model. By using small molecule inhibitors, one can never be a 100% certain about the target specificity. Additionally, the memory phenotype induced *in vitro* needs to be stably imprinted in the T cells, since upon adoptive cell transfer *in vivo*, it is not always feasible to continue drug exposure (e.g., due to systemic toxic effects). Ideally, the genetic approach should be combined with inhibitor experiments to provide, respectively, the specificity and therapeutic applicability of targeting a specific metabolic process for the generation of memory $CD8^+$ T cells.

References

Araki, K., Turner, A. P., Shaffer, V. O., Gangappa, S., Keller, S. A., Bachmann, M. F., et al. (2009). mTOR regulates memory CD8 T-cell differentiation. *Nature*, *460*(7251), 108–112. https://doi.org/10.1038/nature08155.

Balmer, M. L., Ma, E. H., Bantug, G. R., Grahlert, J., Pfister, S., Glatter, T., et al. (2016). Memory CD8(+) T cells require increased concentrations of acetate induced by stress for optimal function. *Immunity*, *44*(6), 1312–1324. https://doi.org/10.1016/j.immuni.2016.03.016.

Chi, H. (2012). Regulation and function of mTOR signalling in T cell fate decisions. *Nature Reviews. Immunology*, *12*(5), 325–338. https://doi.org/10.1038/nri3198.

Crompton, J. G., Sukumar, M., Roychoudhuri, R., Clever, D., Gros, A., Eil, R. L., et al. (2015). Akt inhibition enhances expansion of potent tumor-specific lymphocytes with memory cell characteristics. *Cancer Research*, *75*(2), 296–305. https://doi.org/10.1158/0008-5472.CAN-14-2277.

Divakaruni, A. S., Paradyse, A., Ferrick, D. A., Murphy, A. N., & Jastroch, M. (2014). Analysis and interpretation of microplate-based oxygen consumption and pH data. *Methods in Enzymology*, *547*, 309–354. https://doi.org/10.1016/B978-0-12-801415-8.00016-3.

Ecker, C., Guo, L., Voicu, S., Gil-de-Gomez, L., Medvec, A., Cortina, L., et al. (2018). Differential reliance on lipid metabolism as a salvage pathway underlies functional differences of T cell subsets in poor nutrient environments. *Cell Reports*, *23*(3), 741–755. https://doi.org/10.1016/j.celrep.2018.03.084.

Fraietta, J. A., Nobles, C. L., Sammons, M. A., Lundh, S., Carty, S. A., Reich, T. J., et al. (2018). Disruption of TET2 promotes the therapeutic efficacy of CD19-targeted T cells. *Nature*, *558*(7709), 307–312. https://doi.org/10.1038/s41586-018-0178-z.

Gattinoni, L., Klebanoff, C. A., & Restifo, N. P. (2012). Paths to stemness: Building the ultimate antitumour T cell. *Nature Reviews. Cancer*, *12*(10), 671–684. https://doi.org/10.1038/nrc3322.

Gray, S. M., Amezquita, R. A., Guan, T., Kleinstein, S. H., & Kaech, S. M. (2017). Polycomb repressive complex 2-mediated chromatin repression guides effector CD8(+) T cell terminal differentiation and loss of multipotency. *Immunity*, *46*(4), 596–608. https://doi.org/10.1016/j.immuni.2017.03.012.

Harty, J. T., & Badovinac, V. P. (2008). Shaping and reshaping CD8+ T-cell memory. *Nature Reviews. Immunology*, *8*(2), 107–119. https://doi.org/10.1038/nri2251.

Kaech, S. M., & Cui, W. (2012). Transcriptional control of effector and memory CD8+ T cell differentiation. *Nature Reviews. Immunology*, *12*(11), 749–761. https://doi.org/10.1038/nri3307.

Klebanoff, C. A., Crompton, J. G., Leonardi, A. J., Yamamoto, T. N., Chandran, S. S., Eil, R. L., et al. (2017). Inhibition of AKT signaling uncouples T cell differentiation from expansion for receptor-engineered adoptive immunotherapy. *JCI Insight*, *2*(23). e95103. https://doi.org/10.1172/jci.insight.95103.

Klebanoff, C. A., Gattinoni, L., Palmer, D. C., Muranski, P., Ji, Y., Hinrichs, C. S., et al. (2011). Determinants of successful CD8+ T-cell adoptive immunotherapy for large established tumors in mice. *Clinical Cancer Research*, *17*(16), 5343–5352. https://doi.org/10.1158/1078-0432.CCR-11-0503.

Macintyre, A. N., Finlay, D., Preston, G., Sinclair, L. V., Waugh, C. M., Tamas, P., et al. (2011). Protein kinase B controls transcriptional programs that direct cytotoxic T cell fate but is dispensable for T cell metabolism. *Immunity*, *34*(2), 224–236. https://doi.org/10.1016/j.immuni.2011.01.012.

O'Sullivan, D., van der Windt, G. J., Huang, S. C., Curtis, J. D., Chang, C. H., Buck, M. D., et al. (2014). Memory CD8(+) T cells use cell-intrinsic lipolysis to support the metabolic programming necessary for development. *Immunity*, *41*(1), 75–88. https://doi.org/10.1016/j.immuni.2014.06.005.

Pearce, E. L., Walsh, M. C., Cejas, P. J., Harms, G. M., Shen, H., Wang, L. S., et al. (2009). Enhancing CD8 T-cell memory by modulating fatty acid metabolism. *Nature*, *460*(7251), 103–107. https://doi.org/10.1038/nature08097.

Phan, A. T., Doedens, A. L., Palazon, A., Tyrakis, P. A., Cheung, K. P., Johnson, R. S., et al. (2016). Constitutive glycolytic metabolism supports CD8(+) T cell effector memory differentiation during viral infection. *Immunity*, *45*(5), 1024–1037. https://doi.org/10.1016/j.immuni.2016.10.017.

Raud, B., Roy, D. G., Divakaruni, A. S., Tarasenko, T. N., Franke, R., Ma, E. H., et al. (2018). Etomoxir actions on regulatory and memory T cells are independent of Cpt1a-mediated fatty acid oxidation. *Cell Metabolism*, *28*(3), 504–515. e507 https://doi.org/10.1016/j.cmet.2018.06.002.

Utzschneider, D. T., Charmoy, M., Chennupati, V., Pousse, L., Ferreira, D. P., Calderon-Copete, S., et al. (2016). T cell factor 1-expressing memory-like CD8(+) T cells sustain the immune response to chronic viral infections. *Immunity*, *45*(2), 415–427. https://doi.org/10.1016/j.immuni.2016.07.021.

van der Waart, A. B., van de Weem, N. M., Maas, F., Kramer, C. S., Kester, M. G., Falkenburg, J. H., et al. (2014). Inhibition of Akt signaling promotes the generation of superior tumor-reactive T cells for adoptive immunotherapy. *Blood*, *124*(23), 3490–3500. https://doi.org/10.1182/blood-2014-05-578583.

Youngblood, B., Hale, J. S., Kissick, H. T., Ahn, E., Xu, X., Wieland, A., et al. (2017). Effector CD8 T cells dedifferentiate into long-lived memory cells. *Nature*, *552*(7685), 404–409. https://doi.org/10.1038/nature25144.

Zhang, L., Tschumi, B. O., Lopez-Mejia, I. C., Oberle, S. G., Meyer, M., Samson, G., et al. (2016). Mammalian target of Rapamycin complex 2 controls CD8 T cell memory differentiation in a Foxo1-dependent manner. *Cell Reports*, *14*(5), 1206–1217. https://doi.org/10.1016/j.celrep.2015.12.095.

CHAPTER SIX

Functional characterization of tumor antigen-specific T-cells isolated from the tumor microenvironment of sleeping beauty induced murine glioma models

Mahmoud S. Alghamri[a,b], Felipe J. Núñez[a,b], Neha Kamran[a,b], Stephen Carney[a,b], David Altshuler[a,b], Pedro R. Lowenstein[a,b], Maria G. Castro[a,b,*]

[a]Department of Neurosurgery, University of Michigan Medical School, Ann Arbor, MI, United States
[b]Department of Cell and Developmental Biology, University of Michigan Medical School, Ann Arbor, MI, United States
*Corresponding author: e-mail address: mariacas@umich.edu

Contents

Abstract

Diffuse Gliomas represent 80% of brain tumors with an average survival of the most aggressive form glioblastoma (GBM) 15–22 months from the time of diagnosis. The current standard of care includes tumor resection, chemotherapy and radiation, nevertheless, the incidence of recurrence remains high and there is a critical need for developing new therapeutic strategies. T-cell mediated immunotherapy that triggers an anti-tumor T cell-mediated memory response is a promising approach since it will not only attack the primary tumor but also prevent recurrence. Multiple

Methods in Enzymology, Volume 631
ISSN 0076-6879
https://doi.org/10.1016/bs.mie.2019.05.032

immunotherapeutic strategies against glioma are currently being tested in clinical trials. We have developed an immune-mediated gene therapy (Thymidine kinase plus Fms-like tyrosine kinase 3 ligand: TK/Flt3L) which induces a robust anti-tumor T cell response leading to tumor regression, long-term survival and immunological memory in GBM models. Efficacy of the anti-glioma T cell therapy is determined by anti-tumor specific effector T cells. Therefore, assessing effector T cell activation status and function are critical readouts for determining the effectiveness of the therapy. Here, we detail methodologies to evaluate tumor specific T-cell responses using a genetically engineered Sleeping Beauty transposase-mediated glioma model. We first describe the glioma model and the generation of neurospheres (NS) that express the surrogate antigen cOVA. Then, we describe functional assays to determine anti-tumor T-cell response.

1. Introduction

Gliomas are the most common primary central nervous system neoplasm in adults (Acs, 2013). The current classification of gliomas (reviewed in 2016) are based not only on histopathologic appearance but also on well-established molecular parameters which identified multiple glioma subtypes (Ceccarelli et al., 2016; Louis et al., 2016; Reifenberger, Wirsching, Knobbe-Thomsen, & Weller, 2016). The most common glioma subtype is the most aggressive grade IV astrocytoma, known as glioblastoma (GBM). GBM carries a poor prognosis with a median survival of approximately 15 months (Ostrom et al., 2018; Reifenberger et al., 2016). Lower grade gliomas include grade II and III oligodendrogliomas and astrocytomas, which are now classified as diffuse gliomas. The median age at diagnosis ranges from 43 years for oligodendroglioma, 48 years for low grade astrocytoma, to 65 years for GBM (Ostrom et al., 2018).

Treatment for GBM is relatively standardized and includes maximal surgical resection, when possible and dependent on the location of the tumor, followed by concomitant radiotherapy and temozolomide (Stupp et al., 2005). Surgical resection is often offered if the tumor can be safely removed without risking neurologic injury (Buckner et al., 2016). Ionizing radiation, however, can result in severe cognitive dysfunction and this must be given appropriate consideration particularly for patients with low grade tumors who have relatively favorable prognosis (Racine, Li, Molinaro, Butowski, & Berger, 2015).

In the mid-2000s, large initiatives such as The Cancer Genome Atlas (TCGA) Consortium were established in order to comprehensively define the molecular characteristics of gliomas in a large number of samples (Brat et al., 2015; Verhaak, 2016). These efforts revealed the high heterogeneity of gliomas (Bai et al., 2016; Brat et al., 2015; Michele Ceccarelli et al., 2016). These molecular studies identified a distinct genetic signature found within low grade gliomas defined by mutations in the isocitrate dehydrogenase gene 1 (mIDH1) TP53 and ATRX mutations in astrocytomas as well as chromosomal deletions involving 1p and 19q in oligodendrogliomas (Ceccarelli et al., 2016). Molecular analysis is necessary for the discovery of novel treatments, effective clinical trial design, and clinical decision making (Ceccarelli et al., 2016; Verhaak et al., 2010).

Immunotherapy for glioma is a promising area of active investigation (Dunn-Pirio & Vlahovic, 2017; Kamran et al., 2018; Reardon, Wucherpfennig, & Chiocca, 2017). There are a variety of approaches under investigation in ongoing clinical trials including passive immunotherapy with antibodies, autologous activated lymphocytes, immune-mediated gene therapy, oncolytic viral therapy or active immunotherapy with tumor cell based vaccines, peptides, or dendritic cells (Polivka et al., 2017). Some examples of these therapies include Rindopepimut, a peptide based vaccine that targets the mutant variant of the epidermal growth factor receptor (EGFRvIII) that is present in >40% of GBM (Sampson et al., 2011). Dendritic cell (DC) vaccines loaded with autologous tumor lysate are another example of immunotherapy approaches being studied (Garg et al., 2016). We have developed an immune-mediated gene therapy strategy, which involves two adenoviral vectors (Ad) encoding, Herpes Simplex Type 1-Thymidine kinase (TK) used in combinations with a second Ad encoding Fms-like tyrosine kinase 3 ligand (TK/Flt3L). This gene therapy approach induces a robust anti-tumor T cell response leading to tumor regression, long-term survival and immunological memory in GBM models (Ali et al., 2005; Candolfi et al., 2009, 2014; Curtin et al., 2009; Kamran et al., 2017; Larocque et al., 2010; Yang et al., 2010) that is currently being tested in a Phase I clinical trial at our institution (NCT01811992). Other dendritic cell-based vaccine includes the DCVax-L phase III study (NCT00045968) which is currently ongoing (Dunn-Pirio & Vlahovic, 2017). Another interesting approach include combination of oncolytic virus mediated immune stimulation with the immune checkpoint inhibitors for GBM treatment (Lawler, 2017). This strategy has proven its efficacy in preclinical setting, and it is being tested in multiple clinical trials (NCT03294486, NCT03714334) (Lawler, 2017). Given the success for immune checkpoint inhibitors for other advanced solid tumor types such as melanoma (Gide, Wilmott, Scolyer, & Long, 2018), there is significant interest in this class of immune-based therapy for use in patients with glioma. Current studies involve the CTLA-4 inhibitor, ipilimumab as well as PD-1 inhibitors, nivolumab and pembrolizumab for patients with GBM. For instance, the phase III Check-Mate 143 trial (NCT02017717) is evaluating nivolumab alone and nivolumab plus ipilimumab *versus* the anti-vascular endothelial growth factor antibody, bevacizumab in patients with recurrent GBM. Interestingly, recent clinical trial testing neoadjuvant anti PD-1 therapy in recurrent GBM patients showed that neoadjuvant administration of pembrolizumab extended the overall survival and trigger local and systemic anti glioma immune response (Cloughesy et al., 2019).

The generation of robust animal models for glioma is critical to expand our knowledge of the biology of these tumors and also to develop new therapeutic approaches that could improve patient outcomes in combination with standard treatments. One of the main goals in the generation of animal models is to reproduce the genetics and phenotypic features found in glioma patients, including molecular composition, tumor heterogeneity and tumor microenvironment. We and others have established the Sleeping Beauty (SB)-transposon system (Calinescu et al., 2015; Koschmann et al., 2016; Nunez et al., 2019; Wiesner et al., 2009) which represents a powerful tool to generate genetically engineered glioma models. This system is based on the SB-transposase activity (Paschka et al., 2010), which allows the generation of mouse brain tumors through the insertion of specific genetic lesions into the genome of stem cells present along the lateral ventricle of neonate mice (Calinescu et al., 2015). The advantages of this model is that it allows altering the genome of the somatic cells which reproduces more closely the etiology of human disease and recapitulates histo-pathological features of human GBM (Koschmann et al., 2016; Mendez et al., 2018; Nunez et al., 2019).

Here, we describe the SB glioma model and the method used to generate SB-derived NS. Then, we detail the expression of the surrogate antigen cOVA which will be used to determine antigen specific T cell responses against glioma. Finally, we describe functional assays determine tumor specific T cell activation and proliferation against glioma.

2. Generation of glioma neurospheres derived from sleeping beauty transposon mouse glioma model

Glioma NS represent a valuable source for *in vitro* studies of tumor biology, as well as, to generate orthotropic glioma models by intracranial implantation. Glioma NS can be generated from glioma tissue obtained from patient's biopsies and also from genetically engineered mouse glioma models. Here, we described a protocol to generate NS derived from SB transposase mouse glioma models. It should be noted that SB-mediated mutations integrates in the inter-genomic regions, with no further insertion of passenger mutagenesis (Wiesner et al., 2009). Nevertheless, levels of passenger mutations can be determined by full genomic sequencing.

1. Neonatal pups are injected with the plasmid mix, that will induce the genetic lesions of interest, delivered into the lateral ventricle, as described before (Calinescu et al., 2015).
2. The tumor must grow to ($>10^6$ photons/s/cm^2/sr of luminescence) for an optimal yield of cells (please note that the latency time vary according tumor subtype between 60 and 300 day after plasmid injection). The mice glioma obtained using the SB transposon system express luciferase and they are monitored by bioluminescence, whose signal level is proportional to the tumor size. The luminescence is measured using IVIS® Spectrum *in vivo* imaging system (Caliper Life science). A bioluminescence signal of $>10^6$ photons/s/cm^2/sr is considered enough to proceed with the protocol of glioma NS generation. Animals are injected intraperitoneally (i.p.) with 3 mg luciferin (Gold Biotechnology) dissolved in 100 μL DPBS 5–10 min before IVIS acquisition.
3. Euthanize the mouse harboring the selected tumor using an overdose of isoflurane anesthetic (>5% of isoflurane). After 5–10 min, check absence of pedal reflex of the mouse, decapitate the mouse and dissect the brain as previously described (Calinescu et al., 2015).
4. Once the brain is removed, the tumor mass can be dissected by GFP expression, using a dissecting microscope (Olympus SZX16) equipped with a fluorescent lamp in a sterile environment.
5. For tumor homogenization, place the dissected tissue in a 1.5 mL tube containing 300 μL of NS media (DMEM/F12 (Gibco) with 1 × B27 supplement (Gibco), 1 × N2 supplement (Gibco), 1 × Normocin (InvivoGen), and 1 × Antibiotic/Antimycotic (Gibco) supplemented with human recombinant EGF (Peprotech) and basic-FGF (Peprotech) at a concentration of 20 ng/mL each). Continue with the homogenization, in the same 1.5 mL tube, using a sterile pestle and moving it in up and down motion 20 times under laminar flow hood.
6. The cell clumps must be dissociated by adding 1 mL of enzyme-free dissociation media (Accutase, BioLegend) then incubate for 5 min at 37 °C. Subsequently, filter the cell suspension through a 70 μm cell strainer placed in a 50 mL conical tube and wash the strainer with 10 mL of NS media.

7. To collect the glioma NS, centrifuge the cell suspension contained in the filtered media from step 6 for 5 min at 500 × *g*. Discard the supernatant, re-suspend the cells into 6 mL of NS media and plate into a T-25 (Thermofisher scientific) culture flask and maintain the cells at 37 °C (5% CO_2).
8. After 3 days in culture a mixed population of cells can be identified morphologically as: (i) adherent cells; (ii) dead cells and (iii) glioma NS. Continue collecting the NS population (floating cell clusters), and plate in a new T-25 culture flask in the same conditions as in step 7.
9. After 3 more days, centrifuge glioma NS for 5 min at 500 × *g* and dissociate as step 8. Split the NS in T-25 culture flasks with 6 mL of NS media at a density of 500,000 cells per flask.
10. Freeze aliquots of glioma NS after dissociation (at least one million per aliquot) by re-suspending the pellet in Fetal Bovine Serum (FBS, peak serum) + 10% DMSO (Fisher scientific) and store at −80 for 48 h, then transfer to liquid nitrogen for long term storage.

3. Stable transfection of neurospheres with a surrogate tumor antigen

The plasmid encoding cytoplasmic chicken ovalbumin, pCI-neocOVA, was developed in our lab (Addgene ID:25097) (Yang et al., 2010). NS were transfected with pCl-neocOva, pCl-neo (Promega), or pMax-GFP™ vector (Lonza) using the P3 primary cell 4D-Nucleofector™ X kit (Lonza, Allendale, NJ). The following procedure was followed for NS transfection:

1. Collect the NS from an 80–90% confluent T75 flask by centrifugation at 500 × *g*, and discard the supernatant by decantation.
2. Dissociate the cells using 1 mL of Accutase® cell detachment solution (Biolegend). Incubate for 90s at 37 °C, neutralize with 9 mL HBSS, and centrifuge the NS again at 500 × *g* to separate the supernatant.
3. Resuspend the detached NS in 10 mL HBSS. Take 10 μL to count, then split the NS to have 200,000 cell/tube.
4. Spin the tubes at 90 × *g* (Note: any higher centrifugation speed would impair viability), for 10 min, at room temperature.
5. Resuspend the NS with 20 μL of transfection solution (4D-Nucleofector solution).
6. Add 500 ng of pCl-neocOva, pCl-neo (promega), or pMax-GFP™ (Lonza) (to determine the transfection efficacy) to each tube (duplicate

each condition). Mix gently by pipetting up and down. Keep the cells at room temperature, allow to settle down for 5 min.

7. Transfer the suspended NS in each well to the Nucleocuvette™ Vessel (note: insert the pipet tip all the way to the bottom of the vessel to avoid any air bubbles).
8. Tab the Nucleocuvette™ Vessel gently to get rid of any air bubbles.
9. Place the Nucleocuvette™ Vessels with closed lid into the retainer of the 4D-Nucleofector™ X Unit in the right orientation.
10. Enter the right information regarding the protocol number (Pulse code EN 138) and hit enter to pulse the NS with the plasmid mix. If the pulse is successful you will see a green mark on the screen, otherwise, red mark means either air bubbles interference or failure to pulse due to other reasons.
11. Combine NS from two wells and culture them in one well from a 24-well plate containing 400–600 μg/mL G418 (Fisher scientific) selection antibiotic. (Stock solution of G418 "600 mg/mL, stable at −20 °C" was prepared first, then it was added to the transfected NS flask in 1:1000 dilution.)
12. Remove dead cells and supplement with 20 ng/mL of EGF, FGF growth factors as well as 400–600 μg/mL G418 every other day, for approximately 14 days.
13. Cells are cultured until stable cOva expression is established which can be confirm by western blot using anti ovalbumin antibody (ab181688, Abcam)

4. Assessment of T-cell function

4.1 TK-Flt3L gene therapy

Tumor implantation is done as previously described (Baker, Castro, & Lowenstein, 2015). Briefly, 6–8 weeks old female mice are anesthetized using ketamine and dexmedetomidine prior to stereotactic implantation with 50,000 of cOVA expressing NS in the right striatum. The coordinates for implantation are 0.5 mm anterior and 2.0 mm lateral from the bregma and 3.0 mm ventral from the dura. NS were injected at a rate 1 μL/min. The therapeutic mechanism of the T cell mediated immunotherapy is described in Fig. 1. The procedure for Ad-TK and Ad-Flt3L treatment is as follows:

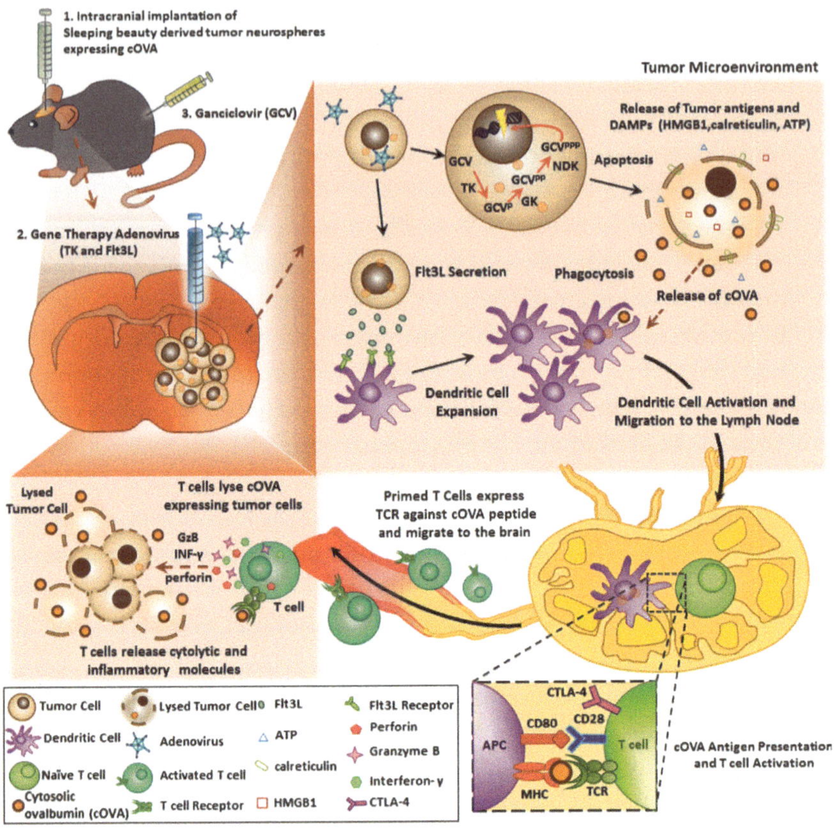

Fig. 1 Depiction of T cell response to tumor specific antigens following TK/Ftlt3 mediated immunotherapy. Neurospheres derived from sleeping beauty induced tumors are implanted intracranially into immunocompetent mice [1]. These neurospheres express cytoplasmic ovalbumin (cOVA) to assess antigen specificity. Two weeks later, animals are treated with adenoviral vectors delivered directly within the tumor mass [2]. The adenoviral vector encodes thymidine kinase (TK) and Fms-like tyrosine kinase 3 ligand (Flt3L). The infected tumor cells secrete the immunostimulatory cytokine Flt3L into circulation, which recruits dendritic cells to the tumor microenvironment. Intraperitoneal injection of ganciclovir (GCV) is administered the day after immunotherapy [3]. Ganciclovir is an inactive prodrug, but can become active by TK. TK initiates the first phosphorylation step followed by a cascade of cellular kinases Guanylate Kinase(GK) and Nucleoside Diphosphokinase (NDK) that further phosphorylate the drug. Activated GCV mimics guanosine and is incorporated into DNA where it induces replication stalling and apoptosis. Tumor cells that undergo apoptosis release damage associated molecular patterns (DAMPs) and cOVA. This stimulates dendritic cell expansion and phagocytosis of cOVA. Dendritic cells migrate to the nearby lymph nodes and present cOVA peptide on major histocompatibility complex (MHC) to naïve T cells. The primed T cells express T cell receptor (TCR) against cOVA peptide and migrate from the lymph node into the bloodstream. The activated T cells then infiltrate the tumor microenvironment and identify cOVA-expressing tumors cells by the formation of immunological synapses. The release of cytolytic (perforin, and granzyme B) and inflammatory (interferon-gamma) molecules by T cells lyses tumor cells.

1. First-generation replication-defective, recombinant adenovirus type 5 (Ad) vectors expressing transgenes (TK or Flt3L) under transcriptional control of the human CMV major intermediate early promoter within the E1 region were developed, purified, characterized, and scaled up, as described previously (Ali et al., 2005; Dewey et al., 1999; Southgate, Kroeger, Liu, Lowenstein, & Castro, 2008).
2. When the tumor reaches a signal of (10^6 photons/s/cm^2/sr of luminescence), each mouse received an intratumoral injection of either saline or the combination of the two viruses: AdV 5×10^8 plaque-forming units (pfu) of Ad-Flt3L and 2×10^8 pfu of Ad-TK in 1.5 μL volume in three locations at 3.5 mm, 3.0 mm, and 2.5 mm ventral from the dura.
3. Ganciclovir (GCV; TSZ Chemicals) was dissolved in DPBS (Gibco) and is given in a dose of 25 mg/kg i.p twice a day for 10 days starting immediately the day after viral injection.
4. By the end of the 10th day of GCV treatment, tumor is harvested and infiltrating immune cells were purified using percoll™ plus lymphocyte purification kit as described before (Kamran et al., 2017).

4.2 Isolation of tumor infiltrating lymphocytes (TILs)

Note that this method can be applied to any immunocompetent murine models of GBM

1. Prepare the following reagents: Complete media: RPMI/DMEM medium supplemented with 10% FBS (Peak Serum), 50 unit/mL penicillin and 50 μg/mL streptomycin (Gibco). 90% stock isotonic percoll (SIP, GE Healthcare)-prepared by mixing 9 parts of percoll and 1 part of 10× HBSS (Gibco). 70% percoll™-prepared by mixing 7 parts of 90% SIP and 3 parts of 1× HBSS.
2. Dissect the tumor carefully avoiding the normal areas of the brain.
3. Place the tumor onto a 70 μm strainer attached to a 50 mL conical tube, filled with 10 mL of complete media (Dulbecco's modified eagle (DMEM) media supplemented with 10% FBS).
4. Using the plunger end of a syringe, carefully mash the tumor through the 70 μm strained (Alkali Scientific Inc.,) on ice. Wash the strained two times with 10 mL complete media for force cell through the strainer and to minimize cell loss.
5. Spin the suspension at 500 × g (Allegra™ 6R centrifuge, Beckman Coulter) for 5 min at room temperature. Discard the supernatant and resuspend the cells in 7 mL of complete media (in 15 mL Falcon tube).

To this, add 3 mL of the 90% SIP (GE Healthcare) and mix well by pipetting up and down 3–5 times. The 30% gradient is now ready.

6. To layer the 70% percoll™ under the 30% percoll™ gradient, fill 1 mL serological pipette with 1 mL of 70% percoll™ and push it to the bottom of the 15 mL Falcon tube containing the homogenized brain tissue. Now slowly release the 1 mL of the 70% percoll™, making sure a clear interface is formed.
7. Repeat steps 2–5 for each mouse in the study.
8. Spin the falcon tubes at 800 × *g* for 20 min at room temperature.
9. After centrifugation, a white band will have formed at the interface containing the immune cells. Discard the supernatant (by pipetting) until about 1 mL from the interface. Collect the PBMCs that have accumulated at the interface between the two density centrifugation media layers using a 1 mL pipette and transfer to a fresh falcon tube. Add an additional 12 mL 1 × HBSS (Gibco) to the isolated cells and spin at 500 × *g* for 5 min to wash out the percoll™.
10. Discard the supernatant and resuspend the cells in an appropriate amount of complete media for counting using Trypan blue (Gibco). After the percoll™ separation, the cells can be stored on ice.
11. After counting, 50,000 cells can be plated per well of a 96 well V-bottom plate for staining for flow cytometry analysis. Cells can be stained with a variety of markers to distinguish the various immune cells as indicated in Table 1. Prior to staining cells should be labeled with viability dye and Fc receptor block (Biolegend) according to the manufacturer's instructions. Antibody staining is performed in Fluorescence-activated cell sorting (FACS) buffer on ice in the dark for 30 min. Post staining the cells are spun down at 500 × *g* for 5 min at 4 °C. Supernatant is discarded and cells are washed with 200 μL of FACS buffer (BD Biosciences).

4.3 Intracellular staining of T cell activation cytokines IFN-γ and GzB

Activated CD8 + T cells lyse cancer cells by secreting Granzyme B (GzB) and interferon gamma (IFN-γ). Methodologies for assessment of GzB and IFN-γ in activated CD8 + T cell are shown in Fig. 2. The detailed procedure is as follows:

1. For intracellular GzB and IFN-γ level, purified tumor infiltrating lymphocytes (TILs) are counted, stimulated *in vitro* with 200 mg/mL lysate from cOVA expressing NS for 24 h in 10% FCS-containing

Table 1 Immune cell markers use to characterize lymphocytes in mice.

Immune cells	**Markers**
CD4 T cells	CD45+, CD3+, CD4+
CD8 T cells	CD45+, CD3+, CD8+
Tregs	CD45+, CD3+, CD4+, FoxP3+, CD25+
Monocytes	CD45+, CD14+
Macrophages	CD45+, CD11b+, F4/80+
M1 Macrophages	CD45+, CD11b+, CD206−
M2 Macrophages	CD45+, CD11b+, CD206+, Arginase+
PMN MDSCs	CD45+, CD11b+, Ly6G+, Ly6Clow
Monocytic MDSCs	CD45+, CD11b+, Ly6G−, Ly6C+
MDSCs	CD45+, CD11b+, Gr-1+
DCs	CD45+, CD11c+, MHCII+
B cells	CD45+, CD19+
NK cells	CD45+, $CD11b^{int}$, NK1.1+

media supplemented with IL-2. (Note: The lysate from cOVA expressing NS is prepared by resuspending 1×10^7 NS in 1 mL RIPA lysis buffer (ThermoFisher Scientific) followed by three cycles of freeze-thawing, 30s each).

2. Six hours before harvesting the TILs, cells are treated with 4 nM Brefeldin and 1 nM monensin to accumulate the cytokines and other secreted proteins, intracellularly.
3. Cells are stained with CD3 (Clone: 17A2, Biolegend) and CD8 (Clone: YTS105.18, Biorad) antibodies, followed by intracellular stains for granzyme B (Clone: NGZB, eBioscience) and IFN-γ (Clone: XMG1.2, BD Biosciences).
4. For the tumor specific CD8 T cells, TILs are stained with Phycoerythrin (PE)-conjugated H2Kb-OVA tetramer (MBL international). Cells are first stained with CD3 (Clone: 17A2, Biolegend) and CD8 (Clone: YTS105.18, Biorad) antibodies.
5. Exhausted T-cells can also be examined by analyzing the expression level of exhausted Tcell markers such as; PD-1 (Clone: 29F.1A12, Biolegend), TIM-3 (Clone: B8.2C12, Biolegend), LAG-3 (Clone: 11C3C65, Biolegend), CTLA-4 (Clone: BNI3, Biolegend).

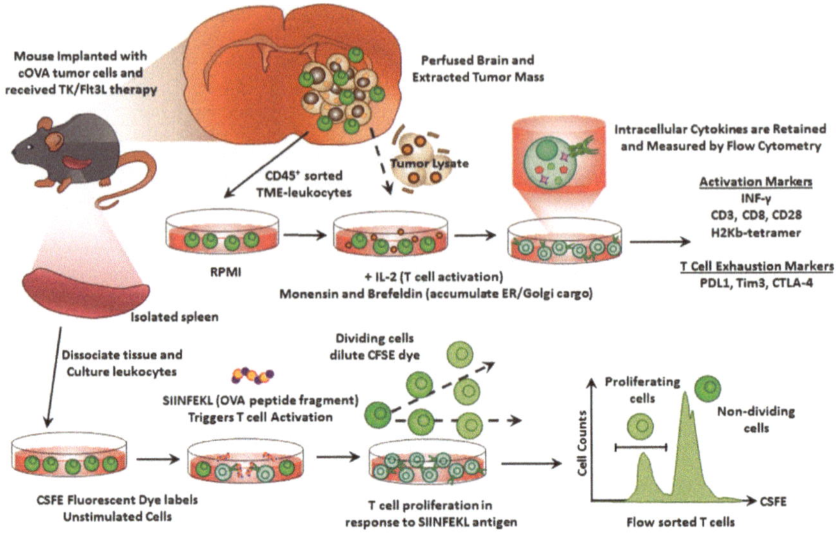

Fig. 2 Functional Assay to evaluate the activation of T cells in response to tumor lysate and stimulation with cOVA peptide *ex vivo*. Following immunotherapy, tumor bearing animals are euthanized to extract $CD45^+$ cells from the tumor mass and leukocytes from the spleen. Leukocytes from the tumor microenvironment (TME) are co-cultured with tumor lysate to stimulate T cell activation *in vitro*. Additionally, interleukin-2 (IL-2) is added to the media to enhance this activation. Monensin and Brefeldin inhibit ER secretory pathway which caused accumulation of intracellular cytokines in order to be measured by flow cytometry. The degree of T cell activation is evaluated based on the levels of activation markers such as INF- γ and GzB. Unstimulated T cells isolated from the spleen are labeled with carboxyfluorescein succinimidyl ester (CFSE). With each cell division, the CFSE dye is diluted signifying the proliferation of the T cells in response to SIINFEKL. Proliferating and non-dividing cells can be assessed by gating $CD3^+/CD8^+$ against CSFE fluorescence. Non-dividing cells will retain high intensity of CSFE, and proliferating T cells will have lower intensity.

6. Antigen-induced T cell proliferation is used to validate the T cell response upon antigen (SIINFEKL) stimulation:
7. Splenocytes from tumor bearing animals are isolated in a single cell suspension.
8. They are then labeled with CFSE dye (fisher scientific) using manufacturer's protocol.
9. Labeled splenocytes are cultured *in vitro* in the presence or absence of a final concentration of 100 nM of cOVA peptide (SIINFEKL) for 4–5 days.
10. Cells are harvested, gated on live/dead, CD3, CD8. The percentage of proliferating T cells are analyzed based on the intensity of CFSE and compared to control (Fig. 2).

5. Summary

Choosing a robust animal model as well as evaluating the immune response in glioma is critical for future implementation of immune-mediated therapeutic strategies in clinical trials. The methods presented here describe in detail the generation of NS-expressing the surrogate tumor antigen, cOVA from SB glioma models. This model can be used to determine and evaluate the efficacy of T cell-mediated anti-glioma immunotherapy induced by a variety of strategies, such as DC mediated immunotherapy. This in turn will have a strong impact on the preclinical evaluation of many immunotherapies in glioma. We also explain detailed methodology to analyze the functional T cell response against glioma. Such methodologies can be applied to other T cell mediated immunotherapy that target a variety of cancers in preclinical animal models.

Acknowledgments

This work was supported by NIH/NINDS Grants, R37-NS094804, R01-NS105556 and 1R21NS107894 to M.G.C.; NIH/NINDS Grants R01-NS076991, and R01-NS096756 to P.R.L.; NIH/NIBIB: R01-EB022563; NIH/NCI U01CA224160; the Department of Neurosurgery, the Rogel Cancer Center, Program in Cancer Hematopoiesis and Immunology (CHI), the ChadTough Foundation and Leah's Happy Hearts to M.G.C. and P.R.L. RNA Biomedicine Grant F046166 to M.G.C.; NIH/NCI T32-CA009676 to M.S.A. and S.C.

References

Acs. (2013). *Cancer facts and figures 2013*. American Cancer Society.

Ali, S., King, G. D., Curtin, J. F., Candolfi, M., Xiong, W., Liu, C., et al. (2005). Combined immunostimulation and conditional cytotoxic gene therapy provide long-term survival in a large glioma model. *Cancer Research*, *65*(16), 7194–7204. https://doi.org/10.1158/0008-5472.CAN-04-3434.

Bai, H., Harmanci, A. S., Erson-Omay, E. Z., Li, J., Coskun, S., Simon, M., et al. (2016). Integrated genomic characterization of IDH1-mutant glioma malignant progression. *Nature Genetics*, *48*(1), 59–66. https://doi.org/10.1038/ng.3457.

Baker, G. J., Castro, M. G., & Lowenstein, P. R. (2015). Isolation and flow cytometric analysis of glioma-infiltrating peripheral blood mononuclear cells. *Journal of Visualized Experiments*, *105*, 53676. https://doi.org/10.3791/53676.

Brat, D. J., Verhaak, R. G., Aldape, K. D., Yung, W. K., Salama, S. R., Cooper, L. A., et al. (2015). Comprehensive, integrative genomic analysis of diffuse lower-grade gliomas. *The New England Journal of Medicine*, *372*(26), 2481–2498. https://doi.org/10.1056/NEJMoa1402121.

Buckner, J. C., Shaw, E. G., Pugh, S. L., Chakravarti, A., Gilbert, M. R., Barger, G. R., et al. (2016). Radiation plus procarbazine, CCNU, and vincristine in low-grade glioma. *The New England Journal of Medicine*, *374*(14), 1344–1355. https://doi.org/10.1056/NEJMoa1500925.

Calinescu, A. A., Nunez, F. J., Koschmann, C., Kolb, B. L., Lowenstein, P. R., & Castro, M. G. (2015). Transposon mediated integration of plasmid DNA into the subventricular zone of neonatal mice to generate novel models of glioblastoma. *Journal of Visualized Experiments*, *96*, e52443. https://doi.org/10.3791/52443.

Candolfi, M., Yagiz, K., Foulad, D., Alzadeh, G. E., Tesarfreund, M., Muhammad, A. K. M. G., et al. (2009). Release of HMGB1 in response to proapoptotic glioma killing strategies: Efficacy and neurotoxicity. *Clinical Cancer Research*, *15*(13), 4401–4414. https://doi.org/10.1158/1078-0432.Ccr-09-0155.

Candolfi, M., Yagiz, K., Wibowo, M., Ahlzadeh, G. E., Puntel, M., Ghiasi, H., et al. (2014). Temozolomide does not impair gene therapy-mediated antitumor immunity in syngeneic brain tumor models. *Clinical Cancer Research*, *20*(6), 1555–1565. https://doi.org/10.1158/1078-0432.ccr-13-2140.

Ceccarelli, M., Barthel, F. P., Malta, T. M., Sabedot, T. S., Salama, S. R., Murray, B. A., et al. (2016). Molecular profiling reveals biologically discrete subsets and pathways of progression in diffuse glioma. *Cell*, *164*(3), 550–563. https://doi.org/10.1016/j.cell.2015.12.028.

Cloughesy, T. F., Mochizuki, A. Y., Orpilla, J. R., Hugo, W., Lee, A. H., Davidson, T. B., et al. (2019). Neoadjuvant anti-PD-1 immunotherapy promotes a survival benefit with intratumoral and systemic immune responses in recurrent glioblastoma. *Nature Medicine*, *25*(3), 477–486. https://doi.org/10.1038/s41591-018-0337-7.

Curtin, J. F., Liu, N., Candolfi, M., Xiong, W., Assi, H., Yagiz, K., et al. (2009). HMGB1 mediates endogenous TLR2 activation and brain tumor regression. *PLoS Medicine*, *6*(1), e10. https://doi.org/10.1371/journal.pmed.1000010.

Dewey, R. A., Morrissey, G., Cowsill, C. M., Stone, D., Bolognani, F., Dodd, N. J., et al. (1999). Chronic brain inflammation and persistent herpes simplex virus 1 thymidine kinase expression in survivors of syngeneic glioma treated by adenovirus-mediated gene therapy: Implications for clinical trials. *Nature Medicine*, *5*(11), 1256–1263. https://doi.org/10.1038/15207.

Dunn-Pirio, A. M., & Vlahovic, G. (2017). Immunotherapy approaches in the treatment of malignant brain tumors: Immunotherapy for malignant brain tumor. *Cancer*, *123*(5), 734–750. https://doi.org/10.1002/cncr.30371.

Garg, A. D., Vandenberk, L., Koks, C., Verschuere, T., Boon, L., Van Gool, S. W., et al. (2016). Dendritic cell vaccines based on immunogenic cell death elicit danger signals and T cell-driven rejection of high-grade glioma. *Science Translational Medicine*, *8*(328), 328ra327. https://doi.org/10.1126/scitranslmed.aae0105.

Gide, T. N., Wilmott, J. S., Scolyer, R. A., & Long, G. V. (2018). Primary and acquired resistance to immune checkpoint inhibitors in metastatic melanoma. *Clinical Cancer Research*, *24*(6), 1260–1270. https://doi.org/10.1158/1078-0432.CCR-17-2267.

Kamran, N., Alghamri, M. S., Nunez, F. J., Shah, D., Asad, A. S., Candolfi, M., et al. (2018). Current state and future prospects of immunotherapy for glioma. *Immunotherapy*, *10*(4), 317–339. https://doi.org/10.2217/imt-2017-0122.

Kamran, N., Kadiyala, P., Saxena, M., Candolfi, M., Li, Y., Moreno-Ayala, M. A., et al. (2017). Immunosuppressive myeloid cells' blockade in the glioma microenvironment enhances the efficacy of immune-stimulatory gene therapy. *Molecular Therapy*, *25*(1), 232–248. https://doi.org/10.1016/j.ymthe.2016.10.003.

Koschmann, C., Calinescu, A. A., Nunez, F. J., Mackay, A., Fazal-Salom, J., Thomas, D., et al. (2016). ATRX loss promotes tumor growth and impairs nonhomologous end joining DNA repair in glioma. *Science Translational Medicine*, *8*(328), 328ra328. https://doi.org/10.1126/scitranslmed.aac8228.

Larocque, D., Sanderson, N. S., Bergeron, J., Curtin, J. F., Girton, J., Wibowo, M., et al. (2010). Exogenous fms-like tyrosine kinase 3 ligand overrides brain immune privilege and facilitates recognition of a neo-antigen without causing autoimmune neuropathology. *Proceedings of the National Academy of Sciences of the United States of America*, *107*(32), 14443–14448. https://doi.org/10.1073/pnas.0913496107.

Lawler, S. E. (2017). Shifting the balance of power? The combination of oncolytic virotherapy and immune checkpoint blockade for glioblastoma treatment. *Neuro-Oncology*, *19*(4), 463–465. https://doi.org/10.1093/neuonc/nox011.

Louis, D. N., Perry, A., Reifenberger, G., von Deimling, A., Figarella-Branger, D., Cavenee, W. K., et al. (2016). The 2016 World Health Organization classification of tumors of the central nervous system: A summary. *Acta Neuropathologica*, *131*(6), 803–820. https://doi.org/10.1007/s00401-016-1545-1.

Mendez, F., Nunez, F., Koschmann, C., Ravindran, R., Dzaman, M., Lowenstein, P., et al. (2018). Gene-34. Mouse model of diffuse intrinsic pontine glioma harboring Acvr1 G328V. *Neuro-Oncology*, *20*(suppl_6), vi110–vi111. https://doi.org/10.1093/neuonc/noy148.460.

Michele Ceccarelli, F. B., Malta, T., Sabedot, T., Salma, S., Murray, B., Morozova, O., et al. (2016). Molecular profiling reveals biologically discrete subsets and pathways of progression in diffuse glioma. *Cell Reports*, *164*, 550–563. https://doi.org/10.1016/j.cell.2015.12.028.

Nunez, F. J., Mendez, F. M., Kadiyala, P., Alghamri, M. S., Savelieff, M. G., Garcia-Fabiani, M. B., et al. (2019). IDH1-R132H acts as a tumor suppressor in glioma via epigenetic up-regulation of the DNA damage response. *Science Translational Medicine*, *11*(479), eaaq1427. https://doi.org/10.1126/scitranslmed.aaq1427. ARTN.

Ostrom, Q. T., Gittleman, H., Truitt, G., Boscia, A., Kruchko, C., & Barnholtz-Sloan, J. S. (2018). CBTRUS statistical report: Primary brain and other central nervous system tumors diagnosed in the United States in 2011–2015. *Neuro-Oncology*, *20*(suppl_4), iv1–iv86. https://doi.org/10.1093/neuonc/noy131.

Paschka, P., Schlenk, R. F., Gaidzik, V. I., Habdank, M., Kronke, J., Bullinger, L., et al. (2010). IDH1 and IDH2 mutations are frequent genetic alterations in acute myeloid leukemia and confer adverse prognosis in cytogenetically normal acute myeloid leukemia with NPM1 mutation without FLT3 internal tandem duplication. *Journal of Clinical Oncology*, *28*(22), 3636–3643. https://doi.org/10.1200/JCO.2010.28.3762.

Polivka, J., Jr., Polivka, J., Holubec, L., Kubikova, T., Priban, V., Hes, O., et al. (2017). Advances in experimental targeted therapy and immunotherapy for patients with glioblastoma multiforme. *Anticancer Research*, *37*(1), 21–33. https://doi.org/10.21873/anticanres.11285.

Racine, C. A., Li, J., Molinaro, A. M., Butowski, N., & Berger, M. S. (2015). Neurocognitive function in newly diagnosed low-grade glioma patients undergoing surgical resection with awake mapping techniques. *Neurosurgery*, 77(3), 371–379. discussion 379. https://doi.org/10.1227/NEU.0000000000000779.

Reardon, D. A., Wucherpfennig, K., & Chiocca, E. A. (2017). Immunotherapy for glioblastoma: On the sidelines or in the game. *Discovery Medicine*, *24*(133), 201–208.

Reifenberger, G., Wirsching, H.-G., Knobbe-Thomsen, C. B., & Weller, M. (2016). Advances in the molecular genetics of gliomas—Implications for classification and therapy. *Nature Reviews. Clinical Oncology*, *14*(7), 434–452. https://doi.org/10.1038/nrclinonc.2016.204.

Sampson, J. H., Aldape, K. D., Archer, G. E., Coan, A., Desjardins, A., Friedman, A. H., et al. (2011). Greater chemotherapy-induced lymphopenia enhances tumor-specific immune responses that eliminate EGFRvIII-expressing tumor cells in patients with glioblastoma. *Neuro-Oncology*, *13*(3), 324–333. https://doi.org/10.1093/neuonc/noq157.

Southgate, T., Kroeger, K. M., Liu, C., Lowenstein, P. R., & Castro, M. G. (2008). Gene transfer into neural cells in vitro using adenoviral vectors. *Current Protocols in Neuroscience*, *45*, 4.23.1–4.23.43. https://doi.org/10.1002/0471142301.ns0423s45. Chapter 4, Unit 4 23.

Stupp, R., Mason, W. P., van den Bent, M. J., Weller, M., Fisher, B., Taphoorn, M. J. B., et al. (2005). Radiotherapy plus concomitant and adjuvant temozolomide for glioblastoma. *New England Journal of Medicine, 352*(10), 987–996. https://doi.org/10.1056/NEJMoa043330.

Verhaak, R. G. W. (2016). Moving the needle: Optimizing classification for glioma. *Science Translational Medicine, 8*(350), 350fs314. 350fs314. https://doi.org/10.1126/scitranslmed.aah4740.

Verhaak, R. G., Hoadley, K. A., Purdom, E., Wang, V., Qi, Y., Wilkerson, M. D., et al. (2010). Integrated genomic analysis identifies clinically relevant subtypes of glioblastoma characterized by abnormalities in PDGFRA, IDH1, EGFR, and NF1. *Cancer Cell, 17*(1), 98–110. https://doi.org/10.1016/j.ccr.2009.12.020.

Wiesner, S. M., Decker, S. A., Larson, J. D., Ericson, K., Forster, C., Gallardo, J. L., et al. (2009). De novo induction of genetically engineered brain tumors in mice using plasmid DNA. *Cancer Research, 69*(2), 431–439. https://doi.org/10.1158/0008-5472.CAN-08-1800.

Yang, J., Sanderson, N. S., Wawrowsky, K., Puntel, M., Castro, M. G., & Lowenstein, P. R. (2010). Kupfer-type immunological synapse characteristics do not predict anti-brain tumor cytolytic T-cell function in vivo. *Proceedings of the National Academy of Sciences of the United States of America, 107*(10), 4716–4721. https://doi.org/10.1073/pnas.0911587107.

CHAPTER SEVEN

Methods to edit T cells for cancer immunotherapy

Francesca Lucibello, Silvia Menegatti, Laurie Menger*
INSERM U932, PSL University, Institut Curie, Paris, France
*Corresponding author: e-mail address: laurie.menger@curie.fr

Contents

Abstract

This chapter summarizes the recent findings regarding the development of gene-editing technologies applied to adoptive T cell immunotherapy for cancer. The emerging tools and strategies described below enable the generation of a tailored system, establishing T cells as an advanced biomaterial with redirected specificities that are both programmable and inducible and establish long-term antitumor potential. These reviewed techniques and strategies are implemented with an optimized protocol for the generation of CAR-T cells using lentiviral gene transfer.

Methods in Enzymology, Volume 631
ISSN 0076-6879
https://doi.org/10.1016/bs.mie.2019.05.048

1. Introduction

The emergence of manipulated T cell-based immunotherapy opens new avenues in medicine, providing a novel class of intelligent living drugs to combat complex diseases such as cancer. With a more specific and continuous control or eradication of pathology, cell-mediated therapy offers an additional effective option to conventional treatments including surgery, chemotherapy or radiotherapy. Hence, the current approach consists of adoptive transfer of T cells (ACT) engineered with chimeric antigen receptors (CARs) or T cell receptors (TCRs) to repurpose cytotoxic T cell activity.

A primary example of the therapeutic potential of these strategies is the redirected specificity against a given disease provided by genetic manipulation technologies ex vivo that allow the introduction of genes into cells using lentivirus or retrovirus transduction (June & Sadelain, 2018; Maus et al., 2014). The adaptability offered by the engineering process can endow T cells with improved functions and simultaneously establish in vivo constraints for safety. For instance, suicide gene systems allow selective ablation of engineered cells (Stavrou et al., 2018) while other Tet-inducible promoter and CAR-RUSH (retention using selective hooks, WO2016012623A1) represent two systems that have been coopted to control the expression of chimeric antigen receptors. Furthermore, contrary to molecular agents that can be degraded or excreted shortly after administration, T cell-based drugs have the ability to persist for several years in patients (Kalos et al., 2011; Maude et al., 2018). Thus, the development of immunotherapeutic strategies favoring the generation of specific T cells subsets (memory or tissue resident memory) or of T cells with specific attributes (stemness) known to be effective in tumor control, represent promising immunotherapeutic strategies (Dumauthioz, Labiano, & Romero, 2018; Malik et al., 2017). Finally, the capacity of T cell intelligent sensing and specific response behaviors can be employed to customized therapeutic responses with a la carte cytokine secretion profiles and differentiation patterns (Roybal et al., 2016). Synthetic receptors, like the SynNotch system can, for instance, supplement and extend the capabilities of the endogenous T cell response, creating a novel transcriptional output by programming a defined cytokine secretion, T cell differentiation, and local delivery of therapeutic antibodies (Morsut et al., 2016; Srivastava et al., 2019).

In this chapter, we review the different methods of T cell engineering, with currently used and emerging tools for the induction of site-specific changes into human and murine T cell genomes, empowering T-cell targeting, functionality and safety. Genetic engineering includes the ablation or modification of endogenous genes, addition of new synthetic genes or transgenes. With this, we provide a general introduction to the detailed CAR T cell generation protocol included below.

2. TCR-engineered T cells, MHC dependency

The principle of clinical TCR gene therapy is straightforward: transfer of antigen-specific TCRαβ genes into T cells followed by ex vivo expansion and infusion of modified T cells into the patient. TCRα and β genes are used as reagents to confer reactivity against tumors expressing the appropriate antigen and HLA restriction element. This concept dates back more than 2 decades, when Z. Dembić and colleagues isolated T-cell receptor α- and β-chain genes from a class I major histocompatibility complex-restricted cytotoxic T-cell clone and transfected it into another cytolytic T-cell clone of different specificity (Dembić et al., 1986). Since then, gene transfer technologies were rapidly improved with retro- and lentiviral platforms generating therapeutic cellular products with a high level of tumor specificity (June, Riddell, & Schumacher, 2015; Maus et al., 2014). This strategy has overcome the relatively low frequency of cancer-reactive T cells that are naturally present in patients.

2.1 TCR gene delivery systems

Most of the TCR gene transfer studies in human are based on viral vector-based expression systems. Gamma-retroviral vectors, mainly represented by the MSGV-based recombinant vector, benefit from long-term gene expression (integration into the host genome), a low intrinsic immunogenicity (Uchida, Cone, Freeman, Mulligan, & Cantor, 1986), a good transduction efficiency and no vector-associated toxicity (Hughes et al., 2005). Initially, TCR α and β chains were assembled in bicistronic vectors with dual promoters and the chains were linked with IRES (internal ribosome entry site) sequences (Morgan et al., 2003). More recent constructs incorporated the 2A peptide to the chains with optimal stoichiometric expression in lentiviral vectors, which confer resistance to gene silencing and a safer integration site profile (Yang et al., 2008). However, most of the studies were performed using retroviral-based gene modification (Johnson et al., 2009; Mitsuyasu et al., 2000; Morgan et al., 2006; Robbins et al., 2011) and a 500 patient-years follow-up study has demonstrated the safety of T cells modified by retroviral gene transfer in clinical application (Scholler et al., 2012). Recent technological advances now permit rapid production of expression-ready retroviral TCR vectors directly from T cell RNA/cDNA (Coren, Jain, Trivett, Ohlen, & Ott, 2015) and the generation of tumor-derived TCR-expressing retroviral libraries from frozen tumor biopsies that contain antigen-specific T cells with potent antitumor activity (Tsuji et al., 2018).

2.2 Antigenic specificity and affinity

TCRs engineering provides broad recognition of epitopes from processed exogenous, membrane-bound and intracellular proteins, while CAR recognition is limited to cell surface structures. For example, most cancer testis antigens which are not expressed on the surface of tumor cells can be efficiently targeted by TCRs, but not by CARs (Obenaus et al., 2015). Hence, given that the majority of oncogenic proteins are intracellular (99%), TCR-engineered T cells offer the possibility to target a great number of molecules from the tumor-associated antigens (TAAs) to the cancer mutagenome (Morris & Stauss, 2016). Most clinical TCRs tested were directed against either differentiation, onco-fetal, or cancer germline antigens. Collectively, these trials have demonstrated significant clinical responses in patients with multiple myeloma (Rapoport et al., 2015), metastatic melanoma, colorectal carcinoma and synovial sarcoma, mainly human leukocyte antigen (HLA-) A1 and A2 restricted (Debets, Donnadieu, Chouaib, & Coukos, 2016; Ping, Liu, & Zhang, 2018). Increasing studies of TCR-T cells targeting leukemia-relative antigens have also been reported. TERT-TCR-T cells demonstrated a capacity to limit the progression of AML and B-ALL (Sandri et al., 2016, 2017). Antileukemia reactivity was observed in xenogeneic mouse models without affecting HLA-A*0201$^+$ normal PBMCs, mitotic PHA-lymphoblasts or cord blood CD34$^+$ cells. CD4 T cells were also redirected to recognize the same HLA-A*0201-restricted AURKA$_{207-215}$ epitope, which could sustain the antileukemia response in vivo after adoptive transfer (Nagai et al., 2012). Finally, the first clinical trial using WT1-TCR-modified T cells for leukemia therapy demonstrated that the cells survived in vivo, retained the capacity to mount immune responses to WT1, and provided transient antileukemic effects in patients with refractory AML (Tawara et al., 2017).

HLA-dependent Ag recognition represents a major disadvantage since therapeutic TCRs can only be used in patients with the appropriate HLA alleles. Therapeutic TCRs restricted by four common HLA class I alleles

(A*0201, A*0301, A*2402, and B*0702) would nevertheless extend the applicability of TCR gene therapy (González-Galarza et al., 2015). Another major concern regarding engineered TCRs is their capacity to mispair with endogenous TCRs and thereby to reduce the desired specificity, increasing the risk to recognize both allo- and auto-Ags in a HLA class I and II restricted fashion (van Loenen et al., 2010). In mouse, native/transduced heterodimers with unknown specificity were able to cause severe graft-versus-host disease (GVHD)-like pathology (Bendle et al., 2010). However, several methods can favor the pairing of transduced TCRs, including the engineering of the constant regions (murine-human hybrid TCRs, adding disulfide bond, mutational inversions, already reviewed in Maus et al., 2014) as well as endogenous TCR editing (Legut, Dolton, Mian, Ottmann, & Sewell, 2018; Mastaglio et al., 2017).

Finally, a thorough understanding of TCR properties and Ag affinity/avidity are mandatory to choose the most appropriate TCR for controlling tumor growth and maintaining durable immunity without any adverse consequences. Despite a large amount of research, controversy still remains as to which TCR properties lead to a safe and effective TCR gene therapy. The uniqueness of a cancer peptide as well as its tissue-expression profiles and tumor-expression levels are critical parameters to consider. It has been observed that T cells expressing highly reactive TCRs can mediate a more robust clinical antitumor response compared to TCRs with lower affinity but can also target rare cognate-Ag-containing cells throughout the body. Melanoma patients treated with T cells transduced with a high-avidity MART-1$_{(27–35)}$ TCR had a significantly higher rate of tumor response than patients treated with T cells transduced with a lower-avidity MART-1$_{(27–35)}$ TCR. However, off-target toxicities, TCR cross-reactivity and autoimmune responses were observed in 55% of the patients who developed uveitis and vitiligo (Borbulevych, Santhanagopolan, Hossain, & Baker, 2011; Johnson et al., 2009). T cell antitumor activity and autoimmunity are closely related and if the affinity of TCRs is ranging between K_d: 10^{-4} and 10^{-6} M, a plateau in the contribution is reached at 10 μM, suggesting that the efficacy of adoptive cell therapy will not be improved utilizing high-affinity TCRs (Schmid et al., 2010; Zhong et al., 2013). In physiological conditions, the immune system appears to maintain a repertoire of flexible low-affinity TCRs at the expense of more effective high-affinity TCRs (Ellis, 2005; Holler, Chlewicki, & Kranz, 2003). Not surprisingly, TCR affinities to cancer antigens (e.g., self-derived peptides) are strikingly lower than TCRs affinities to pathogen-derived antigens (Aleksic et al., 2012), rendering

T cells with a sufficient sensitivity to TAAs difficult to isolate. Recently, a study demonstrated that low-affinity TCRs can also mediate tumor regression and that the functional avidity allows for discrimination between tumor-derived and endogenous antigen (Miller, Bahmanof, Zehn, Cohen, & Schoenberger, 2019). However, endogenous low-affinity TCRs requires higher levels of peptide: ligand interactions, through increased CD8 co-receptor stabilization, DC-derived co-stimulation, and blockade of co-inhibitory receptors (Zhong et al., 2013) to ensure occurrence of activation (Hebeisen et al., 2013). New methods have been developed to concurrently sequence TCRs and measure TCR affinity of individual T cells from humans. This novel ability to isolate TCRs native to humans with a given affinity could enable rapid selection of safer TCR candidates with high potency (Zhang et al., 2016). TCR avidity is also an important component for in vivo T cell efficacy. Enhanced TCR avidity confers a superior efficacy both in vitro and in vivo, promoting earlier and faster recognition of target cells (Derby, Alexander-Miller, Tse, & Berzofsky, 2001; Dutoit et al., 2001). The recent development of technology based on reversible MHC multimer staining (streptamers) allows for an accurate determination of the dissociation of monomeric pMHCs from TCRs on living T cells, providing a measure of the avidity. This highlighted a strong correlation (what type of correlation?) between the k_{off}-rate and protective capacity of transferred T cells (Nauerth et al., 2013). This parameter will allow for selection of the most efficacious T cells for adoptive transfer.

2.3 Identification of antitumoral TCR sequences

Single cell RT-PCR methods are growing rapidly, allowing for rare or heterogeneous cells population profiling with an accurate quantitation of transcriptomes in individual cells that recapitulate bulk transcriptome complexity (Wu et al., 2014). Hence, the paired α and β chains from 91 murine CD4 T cells have been successfully identified (Mahata et al., 2014). Similarly, the intratumoral TCR repertoire in human cancers is intensively being studied, and the first results show that the intrinsic capacity of T cells to recognize adjacent (adjacent or tumor tissue?) tumor tissue can be rare (10%) and variable (Scheper et al., 2019). Therefore, clinical efforts to engineer/reactivate intratumoral T cells will have to benefit from novel technologies that increase the identification and understanding of tumor-infiltrating T cells repertoire. Peptide-MHC multimers labeled with individual DNA barcodes now allow for large scale epitope recognition profiling of tumor-infiltrating lymphocytes and peripheral blood from small clinical samples that have not been expanded in vitro. Moreover, a sorting strategy based on T cell cytokine expression profiles can further elucidate the functional capacity of T-cell specificities (Bentzen et al., 2016). This method can be time efficient and inexpensive when using a library of DNA-barcoded antigen tetramers generated by in vitro transcription and translation. The tetramer-associated T-cell receptors linked to their cognate antigens in single cells are then sequenced to identify patterns in TCR cross-reactivity with cancer neoantigens (Zhang et al., 2018). In this way, neoantigen-specific TCRs that do not cross-react with normal tissue antigens can be rapidly isolated (Zhang et al., 2018). Importantly, this approach emphasized that TCRs recognizing the same epitope can present different recognition motifs leading to potential cross-recognition in the human proteome, another important parameter to consider before selection of TCRs for clinical development (Bentzen et al., 2018; Glanville et al., 2017).

2.4 Gene modifications to enhance TCR-transduced T-cell function

TCR-engineered T cells can be further modified to enhance their functions. Early work aimed at expression of costimulatory ligands (CD80, 4-1BBL) to provide auto- and trans-co-stimulation to provoke potent tumor rejection in the context of tumor cells lacking costimulatory ligands (Stephan et al., 2007). With the persistence of tumor antigen and many distinct immunosuppressive mechanisms, intratumoral T cells display a broad spectrum of dysfunctional states (Wherry, 2011). T cells become less functional with the upregulation of several inhibitory costimulatory molecules such as PD-1, Tim-3, CTLA-4, LAG-3, BTLA, TIGIT (Śledzińska, Menger, Bergerhoff, Peggs, & Quezada, 2015), demonstrating impaired production of effector cytokines, and aberrations in transcription factor expression, chemotaxis, metabolism, cell cycle, and DNA repair (Thommen & Schumacher, 2018). Interestingly, enhancement of T cell antitumor activity was also conducted by genetically diverting PD-1 negative signals into positive ones (CD28 and 4-1BB) leading to increased T-cell proliferation and cytokine secretion (Ankri, Shamalov, Horovitz-Fried, Mauer, & Cohen, 2013). Further, abrogating inhibitory molecules such as Cbl-B (Stromnes et al., 2010), Shp-1 (Stromnes et al., 2012), TGFB (Chou et al., 2012) or PD-1 (Menger et al., 2016) represent alternative strategies to improve the efficacy of adoptive therapies. In a separate context, engineering a selective delivery of cytokine such as IL-12 upon recognition of tumor-specific antigen by TCR can potentiate the antitumor immune response in cell transfer experiments (Zhang et al., 2011). An emerging technology at the interface between TCR-engineered- and CAR-T cells, is the recently described T cell antigen coupler (TAC) technology (chimeric receptor with single chain antibody—scFv). TAC recapitulates the architecture of a TCR-CD3/co-receptor complex to engage natural cellular pathways, and has outperformed the CD28-based CAR therapies in several xenograft models including solid tumors and hematological (Helsen et al., 2018).

3. CAR T-cell, MHC independency

In an attempt to redirect T cell specificity in a major histocompatibility complex (MHC)-restriction independent manner, chimeric TCRs composed of TCR constant domains fused to an antibody (Ab) variable domain (scFv) and the gamma or zeta TCR subunits were transfected into a cytotoxic T-cell hybridoma (Eshhar, Waks, Gross, & Schindler, 1993; Gross, Waks, & Eshhar, 1989). This was the first demonstration that linking recognition elements of an Ab to signaling moieties for T cell activation could imbue T lymphocytes with redirected specificity and functional activity. Co-stimulatory signaling domain CD28 (Finney, Lawson, Bebbington, & Weir, 1998), OX-40 (Pulè et al., 2005) or 41BB (Finney, Akbar, & Lawson, 2004; Johnson et al., 2009; Milone et al., 2009) were then added in between scFv and CD3ζ chain to improve the quality, strength and duration of signals delivered by the chimeric molecules (Kawalekar et al., 2016; van der Stegen, Hamieh, & Sadelain, 2015). However, the optimal signal combination is still debated. Recent findings suggest a promising therapeutic activity in solid tumor models with CD4 and CD8 CARs bearing ICOS and 4-1BB intracellular domains (Guedan et al., 2018). To increase the persistence of engineered cells in the TME, CARs can be endowed with the inducible capacity to secrete IL-12, leading to the recruitment of second wave of immune cells (NK cells, macrophages) (Chmielewski, Kopecky, Hombach, & Abken, 2011; Zhang et al., 2011).

3.1 CAR gene delivery systems

CARs can be introduced into T cells using different vectors and methods which must meet the criteria of efficient delivery, patient safety, and expression efficacy. Both retroviral and lentiviral systems account for excellent gene transduction efficiency but lentiviral is the most frequently used method because of its safer genomic integration profile

(De Palma et al., 2005; McGarrity et al., 2013; Schröder et al., 2002), its insensitivity to transgene silencing (compared to retroviruses) and its ability to infect resting and dividing cells (Ellis, 2005; Kay, 2011; Naldini et al., 1996). The capacity to integrate transgenes in non-replicating cells offers the advantage of obtaining T lymphocytes with a poorly differentiated phenotype which is optimal for therapeutic purposes (June, Blazar, & Riley, 2009). Nevertheless, viral particle production remains laborious with a risk of contamination by infectious agents and the development of innate immune and inflammatory responses to viral vectors (Nayak & Herzog, 2010; Shayakhmetov, Di Paolo, & Mossman, 2010). Among non-viral approaches are the PiggyBac and Sleeping Beauty (SB) transposons that allow integration of large DNA sequences in the host genome through the transposase cut and paste mechanism (Moldt, Yant, Andersen, Kay, & Mikkelsen, 2007; Walisko et al., 2008). These tools offer a reduced cost, a lower immunogenicity and the integration can be enhanced with hyper-active transposase (Hackett, Largaespada, & Cooper, 2010). It was recently showed that CAR T cells can be engineered through SB transposition of CAR genes from minimalistic DNA vectors called minicircles (Monjezi et al., 2017). However, many concerns remain, such as the requirement for several rounds of infusion using transient mRNA transfection and the unknown potential of insertional mutagenesis and remobilization of transposons when using the SB transposon system (Beatty et al., 2014). A major challenge in primary T-cell engineering is the timeframe of multiplex genetic manipulation with high efficiency. Unstimulated primary or naïve T cells are less receptive to exogenous nucleic acid or protein uptake compared with stimulated T cells (Hendel et al., 2015). On the contrary, repeated stimulations can lead to decrease antitumor efficacy due to T-cell exhaustion. Although Crispr is easily implementable at the discovery/screening stage, other technologies such as Zinc-finger nucleases (Doyon et al., 2010; Gaj, Gersbach, & Barbas, 2013; Provasi et al., 2012) and TALEN (Menger et al., 2015; Poirot et al., 2015) have demonstrated similar editing performances and are already in the clinic (Qasim et al., 2017).

Multi-functional universal CAR T cells generated with CRISPR/Cas9 techniques have been developed to incorporate endogenous gene disruption into standard clinical CAR T cell manufacturing processes (Ren et al., 2017). Freshly isolated human T stimulated via anti-CD3/CD28 beads and lentiviral transduction to stably express a CAR transgene are electroporated with RNA-encoding Cas9 and sgRNA to disrupt TCR, HLA-I, and PD1 simultaneously. The editing efficiency indicated >70% CAR transduction efficiency and >60% double-knockout efficiency. These CAR T cells were specific to CD19 targets, resistant to host rejection, and incapable of triggering GVHD. Similar results were also reported using another approach of CRISPR/Cas9 RNPs targeting the same three genes: TCR, B2m, and PD1 (Liu et al., 2017). Another CRISPR system, the CRISPR-Cpf1 (Cas12a) might be of particular interest since it shows multiple inherent advantages over CRISPR-Cas9. For instance, it has a smaller effector gene size, a smaller guide RNA (42 nt), a low off-target activity, the PAM site is preserved, and it has the capacity to generate a staggered cut with a 5′ overhang, in contrast to the blunt ends generated by Cas9 (Garneau et al., 2010; Jinek et al., 2012; Zetsche et al., 2015). This cleavage pattern facilitates non-homologous end joining (NHEJ)-based gene insertion into the mammalian genome and directional gene transfer (Maresca, Lin, Guo, & Yang, 2013). A new method for site-specific CAR gene insertion can therefore be a clear advantage compared to random lentiviral integration. Specifically, in non-dividing cells, in which genome editing via homology-directed repair (HDR) mechanisms is especially challenging (Chan, Hauswirth, Wensel, & Wilson, 2011). Moreover, it was recently reported that a significant portion of humans have pre-existing adaptive immunity to Cas9 proteins (Charlesworth et al., 2019). These observations emphasize the potential advantages of the CRISPR-Cpf1 system. Recently, AAV-based approach combined with site-specific nuclease allow to induce KO (TCR locus for instance) while inserting the CAR specificity (KIKO) (Ehrke-Schulz, Zhang, Gao, & Ehrhardt, 2017; Eyquem et al., 2017; Xu et al., 2019). These strategies can set a new standard in advanced cellular and gene therapies as it presents many advantages: simplicity of gene transduction (by electroporation) and precision, safety for both patients and operators, reduced cost, and less good manufacturing practice requirements.

3.2 Antigenic specificity for solid tumor, affinity and associated toxicities

Heterogeneous and overlapping expression of tumor-associated antigens (TAA) on solid cancers and healthy tissues (Fidler & Hart, 1982) makes it difficult to find suitable targets for CAR T cell therapy. Detailed discussions on new antigenic targets for CAR-T-cell therapy in solid tumors is currently available (Jackson, Rafiq, & Brentjens, 2016; Knochelmann et al., 2018; Martinez & Moon, 2019). They include mesothelin for the treatment of mesothelioma (Adusumilli et al., 2014), ovarian and pancreatic cancers, disialoganglioside GD2 (Singh et al., 2014), EGFRvIII for glioblastoma

(Johnson et al., 2015), and mucin-16 (Koneru, O'Cearbhaill, Pendharkar, Spriggs, & Brentjens, 2015) for ovarian cancer. The on-target, off-tumor toxicities report from these trials is still on-hold as long-term outcomes can be fatal fatal—death attributed to CARs attack of lung epithelial cells expressing HER2 (Morgan et al., 2010) or fatal encephalitis due to high-affinity anti-GD2 CAR for neuroblastoma treatment (Richman et al., 2018). The degree of on-target toxicity relies on the level of Ag expressed on the healthy tissue and its functional importance as well as the CAR affinity and potency. Screening strategies for identifying Ag exclusively expressed by tumors involve bulk- and single-cell deep sequencing technologies such as DNA sequencing (DNA-seq), RNA sequencing (RNA-seq), and tumor exome analysis (Cohen et al., 2015; Srivastava, 2015). These new technologies raise high expectation to better predict the efficacy and toxicity of novel CAR T cells. The level of CAR T cell activation has also to be calibrated, and there is a balance between CAR program efficacy and in vivo persistence. The threshold of stimulation to yield optimal CAR-T cell activation is reached when CARs encode a single immunoreceptor tyrosine-based activation motif (Feucht et al., 2019). It is important to keep in mind that scFvs possess higher affinity (in the nM range, K_d: $10^{-6}-10^{-9}$ M) than native TCRs. Even though, they can similarly inflict sequential killing of multiple tumor cells, suggesting that they dissociate more rapidly (Davenport et al., 2015). Through the active process of trogocytosis, CARs can also provoke reversible antigen loss and transfer to other T cells, thereby decreasing target density on tumor cells and abating T cell activity by promoting fratricide T cell killing (Hamieh et al., 2019). This active and reversible phenomenon that depends on Ag density represent a rational for the development of strategies based on combinatorial targeting.

3.3 Engineering CAR T cells for solid tumors

The demonstrated clinical efficacy using CAR-T-cell therapy is limited to hematological malignancies (June & Sadelain, 2018). Solid tumors present unique challenges not seen in B-ALL. First, their microenvironment can be considerably more immunosuppressive. CAR T cell therapies for solid tumors also face poor trafficking to the tumor site (Pule et al., 2008) and limited persistence/proliferation within the host (Gargett et al., 2016; Hege et al., 2017; Thistlethwaite et al., 2017). To improve T cell trafficking into tumors, inactivating CXCR4 could, through insensitivity to CXCL12 (Feig et al., 2013), enhance T cell infiltration. Another idea would be to target the tumor vascular environment with EC matrix-directed CAR-T cells to destroy the architecture of the neovessels (Wang et al., 2013). CAR-T cells engineered to express catalase can be protected from intratumor high H_2O_2 levels and NK cell oxidative stress (Ligtenberg et al., 2016). Finally, blocking TGF-β signaling via expression of a dominant negative TGF-β receptor II in CAR-T cells has been shown to improve CAR proliferation, cytokine secretion, in vivo persistence, and tumor control in aggressive human prostate cancer mouse models (Kloss et al., 2018).

4. Suicide gene strategies

The efficacy of T cells genetically modified with CAR against cancer is achieved at the expense of enhanced toxicities. These adverse events can be divided into five categories, i.e., on-target on-tumor, on-target off-tumor, off-target, neurotoxicity, and other toxicities (Sun, Hao, Yang, Zhang, & Fu, 2018). The most common toxicity is triggered by excessive cytokine release or tumor cell necrosis (Bonifant, Jackson, Brentjens, & Curran, 2016; Brudno & Kochenderfer, 2016). A direct attack on normal tissues having a shared expression of the targeted antigen can also induce on-target off-tumor toxicity, and fatal cardiac toxicity has been observed (Cameron et al., 2013; Linette et al., 2013). It remains uncertain if the neurotoxicity arises from direct CAR T cell attack on the CNS tissue or generalized cytokine-mediated inflammation, but the most commonly occurring neurological symptoms comprise encephalopathy like cerebral edema, headache, tremor, fever and aphasia (35%) (Rubin et al., 2019). Finally, the lymphodepleting and nonmyeloablative regimen comes along with the well-known toxicities including anemia, coagulopathy, and neutropenic sepsis (Lowe et al., 2018). Genetically encoded elements have been developed to induce a selective destruction of expressing cells in the face of strong toxicity. Several suicide gene strategies have been developed, such as the B-cell antigen CD20 (Serafini et al., 2004), RQR8 (Philip et al., 2014) or huEGFRt (Wang et al., 2011). These surface proteins act as a permanent off switch, activated by therapeutic monoclonal antibodies, rituximab or cetuximab, respectively. These molecules can also serve as in vivo tracking molecules or be used to enrich products in CAR-transduced T cells using biotinylated antibodies and anti-biotin immunomagnetic microbeads. Two other suicide gene strategies that are activated by a pharmaceutical agent have been tested in clinical studies: the highly effective but immunogenic approach using herpes simplex virus thymidine kinase (HSV-TK) with susceptibility to ganciclovir (Ciceri et al., 2009) and the inducible caspase 9 (iCasp9) (Straathof et al., 2005), which is a fusion of a mutated FKBP12 (docking of a small chemical inducer of dimerization) with the catalytic domain of caspase 9. The different methods have been compared. Although CD20, iCasp9, and HSV-TK were equally efficient in destruction of expressing T cells, iCasp9 and CD20 expression resulted in more rapid activity as compared to HSV-TK and mTMPK (Marin et al., 2012). Improvements are still ongoing since there are limitations in therapeutic antibody usage regarding their poor distribution in the brain and some tissues. Another promising method has been recently described using a rapamycin-induced caspase 9 suicide gene version (Stavrou et al., 2018).

5. Genome-wide CRISPR-Cas9 engineering

Pooled screens engineering of primary T cells can be used to rapidly map gene circuits that regulate antitumor immune responses. A recent study used shRNAs targeting known negative regulators to systematically uncover the regulatory switches controlling T-cell function in an immunosuppressive tumor microenvironment (Zhou et al., 2014). Using this elegant approach, they identified Ppp2r2d as an important regulator of T-cell apoptosis, proliferation and cytokine production. In another study, a genome-scale CRISPR–Cas9 library was used to profile genes whose loss in tumor cells impaired the effector function of $CD8^+$ T cells. These experiments unveiled the apelin receptor (APLNR) as an important modulator of interferon-γ responses in tumors for which functional loss reduces the efficacy of adoptive cell transfer and checkpoint blockade immunotherapies (Patel et al., 2017). Perturbational strategies have been challenging to apply at the genome level in primary immune cells due to…? However, Cas9-expressing transgenic mice enabled the dissection of TLR(definition?) signaling modulators in infected DCs with a genome-wide library of lenti-based sgRNAs (Parnas et al., 2015). A validated retroviral version of a genome-wide CRISPR knockout library is now available, and it enabled the dissection of regulatory circuitry governing murine Th2 cell differentiation (Henriksson et al., 2019). CRISPR–Cas9 genome-wide interrogation of gene function is also now achievable in human primary T cells. This method is based on the finding that editing by lentiviral delivery of targeted gRNAs occurs efficiently with cas9 mRNA (Hendel et al., 2015) or using ribonuleoproteins (RNPs) (Ting et al., 2018). Genome-wide loss-of-function screens identified essential, intrinsic, and non-redundant T cell receptor signaling components as well as genes that negatively modulate proliferation such as Cbl-B, RASA2, CD5 and SOCS-1. Importantly, targeted ablation of these candidate genes demonstrated enhanced cancer cell killing in vitro (Shifrut et al., 2018).

6. Generation of CAR T cells using lentiviral strategy

The following section presents a general and condensed protocol for generating CAR-T cells through lentiviral transduction. It is based on the expertise we have developed along with published protocols from the literature.

6.1 Equipment

1. Beckman Coulter Optima L-100XP ultracentrifuge with a SW32Ti rotor
2. Laminar flow hood

6.2 Lentiviral production

- Day 0: seed 15×10^6 HEK 293T/17 cells (ATCC No. CRL-11268) in DMEM (Gibco) supplemented with 10% fetal calf serum (FCS, Gibco), MEM Non-Essential Amino Acids Solution, HEPES, L-Glutamine (Thermofisher) at 37 °C, 5% CO_2 into dish $145 \times 200\,mm^2$ (Greiner CELLSTAR® 639160) and incubate the cells overnight at 37 °C in a humidified 5% CO_2 incubator.
- Day 1: Cells should be 70–80% confluent. Prepare the transfection complex in two 15-mL tubes. Plasmids must be prepared using the Endofree Plasmid Maxi kit (Qiagen). A recent study demonstrated that the titer of virus packaged with psPAX2/pMD2.G system (2.46×10^6 TU/mL) was significantly higher than that packaged with pCMV-dR8.91/pVSV-G (0.71×10^6 TU/mL) and that the pLVX-EF1a-CAR construct expressed the highest level of CAR in human peripheral T cells compared to pLVX-CMV-CAR and pCDH-EF1a-CAR (Yuan et al., 2018). The recommended ratios are 4:2:1 for lentiviral vector plasmid:packaging plasmid (psPAX2):envelope plasmid (pMD2G), respectively.
- Tube A: dilute 20 µg of plasmid DNA, 10 µg of pPAX2, 5 µg of pMD2.G in 1 mL of Opti-MEM (room temperature) and mix gently. Tube B: dilute 120 µL of TransIT-293 (Mirus) in 1 mL of Opti-MEM and mix gently. Pipet the contents of tube B into tube A and incubate the tube at RT for 30 min.

- Using a pipette, add contents dropwise into the dish of 293T cells to disperse the transfection complex and incubate overnight at 37 °C in a humidified 5% CO_2 incubator. Gently mix of the contents.
- Day 2: remove the culture medium from the dish and slowly add 16 mL of fresh DMEM culture medium and incubate the cells overnight at 37 °C in a humidified 5% CO_2 incubator.
- Day 3 a.m.: harvest the medium into a 50-mL centrifuge tube and slowly add 16 mL of fresh DMEM culture medium. Keep the tube at 4 °C.
- Day 3 p.m. (10 h later): harvest the medium into a 50-mL centrifuge tube and slowly add 16 mL of fresh DMEM culture medium. Keep the tube at 4 °C.
- Day 4 a.m.: harvest the medium into a 50-mL centrifuge tube and slowly add 16 mL of fresh DMEM culture medium. Keep the tube at 4 °C.
- Pool the 50-mL centrifuge tubes from the tree collection and centrifuge at 2000 × *g* for 30 min at 4 °C to pellet the cells and debris.
- Filter supernatant through a 0.45 μm pore PVDF Millex-HV filter (Millipore).
- Perform ultracentrifugation with a 20% sucrose cushion at 20,000 × *g* for 2 h at 4 °C, using L-100XP ultracentrifuge with a SW32Ti rotor (Cribbs, Kennedy, Gregory, & Brennan, 2013; Yuan et al., 2018).
- Pour the supernatant from the tube and remove the excess liquid by blotting onto a paper towel.
- Add 100 μL ofDMEM directly onto the viral pellet and place the tube on ice for 15 min. Once on ice, carefully pipet up and down multiple times. Avoid generating bubbles (Xu et al., 2019).
- Aliquot the lentiviral stock into pre-chilled sterile tubes and store in a −80 °C freezer.

6.3 Viral titration

The use of a certain MOI is intimately dependent of which titration method was used. The methods are classified in functional and non-functional. The functional titration is more accurate and is determined by the direct transduction of cells of interest and subsequent evaluation of reporter protein activity, (e.g., beta-galactosidase positive cells) or by assessment of the number of colonies forming units following antibiotic selection (Geraerts, Willems, Baekelandt, Debyser, & Gijsbers, 2006). The non-functional titration tends to overestimate the functional vector titer and includes p24 antigen ELISA (quantification of free p24), assessment of the reverse transcriptase activity (assess defective particles) and determination of the genomic RNA concentration in vector preparations by quantitative RT-PCR using custom Taqman assays (Thermo Fisher) targeting the U6 promoter or using a Lenti-X kit (Clontech). From one 150-mm dish, the viral titration ranges between 10^9 and 10^{10} viral copies/mL. Alternatively, the titers can be assessed by the formula: Titer (TU/mL) = (P × N)/(D × V). P: percentage of GFP^+ cells, N: number of cells/well, D: dilution factor of virus, V: volume of virus/well.

- Plate 293T cells (7×10^4/well) in a 24-well plate to reach 30–40% confluency the following day.
- Remove medium from 293T cells.
- Add 200 μL of dilutions ranging from 10^{-2} to 10^{-5} of each viral sample, keeping one un-transduced control well.
- 24 h after transduction, remove medium containing virus and replace with fresh medium.
- 72 h after transduction, evaluate the percentage of GFP-positive cells by flow cytometry.

6.4 T cell activation, transduction, and expansion conditions

- Isolate T cells from fresh PBMCs using EasySep™ Human T Cell Enrichment Kit (Stemcell technologies).
- Culture T cells in CAR-T cell medium: X-VIVO media (Lonza) with 5% human AB (blood type) serum and 30–100 IU/mL rIL-2 (Peprotech), addition of 10–20 mM of the antioxidant *N*-acetyl cysteine (NAC) can improve the quality and therapeutic efficacy of adoptive T cells (Pilipow et al., 2018; Scheffel et al., 2016). NAC (A7250, Sigma) is dissolved in H_2O, and the pH is adjusted to 7.4 with NaOH.
- Wash CD3/CD28 Dynabeads (Thermo Fisher) several times with PBS containing 0.1% BSA.
- T cells are activated with a 1:1 ratio of human anti-CD3/anti-CD28 beads for 2 days in a non-treated 24-well plate at 10^6 cells/mL.
- Harvest the cells and remove Dynabeads with magnetic separation rack.
- Thaw an aliquot of the lentiviral stock on ice.
- Seed T cells into 24-well plates at 5×10^5 cells/500 μL/well.
- Add TransPlus (Alstem) or 6–8 μg/mL polybrene (Sigma) into each well and mix.

- Pipet the lentivirus at a multiplicity of infection (MOI) of 5 (Xu et al., 2019) or 40 (Yuan et al., 2018) or 50 (Cribbs et al., 2013) for optimal transduction efficiency and cell viability.
- Centrifuge the 24-well plate at 1200 × *g* for 2h at 32°C with slow acceleration and slow deceleration, and culture for 10–14h before replenishing with fresh medium.
- To improve efficiency, a second round of transduction can be performed, preferably 24h after the first round.
- Count cells in the wells every 2–3 days and add fresh CAR-T medium to maintain the cell density below 3×10^6 cells/mL.
- Bioreactor culture systems can also be used to expand CAR-T cells number: WAVE Bioreactor, G-Rex and CliniMACS Prodigy (Bajgain et al., 2014).
- Measure CAR Expression by Flow Cytometry after 72h.

A major challenge in the CAR detection is the lack of a fast, economical, sensitive, and robust assay. Generally, expression of CARs on T cells is detected by flow cytometry using fluorochrome-tagged antibodies or ligands that bind to the extracellular domain of the CAR (Table 1).

Recently a new assay based on the recombinant fusion (Topanga reagent) of a CAR-target in frame with a small luciferase gene allowed for detection by both bioluminescence and flow cytometry of CARs comprising either immunoglobulin- or non-immunoglobulin-based antigen binding domains (Gopalakrishnan et al., 2019). In addition, basic procedures to detect and assess the killing effect of CAR-T cells have been recently reviewed in Mihara, Yoshida, and Bhattacharyya (2019).

Table 1 CAR expression assessment options by flow cytometry.

	Binding domain	**Advantages**	**Limitations**
Protein L (Zheng, Chinnasamy, & Morgan, 2012)	kappa light chain of immunoglobulin	General and low cost	High background Do not detect the anti-lambda light chain CAR
Anti-Fab antibody	Fab portion of immunoglobulin	General and low cost	High background
Target antigens (De Oliveira et al., 2013; Jena et al., 2013)	Antigen-binding domains of CARs	High specificity Minimal background	High cost Unique to CAR

7. Summary

Concerns about cost, safety and development of resistance are tempering the optimism about T cell cancer therapies. Understanding T cell mechanisms of regulation is an essential pre-requisite to predict how new T cell therapies will function in vivo and how they can be empowered. Efforts are currently ongoing and the use of gene-editing technologies may help to overcome many of these limitations. Envisioning the next generation of cancer immunotherapies is an exciting challenge. There is a need to push the boundaries of traditional medicine, to revolutionize cancer treatment, and develop an a la carte T cell product that can be extended to other diseases.

References

Adusumilli, P. S., Cherkassky, L., Villena-Vargas, J., Colovos, C., Servais, E., Plotkin, J., et al. (2014). Regional delivery of mesothelin-targeted CAR T cell therapy generates potent and long-lasting CD4-dependent tumor immunity. *Science Translational Medicine*, *6*, 261ra151. https://doi.org/10.1126/scitranslmed.3010162.

Aleksic, M., Liddy, N., Molloy, P. E., Pumphrey, N., Vuidepot, A., Chang, K.-M., et al. (2012). Different affinity windows for virus and cancer-specific T-cell receptors: Implications for therapeutic strategies. *European Journal of Immunology*, *42*, 3174–3179. https://doi.org/10.1002/eji.201242606.

Ankri, C., Shamalov, K., Horovitz-Fried, M., Mauer, S., & Cohen, C. J. (2013). Human T cells engineered to express a programmed death 1/28 costimulatory retargeting molecule display enhanced antitumor activity. *Journal of Immunology*, *191*, 4121–4129. https://doi.org/10.4049/jimmunol.1203085.

Bajgain, P., Mucharla, R., Wilson, J., Welch, D., Anurathapan, U., Liang, B., et al. (2014). Optimizing the production of suspension cells using the G-rex "M" series. *Molecular Therapy. Methods & Clinical Development*, *1*, 14015. https://doi.org/10.1038/mtm.2014.15.

Beatty, G. L., Haas, A. R., Maus, M. V., Torigian, D. A., Soulen, M. C., Plesa, G., et al. (2014). Mesothelin-specific chimeric antigen receptor mRNA-engineered T cells induce anti-tumor activity in solid malignancies. *Cancer Immunology Research*, *2*, 112–120. https://doi.org/10.1158/2326-6066.CIR-13-0170.

Bendle, G. M., Linnemann, C., Hooijkaas, A. I., Bies, L., de Witte, M. A., Jorritsma, A., et al. (2010). Lethal graft-versus-host disease in mouse models of T cell receptor gene therapy. *Nature Medicine*, *16*, 565–570. 1p following 570. https://doi.org/10.1038/nm.2128.

Bentzen, A. K., Marquard, A. M., Lyngaa, R., Saini, S. K., Ramskov, S., Donia, M., et al. (2016). Large-scale detection of antigen-specific T cells using peptide-MHC-I multimers labeled with DNA barcodes. *Nature Biotechnology*, *34*, 1037–1045. https://doi.org/10.1038/nbt.3662.

Bentzen, A. K., Such, L., Jensen, K. K., Marquard, A. M., Jessen, L. E., Miller, N. J., et al. (2018). T cell receptor fingerprinting enables in-depth characterization of the interactions governing recognition of peptide-MHC complexes. *Nature Biotechnology*, *36*(12), 1191–1196. https://doi.org/10.1038/nbt.4303.

Bonifant, C. L., Jackson, H. J., Brentjens, R. J., & Curran, K. J. (2016). Toxicity and management in CAR T-cell therapy. *Molecular Therapy Oncolytics*, *3*, 16011. https://doi.org/10.1038/mto.2016.11.

Borbulevych, O. Y., Santhanagopolan, S. M., Hossain, M., & Baker, B. M. (2011). TCRs used in cancer gene therapy cross-react with MART-1/Melan-a tumor antigens via distinct mechanisms. *Journal of Immunology*, *187*, 2453–2463. https://doi.org/10.4049/jimmunol.1101268.

Brudno, J. N., & Kochenderfer, J. N. (2016). Toxicities of chimeric antigen receptor T cells: Recognition and management. *Blood*, *127*, 3321–3330. https://doi.org/10.1182/blood-2016-04-703751.

Cameron, B. J., Gerry, A. B., Dukes, J., Harper, J. V., Kannan, V., Bianchi, F. C., et al. (2013). Identification of a titin-derived HLA-A1-presented peptide as a cross-reactive target for engineered MAGE A3-directed T cells. *Science Translational Medicine*, *5*, 197ra103. https://doi.org/10.1126/scitranslmed.3006034.

Chan, F., Hauswirth, W. W., Wensel, T. G., & Wilson, J. H. (2011). Efficient mutagenesis of the rhodopsin gene in rod photoreceptor neurons in mice. *Nucleic Acids Research*, *39*, 5955–5966. https://doi.org/10.1093/nar/gkr196.

Charlesworth, C. T., Deshpande, P. S., Dever, D. P., Camarena, J., Lemgart, V. T., Cromer, M. K., et al. (2019). Identification of preexisting adaptive immunity to Cas9 proteins in humans. *Nature Medicine*, *25*, 249–254. https://doi.org/10.1038/s41591-018-0326-x.

Chmielewski, M., Kopecky, C., Hombach, A. A., & Abken, H. (2011). IL-12 release by engineered T cells expressing chimeric antigen receptors can effectively muster an antigen-independent macrophage response on tumor cells that have shut down tumor antigen expression. *Cancer Research*, *71*, 5697–5706. https://doi.org/10.1158/0008-5472.CAN-11-0103.

Chou, C. K., Schietinger, A., Liggitt, H. D., Tan, X., Funk, S., Freeman, G. J., et al. (2012). Cell-intrinsic abrogation of TGF-β signaling delays but does not prevent dysfunction of self/tumor-specific CD8 T cells in a murine model of autochthonous prostate cancer. *Journal of Immunology*, *189*, 3936–3946. https://doi.org/10.4049/jimmunol.1201415.

Ciceri, F., Bonini, C., Stanghellini, M. T. L., Bondanza, A., Traversari, C., Salomoni, M., et al. (2009). Infusion of suicide-gene-engineered donor lymphocytes after family haploidentical haemopoietic stem-cell transplantation for leukaemia (the TK007 trial): A non-randomised phase I–II study. *The Lancet Oncology*, *10*, 489–500. https://doi.org/10.1016/S1470-2045(09)70074-9.

Cohen, C. J., Gartner, J. J., Horovitz-Fried, M., Shamalov, K., Trebska-McGowan, K., Bliskovsky, V. V., et al. (2015). Isolation of neoantigen-specific T cells from tumor and peripheral lymphocytes. *The Journal of Clinical Investigation*, *125*, 3981–3991. https://doi.org/10.1172/JCI82416.

Coren, L. V., Jain, S., Trivett, M. T., Ohlen, C., & Ott, D. E. (2015). Production of retroviral constructs for effective transfer and expression of T-cell receptor genes using Golden Gate cloning. *BioTechniques*, *58*, 135–139. https://doi.org/10.2144/000114265.

Cribbs, A. P., Kennedy, A., Gregory, B., & Brennan, F. M. (2013). Simplified production and concentration of lentiviral vectors to achieve high transduction in primary human T cells. *BMC Biotechnology*, *13*, 98. https://doi.org/10.1186/1472-6750-13-98.

Davenport, A. J., Jenkins, M. R., Cross, R. S., Yong, C. S., Prince, H. M., Ritchie, D. S., et al. (2015). CAR-T cells inflict sequential killing of multiple tumor target cells. *Cancer Immunology Research*, *3*, 483–494. https://doi.org/10.1158/2326-6066.CIR-15-0048.

De Oliveira, S. N., Wang, J., Ryan, C., Morrison, S. L., Kohn, D. B., & Hollis, R. P. (2013). A CD19/fc fusion protein for detection of anti-CD19 chimeric antigen receptors. *Journal of Translational Medicine*, *11*, 23. https://doi.org/10.1186/1479-5876-11-23.

De Palma, M., Montini, E., Santoni de Sio, F. R., Benedicenti, F., Gentile, A., Medico, E., et al. (2005). Promoter trapping reveals significant differences in integration site selection between MLV and HIV vectors in primary hematopoietic cells. *Blood*, *105*, 2307–2315. https://doi.org/10.1182/blood-2004-03-0798.

Debets, R., Donnadieu, E., Chouaib, S., & Coukos, G. (2016). TCR-engineered T cells to treat tumors: Seeing but not touching? *Seminars in Immunology*, *28*, 10–21. https://doi.org/10.1016/j.smim.2016.03.002.

Dembić, Z., Haas, W., Weiss, S., McCubrey, J., Kiefer, H., von Boehmer, H., et al. (1986). Transfer of specificity by murine alpha and beta T-cell receptor genes. *Nature*, *320*, 232–238. https://doi.org/10.1038/320232a0.

Derby, M., Alexander-Miller, M., Tse, R., & Berzofsky, J. (2001). High-avidity CTL exploit two complementary mechanisms to provide better protection against viral infection than low-avidity CTL. *Journal of Immunology*, *166*, 1690–1697.

Doyon, Y., Choi, V. M., Xia, D. F., Vo, T. D., Gregory, P. D., & Holmes, M. C. (2010). Transient cold shock enhances zinc-finger nuclease-mediated gene disruption. *Nature Methods*, *7*, 459–460. https://doi.org/10.1038/nmeth.1456.

Dumauthioz, N., Labiano, S., & Romero, P. (2018). Tumor resident memory T cells: New players in immune surveillance and therapy. *Frontiers in Immunology*, *9*, 2076. https://doi.org/10.3389/fimmu.2018.02076.

Dutoit, V., Rubio-Godoy, V., Dietrich, P. Y., Quiqueres, A. L., Schnuriger, V., Rimoldi, D., et al. (2001). Heterogeneous T-cell response to MAGE-A10(254-262): High avidity-specific cytolytic T lymphocytes show superior antitumor activity. *Cancer Research*, *61*, 5850–5856.

Ehrke-Schulz, E., Zhang, W., Gao, J., & Ehrhardt, A. (2017). Recent advances in preclinical developments using adenovirus hybrid vectors. *Human Gene Therapy*, *28*, 833–841. https://doi.org/10.1089/hum.2017.140.

Ellis, J. (2005). Silencing and variegation of gammaretrovirus and lentivirus vectors. *Human Gene Therapy*, *16*, 1241–1246. https://doi.org/10.1089/hum.2005.16.1241.

Eshhar, Z., Waks, T., Gross, G., & Schindler, D. G. (1993). Specific activation and targeting of cytotoxic lymphocytes through chimeric single chains consisting of antibody-binding domains and the gamma or zeta subunits of the immunoglobulin and T-cell receptors. *Proceedings of the National Academy of Sciences of the United States of America*, *90*, 720–724.

Eyquem, J., Mansilla-Soto, J., Giavridis, T., van der Stegen, S. J. C., Hamieh, M., Cunanan, K. M., et al. (2017). Targeting a CAR to the TRAC locus with CRISPR/Cas9 enhances tumour rejection. *Nature*, *543*, 113–117. https://doi.org/10.1038/nature21405.

Feig, C., Jones, J. O., Kraman, M., Wells, R. J. B., Deonarine, A., Chan, D. S., et al. (2013). Targeting CXCL12 from FAP-expressing carcinoma-associated fibroblasts synergizes with anti-PD-L1 immunotherapy in pancreatic cancer. *Proceedings of the National Academy of Sciences of the United States of America*, *110*, 20212–20217. https://doi.org/10.1073/pnas.1320318110.

Feucht, J., Sun, J., Eyquem, J., Ho, Y.-J., Zhao, Z., Leibold, J., et al. (2019). Publisher correction: Calibration of CAR activation potential directs alternative T cell fates and therapeutic potency. *Nature Medicine*, *25*, 530. https://doi.org/10.1038/s41591-019-0360-3.

Fidler, I. J., & Hart, I. R. (1982). Biological diversity in metastatic neoplasms: Origins and implications. *Science*, *217*, 998–1003.

Finney, H. M., Akbar, A. N., & Lawson, A. D. G. (2004). Activation of resting human primary T cells with chimeric receptors: Costimulation from CD28, inducible costimulator, CD134, and CD137 in series with signals from the TCR zeta chain. *Journal of Immunology*, *172*, 104–113.

Finney, H. M., Lawson, A. D., Bebbington, C. R., & Weir, A. N. (1998). Chimeric receptors providing both primary and costimulatory signaling in T cells from a single gene product. *Journal of Immunology*, *161*, 2791–2797.

Gaj, T., Gersbach, C. A., & Barbas, C. F. (2013). ZFN, TALEN, and CRISPR/Cas-based methods for genome engineering. *Trends in Biotechnology*, *31*, 397–405. https://doi.org/10.1016/j.tibtech.2013.04.004.

Gargett, T., Yu, W., Dotti, G., Yvon, E. S., Christo, S. N., Hayball, J. D., et al. (2016). GD2-specific CAR T cells undergo potent activation and deletion following antigen encounter but can be protected from activation-induced cell death by PD-1 blockade. *Molecular Therapy*, *24*, 1135–1149. https://doi.org/10.1038/mt.2016.63.

Garneau, J. E., Dupuis, M.-È., Villion, M., Romero, D. A., Barrangou, R., Boyaval, P., et al. (2010). The CRISPR/Cas bacterial immune system cleaves bacteriophage and plasmid DNA. *Nature*, *468*, 67–71. https://doi.org/10.1038/nature09523.

Geraerts, M., Willems, S., Baekelandt, V., Debyser, Z., & Gijsbers, R. (2006). Comparison of lentiviral vector titration methods. *BMC Biotechnology*, *6*, 34. https://doi.org/10.1186/1472-6750-6-34.

Glanville, J., Huang, H., Nau, A., Hatton, O., Wagar, L. E., Rubelt, F., et al. (2017). Identifying specificity groups in the T cell receptor repertoire. *Nature*, *547*, 94–98. https://doi.org/10.1038/nature22976.

González-Galarza, F. F., Takeshita, L. Y. C., Santos, E. J. M., Kempson, F., Maia, M. H. T., da Silva, A. L. S., et al. (2015). Allele frequency net 2015 update: New features for HLA epitopes, KIR and disease and HLA adverse drug reaction associations. *Nucleic Acids Research*, *43*, D784–D788. https://doi.org/10.1093/nar/gku1166.

Gopalakrishnan, R., Matta, H., Choi, S., Natarajan, V., Prins, R., Gong, S., et al. (2019). A novel luciferase-based assay for the detection of chimeric antigen receptors. *Scientific Reports*, *9*, 1957. https://doi.org/10.1038/s41598-018-38258-z.

Gross, G., Waks, T., & Eshhar, Z. (1989). Expression of immunoglobulin-T-cell receptor chimeric molecules as functional receptors with antibody-type specificity. *Proceedings of the National Academy of Sciences of the United States of America*, *86*, 10024–10028.

Guedan, S., Posey, A. D., Shaw, C., Wing, A., Da, T., Patel, P. R., et al. (2018). Enhancing CAR T cell persistence through ICOS and 4-1BB costimulation. *JCI Insight*, *3* https://doi.org/10.1172/jci.insight.96976.

Hackett, P. B., Largaespada, D. A., & Cooper, L. J. N. (2010). A transposon and transposase system for human application. *Molecular Therapy*, *18*, 674–683. https://doi.org/10.1038/mt.2010.2.

Hamieh, M., Dobrin, A., Cabriolu, A., van der Stegen, S. J. C., Giavridis, T., Mansilla-Soto,- J., et al. (2019). CAR T cell trogocytosis and cooperative killing regulate tumour antigen escape. *Nature*, *568*, 112–116. https://doi.org/10.1038/s41586-019-1054-1.

Hebeisen, M., Oberle, S. G., Presotto, D., Speiser, D. E., Zehn, D., & Rufer, N. (2013). Molecular insights for optimizing T cell receptor specificity against cancer. *Frontiers in Immunology*, *4*, 154. https://doi.org/10.3389/fimmu.2013.00154.

Hege, K. M., Bergsland, E. K., Fisher, G. A., Nemunaitis, J. J., Warren, R. S., McArthur, J. G., et al. (2017). Safety, tumor trafficking and immunogenicity of chimeric antigen receptor (CAR)-T cells specific for TAG-72 in colorectal cancer. *Journal for Immunotherapy of Cancer*, *5*, 22. https://doi.org/10.1186/s40425-017-0222-9.

Helsen, C. W., Hammill, J. A., Lau, V. W. C., Mwawasi, K. A., Afsahi, A., Bezverbnaya, K., et al. (2018). The chimeric TAC receptor co-opts the T cell receptor yielding robust anti-tumor activity without toxicity. *Nature Communications*, *9*, 3049. https://doi.org/10.1038/s41467-018-05395-y.

Hendel, A., Bak, R. O., Clark, J. T., Kennedy, A. B., Ryan, D. E., Roy, S., et al. (2015). Chemically modified guide RNAs enhance CRISPR-Cas genome editing in human primary cells. *Nature Biotechnology*, *33*, 985–989. https://doi.org/10.1038/nbt.3290.

Henriksson, J., Chen, X., Gomes, T., Ullah, U., Meyer, K. B., Miragaia, R., et al. (2019). Genome-wide CRISPR screens in T helper cells reveal pervasive crosstalk between activation and differentiation. *Cell*, *176* 882–896.e18. https://doi.org/10.1016/j.cell.2018.11.044.

Holler, P. D., Chlewicki, L. K., & Kranz, D. M. (2003). TCRs with high affinity for foreign pMHC show self-reactivity. *Nature Immunology*, *4*, 55–62. https://doi.org/10.1038/ni863.

Hughes, M. S., Yu, Y. Y. L., Dudley, M. E., Zheng, Z., Robbins, P. F., Li, Y., et al. (2005). Transfer of a TCR gene derived from a patient with a marked antitumor response conveys highly active T-cell effector functions. *Human Gene Therapy*, *16*, 457–472. https://doi.org/10.1089/hum.2005.16.457.

Jackson, H. J., Rafiq, S., & Brentjens, R. J. (2016). Driving CAR T-cells forward. *Nature Reviews. Clinical Oncology*, *13*, 370–383. https://doi.org/10.1038/nrclinonc.2016.36.

Jena, B., Maiti, S., Huls, H., Singh, H., Lee, D. A., Champlin, R. E., et al. (2013). Chimeric antigen receptor (CAR)-specific monoclonal antibody to detect CD19-specific T cells in clinical trials. *PLoS One*, *8*, e57838 https://doi.org/10.1371/journal.pone.0057838.

Jinek, M., Chylinski, K., Fonfara, I., Hauer, M., Doudna, J. A., & Charpentier, E. (2012). A programmable dual-RNA-guided DNA endonuclease in adaptive bacterial immunity. *Science*, *337*, 816–821. https://doi.org/10.1126/science.1225829.

Johnson, L. A., Morgan, R. A., Dudley, M. E., Cassard, L., Yang, J. C., Hughes, M. S., et al. (2009). Gene therapy with human and mouse T-cell receptors mediates cancer regression and targets normal tissues expressing cognate antigen. *Blood*, *114*, 535–546. https://doi.org/10.1182/blood-2009-03-211714.

Johnson, L. A., Scholler, J., Ohkuri, T., Kosaka, A., Patel, P. R., McGettigan, S. E., et al. (2015). Rational development and characterization of humanized anti-EGFR variant III chimeric antigen receptor T cells for glioblastoma. *Science Translational Medicine*, *7*, 275ra22. https://doi.org/10.1126/scitranslmed.aaa4963.

June, C. H., Blazar, B. R., & Riley, J. L. (2009). Engineering lymphocyte subsets: Tools, trials and tribulations. *Nature Reviews. Immunology*, *9*, 704–716. https://doi.org/10.1038/nri2635.

June, C. H., Riddell, S. R., & Schumacher, T. N. (2015). Adoptive cellular therapy: A race to the finish line. *Science Translational Medicine*, *7*, 280ps7. https://doi.org/10.1126/scitranslmed.aaa3643.

June, C. H., & Sadelain, M. (2018). Chimeric antigen receptor therapy. *The New England Journal of Medicine*, *379*, 64–73. https://doi.org/10.1056/NEJMra1706169.

Kalos, M., Levine, B. L., Porter, D. L., Katz, S., Grupp, S. A., Bagg, A., et al. (2011). T cells with chimeric antigen receptors have potent antitumor effects and can establish memory in patients with advanced leukemia. *Science Translational Medicine*, *3*, 95ra73. https://doi.org/10.1126/scitranslmed.3002842.

Kawalekar, O. U., O'Connor, R. S., Fraietta, J. A., Guo, L., McGettigan, S. E., Posey, A. D., et al. (2016). Distinct signaling of coreceptors regulates specific metabolism pathways and impacts memory development in CAR T cells. *Immunity*, *44*, 380–390. https://doi.org/10.1016/j.immuni.2016.01.021.

Kay, M. A. (2011). State-of-the-art gene-based therapies: The road ahead. *Nature Reviews. Genetics*, *12*, 316–328. https://doi.org/10.1038/nrg2971.

Kloss, C. C., Lee, J., Zhang, A., Chen, F., Melenhorst, J. J., Lacey, S. F., et al. (2018). Dominant-negative TGF-β receptor enhances PSMA-targeted human CAR T cell proliferation and augments prostate cancer eradication. *Molecular Therapy*, *26*, 1855–1866. https://doi.org/10.1016/j.ymthe.2018.05.003.

Knochelmann, H. M., Smith, A. S., Dwyer, C. J., Wyatt, M. M., Mehrotra, S., & Paulos, C. M. (2018). CAR T cells in solid tumors: Blueprints for building effective therapies. *Frontiers in Immunology*, *9*, 1740. https://doi.org/10.3389/fimmu.2018.01740.

Koneru, M., O'Cearbhaill, R., Pendharkar, S., Spriggs, D. R., & Brentjens, R. J. (2015). A phase I clinical trial of adoptive T cell therapy using IL-12 secreting MUC-16(ecto) directed chimeric antigen receptors for recurrent ovarian cancer. *Journal of Translational Medicine*, *13*, 102. https://doi.org/10.1186/s12967-015-0460-x.

Legut, M., Dolton, G., Mian, A. A., Ottmann, O. G., & Sewell, A. K. (2018). CRISPR-mediated TCR replacement generates superior anticancer transgenic T cells. *Blood*, *131*, 311–322. https://doi.org/10.1182/blood-2017-05-787598.

Ligtenberg, M. A., Mougiakakos, D., Mukhopadhyay, M., Witt, K., Lladser, A., Chmielewski, M., et al. (2016). Coexpressed catalase protects chimeric antigen receptor-redirected T cells as well as bystander cells from oxidative stress-induced loss of antitumor activity. *Journal of Immunology*, *196*, 759–766. https://doi.org/10.4049/jimmunol.1401710.

Linette, G. P., Stadtmauer, E. A., Maus, M. V., Rapoport, A. P., Levine, B. L., Emery, L., et al. (2013). Cardiovascular toxicity and titin cross-reactivity of affinity-enhanced T cells in myeloma and melanoma. *Blood*, *122*, 863–871. https://doi.org/10.1182/blood-2013-03-490565.

Liu, X., Zhang, Y., Cheng, C., Cheng, A. W., Zhang, X., Li, N., et al. (2017). CRISPR-Cas9-mediated multiplex gene editing in CAR-T cells. *Cell Research*, *27*, 154–157. https://doi.org/10.1038/cr.2016.142.

Lowe, K. L., Mackall, C. L., Norry, E., Amado, R., Jakobsen, B. K., & Binder, G. (2018). Fludarabine and neurotoxicity in engineered T-cell therapy. *Gene Therapy*, *25*, 176–191. https://doi.org/10.1038/s41434-018-0019-6.

Mahata, B., Zhang, X., Kolodziejczyk, A. A., Proserpio, V., Haim-Vilmovsky, L., Taylor, A. E., et al. (2014). Single-cell RNA sequencing reveals T helper cells synthesizing steroids de novo to contribute to immune homeostasis. *Cell Reports*, 7, 1130–1142. https://doi.org/10.1016/j.celrep.2014.04.011.

Malik, B. T., Byrne, K. T., Vella, J. L., Zhang, P., Shabaneh, T. B., Steinberg, S. M., et al. (2017). Resident memory T cells in the skin mediate durable immunity to melanoma. *Science Immunology*, *2* https://doi.org/10.1126/sciimmunol.aam6346.

Maresca, M., Lin, V. G., Guo, N., & Yang, Y. (2013). Obligate ligation-gated recombination (ObLiGaRe): Custom-designed nuclease-mediated targeted integration through non-homologous end joining. *Genome Research*, *23*, 539–546. https://doi.org/10.1101/gr.145441.112.

Marin, V., Cribioli, E., Philip, B., Tettamanti, S., Pizzitola, I., Biondi, A., et al. (2012). Comparison of different suicide-gene strategies for the safety improvement of genetically manipulated T cells. *Human Gene Therapy Methods*, *23*, 376–386. https://doi.org/10.1089/hgtb.2012.050.

Martinez, M., & Moon, E. K. (2019). CAR T cells for solid tumors: New strategies for finding, infiltrating, and surviving in the tumor microenvironment. *Frontiers in Immunology*, *10*, 128. https://doi.org/10.3389/fimmu.2019.00128.

Mastaglio, S., Genovese, P., Magnani, Z., Ruggiero, E., Landoni, E., Camisa, B., et al. (2017). NY-ESO-1 TCR single edited stem and central memory T cells to treat multiple myeloma without graft-versus-host disease. *Blood*, *130*, 606–618. https://doi.org/10.1182/blood-2016-08-732636.

Maude, S. L., Laetsch, T. W., Buechner, J., Rives, S., Boyer, M., Bittencourt, H., et al. (2018). Tisagenlecleucel in children and young adults with B-cell lymphoblastic leukemia. *The New England Journal of Medicine*, *378*, 439–448. https://doi.org/10.1056/NEJMoa1709866.

Maus, M. V., Fraietta, J. A., Levine, B. L., Kalos, M., Zhao, Y., & June, C. H. (2014). Adoptive immunotherapy for cancer or viruses. *Annual Review of Immunology*, *32*, 189–225. https://doi.org/10.1146/annurev-immunol-032713-120136.

McGarrity, G. J., Hoyah, G., Winemiller, A., Andre, K., Stein, D., Blick, G., et al. (2013). Patient monitoring and follow-up in lentiviral clinical trials. *The Journal of Gene Medicine*, *15*, 78–82. https://doi.org/10.1002/jgm.2691.

Menger, L., Gouble, A., Marzolini, M. A. V., Pachnio, A., Bergerhoff, K., Henry, J. Y., et al. (2015). TALEN-mediated genetic inactivation of the glucocorticoid receptor in cytomegalovirus-specific T cells. *Blood*, *126*, 2781–2789. https://doi.org/10.1182/blood-2015-08-664755.

Menger, L., Sledzinska, A., Bergerhoff, K., Vargas, F. A., Smith, J., Poirot, L., et al. (2016). TALEN-mediated inactivation of PD-1 in tumor-reactive lymphocytes promotes intratumoral T-cell persistence and rejection of established tumors. *Cancer Research, 76*, 2087–2093. https://doi.org/10.1158/0008-5472.CAN-15-3352.

Mihara, K., Yoshida, T., & Bhattacharyya, J. (2019). Basic procedures for detection and cytotoxicity of chimeric antigen receptors. *Methods in Molecular Biology, 1904*, 299–306. https://doi.org/10.1007/978-1-4939-8958-4_13.

Miller, A. M., Bahmanof, M., Zehn, D., Cohen, E. E. W., & Schoenberger, S. P. (2019). Leveraging TCR affinity in adoptive immunotherapy against shared tumor/self-antigens. *Cancer Immunology Research*, 7, 40–49. https://doi.org/10.1158/2326-6066. CIR-18-0371.

Milone, M. C., Fish, J. D., Carpenito, C., Carroll, R. G., Binder, G. K., Teachey, D., et al. (2009). Chimeric receptors containing CD137 signal transduction domains mediate enhanced survival of T cells and increased antileukemic efficacy in vivo. *Molecular Therapy, 17*, 1453–1464. https://doi.org/10.1038/mt.2009.83.

Mitsuyasu, R. T., Anton, P. A., Deeks, S. G., Scadden, D. T., Connick, E., Downs, M. T., et al. (2000). Prolonged survival and tissue trafficking following adoptive transfer of CD4zeta gene-modified autologous CD4(+) and CD8(+) T cells in human immunodeficiency virus-infected subjects. *Blood, 96*, 785–793.

Moldt, B., Yant, S. R., Andersen, P. R., Kay, M. A., & Mikkelsen, J. G. (2007). Cis-acting gene regulatory activities in the terminal regions of sleeping beauty DNA transposon-based vectors. *Human Gene Therapy, 18*, 1193–1204. https://doi.org/10.1089/hum. 2007.099.

Monjezi, R., Miskey, C., Gogishvili, T., Schleef, M., Schmeer, M., Einsele, H., et al. (2017). Enhanced CAR T-cell engineering using non-viral sleeping beauty transposition from minicircle vectors. *Leukemia, 31*, 186–194. https://doi.org/10.1038/leu.2016.180.

Morgan, R. A., Dudley, M. E., Wunderlich, J. R., Hughes, M. S., Yang, J. C., Sherry, R. M., et al. (2006). Cancer regression in patients after transfer of genetically engineered lymphocytes. *Science, 314*, 126–129. https://doi.org/10.1126/science. 1129003.

Morgan, R. A., Dudley, M. E., Yu, Y. Y. L., Zheng, Z., Robbins, P. F., Theoret, M. R., et al. (2003). High efficiency TCR gene transfer into primary human lymphocytes affords avid recognition of melanoma tumor antigen glycoprotein 100 and does not alter the recognition of autologous melanoma antigens. *Journal of Immunology, 171*, 3287–3295.

Morgan, R. A., Yang, J. C., Kitano, M., Dudley, M. E., Laurencot, C. M., & Rosenberg, S. A. (2010). Case report of a serious adverse event following the administration of T cells transduced with a chimeric antigen receptor recognizing ERBB2. *Molecular Therapy, 18*, 843–851. https://doi.org/10.1038/mt.2010.24.

Morris, E. C., & Stauss, H. J. (2016). Optimizing T-cell receptor gene therapy for hematologic malignancies. *Blood, 127*, 3305–3311. https://doi.org/10.1182/blood-2015-11-629071.

Morsut, L., Roybal, K. T., Xiong, X., Gordley, R. M., Coyle, S. M., Thomson, M., et al. (2016). Engineering customized cell sensing and response behaviors using synthetic notch receptors. *Cell, 164*, 780–791. https://doi.org/10.1016/j.cell.2016.01.012.

Nagai, K., Ochi, T., Fujiwara, H., An, J., Shirakata, T., Mineno, J., et al. (2012). Aurora kinase A-specific T-cell receptor gene transfer redirects T lymphocytes to display effective antileukemia reactivity. *Blood, 119*, 368–376. https://doi.org/10.1182/blood-2011-06-360354.

Naldini, L., Blömer, U., Gallay, P., Ory, D., Mulligan, R., Gage, F. H., et al. (1996). In vivo gene delivery and stable transduction of nondividing cells by a lentiviral vector. *Science, 272*, 263–267.

Nauerth, M., Weißbrich, B., Knall, R., Franz, T., Dössinger, G., Bet, J., et al. (2013). TCR-ligand koff rate correlates with the protective capacity of antigen-specific CD8+ T cells for adoptive transfer. *Science Translational Medicine*, *5*, 192ra87. https://doi.org/10.1126/scitranslmed.3005958.

Nayak, S., & Herzog, R. W. (2010). Progress and prospects: Immune responses to viral vectors. *Gene Therapy*, *17*, 295–304. https://doi.org/10.1038/gt.2009.148.

Obenaus, M., Leitão, C., Leisegang, M., Chen, X., Gavvovidis, I., van der Bruggen, P., et al. (2015). Identification of human T-cell receptors with optimal affinity to cancer antigens using antigen-negative humanized mice. *Nature Biotechnology*, *33*, 402–407. https://doi.org/10.1038/nbt.3147.

Parnas, O., Jovanovic, M., Eisenhaure, T. M., Herbst, R. H., Dixit, A., Ye, C. J., et al. (2015). A genome-wide CRISPR screen in primary immune cells to dissect regulatory networks. *Cell*, *162*, 675–686. https://doi.org/10.1016/j.cell.2015.06.059.

Patel, S. J., Sanjana, N. E., Kishton, R. J., Eidizadeh, A., Vodnala, S. K., Cam, M., et al. (2017). Identification of essential genes for cancer immunotherapy. *Nature*, *548*, 537–542. https://doi.org/10.1038/nature23477.

Philip, B., Kokalaki, E., Mekkaoui, L., Thomas, S., Straathof, K., Flutter, B., et al. (2014). A highly compact epitope-based marker/suicide gene for easier and safer T-cell therapy. *Blood*, *124*, 1277–1287. https://doi.org/10.1182/blood-2014-01-545020.

Pilipow, K., Scamardella, E., Puccio, S., Gautam, S., De Paoli, F., Mazza, E. M., et al. (2018). Antioxidant metabolism regulates CD8+ T memory stem cell formation and antitumor immunity. *JCI Insight*, *3*. https://doi.org/10.1172/jci.insight.122299.

Ping, Y., Liu, C., & Zhang, Y. (2018). T-cell receptor-engineered T cells for cancer treatment: Current status and future directions. *Protein & Cell*, *9*, 254–266. https://doi.org/10.1007/s13238-016-0367-1.

Poirot, L., Philip, B., Schiffer-Mannioui, C., Le Clerre, D., Chion-Sotinel, I., Derniame, S., et al. (2015). Multiplex genome-edited T-cell manufacturing platform for "off-the-shelf" adoptive T-cell immunotherapies. *Cancer Research*, *75*, 3853–3864. https://doi.org/10.1158/0008-5472.CAN-14-3321.

Provasi, E., Genovese, P., Lombardo, A., Magnani, Z., Liu, P.-Q., Reik, A., et al. (2012). Editing T cell specificity towards leukemia by zinc finger nucleases and lentiviral gene transfer. *Nature Medicine*, *18*, 807–815. https://doi.org/10.1038/nm.2700.

Pule, M. A., Savoldo, B., Myers, G. D., Rossig, C., Russell, H. V., Dotti, G., et al. (2008). Virus-specific T cells engineered to coexpress tumor-specific receptors: Persistence and antitumor activity in individuals with neuroblastoma. *Nature Medicine*, *14*, 1264–1270. https://doi.org/10.1038/nm.1882.

Pulè, M. A., Straathof, K. C., Dotti, G., Heslop, H. E., Rooney, C. M., & Brenner, M. K. (2005). A chimeric T cell antigen receptor that augments cytokine release and supports clonal expansion of primary human T cells. *Molecular Therapy*, *12*, 933–941. https://doi.org/10.1016/j.ymthe.2005.04.016.

Qasim, W., Zhan, H., Samarasinghe, S., Adams, S., Amrolia, P., Stafford, S., et al. (2017). Molecular remission of infant B-ALL after infusion of universal TALEN gene-edited CAR T cells. *Science Translational Medicine*, *9*(374). https://doi.org/10.1126/scitranslmed.aaj2013.

Rapoport, A. P., Stadtmauer, E. A., Binder-Scholl, G. K., Goloubeva, O., Vogl, D. T., Lacey, S. F., et al. (2015). NY-ESO-1-specific TCR-engineered T cells mediate sustained antigen-specific antitumor effects in myeloma. *Nature Medicine*, *21*, 914–921. https://doi.org/10.1038/nm.3910.

Ren, J., Liu, X., Fang, C., Jiang, S., June, C. H., & Zhao, Y. (2017). Multiplex genome editing to generate universal CAR T cells resistant to PD1 inhibition. *Clinical Cancer Research*, *23*, 2255–2266. https://doi.org/10.1158/1078-0432.CCR-16-1300.

Richman, S. A., Nunez-Cruz, S., Moghimi, B., Li, L. Z., Gershenson, Z. T., Mourelatos, Z., et al. (2018). High-affinity GD2-specific CAR T cells induce fatal encephalitis in a preclinical neuroblastoma model. *Cancer Immunology Research*, *6*, 36–46. https://doi.org/10.1158/2326-6066.CIR-17-0211.

Robbins, P. F., Morgan, R. A., Feldman, S. A., Yang, J. C., Sherry, R. M., Dudley, M. E., et al. (2011). Tumor regression in patients with metastatic synovial cell sarcoma and melanoma using genetically engineered lymphocytes reactive with NY-ESO-1. *Journal of Clinical Oncology*, *29*, 917–924. https://doi.org/10.1200/JCO.2010.32.2537.

Roybal, K. T., Williams, J. Z., Morsut, L., Rupp, L. J., Kolinko, I., Choe, J. H., et al. (2016). Engineering T cells with customized therapeutic response programs using synthetic notch receptors. *Cell*, *167* 419–432.e16. https://doi.org/10.1016/j.cell.2016.09.011.

Rubin, D. B., Danish, H. H., Ali, A. B., Li, K., LaRose, S., Monk, A. D., et al. (2019). Neurological toxicities associated with chimeric antigen receptor T-cell therapy. *Brain*, *142*(5), 1334–1348. https://doi.org/10.1093/brain/awz053.

Sandri, S., Bobisse, S., Moxley, K., Lamolinara, A., De Sanctis, F., Boschi, F., et al. (2016). Feasibility of telomerase-specific adoptive T-cell therapy for B-cell chronic lymphocytic leukemia and solid malignancies. *Cancer Research*, *76*, 2540–2551. https://doi.org/10.1158/0008-5472.CAN-15-2318.

Sandri, S., De Sanctis, F., Lamolinara, A., Boschi, F., Poffe, O., Trovato, R., et al. (2017). Effective control of acute myeloid leukaemia and acute lymphoblastic leukaemia progression by telomerase specific adoptive T-cell therapy. *Oncotarget*, *8*, 86987–87001. https://doi.org/10.18632/oncotarget.18115.

Scheffel, M. J., Scurti, G., Simms, P., Garrett-Mayer, E., Mehrotra, S., Nishimura, M. I., et al. (2016). Efficacy of adoptive T-cell therapy is improved by treatment with the antioxidant N-acetyl cysteine, which limits activation-induced T-cell death. *Cancer Research*, *76*, 6006–6016. https://doi.org/10.1158/0008-5472.CAN-16-0587.

Scheper, W., Kelderman, S., Fanchi, L. F., Linnemann, C., Bendle, G., de Rooij, M. A. J., et al. (2019). Low and variable tumor reactivity of the intratumoral TCR repertoire in human cancers. *Nature Medicine*, *25*, 89–94. https://doi.org/10.1038/s41591-018-0266-5.

Schmid, D. A., Irving, M. B., Posevitz, V., Hebeisen, M., Posevitz-Fejfar, A., Sarria, J.-C. F., et al. (2010). Evidence for a TCR affinity threshold delimiting maximal CD8 T cell function. *Journal of Immunology*, *184*, 4936–4946. https://doi.org/10.4049/jimmunol.1000173.

Scholler, J., Brady, T. L., Binder-Scholl, G., Hwang, W.-T., Plesa, G., Hege, K. M., et al. (2012). Decade-long safety and function of retroviral-modified chimeric antigen receptor T cells. *Science Translational Medicine*, *4*, 132ra53. https://doi.org/10.1126/scitranslmed.3003761.

Schröder, A. R. W., Shinn, P., Chen, H., Berry, C., Ecker, J. R., & Bushman, F. (2002). HIV-1 integration in the human genome favors active genes and local hotspots. *Cell*, *110*, 521–529.

Serafini, M., Manganini, M., Borleri, G., Bonamino, M., Imberti, L., Biondi, A., et al. (2004). Characterization of CD20-transduced T lymphocytes as an alternative suicide gene therapy approach for the treatment of graft-versus-host disease. *Human Gene Therapy*, *15*, 63–76. https://doi.org/10.1089/10430340460732463.

Shayakhmetov, D. M., Di Paolo, N. C., & Mossman, K. L. (2010). Recognition of virus infection and innate host responses to viral gene therapy vectors. *Molecular Therapy*, *18*, 1422–1429. https://doi.org/10.1038/mt.2010.124.

Shifrut, E., Carnevale, J., Tobin, V., Roth, T. L., Woo, J. M., Bui, C. T., et al. (2018). Genome-wide CRISPR screens in primary human T cells reveal key regulators of immune function. *Cell*, *175* 1958–1971.e15. https://doi.org/10.1016/j.cell.2018.10.024.

Singh, N., Liu, X., Hulitt, J., Jiang, S., June, C. H., Grupp, S. A., et al. (2014). Nature of tumor control by permanently and transiently modified GD2 chimeric antigen receptor T cells in xenograft models of neuroblastoma. *Cancer Immunology Research*, *2*, 1059–1070. https://doi.org/10.1158/2326-6066.CIR-14-0051.

Śledzińska, A., Menger, L., Bergerhoff, K., Peggs, K. S., & Quezada, S. A. (2015). Negative immune checkpoints on T lymphocytes and their relevance to cancer immunotherapy. *Molecular Oncology*, *9*, 1936–1965. https://doi.org/10.1016/j.molonc.2015.10.008.

Srivastava, P. K. (2015). Neoepitopes of cancers: Looking back, looking ahead. *Cancer Immunology Research*, *3*, 969–977. https://doi.org/10.1158/2326-6066.CIR-15-0134.

Srivastava, S., Salter, A. I., Liggitt, D., Yechan-Gunja, S., Sarvothama, M., Cooper, K., et al. (2019). Logic-gated ROR1 chimeric antigen receptor expression rescues T cell-mediated toxicity to normal tissues and enables selective tumor targeting. *Cancer Cell*, *35* 489–503.e8. https://doi.org/10.1016/j.ccell.2019.02.003.

Stavrou, M., Philip, B., Traynor-White, C., Davis, C. G., Onuoha, S., Cordoba, S., et al. (2018). A rapamycin-activated caspase 9-based suicide gene. *Molecular Therapy*, *26*, 1266–1276. https://doi.org/10.1016/j.ymthe.2018.03.001.

Stephan, M. T., Ponomarev, V., Brentjens, R. J., Chang, A. H., Dobrenkov, K. V., Heller, G., et al. (2007). T cell-encoded CD80 and 4-1BBL induce auto- and trans-costimulation, resulting in potent tumor rejection. *Nature Medicine*, *13*, 1440–1449. https://doi.org/10.1038/nm1676.

Straathof, K. C. M., Bollard, C. M., Popat, U., Huls, M. H., Lopez, T., Morriss, M. C., et al. (2005). Treatment of nasopharyngeal carcinoma with Epstein-Barr virus-specific T lymphocytes. *Blood*, *105*, 1898–1904. https://doi.org/10.1182/blood-2004-07-2975.

Stromnes, I. M., Blattman, J. N., Tan, X., Jeevanjee, S., Gu, H., & Greenberg, P. D. (2010). Abrogating Cbl-b in effector CD8(+) T cells improves the efficacy of adoptive therapy of leukemia in mice. *The Journal of Clinical Investigation*, *120*, 3722–3734. https://doi.org/10.1172/JCI41991.

Stromnes, I. M., Fowler, C., Casamina, C. C., Georgopolos, C. M., McAfee, M. S., Schmitt, T. M., et al. (2012). Abrogation of SRC homology region 2 domain-containing phosphatase 1 in tumor-specific T cells improves efficacy of adoptive immunotherapy by enhancing the effector function and accumulation of short-lived effector T cells in vivo. *Journal of Immunology*, *189*, 1812–1825. https://doi.org/10.4049/jimmunol.1200552.

Sun, S., Hao, H., Yang, G., Zhang, Y., & Fu, Y. (2018). Immunotherapy with CAR-modified T cells: Toxicities and overcoming strategies. *Journal of Immunology Research*, *2018*, 2386187. https://doi.org/10.1155/2018/2386187.

Tawara, I., Kageyama, S., Miyahara, Y., Fujiwara, H., Nishida, T., Akatsuka, Y., et al. (2017). Safety and persistence of WT1-specific T-cell receptor gene-transduced lymphocytes in patients with AML and MDS. *Blood*, *130*, 1985–1994. https://doi.org/10.1182/blood-2017-06-791202.

Thistlethwaite, F. C., Gilham, D. E., Guest, R. D., Rothwell, D. G., Pillai, M., Burt, D. J., et al. (2017). The clinical efficacy of first-generation carcinoembryonic antigen (CEACAM5)-specific CAR T cells is limited by poor persistence and transient pre-conditioning-dependent respiratory toxicity. *Cancer Immunology, Immunotherapy*, *66*, 1425–1436. https://doi.org/10.1007/s00262-017-2034-7.

Thommen, D. S., & Schumacher, T. N. (2018). T cell dysfunction in cancer. *Cancer Cell*, *33*, 547–562. https://doi.org/10.1016/j.ccell.2018.03.012.

Ting, P. Y., Parker, A. E., Lee, J. S., Trussell, C., Sharif, O., Luna, F., et al. (2018). Guide swap enables genome-scale pooled CRISPR-Cas9 screening in human primary cells. *Nature Methods*, *15*, 941–946. https://doi.org/10.1038/s41592-018-0149-1.

Tsuji, T., Yoneda, A., Matsuzaki, J., Miliotto, A., Ryan, C., Koya, R. C., et al. (2018). Rapid construction of antitumor T-cell receptor vectors from frozen tumors for engineered T-cell therapy. *Cancer Immunology Research*, *6*, 594–604. https://doi.org/10.1158/2326-6066.CIR-17-0434.

Uchida, N., Cone, R. D., Freeman, G. J., Mulligan, R. C., & Cantor, H. (1986). High efficiency gene transfer into murine T cell clones using a retroviral vector. *Journal of Immunology*, *136*, 1876–1879.

van der Stegen, S. J. C., Hamieh, M., & Sadelain, M. (2015). The pharmacology of second-generation chimeric antigen receptors. *Nature Reviews. Drug Discovery*, *14*, 499–509. https://doi.org/10.1038/nrd4597.

van Loenen, M. M., de Boer, R., Amir, A. L., Hagedoorn, R. S., Volbeda, G. L., Willemze, R., et al. (2010). Mixed T cell receptor dimers harbor potentially harmful neoreactivity. *Proceedings of the National Academy of Sciences of the United States of America*, *107*, 10972–10977. https://doi.org/10.1073/pnas.1005802107.

Walisko, O., Schorn, A., Rolfs, F., Devaraj, A., Miskey, C., Izsvák, Z., et al. (2008). Transcriptional activities of the sleeping beauty transposon and shielding its genetic cargo with insulators. *Molecular Therapy*, *16*, 359–369. https://doi.org/10.1038/sj.mt.6300366.

Wang, X., Chang, W.-C., Wong, C. W., Colcher, D., Sherman, M., Ostberg, J. R., et al. (2011). A transgene-encoded cell surface polypeptide for selection, in vivo tracking, and ablation of engineered cells. *Blood*, *118*, 1255–1263. https://doi.org/10.1182/blood-2011-02-337360.

Wang, W., Ma, Y., Li, J., Shi, H.-S., Wang, L.-Q., Guo, F.-C., et al. (2013). Specificity redirection by CAR with human VEGFR-1 affinity endows T lymphocytes with tumor-killing ability and anti-angiogenic potency. *Gene Therapy*, *20*, 970–978. https://doi.org/10.1038/gt.2013.19.

Wherry, E. J. (2011). T cell exhaustion. *Nature Immunology*, *12*, 492–499.

Wu, A. R., Neff, N. F., Kalisky, T., Dalerba, P., Treutlein, B., Rothenberg, M. E., et al. (2014). Quantitative assessment of single-cell RNA-sequencing methods. *Nature Methods*, *11*, 41–46. https://doi.org/10.1038/nmeth.2694.

Xu, Q., Harto, H., Berahovich, R., Xu, S., Zhou, H., Golubovskaya, V., et al. (2019). Generation of CAR-T cells for cancer immunotherapy. *Methods in Molecular Biology*, *1884*, 349–360. https://doi.org/10.1007/978-1-4939-8885-3_24.

Yang, S., Cohen, C. J., Peng, P. D., Zhao, Y., Cassard, L., Yu, Z., et al. (2008). Development of optimal bicistronic lentiviral vectors facilitates high-level TCR gene expression and robust tumor cell recognition. *Gene Therapy*, *15*, 1411–1423. https://doi.org/10.1038/gt.2008.90.

Yuan, W., Chen, J., Cao, Y., Yang, L., Shen, L., Bian, Q., et al. (2018). Comparative analysis and optimization of protocols for producing recombinant lentivirus carrying the anti-Her2 chimeric antigen receptor gene. *The Journal of Gene Medicine*, *20* e3027 https://doi.org/10.1002/jgm.3027.

Zetsche, B., Gootenberg, J. S., Abudayyeh, O. O., Slaymaker, I. M., Makarova, K. S., Essletzbichler, P., et al. (2015). Cpf1 is a single RNA-guided endonuclease of a class 2 CRISPR-Cas system. *Cell*, *163*, 759–771. https://doi.org/10.1016/j.cell.2015.09.038.

Zhang, L., Kerkar, S. P., Yu, Z., Zheng, Z., Yang, S., Restifo, N. P., et al. (2011). Improving adoptive T cell therapy by targeting and controlling IL-12 expression to the tumor environment. *Molecular Therapy*, *19*, 751–759. https://doi.org/10.1038/mt.2010.313.

Zhang, S.-Q., Ma, K.-Y., Schonnesen, A. A., Zhang, M., He, C., Sun, E., et al. (2018). High-throughput determination of the antigen specificities of T cell receptors in single cells. *Nature Biotechnology*, *36*(12), 1156–1159. https://doi.org/10.1038/nbt.4282.

Zhang, S.-Q., Parker, P., Ma, K.-Y., He, C., Shi, Q., Cui, Z., et al. (2016). Direct measurement of T cell receptor affinity and sequence from naïve antiviral T cells. *Science Translational Medicine*, *8*, 341ra77. https://doi.org/10.1126/scitranslmed.aaf1278.

Zheng, Z., Chinnasamy, N., & Morgan, R. A. (2012). Protein L: A novel reagent for the detection of chimeric antigen receptor (CAR) expression by flow cytometry. *Journal of Translational Medicine, 10*, 29. https://doi.org/10.1186/1479-5876-10-29.

Zhong, S., Malecek, K., Johnson, L. A., Yu, Z., Vega-Saenz de Miera, E., Darvishian, F., et al. (2013). T-cell receptor affinity and avidity defines antitumor response and autoimmunity in T-cell immunotherapy. *Proceedings of the National Academy of Sciences of the United States of America, 110*, 6973–6978. https://doi.org/10.1073/pnas.1221609110.

Zhou, P., Shaffer, D. R., Alvarez Arias, D. A., Nakazaki, Y., Pos, W., Torres, A. J., et al. (2014). In vivo discovery of immunotherapy targets in the tumour microenvironment. *Nature, 506*, 52–57. https://doi.org/10.1038/nature12988.

CHAPTER EIGHT

Generating stem-like memory T cells with antioxidants for adoptive cell transfer immunotherapy of cancer

Karolina Pilipow[a,†], Eloise Scamardella[a,†], Enrico Lugli[a,b,*]
[a]Laboratory of Translational Immunology, Humanitas Clinical and Research Center, Rozzano, Milan, Italy
[b]Flow Cytometry Core, Humanitas Clinical and Research Center, Rozzano, Milan, Italy
*Corresponding author: e-mail address: enrico.lugli@humanitasresearch.it

Contents

Abstract

Among the multiple factors that are responsible for the success of adoptive cell transfer (ACT) immunotherapy for cancer, the differentiation status of the in vitro expanded T cell product at the time of transfer seems to play a major role. In particular, less differentiated memory $CD8^+$ T cells endowed with self-renewing capacity and multipotency exert the most potent antitumor activity. To this aim, expansion protocols that generate sufficient numbers of tumor-specific $CD8^+$ T cells with superior capacity to

† Equal contribution.

Methods in Enzymology, Volume 631
ISSN 0076-6879
https://doi.org/10.1016/bs.mie.2019.08.016

persist in vivo following ACT are needed. We describe a procedure for the differentiation of TCF-1$^+$ stem-like CD8$^+$ memory T cells from peripheral blood naïve precursors that takes advantage of the use of antioxidants, in particular *N*-acetylcysteine (NAC), in combination with T cell receptor stimulation and proinflammatory cytokines. We additionally describe how to conduct in vitro assays to test the stem-like features of the generated cells at the phenotypic, functional and metabolic level. Balancing the oxidative metabolism by the addition of antioxidants during in vitro manipulation of CD8$^+$ T cells results in the generation of cell products with potent antitumor characteristics following ACT.

1. Introduction

1.1 Successful adoptive cell transfer (ACT) immunotherapy relies on early differentiated CD8$^+$ T cell subsets

Adoptive cell transfer (ACT) of tumor-reactive T lymphocytes is a potent immunotherapy against solid and hematological malignancies (Rosenberg & Restifo, 2015). Together with immune checkpoint blockade, ACT has demonstrated that the immune system can effectively reject established tumors in humans. Originally, ACT protocols required the isolation of tumor-infiltrating lymphocytes (TILs), their extensive expansion in vitro to huge numbers with the autologous tumor and cytokines, followed by infusion in the patient. More recently, the specificity of autologous or, in some cases, third party T cells from the peripheral blood has been redirected by lentivirally transducing T cell receptors (TCR) or chimeric antigen receptors (CAR) specific for antigens overexpressed by tumor cells (June et al., 2014). ACT of CAR-transduced T cells recognizing CD19 (CART-19), the antigen that is present on some B cell neoplasia in addition to normal B cells, has been approved by the Food and Drug Administration (FDA) for the treatment of relapsed/refractory acute lymphoblastic leukemia and diffuse large B cell lymphoma. Fostered by its success, the CAR-T cell approach is currently being developed or tested for a wide range of hematological and solid tumor targets, including, but not limited to B cell maturation antigen (BCMA) for multiple myeloma, CD20 for relapse or refractory B-cell non-Hodgkin lymphoma, mesothelin (MSLN) for non-small cell lung cancer, mesothelioma and pancreatic ductal adenocarcinoma (which have abundant MSLN expression), and human epidermal growth factor receptor 2 (Her-2) for Her-2-positive breast cancer or gliomas. These targets have been tested at the preclinical level, yielding promising results (Forsberg et al., 2019; Ye, Lou, Lu, & Fan, 2019; Yu et al., 2017; Zhang et al., 2016).

Although potent T cell immunity upon ACT is dependent on multiple mechanisms, persistence of the transferred cells in vivo is one of the major determinants of the success of the treatment (Pilipow et al., 2015). A plethora of data has shown that durability of antitumor function largely depends on the state of differentiation at the time of transfer, where less differentiated memory T cells such as stem cell memory (T_{SCM}) and central memory (T_{CM}) T cells are the most potent in this regard. These subsets are endowed with heightened longevity (Biasco et al., 2015; Graef et al., 2014; Ladell et al., 2008; Lugli, Dominguez, et al., 2013), superior reconstitution capacity (Cieri et al., 2015; Gattinoni et al., 2011; Roberto et al., 2015), which might be linked to their long telomere length and in vivo dynamic proliferative nature (Ahmed et al., 2016), their potential to differentiate into multiple subsets of T cells (Biasco et al., 2015; Cieri et al., 2015; Oliveira et al., 2015; Roberto et al., 2015) and to mediate potent antitumor activity (Gattinoni et al., 2011; Klebanoff et al., 2005) when compared to more differentiated effector memory (T_{EM}) and terminal effector (T_{TE}) T cells (reviewed in Mahnke, Brodie, Sallusto, Roederer, & Lugli, 2013). These properties have been linked to the enhanced expression of a network of transcription factors as well as production of metabolites that couple extensive proliferation with self-renewal capacity. However, the frequency of stem-like T cells is low ex vivo, thus limiting potential translational applications. To this aim, several new strategies have been devised for their generation from the more abundant, uncommitted naïve T (T_N) cells.

1.2 Protocols for the generation of $CD8^+$ T_{SCM} cells

Over the last decade, several protocols have been described to enrich for the presence of stem-like T cells from total peripheral blood mononuclear cells (PBMCs) or T_N precursors (Fig. 1A) (reviewed in Gattinoni, Speiser, Lichterfeld, & Bonini, 2017). These include the use of small molecules such as inhibitors of glycogen synthase kinase-3β (GSK-3β), the most popular being TWS119 (Forget et al., 2012; Gattinoni et al., 2011; Sabatino et al., 2016), of AKT (Crompton et al., 2015; Mousset et al., 2018; van der Waart et al., 2014) or of BET bromodomain (Kagoya et al., 2016) in T cells activated with anti-CD3/CD28 antibody-conjugated beads mimicking T cell receptor (TCR) signaling (Fig. 1A and B). Alternatively, IL-2, a cytokine capable to expand T cells but promoting terminal differentiation and apoptosis, has been replaced by homeostatic cytokines such as IL-7, IL-15 (Cieri et al., 2015; Lugli, Gattinoni, et al., 2013) or IL-21 (Alvarez-Fernandez, Escriba-Garcia, Vidal, Sierra, & Briones, 2016;

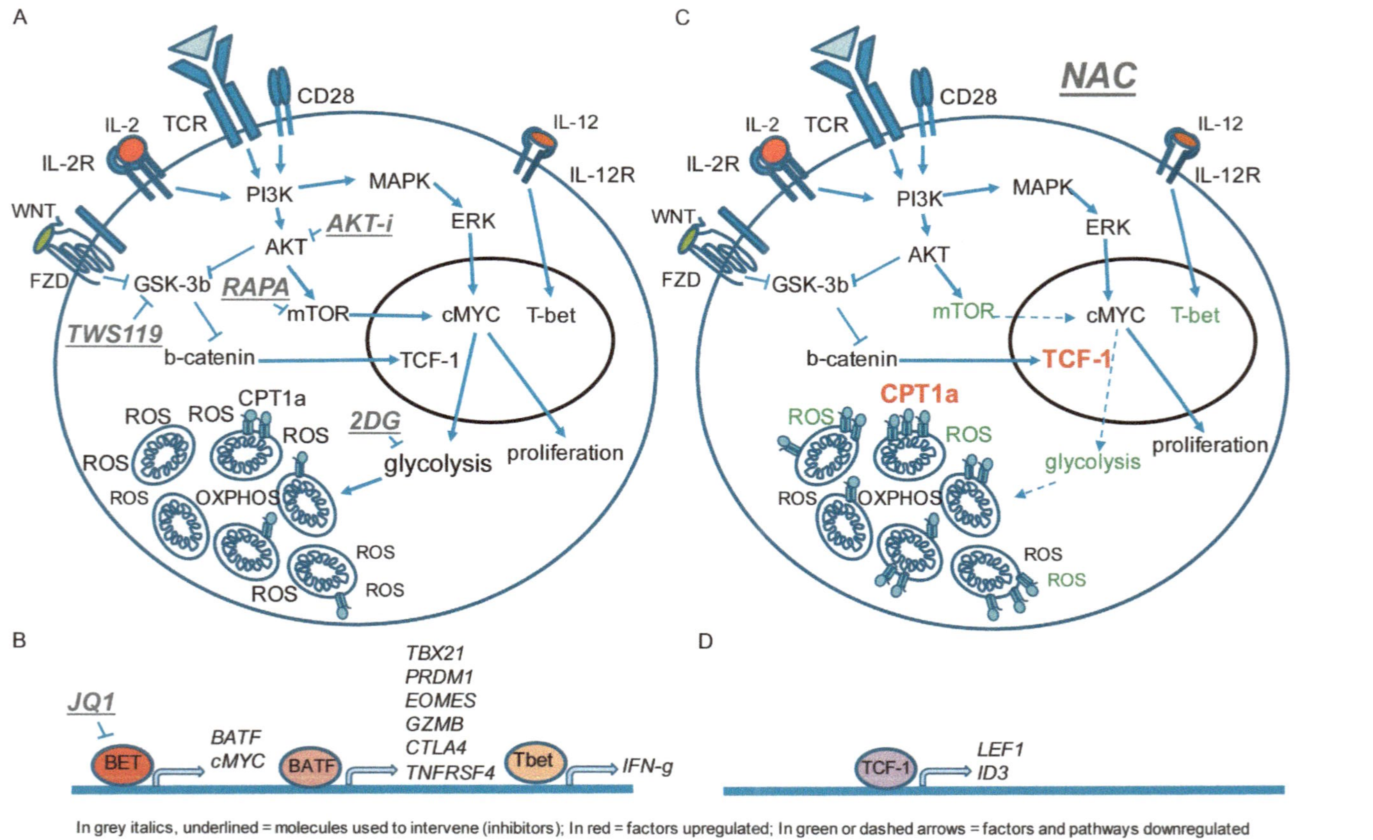

Fig. 1 See legend on opposite page.

Sabatino et al., 2016), which instead favor the formation of long-lived memory T cells. Additional mechanisms were reported to regulate T_{SCM} cell formation such as diminished TCR signaling (Zanon et al., 2017) or stimulation via Notch (Kondo et al., 2017). Finally, modification of the metabolic status has been shown to play a role in this regard (Gattinoni et al., 2017). It has become clear that cellular metabolism is implicated in $CD8^+$ T cell differentiation: while T_N and memory cells mainly rely on fatty acid oxidation (FAO) and oxidative phosphorylation (OXPHOS) to meet their energy demands (Pearce, Poffenberger, Chang, & Jones, 2013), effector T cells such as those activated by cognate antigen preferentially use glycolysis to support their rapid growth, proliferation and effector functions (Fig. 1A) (Chang et al., 2013). Different metabolic modulators have been used in vitro in order to obtain $CD8^+$ T cells with memory-like properties and potent antitumor activity, including inhibition of hexokinase activity by 2-deoxyglucose (2-DG; a glucose molecule that has the 2-hydroxyl group replaced by hydrogen) (Sukumar et al., 2013), by increasing the plasma membrane cholesterol level through blockade of the key cholesterol esterification enzyme ACAT1 (Yang et al., 2016) or by supplementing culture medium with L-arginine which promotes mitochondrial oxidative phosphorylation (Geiger et al., 2016; Sukumar et al., 2013; Yang et al., 2016). A similar phenotype can also be obtained by altering mitochondria fusion and fission, a process collectively known as mitochondrial dynamics.

Fig. 1 Metabolic and transcriptional reprograming during $CD8^+$ T_N cell activation with different protocols compared to *N*-acetylcysteine (NAC) treatment. (A) Following stimulation via the TCR and CD28 in the presence of IL-2 and IL-12, $CD8^+$ T_N cells activate an effector-differentiation program that ensures pathogen/tumor elimination. MAPK and mTOR signaling pathways induce cell proliferation supported by increased metabolic activity including glycolysis and OXPHOS, the latter resulting in augmented ROS production. Exacerbated levels of ROS induce terminal differentiation and apoptosis, thereby limiting the antitumor effect of $CD8^+$ T cells upon ACT. Inhibitors reported to induce the formation of stem-like T cells are indicated in gray (italic underlined). AKTi, inhibitors of AKT; RAPA, rapamycin; 2DG, 2-deoxyglucose. (B) Upregulation of crucial transcription factors following T cell activation such as BET, BATF and T-bet regulate effector differentiation by inducing target genes (selected genes are listed). JQ1, a thienotriazolodiazepine, inhibitor of BET. (C) When $CD8^+$ T_N cells are activated as in A but in the presence of NAC, formation of stem-like memory T cells is favored. NAC exerts global changes at the functional, metabolic and transcriptional level. NAC-treated T cells display reduced levels of ROS as well as of T-bet and reduced mTOR and glycolytic activity (in green), but increased levels of CPT1α and TCF1 expression (in red). Although the exact molecular mechanism of NAC function is poorly understood, induction of TCF1 and its downstream target genes ID3 and LEF1 are likely responsible for the enhanced persistence of stem-like memory T cells in the long term (D).

Specifically, promoting mitochondrial fusion by increased expression of optic atrophy 1 (Opa1), an inner mitochondrial membrane fusion protein, or pharmacological treatment with a "fusion promoter" M1 molecule together with fission inhibitor (Mdivi-1) in activated effector cells boosted effector cytokine production, memory formation as well as enhanced longevity and persistence upon ACT (Buck et al., 2016).

1.3 Reactive oxygen species (ROS) metabolism in CD8$^+$ T cells

ROS refer to a set of metabolites derived from oxygen metabolism that act as second messengers in and between cells. ROS are generated mainly by xanthine oxidase-dependent, respiratory chain and NADPH oxidase- (NOX-) dependent pathways. The majority of ROS in the cell comes from mitochondrial respiration, and the complex I and III of the electron transport system are major sources of ROS, specifically superoxide anion (O_2^-) (Turrens, 2003). O_2^- is converted immediately into hydrogen peroxide (H_2O_2) that is more stable compared to other ROS, can diffuse freely through cell membranes and can be further converted into hydroxyl radicals (Turrens, 2003). Furthermore, it reacts with thiol groups at the physiological concentrations and forms disulfide bonds on the proteins modulating their structures and activity. Overall, increased mitochondrial respiration results in increased ROS levels that have broad consequences on cell function. Importantly, ROS levels that are too high induce oxidative stress that is detrimental for cell survival as it might result in DNA damage and, ultimately, apoptosis (Turrens, 2003). Therefore, ROS levels must be tightly controlled by antioxidant mechanisms, the most abundant of which is represented by intracellular reduced glutathione (GSH), a tripeptide synthesized from cysteine, glutamic acid and glycine in a two-step, ATP-dependent reaction (Ferraresi et al., 2005).

Upon activation, T cell experience a rapid increase in ROS production (Fig. 1A) that is coupled to an endogenous antioxidant response. Increased levels of reduced glutathione (GSH) are required for buffering the harmful ROS levels as well as for optimal activation of mechanistic/mammalian target of Rapamycin (mTOR) (Mak et al., 2017a). In fact, genetic deletion of the catalytic subunit of glutamate cysteine ligase, *Gclc*, involved in GSH biosynthesis, leads to reduced mTOR and NFAT-dependent Myc regulation of glycolysis and glutaminolysis and, as a consequence, to the loss of T cells due to their inability to fuel the metabolic demand required after activation (Mak et al., 2017b). At the same time, deletion of *Uqcrfs1*, a component of mitochondrial complex III, results in ROS abrogation and in the incapability to develop antigen-specific $CD4^+$ T cell responses upon infection (Sena et al., 2013). These two papers confirm that ROS levels must be within a given range to generate optimal T cell responses.

More recent data have shown that ROS levels are differentially associated with effector and memory T cells (Sukumar et al., 2016). In fact, following activation, $CD8^+$ T cells that are memory precursors display reduced mitochondrial membrane potential (Ψ_m) as well as reduced ROS compared to those that are short-lived effectors, therefore Ψ_m can be used as a marker to isolate those cells with improved persistence potential and anti-tumor immunity upon adoptive transfer (Sukumar et al., 2016). We have additionally shown that modulating ROS metabolism during $CD8^+$ T cell activation through the use of antioxidants blocks terminal differentiation of CART-19-redirected T cells in vitro by decreasing mTOR activity and by inhibiting the effector-like transcriptomic program, and promotes the generation of memory T cells with stem-like properties and enhanced anti-leukemic immunity in a preclinical model of acute lymphoblastic leukemia (ALL) (Fig. 1C and D) (Pilipow et al., 2018).

1.4 Modulation of oxidative metabolism by antioxidants as a feasible and inexpensive method to generate $CD8^+$ T_{SCM} cells in vitro

In vitro expansion is required to obtain sufficient numbers of therapeutic $CD8^+$ T cells for ACT. However, high proliferation rates are generally associated with terminal differentiation, one of the major determinants of the failure of the treatment. The simple addition of an inexpensive antioxidant, i.e., *N*-acetylcysteine (NAC), to the manufacturing procedure of CART-19 cells inhibits effector differentiation while allowing for extensive expansion and acquisition of stem-like $CD8^+$ T cell properties (Pilipow et al., 2018). Additional antioxidants such as GSH and apocyanin inhibited terminal differentiation, but only NAC generated T cells with a $CCR7^+CD45RO^-CD45RA^{+/int}$ $CD95^{dim}CXCR3^{bright}$ immunophenotype, resembling naturally occurring T_{SCM} cells (Lugli, Gattinoni, et al., 2013). NAC allows extensive T cell proliferation when used in the range of 10–20 mM, while becoming immunosuppressive at increased concentrations. The $CD8^+$ T_{SCM} cells generated in vitro in the presence of NAC are characterized by transcriptional and metabolic programs reminiscent of self-renewing memory T cells (Fig. 1C and D) (Pilipow et al., 2018).

1.5 Consequences of NAC treatment in T cells

NAC has been used in the clinic for several decades. The most common application is as mucolytic agent. Additional applications include the cure of acute respiratory distress syndrome, bronchitis as well as acetaminophen, heavy metal or chemotherapy-induced toxicity, among others (Kelly, 1998). The mechanisms of its therapeutic and clinical applications are complex and still poorly understood. Based on its chemical properties, NAC reacts slowly with free radicals such as O_2^-, H_2O_2 and peroxynitrite (NO_3^-), thus under physiological conditions direct ROS scavenging by NAC is questionable (Samuni, Goldstein, Dean, & Berk, 2013). The most probable biologically relevant mechanism of action of NAC is the reduction of disulfide bonds in proteins, which alters their structures and functions (Samuni et al., 2013). Moreover, NAC is a precursor of cysteine for GSH synthesis, thereby allowing to maintain high levels of this dominant cellular antioxidant.

When $CD8^+$ T_N cells are stimulated in the presence of NAC, a significant increase in the intracellular GSH and decrease in cellular ROS levels can be found 18h post activation (Pilipow et al., 2018). NAC has been shown to reduce several cellular proteins, whose function can be induced or repressed depending on the specific target. Moreover, protein function can be influenced by direct glutathionylation (Fratelli et al., 2002). Therefore, NAC may regulate cellular behavior in many ways. We have shown that treatment with NAC broadly affects gene expression following TCR activation, specifically it decreases the expression of several effector-related genes including transcription factors (TFs; *IRF4*, *BATF*, *ID2*, *PRDM1*, *TBX21*), effector mediators (*IFNG*, *GZMA*, *GZMB*, *GZMH* and *GZMK*), chemokines (*CCL1*, *CCL3*, *CCL4* and *CCL22*), molecules associated with T cell activation (*IL2RA*, *CD40*, *ICOS*, *HAVCR2* (encoding TIM-3), *LAG3* and *CTLA4*) and surface molecules involved in tissue migration or retention (*CCR3, CCR4, CCR5, ITGAE*) while it increases master regulators of T cell persistence (*TCF7*, *LEF1* and *ID3*) (Pilipow et al., 2018) (Fig. 1C and D). Of note, NAC-treated cells are characterized by decreased activity of mTOR (Fig. 1C), a kinase promoting effector differentiation (Araki et al., 2009). Furthermore, mTOR is a master regulator of multiple metabolic processes, including glycolysis in T cells (Salmond, 2018). According to lower mTOR activity, multiple glycolytic transcripts are decreased in NAC-treated cells, which also show decreased uptake of the fluorescent glucose analog 2-(*N*-(7-Nitrobenz-2-oxa-1,3-diazol-4-yl) Amino)-2-Deoxyglucose (2-NBDG). Moreover, NAC-treated $CD8^+$ T cells are characterized by increased levels of carnitine palmitoyltransferase1a

(*CPT1a*) and lysosomal hydrolase (*LAL*) transcripts, the former encoding a rate-limiting transporter of fatty acids from cytoplasm to mitochondria, the latter a lipase involved in cell-intrinsic lipolysis to mobilize fatty acids for oxidation (FAO) (O'Sullivan et al., 2014; van der Windt et al., 2012). Both CPT1α and LAL are necessary to fuel OXPHOS, in turn required to support the metabolic program necessary for the development of memory $CD8^+$ T cells. Overall, these results indicate that $CD8^+$ T cells expanded ex vivo in the presence of NAC maintain a less differentiated memory phenotype also at the metabolic level.

2. Protocol

2.1 Equipment

1. pH meter
2. Weight scale
3. Water bath (set to 37 °C)
4. Bench top centrifuge
5. CO_2 Incubator (5% CO_2, 37 °C)
6. Disposable, conical sterile centrifuge Falcon tubes (50 mL, 15 mL)
7. Round-bottom tubes (14 mL)
8. Polystyrene round-bottom tubes (5 mL)
9. 1.5 mL Eppendorf vials
10. U-bottom 96-multiwell plate
11. Cell strainer (40 μM)
12. Filter (0.22 μm)
13. Magnet for magnetic beads (Invitrogen)
14. MojoSort Magnet (Biolegend)
15. Ice
16. Multiparameter flow cytometer (FACS) capable to collect at least eight colors
17. Flow cytometry analysis software, such as FlowJo (BD Biosciences)

2.2 Materials

1. Peripheral blood mononuclear cells (PBMCs) from healthy individuals (50 mL of blood) or from a buffy coat isolated by using standard Ficoll gradient separation, either fresh or previously frozen in liquid nitrogen in a solution containing fetal bovine serum (FBS) + 10% DMSO.
2. Phosphate buffered saline without calcium and magnesium (PBS −/−).
3. Hank's balanced salt solution (HBSS −/−).
4. Sodium Hydroxide (NaOH).

5. Trypan Blue solution for cell counting: dilute 0.4% stock solution to 0.04% working solution in PBS −/−.
6. Complete culture medium (R10; RPMI 1640 medium with phenol red supplemented with 10% FBS, 1% Penicillin/Streptomycin, and 2 mM L-Glutamine).
7. Recombinant human cytokines: IL-2 and IL-12 (Peprotech).
8. Human Dynabeads T-ACT CD3/CD28 (Life Technologies).
9. Human $CD8^+$ T_{Naive} isolation kit, MojoSort (BioLegend).
10. Antioxidants: *N*-acetylcysteine (NAC, Sigma-Aldrich), L-Glutathione reduced (GSH, Sigma-Aldrich), L-Ascorbic acid (vitC, Sigma-Aldrich), Apocyanin (Apo, Sigma-Aldrich).
11. Cell separation buffer (Stemcell Technologies).
12. FACS staining buffer: 2% (vol/vol) FBS in HBSS.
13. Staining buffer for phospho-proteins (Bovine Serum Albumine, BSA 1% (vol/vol) in PBS −/−).
14. Formalin (Tousimis): dilute to 1% (vol/vol) working solution in PBS −/−.
15. Paraformaldehyde 4% in PBS −/−.
16. Triton X-100.
17. Methanol 50% in PBS −/−.
18. Fluorescent dead cell discriminator such as ZOMBIE dye (BioLegend).
19. CellTrace™, Carboxyfluorescein succinimidyl ester (CFSE) Cell Proliferation Kit (Invitrogen), 5 mM aliquot (previously resuspended in DMSO and stored at −20 °C).
20. Fluorescently conjugated monoclonal anti-human antibodies: αCCR7 clone: 150503 PE-CF594 BD; αCD3 clone: OKT3 BV650 BioLegend; αCD45RO clone: UCHL1 PerCP-Cy5.5 BioLegend; αCD8 clone: RPA-T8 BV786 BD; αpS6 Ser235/236 clone: D57.2.2E AF647 CellSignaling; αpS6 Ser240/244 clone: D68F8 PE CellSignaling.
21. Fluorescent probes: Monochlorobimane (mBCI), CellROX® Deep Red Reagent (Invitrogen).

2.3 Methods

2.3.1 Thawing frozen cells

Note: the thawing procedure should be carried out as fast as possible to ensure high viability.

1. Remove the cryovial containing frozen PBMCs from the liquid nitrogen storage and immediately place it into a 37 °C water bath.
2. Quickly thaw the cells (<1 min) by gently swirling the vial in the 37 °C water bath until a pea-size piece of ice is left in the vial.
3. Add 1 mL prewarmed R10 dropwise into the cryovial.

4. Gently resuspend the cells in R10, and transfer them into a conical centrifuge tube with the desired amount of prewarmed R10 appropriate for the number of your cells (e.g., 10 mL for up to 10^7 cells, 20 mL for up to 5×10^7 cells).
5. Centrifuge the cell suspension at $400 \times g$ for 5 min.
6. After the centrifugation, aseptically decant the supernatant.
7. Repeat steps 4–6.
8. Gently resuspend the cells in R10.

2.3.2 Human CD8$^+$ naïve T (T_N) cell isolation by magnetic separation

Note: CD8$^+$ T_N cells are enriched by magnetic negative selection. Memory T cells are removed to facilitate T_{SCM} differentiation from T_N precursors. In fact, the presence of terminally differentiated T cell subsets hastens their differentiation into T_{EM} and T_{TE} cells (Klebanoff et al., 2016).

1. Determine cell number by Burker chamber in a Trypan blue solution and pellet cells at $400 \times g$ for 5 min at room temperature (RT).
2. Eliminate supernatant and thoroughly resuspend cells in cell separation buffer (proportion: 100×10^6 cells in 1 mL of buffer).
3. Filter cells in round-bottom tubes using a 40 μM cell strainer in order to eliminate aggregates.
4. Vortex the vial containing the mAbs (MojoSort kit) and resuspend according to the following proportion: 20 μL of mAbs for 100×10^6 cells.
5. Vortex and incubate 15 min in ice.
6. Vortex 30 s the vial containing the magnetic beads (MojoSort kit) and add beads following the same proportion of mAbs indicated in step 4, vortex thoroughly and incubate 15 min in ice. Vortex every 3 min to prevent the beads to precipitate.
7. Top up with cell separation buffer and put the tubes containing the cells in the magnets and incubate 7 min.
8. Pour out and collect the liquid in a 15 mL Falcon tube while maintaining the tube in the magnet. These are your cells of interest.
9. Repeat steps 7 and 8.
10. Recover the supernatant containing the CD8$^+$ T_N cells and put in a 15 mL Falcon tube.

Optional: take a small aliquot of PBMCs before incubating with MojoSort mAbs and after CD8$^+$ T_N enrichment to monitor purity and yield. We recommend to use cells only when purity after isolation is >95%, as determined by flow cytometry analysis of CCR7 and CD45RO. Important: this procedure will not remove naturally occurring CCR7$^+$CD45RO$^-$CD45RA$^+$CD27$^+$CD95$^+$ T_{SCM} cells.

2.3.3 Assessment of cell proliferation by CFSE dilution assay

Note: avoid concentrations of CFSE that are too high because it may inhibit cell proliferation and/or cause toxicity. CFSE works well in the 1–10 μM range, however, optimal concentration should be determined by preliminary titration experiments.

1. Start from human $CD8^+$ T_N cells obtained by enrichment with MojoSort.
2. Pellet $CD8^+$ T_N cells by centrifugation (5 min at 400 × *g*).
3. Wash cells in PBS −/− to remove cell separation buffer.
4. In the meantime, prepare 2 μM CFSE (0.4 μL from 5 mM stock in 1 mL PBS −/−) and warm at 37 °C.
5. Resuspend cells in CFSE solution (use 1 mL for 10^7 cells, minimum 200 μL for as few as 5×10^3 cells).
6. Vortex and make sure that the cells are resuspended.
7. Incubate 8 min at 37 °C in a water bath or incubator.
8. Add cold FBS (1 mL for 200 μL of CFSE solution, 5 mL for 1 mL) to stop the reaction.
9. Vortex and top up with cold R10.
10. Vortex and pellet cells by centrifugation (5 min at 400 × *g*).
11. Count and plate.

2.3.4 Human $CD8^+$ T_N cell differentiation and expansion in the presence of NAC

Note: the expansion is favored when cells are plated in U-bottomed instead of flat-bottomed plates and when cell density is 50,000/200 μL of total volume, presumably due to increased cell-cell contact. In this protocol, human $CD8^+$ T_N cells are stimulated with αCD3/CD28-coated Dynabeads in 96 U-bottomed well plates for 8 days in the presence of human cytokines IL-2 and IL-12 at a final concentration of 10 ng/mL each, as described previously (Zanon et al., 2017).

1. Start from human $CD8^+$ T_N cells obtained by enrichment with MojoSort, previously stained or not with CFSE.
2. Pellet purified $CD8^+$ T_N cells by centrifugation (5 min at 400 × *g*).
3. Resuspend cell pellet in R10 to a concentration of 1×10^6 cells/mL.
4. Plate 50 μL of cells in R10 per well into a U-bottom 96-well plate (50,000 cells/well).
5. Place the plate in 37 °C incubator.
6. Prepare stimuli as described

   Cytokines: R10 supplemented with 40 ng/mL IL-2 and IL-12 (4 × intermediate solution; final concentration in culture: 10 ng/mL).

Dynabeads: human T-ACT CD3/CD28 at a bead:cell ratio of 1:2. Note: 1 μL Dynabeads contain 40,000 beads, capable to stimulate 80,000 cells.

a. Vortex the Dynabeads vigorously and transfer the desired volume of Dynabeads in the 1.5 mL Eppendorf containing 1 mL R10.
b. Place the Eppendorf containing the Dynabeads in the magnet.
c. Incubate for 1 min and discard the supernatant by vacuum aspiration (maintaining the vial on the magnet).
d. Add 1 mL R10 to the Dynabeads, incubate 1 min and eliminate the supernatant by vacuum aspiration.
e. Remove the Eppendorf tube from the magnet and resuspend the Dynabeads in appropriate volume of R10 and IL-2/IL-12 (at the concentration 4 ×).
f. Add 50 μL of the Dynabeads suspension to wells containing cells.

NAC

a. Dissolve NAC powder in distilled H_2O. *Note: check the pH before adjusting to the final volume.*
b. Critical step: adjust pH to 7.4 by adding NaOH.
c. Filter through a 0.22 μm filter before use. *Note: aliquots are stored at −20 °C for up to 1 month.*
d. Prepare an intermediate concentration of NAC at 40 mM in R10.
e. Add 100 μL of NAC to desired wells. The final concentration of NAC in the culture is 20 mM.

7. Place the plate in a 37 °C incubator.
8. Replace culture medium every 3 days.

Note: once dissolved, NAC solution has been reported to be unstable. However, we noted no difference when culture medium was changed every day or every 3 days in terms of proliferation and differentiation evaluated by CFSE dilution and CCR7 and CD45RO expression, respectively. Thus, to minimize manipulation of the cells in culture and to simplify the protocol we recommend to supplement cell culture with fresh NAC and cytokines every 3 days. For this purpose:

- Collect the cells in an appropriate Falcon tube.
- Centrifuge 400 × *g* for 5 min.
- Eliminate the supernatant by vacuum aspiration, resuspend cells in 1 mL of fresh R10 + 10 ng/mL IL-2/12 ± 20 mM NAC and count by Trypan blue solution.
- Adjust the volume of cells in order to plate 50,000 cells in 200 μL/well in a 96-U bottom plate.
- Incubate cells at 37 °C.

2.3.5 Preparation of chemical compounds

1. Dissolve GSH and vitC in distilled H_2O. Dissolve Apo in DMSO.
2. Pass the prepared solution through a 0.22 μm filter before use.
3. Add the compounds to the culture medium at the time of cell seeding.

Note: although we recommend specific concentrations (GSH 4 mM, vitC 100 μM, Apo 0.4 mM), compounds should be titrated before use.

2.3.6 Staining with fluorescent probes

Note: GSH and total cellular ROS are measured by incubating cells with mBCI and CellROX fluorescent probes, respectively, followed by FACS analysis. Both are non-fluorescent, membrane permeant compounds and exhibit bright fluorescence when conjugated to intracellular thiols such as GSH or when oxidized by intracellular ROS. The CellROX probe is ready to use, whereas mBCI should be dissolved in DMSO at a stock concentration of 40 μM. Staining with CellROX can be performed in the presence of R10, whereas staining with mBCI must be performed on cells washed with PBS −/−. In fact, the presence of FBS or other proteins may interfere with the reaction.

1. Collect 0.5×10^6 cells resuspended in R10 and stain with CellROX probe at the final concentration of 1 μM in 100 μL for 30 min at 37 °C in the dark.
2. Wash cells in PBS −/− by centrifugation at RT.

Optional: cells can be additionally stained with Zombie Live/Dead probe or surface markers Abs to exclude dead cells and identify specific leukocyte subsets, respectively.

3. Fix cells with 100 μL 1% formalin for 5 min at RT.
4. Wash cells in PBS −/−. The cells are ready for acquisition on a flow cytometer.
5. For mBCI staining, collect 0.5×10^6 cells and wash with PBS −/− by centrifugation at RT.
6. Incubate cells with mBCI probe at the final concentration of 10 μM in 100 μL PBS −/− for 10 min at 37 °C in the dark.
7. Wash cells in PBS −/− by centrifugation at 4 °C for 5 min at $400 \times g$.

Note: mBCI is not formaldehyde-fixable.

2.3.7 pS6 intracellular staining

Note: phosphorylation of S6 ribosomal protein (pS6) is downstream of mTOR activation and is a reliable surrogate marker of mTOR activity. Differently from other phosphorylated targets, pS6 staining is extremely bright, thus enabling the precise measurement of the cells with active mTOR signaling.

1. Collect 2×10^6 cells cultured in the absence or presence of NAC for 8 days.
2. Wash cells with PBS −/− by centrifugation for 5 min at $400 \times g$.

3. Immediately fix cells with 200 μL of 4% paraformaldehyde for 15 min at RT.
4. Wash cells with 1% BSA in PBS −/− by centrifugation for 5 min at 400 × *g*.
5. Incubate with 200 μL of Triton X-100 for 30 min at RT.
6. Wash twice with 1% BSA in PBS −/− by centrifugation for 5 min at 400 × *g*.
7. Permeabilize for 10 min in ice with ice-cold methanol (50% in PBS −/−).
8. Wash twice with 1% BSA in PBS −/− by centrifugation for 5 min at 400 × *g*.
9. Stain cells with mAbs specific for CD3, CD8, and phosphorylated S6 (pS6) Ser240/244 in 1% BSA PBS −/− buffer for 20 min at RT in the dark.

Note: we have noticed that phosphorylation of S6 at the Ser235/236 site is highly overlapping with Ser240/244, so there is no need to monitor the phosphorylation of the two different sites simultaneously.

10. Wash with 1% BSA in PBS −/− by centrifugation for 5 min at 400 × *g*.
11. The cells are ready for acquisition on a flow cytometer.

3. Analysis

ACT with CART-19 has demonstrated incredible efficacy in blood cancers and is currently being tested for some solid tumors. Still, some patients do not respond, or respond only poorly to the treatment, therefore the development of more potent cell products to be used in ACT is highly desirable. By simple addition of inexpensive antioxidants to the standard manufacturing procedure, we describe the generation of stem-like memory T cells as reported in Pilipow et al. (2018). Herein, we also describe flow cytometric assays to define their surface phenotype, ROS and GSH content and mTOR activity.

We used different antioxidants to modulate the ROS levels in culture, although only NAC is capable to generate T cells with a naturally occurring T_{SCM} phenotype (Pilipow et al., 2018). This is evaluated by combining the measurement of a selection of markers on the cell surface, e.g., CD45RO, CCR7, CD27 and CD95, that are informative of the stage of naïve and memory differentiation (Lugli, Gattinoni, et al., 2013). Several, additional markers can now be investigated simultaneously by 30-parameter FACS, especially TFs that are indicative of the long-term persistence potential of the cells, such as LEF1 and TCF1, in order to obtain a comprehensive characterization of the proliferative, exhausted and functional state (Brummelman et al., 2019; Mazza et al., 2018). The addition of NAC, GSH and Apo, but not vitC to the culture condition blocks the generation of effector memory T cells (T_{EM}) while favoring the accumulation of less differentiated memory T cell phenotypes. Among all antioxidants tested, only NAC preserves a $CCR7^+CD45RO^-$ phenotype while allowing substantial proliferation (Pilipow et al., 2018). This is important since previous reports have demonstrated that NAC can inhibit proliferation (reviewed in Samuni et al., 2013). Indeed, in our hands, NAC can partially or completely inhibit proliferation when used at 40 mM and 80 mM, respectively, while no effect is seen at 20 mM (Pilipow et al., 2018). It is therefore of pivotal importance to correctly titrate NAC concentrations to obtain the desired effect of inhibition of effector differentiation.

To ensure that NAC is working properly, a rapid test can be performed as soon as 18 h post-stimulation. In this regard, the oxidative status of NAC-treated cells can be evaluated by the use of specific fluorescent probes followed by FACS analysis. ROS levels are evaluated after activation by staining with CellROX fluorescent probe while GSH levels are evaluated by staining with mBCI fluorescent probe. NAC treated cells display a significant increase in GSH content and concomitant decrease in total ROS content at 18 h after stimulation with αCD3/CD28 Dynabeads and IL-2/IL-12 cytokine cocktail (Pilipow et al., 2018).

Terminal T cell differentiation is induced by increased activity of mTOR at the expense of long-lived memory T cell development (Araki et al., 2009). mTOR activity influences several intracellular pathways including cell survival, proliferation and metabolism, among others (Howie, Waldmann, & Cobbold, 2014). The activity of mTOR is downmodulated in NAC-treated cells compared to the control condition (Pilipow et al., 2018). The frequency of $CD8^+$ T cells with detectable phosphorylation of S6 protein on mTOR-dependent site Ser240/244 is decreased after 8 days of treatment with NAC. This possibly explains, at least in part, the memory-like phenotype and transcriptional program of $CD8^+$ T cells generated in the presence of NAC (Scholz et al., 2016). Importantly, suppression of the mTOR activity is only transient as removal of NAC permits T cell activation and differentiation of the cells into potent effectors following adoptive transfer (Pilipow et al., 2018).

4. Summary

In conclusion, tightly balanced ROS metabolism influences effector and memory differentiation and benefits adoptive T cell transfer approaches for the treatment of cancer. Current immunotherapy protocols of T cell expansion generally involve TCR stimulation and high dose IL-2. This

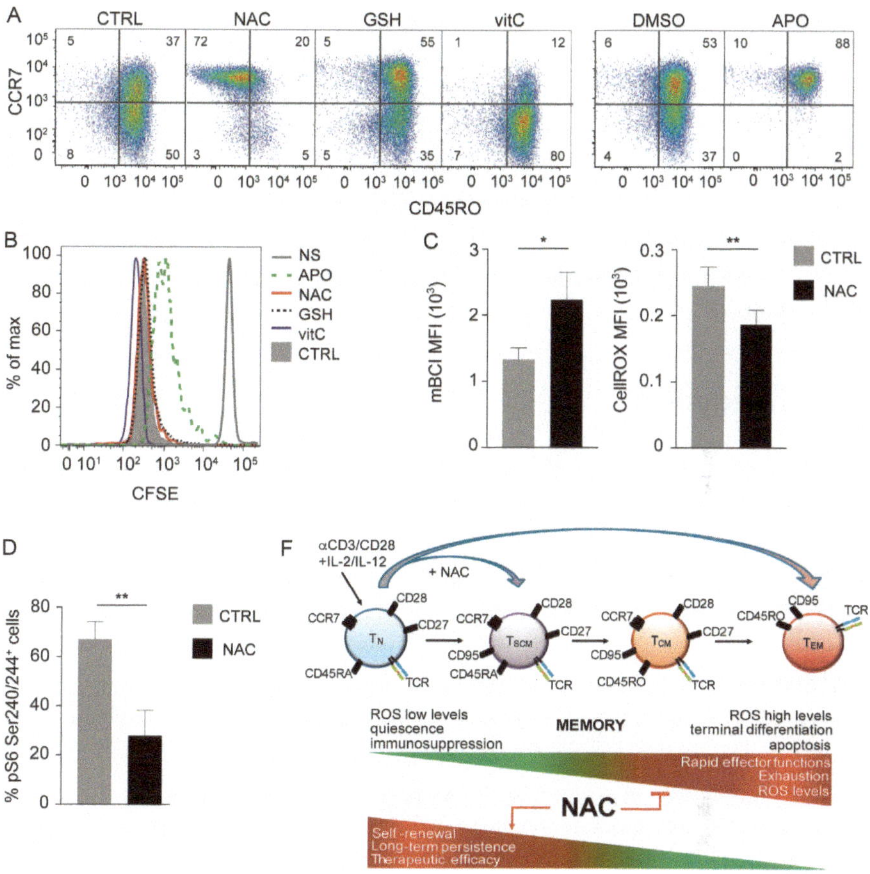

Fig. 2 Model of peripheral T cell differentiation according to ROS levels. Low ROS amounts characterize quiescent T_N cells. Following T cell activation, ROS levels increase and are required for optimal effector T cell differentiation and functions. Excessive ROS amounts accumulate in terminal differentiated effectors resulting in oxidative stress and apoptosis. *N*-acetylcysteine (NAC) balances ROS levels in $CD8^+$ T cells by inhibiting terminal differentiation while inducing T cells with stem-like qualities, even in the presence of strong pro-inflammatory stimuli such as IL-2 and IL-12. These stem-like T cells retain enhanced persistence and therapeutic efficacy against tumors upon adoptive transfer.

results in optimal T cell activation and expansion in vitro, but is associated with terminal differentiation and apoptosis (Fig. 2). Once transferred in vivo, these cells may rapidly lose their effector capacity, become exhausted and persist poorly in the long-term. The addition of antioxidants prevents activation-induced T-cell death (Scheffel et al., 2016), limits terminal differentiation as well as exhaustion (Scheffel et al., 2018) and generates long-lived, stem-like cells with improved persistent capacity, likely mediated by master regulators of T cell self-renewal LEF1 (Gattinoni et al., 2009) and TCF1 (Brummelman et al., 2018; Utzschneider et al., 2016). Recently, $TCF1^+$ T cells have been found in tumors, and are of high interest as potential sources of effective $CD8^+$ T cell responses upon immunotherapy (Brummelman et al., 2018; Kurtulus et al., 2019; Miller et al., 2019; Siddiqui et al., 2019). We anticipate that the addition of NAC to the manufacturing procedure of CART-19 or expansion of $LEF1^+/TCF1^+$ tumor-specific TILs will benefit the persistence of transferred T cells in ACT trials.

Acknowledgments

We wish to thank the members of the Laboratory of Translational Immunology at the Humanitas Clinical and Research Center for critical discussions. This work was supported by grants from the Humanitas Clinical and Research Center to E.L.. K.P. is recipient of the 2018 and 2019 Fondazione Umberto Veronesi postdoctoral fellowship. E.S. is recipient of the Associazione Italiana per la Ricerca sul Cancro (AIRC) 3-year fellowship. There are no conflicts of interest to declare.

References

Ahmed, R., Roger, L., Costa Del Amo, P., Miners, K. L., Jones, R. E., Boelen, L., et al. (2016). Human stem cell-like memory T cells are maintained in a state of dynamic flux. *Cell Reports*, *17*, 2811–2818.

Alvarez-Fernandez, C., Escriba-Garcia, L., Vidal, S., Sierra, J., & Briones, J. (2016). A short CD3/CD28 costimulation combined with IL-21 enhance the generation of human memory stem T cells for adoptive immunotherapy. *Journal of Translational Medicine*, *14*, 214.

Araki, K., Turner, A. P., Shaffer, V. O., Gangappa, S., Keller, S. A., Bachmann, M. F., et al. (2009). mTOR regulates memory CD8 T-cell differentiation. *Nature*, *460*, 108–112.

Biasco, L., Scala, S., Basso Ricci, L., Dionisio, F., Baricordi, C., Calabria, A., et al. (2015). In vivo tracking of T cells in humans unveils decade-long survival and activity of genetically modified T memory stem cells. *Science Translational Medicine*, *7*, 273ra213.

Brummelman, J., Haftmann, C., Nunez, N. G., Alvisi, G., Mazza, E. M. C., Becher, B., et al. (2019). Development, application and computational analysis of high-dimensional fluorescent antibody panels for single-cell flow cytometry. *Nature Protocols*, *14*, 1946–1969.

Brummelman, J., Mazza, E. M. C., Alvisi, G., Colombo, F. S., Grilli, A., Mikulak, J., et al. (2018). High-dimensional single cell analysis identifies stem-like cytotoxic CD8(+) T cells infiltrating human tumors. *Journal of Experimental Medicine*, *215*, 2520–2535.

Buck, M. D., O'Sullivan, D., Klein Geltink, R. I., Curtis, J. D., Chang, C. H., Sanin, D. E., et al. (2016). Mitochondrial dynamics controls T cell fate through metabolic programming. *Cell*, *166*, 63–76.

Chang, C. H., Curtis, J. D., Maggi, L. B., Jr., Faubert, B., Villarino, A. V., O'Sullivan, D., et al. (2013). Posttranscriptional control of T cell effector function by aerobic glycolysis. *Cell*, *153*, 1239–1251.

Cieri, N., Oliveira, G., Greco, R., Forcato, M., Taccioli, C., Cianciotti, B., et al. (2015). Generation of human memory stem T cells after haploidentical T-replete hematopoietic stem cell transplantation. *Blood*, *125*, 2865–2874.

Crompton, J. G., Sukumar, M., Roychoudhuri, R., Clever, D., Gros, A., Eil, R. L., et al. (2015). Akt inhibition enhances expansion of potent tumor-specific lymphocytes with memory cell characteristics. *Cancer Research*, *75*, 296–305.

Ferraresi, R., Troiano, L., Roat, E., Lugli, E., Nemes, E., Nasi, M., et al. (2005). Essential requirement of reduced glutathione (GSH) for the anti-oxidant effect of the flavonoid quercetin. *Free Radical Research*, *39*, 1249–1258.

Forget, M. A., Huon, Y., Reuben, A., Grange, C., Liberman, M., Martin, J., et al. (2012). Stimulation of Wnt/ss-catenin pathway in human CD8 + T lymphocytes from blood and lung tumors leads to a shared young/memory phenotype. *PLoS One*, *7*, e41074.

Forsberg, E. M. V., Lindberg, M. F., Jespersen, H., Alsen, S., Bagge, R. O., Donia, M., et al. (2019). HER2 CAR-T cells eradicate uveal melanoma and T-cell therapy-resistant human melanoma in IL2 transgenic NOD/SCID IL2 receptor knockout mice. *Cancer Research*, *79*, 899–904.

Fratelli, M., Demol, H., Puype, M., Casagrande, S., Eberini, I., Salmona, M., et al. (2002). Identification by redox proteomics of glutathionylated proteins in oxidatively stressed human T lymphocytes. *Proceedings of the National Academy of Sciences of the United States of America*, *99*, 3505–3510.

Gattinoni, L., Lugli, E., Ji, Y., Pos, Z., Paulos, C. M., Quigley, M. F., et al. (2011). A human memory T cell subset with stem cell-like properties. *Nature Medicine*, *17*, 1290–1297.

Gattinoni, L., Speiser, D. E., Lichterfeld, M., & Bonini, C. (2017). T memory stem cells in health and disease. *Nature Medicine*, *23*, 18–27.

Gattinoni, L., Zhong, X. S., Palmer, D. C., Ji, Y., Hinrichs, C. S., Yu, Z., et al. (2009). Wnt signaling arrests effector T cell differentiation and generates CD8 + memory stem cells. *Nature Medicine*, *15*, 808–813.

Geiger, R., Rieckmann, J. C., Wolf, T., Basso, C., Feng, Y., Fuhrer, T., et al. (2016). L-arginine modulates T cell metabolism and enhances survival and anti-tumor activity. *Cell*, *167*(829–842), e813.

Graef, P., Buchholz, V. R., Stemberger, C., Flossdorf, M., Henkel, L., Schiemann, M., et al. (2014). Serial transfer of single-cell-derived immunocompetence reveals stemness of CD8(+) central memory T cells. *Immunity*, *41*, 116–126.

Howie, D., Waldmann, H., & Cobbold, S. (2014). Nutrient sensing via mTOR in T cells maintains a tolerogenic microenvironment. *Frontiers in Immunology*, *5*, 409.

June, C. H., Maus, M. V., Plesa, G., Johnson, L. A., Zhao, Y., Levine, B. L., et al. (2014). Engineered T cells for cancer therapy. *Cancer Immunology, Immunotherapy*, *63*, 969–975.

Kagoya, Y., Nakatsugawa, M., Yamashita, Y., Ochi, T., Guo, T., Anczurowski, M., et al. (2016). BET bromodomain inhibition enhances T cell persistence and function in adoptive immunotherapy models. *Journal of Clinical Investigation*, *126*, 3479–3494.

Kelly, G. S. (1998). Clinical applications of N-acetylcysteine. *Alternative Medicine Review*, *3*, 114–127.

Klebanoff, C. A., Gattinoni, L., Torabi-Parizi, P., Kerstann, K., Cardones, A. R., Finkelstein, S. E., et al. (2005). Central memory self/tumor-reactive CD8 + T cells confer superior antitumor immunity compared with effector memory T cells. *Proceedings of the National Academy of Sciences of the United States of America*, *102*, 9571–9576.

Klebanoff, C. A., Scott, C. D., Leonardi, A. J., Yamamoto, T. N., Cruz, A. C., Ouyang, C., et al. (2016). Memory T cell-driven differentiation of naive cells impairs adoptive immunotherapy. *Journal of Clinical Investigation*, *126*, 318–334.

Kondo, T., Morita, R., Okuzono, Y., Nakatsukasa, H., Sekiya, T., Chikuma, S., et al. (2017). Notch-mediated conversion of activated T cells into stem cell memory-like T cells for adoptive immunotherapy. *Nature Communications*, *8*, 15338.

Kurtulus, S., Madi, A., Escobar, G., Klapholz, M., Nyman, J., Christian, E., et al. (2019). Checkpoint blockade immunotherapy induces dynamic changes in PD-1(−)CD8(+) tumor-infiltrating T cells. *Immunity*, *50*(181–194), e186.

Ladell, K., Hellerstein, M. K., Cesar, D., Busch, R., Boban, D., & McCune, J. M. (2008). Central memory CD8 + T cells appear to have a shorter lifespan and reduced abundance as a function of HIV disease progression. *Journal of Immunology*, *180*, 7907–7918.

Lugli, E., Dominguez, M. H., Gattinoni, L., Chattopadhyay, P. K., Bolton, D. L., Song, K., et al. (2013). Superior T memory stem cell persistence supports long-lived T cell memory. *Journal of Clinical Investigation*, *123*, 594–599.

Lugli, E., Gattinoni, L., Roberto, A., Mavilio, D., Price, D. A., Restifo, N. P., et al. (2013). Identification, isolation and in vitro expansion of human and nonhuman primate T stem cell memory cells. *Nature Protocols*, *8*, 33–42.

Mahnke, Y. D., Brodie, T. M., Sallusto, F., Roederer, M., & Lugli, E. (2013). The who's who of T-cell differentiation: Human memory T-cell subsets. *European Journal of Immunology*, *43*, 2797–2809.

Mak, T. W., Grusdat, M., Duncan, G. S., Dostert, C., Nonnenmacher, Y., Cox, M., et al. (2017a). Glutathione primes T cell metabolism for inflammation. *Immunity*, *46*, 675–689.

Mak, T. W., Grusdat, M., Duncan, G. S., Dostert, C., Nonnenmacher, Y., Cox, M., et al. (2017b). Glutathione primes T cell metabolism for inflammation. *Immunity*, *46*, 1089–1090.

Mazza, E. M. C., Brummelman, J., Alvisi, G., Roberto, A., De Paoli, F., Zanon, V., et al. (2018). Background fluorescence and spreading error are major contributors of variability in high-dimensional flow cytometry. data visualization by t-distributed stochastic neighboring embedding. *Cytometry. Part A*, *93*, 785–792.

Miller, B. C., Sen, D. R., Al Abosy, R., Bi, K., Virkud, Y. V., LaFleur, M. W., et al. (2019). Subsets of exhausted CD8(+) T cells differentially mediate tumor control and respond to checkpoint blockade. *Nature Immunology*, *20*, 326–336.

Mousset, C. M., Hobo, W., Ji, Y., Fredrix, H., De Giorgi, V., Allison, R. D., et al. (2018). Ex vivo AKT-inhibition facilitates generation of polyfunctional stem cell memory-like CD8(+) T cells for adoptive immunotherapy. *Oncoimmunology*, *7*, e1488565.

Oliveira, G., Ruggiero, E., Stanghellini, M. T., Cieri, N., D'Agostino, M., Fronza, R., et al. (2015). Tracking genetically engineered lymphocytes long-term reveals the dynamics of T cell immunological memory. *Science Translational Medicine*, 7, 317ra198.

O'Sullivan, D., van der Windt, G. J., Huang, S. C., Curtis, J. D., Chang, C. H., Buck, M. D., et al. (2014). Memory CD8(+) T cells use cell-intrinsic lipolysis to support the metabolic programming necessary for development. *Immunity*, *41*, 75–88.

Pearce, E. L., Poffenberger, M. C., Chang, C. H., & Jones, R. G. (2013). Fueling immunity: Insights into metabolism and lymphocyte function. *Science*, *342*, 1242454.

Pilipow, K., Roberto, A., Roederer, M., Waldmann, T. A., Mavilio, D., & Lugli, E. (2015). IL15 and T-cell stemness in T-cell-based cancer immunotherapy. *Cancer Research*, *75*, 5187–5193.

Pilipow, K., Scamardella, E., Puccio, S., Gautam, S., De Paoli, F., Mazza, E. M., et al. (2018). Antioxidant metabolism regulates CD8 + T memory stem cell formation and antitumor immunity. *JCI Insight*, *3* e122299.

Roberto, A., Castagna, L., Zanon, V., Bramanti, S., Crocchiolo, R., McLaren, J. E., et al. (2015). Role of naive-derived T memory stem cells in T-cell reconstitution following allogeneic transplantation. *Blood*, *125*, 2855–2864.

Rosenberg, S. A., & Restifo, N. P. (2015). Adoptive cell transfer as personalized immunotherapy for human cancer. *Science*, *348*, 62–68.

Sabatino, M., Hu, J., Sommariva, M., Gautam, S., Fellowes, V., Hocker, J. D., et al. (2016). Generation of clinical-grade CD19-specific CAR-modified CD8+ memory stem cells for the treatment of human B-cell malignancies. *Blood*, *128*, 519–528.

Salmond, R. J. (2018). mTOR regulation of glycolytic metabolism in T cells. *Frontiers in Cell and Development Biology*, *6*, 122.

Samuni, Y., Goldstein, S., Dean, O. M., & Berk, M. (2013). The chemistry and biological activities of N-acetylcysteine. *Biochimica et Biophysica Acta*, *1830*, 4117–4129.

Scheffel, M. J., Scurti, G., Simms, P., Garrett-Mayer, E., Mehrotra, S., Nishimura, M. I., et al. (2016). Efficacy of adoptive T-cell therapy is improved by treatment with the antioxidant N-acetyl cysteine, which limits activation-induced T-cell death. *Cancer Research*, *76*, 6006–6016.

Scheffel, M. J., Scurti, G., Wyatt, M. M., Garrett-Mayer, E., Paulos, C. M., Nishimura, M. I., et al. (2018). N-acetyl cysteine protects anti-melanoma cytotoxic T cells from exhaustion induced by rapid expansion via the downmodulation of Foxo1 in an Akt-dependent manner. *Cancer Immunology, Immunotherapy*, *67*, 691–702.

Scholz, G., Jandus, C., Zhang, L., Grandclement, C., Lopez-Mejia, I. C., Soneson, C., et al. (2016). Modulation of mTOR signalling triggers the formation of stem cell-like memory T cells. *eBioMedicine*, *4*, 50–61.

Sena, L. A., Li, S., Jairaman, A., Prakriya, M., Ezponda, T., Hildeman, D. A., et al. (2013). Mitochondria are required for antigen-specific T cell activation through reactive oxygen species signaling. *Immunity*, *38*, 225–236.

Siddiqui, I., Schaeuble, K., Chennupati, V., Fuertes Marraco, S. A., Calderon-Copete, S., Pais Ferreira, D., et al. (2019). Intratumoral Tcf1(+)PD-1(+)CD8(+) T cells with stem-like properties promote tumor control in response to vaccination and checkpoint blockade immunotherapy. *Immunity*, *50*(195–211), e110.

Sukumar, M., Liu, J., Ji, Y., Subramanian, M., Crompton, J. G., Yu, Z., et al. (2013). Inhibiting glycolytic metabolism enhances CD8+ T cell memory and antitumor function. *Journal of Clinical Investigation*, *123*, 4479–4488.

Sukumar, M., Liu, J., Mehta, G. U., Patel, S. J., Roychoudhuri, R., Crompton, J. G., et al. (2016). Mitochondrial membrane potential identifies cells with enhanced stemness for cellular therapy. *Cell Metabolism*, *23*, 63–76.

Turrens, J. F. (2003). Mitochondrial formation of reactive oxygen species. *Journal of Physiology*, *552*, 335–344.

Utzschneider, D. T., Charmoy, M., Chennupati, V., Pousse, L., Ferreira, D. P., Calderon-Copete, S., et al. (2016). T cell factor 1-expressing memory-like CD8(+) T cells sustain the immune response to chronic viral infections. *Immunity*, *45*, 415–427.

van der Waart, A. B., van de Weem, N. M., Maas, F., Kramer, C. S., Kester, M. G., Falkenburg, J. H., et al. (2014). Inhibition of Akt signaling promotes the generation of superior tumor-reactive T cells for adoptive immunotherapy. *Blood*, *124*, 3490–3500.

van der Windt, G. J., Everts, B., Chang, C. H., Curtis, J. D., Freitas, T. C., Amiel, E., et al. (2012). Mitochondrial respiratory capacity is a critical regulator of CD8+ T cell memory development. *Immunity*, *36*, 68–78.

Yang, W., Bai, Y., Xiong, Y., Zhang, J., Chen, S., Zheng, X., et al. (2016). Potentiating the antitumour response of CD8(+) T cells by modulating cholesterol metabolism. *Nature*, *531*, 651–655.

Ye, L., Lou, Y., Lu, L., & Fan, X. (2019). Mesothelin-targeted second generation CAR-T cells inhibit growth of mesothelin-expressing tumors in vivo. *Experimental and Therapeutic Medicine*, *17*, 739–747.

Yu, S., Li, A., Liu, Q., Li, T., Yuan, X., Han, X., et al. (2017). Chimeric antigen receptor T cells: A novel therapy for solid tumors. *Journal of Hematology & Oncology*, *10*, 78.

Zanon, V., Pilipow, K., Scamardella, E., De Paoli, F., De Simone, G., Price, D. A., et al. (2017). Curtailed T-cell activation curbs effector differentiation and generates CD8 (+) T cells with a naturally-occurring memory stem cell phenotype. *European Journal of Immunology*, *47*, 1468–1476.
Zhang, W. Y., Wang, Y., Guo, Y. L., Dai, H. R., Yang, Q. M., Zhang, Y. J., et al. (2016). Treatment of CD20-directed chimeric antigen receptor-modified T cells in patients with relapsed or refractory B-cell non-Hodgkin lymphoma: An early phase IIa trial report. *Signal Transduction and Targeted Therapy*, *1*, 16002.

CHAPTER NINE

Reverse immunology: From peptide sequence to tumor-killing human T-cell clones

Christophe Vanhaver[a,*], Monica Gordon-Alonso[a,b], Alexandre Bayard[a], Maria Teresa Catanese[c], Didier Colau[a], Pierre van der Bruggen[a,d,*], Annika M. Bruger[a]

[a]De Duve Institute, Université catholique de Louvain, Brussels, Belgium
[b]F-star Biotechnology, Babraham Research Campus, Cambridge, United Kingdom
[c]Nouscom S.R.L., Rome, Italy
[d]WELBIO, Brussels, Belgium
*Corresponding authors: e-mail address: christophe.vanhaver@uclouvain.be; pierre.vanderbruggen@uclouvain.be

Contents

Methods in Enzymology, Volume 631
ISSN 0076-6879
https://doi.org/10.1016/bs.mie.2019.05.033

Abstract

Recent advances in next generation sequencing expanded the availability of tumor mutanome data that list the mutations present in cancer cells. Mutated proteins are an interesting source of neoantigens that can be used to specifically target tumor cells in the context of immunotherapy. However, identifying new antigenic peptides from mutated proteins remains challenging. In this chapter, we present *Reverse Immunology* as an approach to identify potential antigens from any given polypeptide sequence. First, we explain the rationale behind the identification of candidate HLA-binding peptides through mass spectrometry or *in silico* approaches. Then, we describe the isolation of low-frequency T-cell precursors specific for the candidate peptides using peptide-HLA multimers. Finally, we discuss validation steps leading to the identification of a T-cell clone recognizing tumor cells that endogenously process the candidate peptide. We also present approaches to study the impact of the proteasome complex on candidate peptide processing.

Abbreviations

aa	amino acid
CTL	cytotoxic T-lymphocyte
DC	dendritic cell
EBV-B cell	Epstein-Barr virus transformed B cell
FACS	fluorescence-activated cell sorting
FCS	fetal calf serum
HPLC	high-performance liquid chromatography
HS	human serum
IFN-γ	interferon-gamma
MACS	magnetic-activated cell sorting
PBMC	peripheral blood mononuclear cell
PBS	phosphate buffered saline
pHLA	peptide-HLA complex
RBC	red blood cells
RT	room temperature
shRBC	sheep red blood cells
WBC	white blood cells

1. Introduction

Since the 1970s, many efforts were made to identify genes associated with antigenic peptides specifically expressed by tumor cells (Coulie, Van den Eynde, van der Bruggen, & Boon, 2014). The "direct" or "genetic" approach was the first method used to identify tumor-specific antigens recognized by T-cells, particularly $CD8^+$ T-cells. This approach uses tumor-specific cytolytic T-lymphocytes (CTL) isolated from peripheral blood mononuclear cells (PBMC) of cancer patients after several rounds of stimulation with autologous tumor cells. The specific DNA sequence encoding the tumor antigen is further identified by screening tumor-reactive CTL against recipient cells (such as COS cells) that have been transfected with cDNA libraries constructed from the genetic material isolated from tumors (De Plaen et al., 1997). Functional assays such as cytokine secretion upon activation are used as read-outs during this screening step. This direct approach enabled the identification of several tumor antigens recognized by $CD8^+$ T-cells, and some $CD4^+$ T-cells (Topalian et al., 1996; van der Bruggen et al., 1991). A database referencing the different tumor antigens identified is available at https://caped.icp.ucl.ac.be/.

However, the direct approach requires the isolation of T-cells from a cancer patient and the availability of an autologous tumor cell line. Therefore, the method is impractical to identify large numbers of antigenic peptides presented by different HLA class I and class II molecules. Moreover, the approach is rarely suitable for the identification of new antigenic peptides from tumor mutanome data.

In this chapter, we describe an approach known as *Reverse Immunology* (Fig. 1) (Boon & van der Bruggen, 1996; Viatte, Alves, & Romero, 2006).

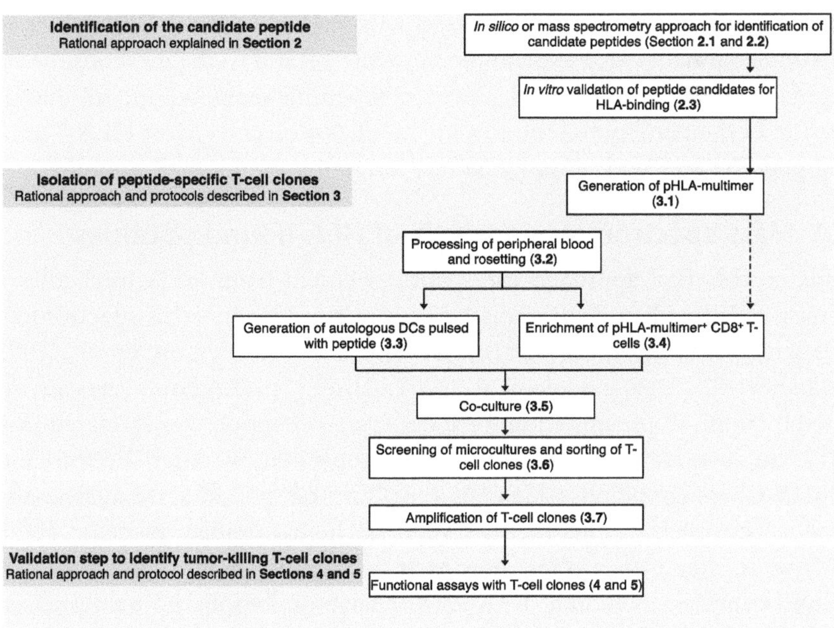

Fig. 1 Overview of the identification process for T-cell clones targeting tumor antigens.

This aims to use gene sequences to identify antigenic peptides presented by tumor cells that are recognized by T-cells. The approach was used to successfully discover the antigenic peptides encoded by genes of the MAGE family (van der Bruggen et al., 2002). We currently use *Reverse Immunology* to identify tumor antigenic peptides encoded by genes mutated in cancer cells. First, we discuss the different approaches used to identify candidate peptides binding to HLA molecules (Section 2). These peptides can be selected from any polypeptide sequence including mutated genes expressed by tumor cells. Next, we provide an overview of the procedure to generate fluorescent peptide-HLA (pHLA)-multimers with the candidate peptides (Section 3.1). Then, we outline the isolation of T-cell clones from donor blood (Sections 3.2–3.7). Finally, we describe the different validation steps suggested to verify that the isolated T-cell clones recognize the candidate peptide, and that the peptide is indeed endogenously processed and presented by tumor cells (Sections 4 and 5).

2. Identification of peptides binding to HLA molecules

The starting point of this approach is the identification of a candidate peptide from a polypeptide sequence. This sequence could be, e.g., a protein containing a point mutation or a frameshift mutation resulting in a modification of the C-terminus of a protein (Mlecnik et al., 2016; Williams et al., 2010). Sections 2.1–2.3 explain the rationale behind (i) the methods used to select candidate peptides from a given polypeptide sequence and (ii) the validation of their ability to bind to a given HLA molecule. Most HLA-I binding peptides are 8–11 amino acid (aa) long.

2.1 Mass spectrometry analysis of HLA-bound peptides

This biochemical approach uses peptides eluted from HLA molecules of tumor cells to identify potential antigenic peptides by mass spectrometry (Bassani-Sternberg et al., 2016; Cox et al., 1994; Hunt et al., 1992; Kalaora et al., 2016; Pascolo et al., 2001). Briefly, pHLA complexes are isolated by immunoprecipitation from cell lysates. Peptides are extracted from HLA molecules by acidic elution and separated by High-Performance Liquid Chromatography (HPLC). Finally, these peptides are analyzed by mass spectrometry. This method requires high numbers of cells (10^8 to 10^9). It is, however, a direct approach that allows identification of HLA-bound peptides, including the ones that have undergone post-translational modifications (Dalet et al., 2011; Skipper et al., 1996).

Although most laboratories use one of the *in silico* approaches detailed next, the analysis of the peptidome by mass spectrometry is an undeniably useful tool to directly confirm that a specific target peptide is indeed present on the tumor surface.

2.2 *In silico* approach

2.2.1 HLA-binding prediction

Several algorithms predicting peptide binding to HLA molecules have been developed. These algorithms screen polypeptide sequences to find short peptides that are likely to bind to a given HLA molecule. The two most commonly used methods are (i) position-specific scoring matrix and (ii) machine learning. While a third approach that links chemical structure and biological function is emerging, the speed and accuracy of its prediction method heavily relies on the availability of pHLA crystal structure data (Jojic, Reyes-gomez, Heckerman, Kadie, & Schueler-furman, 2006; Zhang et al., 2010). This structure-based approach will not be described further.

2.2.1.1 Position-specific scoring matrix

This method uses matrices that assign an HLA-binding score to each peptide residue's position. These matrices are generated using databases that compile sequences of known eluted peptides from HLA molecules, and count the frequency of each amino acid for each position of the peptide. Some of these algorithms are available online, such as the epitope prediction tool SYFPEITHI on http://www.syfpeithi.de (Bachmann, Alexander, & Stefan, 1999). Another tool called BIMAS used HLA-peptide binding data to generate a matrix that gives a score based on the contribution of each residue's position to the overall binding of the peptide to HLA (Parker, Bednarek, & Coligan, 1994). However, this tool is no longer available online.

These algorithms screen the polypeptide sequence of interest and calculate a binding score for all nonamers (peptides of 9 aa). More than just summarizing the score from each residue's position, these algorithms also consider the anchor positions, which are weighted differently in the matrix. However, while these methods score the independent contribution of each amino acid to the overall binding, they disregard the overall peptide structure. Furthermore, these position-specific scoring matrices are usually designed to identify peptides binding to frequently studied HLA alleles (such as HLA-A2) and therefore are less efficient at predicting peptide-binding to rare HLA alleles. Additionally, they were developed at a time when limited structural data were available, and now tend to be replaced by faster and more accurate machine-learning methods (Nielsen et al., 2003; Paul et al., 2013).

2.2.1.2 Machine learning-methods

These approaches use large datasets and artificial intelligence to predict the likelihood of peptide-HLA binding. There are several methods based on machine learning, such as Hidden Markov Model, Decision Tree and Artificial Neural Network. Describing them all in detail is not the goal of this chapter. The most referenced tools available online are *NetMHC 4.0* and *NetMHCpan* (available on http://www.cbs.dtu.dk/services/), which use position-specific scoring matrices generated from eluted peptide databases and HLA-allele sequences. NetMHC is a method based on an artificial neural network that is HLA-allele specific, and therefore can predict peptide binding only to HLA alleles that it was trained for. NetMHCpan is also based on artificial neural networks but, unlike NetMHC, it can predict peptide binding to classes of HLA-molecules for which little or no empirical data are available. The efficiency of these two tools was compared by Karosiene and colleagues who showed that using them in combination resulted in the best predictions (Karosiene, Lundegaard, Lund, & Nielsen, 2012). Therefore, a third software combining NetMHC and NetMHCpan algorithms, called NetMHCcons, was created (http://www.cbs.dtu.dk/services/-NetMHCcons). Several other tools using different machine-learning methods are also available online and are reviewed elsewhere (Luo et al., 2015; Tong, Tan, & Ranganathan, 2006).

2.2.2 Peptide-cleavage prediction

Most antigenic peptides are cleaved from a protein/polypeptide by a multi-catalytic protease called the proteasome. Thus, it is important to consider proteasome cleavage sites when identifying HLA-binding peptides, since peptides with internal proteasome cleavage sites will be degraded rather than incorporated in pHLA complexes. NetChop 3.0 tool (http://www.cbs.dtu.dk/services/NetChop/) uses artificial neural network methods to predict proteasome cleavage sites in a given polypeptide sequence. NetChop can be used with two prediction methods: the first one, NetChop 20S is trained on data obtained from *in vitro* proteolysis of peptide precursors, while the second NetChop Cterm uses the HLA-I ligand dataset. It has been shown that *in vivo* proteolysis was best predicted using NetChop Cterm methods (Calis, Reinink, Keller, Kloetzel, & Kesmir, 2015). However, the reliability of these prediction tools is far from established. Ideally, candidate peptides should be screened by *in vitro* proteolysis with purified proteasome complexes (Vigneron & Abi, 2015).

2.3 Validation of HLA binding

The goal of this step is to validate whether *in silico* predicted peptides bind to HLA *in vitro*. A classical approach is to perform a peptide-HLA binding assay with either cell-bound or solubilized HLA molecules.

2.3.1 Cell-based validation

To confirm that candidate peptides indeed bind to HLA-A2, the T2 cell line is often used. This cell line is an Epstein-Barr Virus-transformed lymphoblastoid (EBV-B) cell line deficient for the *transporter associated with antigen processing* (TAP) protein, which is required for endogenous antigen processing (Kelly et al., 1992; Li, Wang, Chen, Wu, & Chen, 2005). As a result, T2 cells mainly express unstable empty HLA class I molecules on their surface. T2 cells are incubated with β2-microglobulin and different concentrations of the peptide of interest. The peptide binds to the empty HLA molecules, stabilizing the pHLA-β2-microglobulin complex (Ljunggren et al., 1990; Pogue, Eron, Frelinger, & Matsui, 1995). Fluorescent antibodies specific for these stable pHLA-β2-microglobulin complexes, such as W6/32, are subsequently used to quantify peptide binding to a given HLA class I molecule. Other methods are not restricted to HLA-A2. It is possible to naturally elute processed peptides from the surface of cells by cooling them for 5 min on ice and adding cold citric acid (Kessler et al., 2003). The peptide-stripped cells are then incubated with different concentrations of the peptide of interest and a fluorescent reference peptide. The competition for binding to HLA between the peptide of interest and the reference peptide allows comparing the peptides' affinity for a given HLA molecule.

2.3.2 Cell-free validation

This approach relies on the ability of a peptide to associate *in vitro* with purified, unfolded class I HLA heavy chain and β2-microglobulin (Elvin, Potter, Elliott, Cerundolo, & Townsend, 1993). Similarly to the cell-based competition assay, the cell-free method consists of monitoring the ability of the selected peptide to compete with a fluorescent peptide to form pHLA complexes (Tan, Geluk, Toebes, Ottenhoff, & Drijfhout, 1997). In 2005, Buchli et al. proposed to use fluorescence polarization to quantify pHLA association (Buchli et al., 2005). Indeed small molecules, such as free fluorescent peptides, rotate faster than larger ones (e.g., fluorescent peptides bound to soluble HLA molecules). As a decrease in rotation rate results in an increase in light polarization, peptide binding to HLA can be measured. This approach is amenable to high-throughput screening.

3. Isolation of peptide-specific T-cell clones

We have used several strategies in the past to isolate T-cell clones specific for a candidate antigenic peptide (van der Bruggen et al., 2002). We always start with T-cells isolated from non-cancerous donor's blood. The advantage of such an approach is the relative ease of obtaining large volumes of blood. A major challenge is the low precursor frequency of $CD8^+$ T-cells recognizing a specific peptide. This is expected to be in the range of 1×10^{-7} to 3×10^{-7} of the $CD8^+$ population, corresponding approximately to one specific T-cell per 30–100 mL of peripheral blood (Chaux, Vantomme, Coulie, Boon, & van der Bruggen, 1998). Here, we first label T-cells from PBMC with an HLA-multimer folded with the candidate peptide (pHLA-multimer), and enrich the T-cell population for pHLA-multimer$^+$ cells by magnetic sorting (Fig. 2). This enrichment step is followed by an *in vitro*

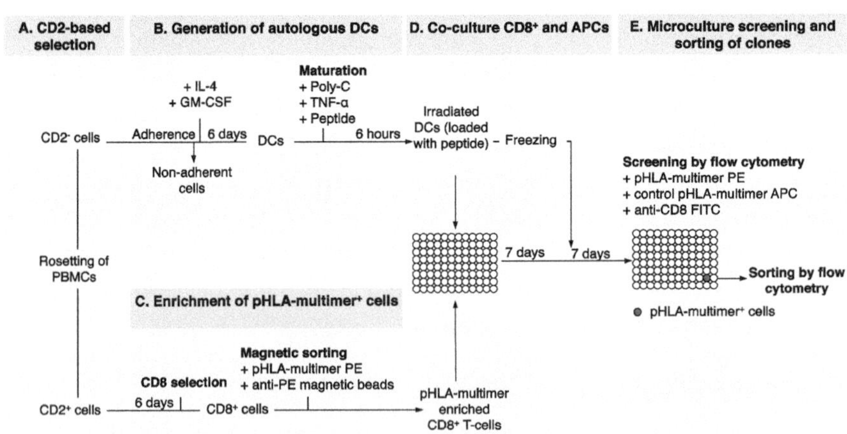

Fig. 2 Overview of the procedure to isolate antigen-specific T-cell clones from PBMCs. Peripheral blood mononuclear cells (PBMCs) are isolated from donor blood by density gradient separation. $CD2^+$ are separated from $CD2^-$ cells by rosetting with sheep red blood cells (shRBC). HPLC-purified candidate peptides are used to produce peptide-HLA (pHLA)-multimers. pHLA-multimer-enriched $CD8^+$ cells and autologous dendritic cells (DCs) loaded with peptides are co-cultured at a ratio of 1:1. Approximately 1/3 of each microculture (around 10,000 cells) is used to screen for pHLA-multimer$^+$ cells. pHLA-multimers produced with irrelevant peptides are used as staining controls.

amplification step by co-culturing T-cells in microwells with autologous mature dendritic cells (DC) loaded with the candidate peptide. After two to three rounds of stimulation, we screen the microwells and sort the pHLA-multimer^{+} cells by flow cytometry. This approach enables the isolation of multiple peptide-specific human CD8^{+} T-cells clones from one blood sample (Ottaviani, Zhang, Boon, & Van Der Bruggen, 2005; van der Bruggen et al., 2002).

3.1 Production of pHLA-multimers

Fluorescent pHLA-multimers are used to stain and sort T-cells recognizing a candidate peptide. We use complexes made in-house by multimerizing soluble biotinylated pHLA complexes with fluorophore-coupled streptavidin (Fig. 3). The fluorophores typically used are phycoerythrin (PE) or allophycocyanin (APC). We provide a concise protocol in this section and refer you to the NIH Multimer Core Facility website (http://multimer.yerkes.emory.edu/) for further details.

a. Soluble extracellular domains of HLA class I heavy chain and β2-microglobulin are produced as recombinant proteins by *Escherichia coli* (Fig. 3A). A biotinylation sequence is inserted at the end of the α3 extracellular domain of the HLA class I heavy chain. After induction of their expression, recombinant proteins are produced as insoluble inclusion bodies. The bacteria are lysed and the inclusion bodies are enriched by successive centrifugation and soft detergent washes. Inclusion bodies are solubilized with urea or guanidine chloride.

b. The pHLA complex is refolded in the presence of high concentrations of synthetic peptide of great purity (>99%) (Fig. 3B). We purified our peptide by HPLC, as peptide purity is crucial to avoid generating fluorescent pHLA-multimers that unspecifically bind to T-cells. Refolding is performed in the presence of reduced and oxidized glutathione to form disulfide bonds on the heavy chains and β2-microglobulin. The pHLA complexes are then purified by size-exclusion chromatography.

c. pHLA complexes are biotinylated by incubation with biotin-protein ligase BirA, biotin and ATP (Fig. 3C). Biotinylated pHLA complexes are then purified by gel filtration chromatography to remove any residual biotin that could interfere with the multimerization step.

d. Soluble biotinylated pHLA complexes are multimerized by incubation with streptavidin-fluorophore-conjugates (Fig. 3D). We use either PE or APC as fluorophores. Our colleagues who produce the

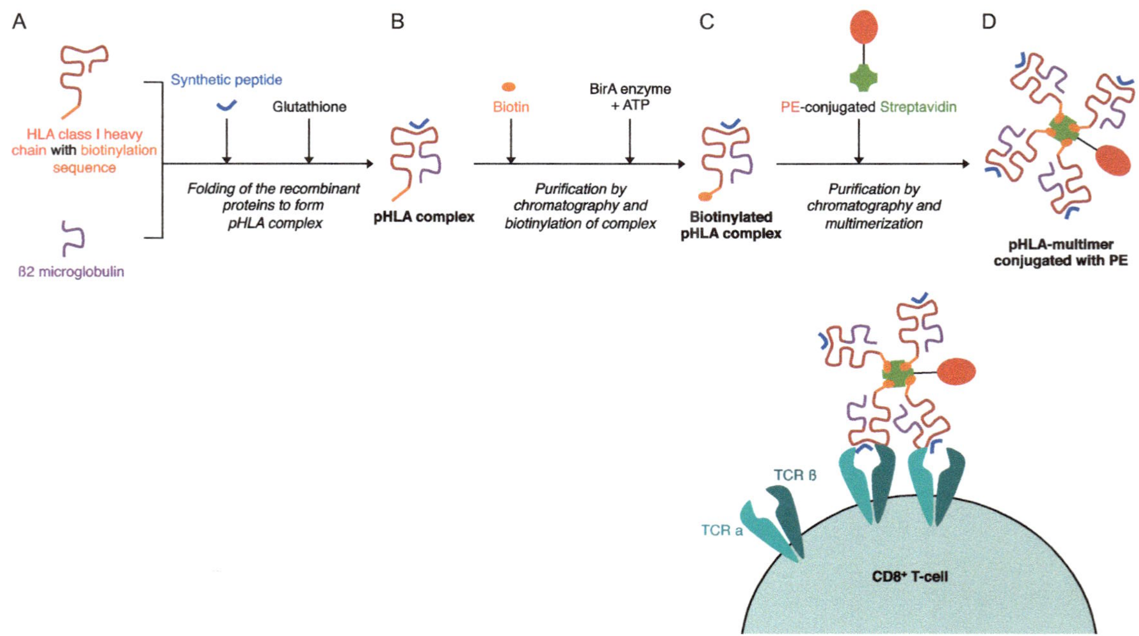

Fig. 3 Scheme for the production of peptide-HLA-multimers. Fluorescent peptide-HLA (pHLA)-multimers are generated in four main steps: (A) Production and purification of HLA class-I recombinant proteins; (B) folding of the recombinant HLA heavy chain and β2 microglobulin with a synthetic peptide; (C) biotinylation of the resulting pHLA complexes; (D) multimerization of the biotinylated pHLA complexes.

pHLA-multimer first incubate the biotinylated pHLA-complexes with a limiting amount of streptavidin. Next, the concentration of unbound pHLA-complexes is measured by ELISA using Streptawell 96-well plates (Roche) and anti-β2-microglobulin antibodies. Finally, an adequate amount of streptavidin is added. The multivalent streptavidin binds to and multimerizes the biotinylated pHLA. The multimerization buffer is PBS-20 mM Tris-50 mM NaCl (pH 8). Fluorescent pHLA-multimers are stored overnight at 4 °C. The complexes are either used over the following days or stored at −20 °C for later use. Concentrations are expressed in moles of pHLA molecules.

3.2 Processing of blood samples and CD2-based cell sorting

Due to the low frequency of $CD8^+$ T-cells specific for a given peptide (3×10^{-7}), we recommend processing large volumes of PBMCs per donor. Specifically, our laboratory frequently obtains 500 mL of peripheral blood from hemochromatosis patients who have to be bled regularly for medical reasons.

3.2.1 Mononuclear cell isolation from blood

a. The following protocol assumes reception of a 500 mL blood bag.

b. Concentrate the white blood cells (WBC) by spinning the blood bag at 3720 × *g* for 20 min at room temperature (RT). Serum and red blood cells (RBC) will be separated from the WBC layer. Collect the serum in sterile conditions and keep it for de-complementation (step c). Collect the WBC in a new sterile blood bag.

c. De-complement the serum by heating it to 56 °C for 20 min. Transfer it into 50-mL Falcon tubes and centrifuge for 10 min at 1830 × *g*. Filter the supernatant through a 70-μm filter and store it at 4 °C (for 1 day) or −20 °C (for longer storage).

d. Dilute the WBC fraction at least twice in cold sterile PBS 1 mM EDTA. Prepare a 50-mL Falcon tube containing 15 mL of Ficoll (Sigma) or Lymphoprep™ (Alere Technologies) solution for density gradient separation (one Falcon tube per 30 mL of diluted WBC) at RT. Lean the Falcon containing the Lymphoprep™ solution and very slowly layer the diluted WBC fraction on top. Do not mix diluted WBC and Lymphoprep™ solution.

e. Centrifuge the cells at 920 × *g* for 20 min without acceleration and brakes to avoid disrupting the density gradient established during centrifugation.

f. Collect, with a pipette, the PBMCs found at the interface below plasma and above Lymphoprep and RBC pellet (Fig. 4).

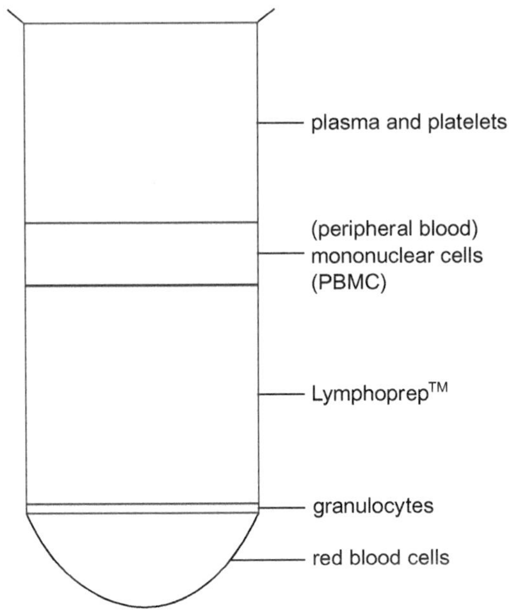

Fig. 4 Density gradient separation of blood. Whole blood is diluted with PBS 1 mM EDTA, and 30 mL are carefully layered onto 15 mL Lymphoprep™ solution in a 50-mL Falcon tube. Density separation is achieved by centrifugation at $920 \times g$ for 20 min without acceleration or brakes. Shown is a sketch of how the blood components separate.

g. Transfer the PBMCs to a 50-mL Falcon tube. The tube is pre-coated with fetal calf serum (FCS) to prevent monocytes from adhering to the plastic.

h. Wash the PBMCs with cold sterile PBS 1 mM EDTA and centrifuge for 10 min at $400 \times g$. This first wash is performed at higher speed to ensure that all cells pellet in case some Lymphoprep™ remains.

i. Wash the PBMCs with cold PBS 1 mM EDTA a further three times. EDTA is a Ca^{+} chelating agent that prevents platelet aggregation. Count the PBMCs and make sure that there are no platelets remaining as they may interfere with the differentiation of monocytes into DCs. Resuspend the cells at 10^7 cells/mL in warm plain RPMI.

3.2.2 Sorting of CD2⁺ and CD2⁻ cells by sheep red blood cell rosetting

Sheep red blood cell (shRBC) rosetting is used to separate the blood mononuclear cells into $CD2^+$ and $CD2^-$ fractions. This is an effective technique that, unlike FACS or MACS, keeps the T-cell surface free of antibodies, allowing further sorting steps at later stages. CD2 is an adhesion molecule present on T-cells and NK cells only. CD2 binds spontaneously to the LFA3-homolog on shRBC (Ocklind, 1988). The binding event triggers the aggregation of NK and T-cells around the shRBC. Applying a density gradient the $CD2^+$ cells pellet together with the shRBC, while the monocytes and B-cells settle in the mononuclear cell layer (Fig. 5). Alternatively, a Percoll gradient could be used (Gmelig-myeling & Waldmann, 1980).

3.2.2.1 Preparation of shRBC treated with AET

a. Use reagents equilibrated to RT. We use shRBC in a 50% Alsever's suspension provided by Labor Dr. Merck (E-400). 2-Aminoethylisothiouronium bromide (AET) is provided by Sigma (A54601). Pre-treatment with AET removes negative charges on shRBC and avoids their mutual repulsion.

b. Check hematocrit (% erythrocytes) of shRBC using the provider's website: https://labormerk.com/manufacturing/blood_products/

c. Referring to the hematocrit value, prepare the AET solution by dissolving **X** mg of AET in **Y** mL of milli-Q water as detailed by Table 1. Add 1 M NaOH and ensure to bring the solution to pH 9 exactly.

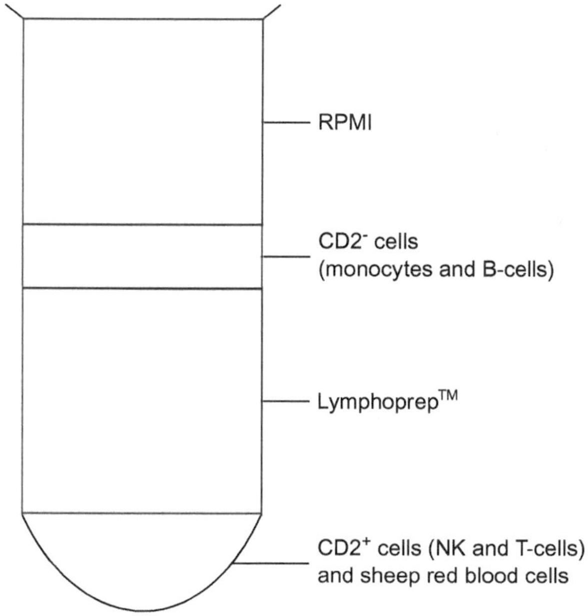

Fig. 5 Density gradient separation of $CD2^+$ and $CD2^-$ cells. Sheep red blood cells (shRBC) express an LFA3-homolog that binds to human CD2, causing spontaneous aggregation of human NK and T-cells (Ocklind, 1988). The sketch indicates how $CD2^+$ and $CD2^-$ cells separate by density gradient centrifugation.

d. Filter the AET suspension with a 0.22 μm Steriflip (Millipore, SCGP00525).

e. Resuspend shRBC by shaking and/or pipetting, and collect 12.5 mL of the shRBC suspension. Wash the suspension by adding 37.5 mL PBS and centrifuge at 910 × *g* for 8 min.

f. Aspirate supernatant. Leave some PBS (1 or 2 mL) to avoid aspirating any shRBC. Add the AET solution and resuspend the shRBC. Incubate for 15 min at 37 °C.

g. Pellet the shRBC suspension by centrifugation at 910 × *g* for 8 min. Remove AET supernatant. Resuspend in 50 mL PBS. The shRBC pellet should be of dense consistency and require repeated gentle agitation for complete resuspension.

h. Wash the cells three times with 50 mL PBS and centrifuge at 910 × *g* for 8 min. The pellet's consistency should become less dense with each washing step. Perform an additional washing step if the supernatant is quite red which indicates RBC lysis.

i. Resuspend the cells in **Z** mL of RPMI (Table 1). Store the cells at 4 °C for up to 15 days. Test the shRBC on donor blood before use, and control the efficiency of the rosetting by flow cytometry.

Table 1 Reference table for the resuspension of shRBC in AET and RPMI 1640 according to the hematocrit content.

% hematocrit	X: A.E.T. (mg)	Y: H_20 volume for AET (mL)	Z: RPMI final volume (mL)
10	625	13.5	15.6
11	688	14.7	17.2
12	750	16.2	18.7
13	813	17.3	20.3
14	875	18.9	21.9
15	938	20.0	23.4
16	1000	21.6	25.0
17	1063	22.7	26.6
18	1125	24.3	28.0
19	1188	25.7	29.7
20	1250	27.0	31.2

3.2.2.2 Rosetting

a. Resuspend the PBMC in warm RPMI at 10^7 cells/mL or in at least 10 mL.

b. Add 1 mL of shRBC pre-treated with AET per 10^8 PBMCs. Incubate 5–10 min at RT. During this time, prepare 50-mL Falcon tube(s) with 15 mL Lymphoprep™ (one tube per 30 mL of cell suspension).

c. Carefully transfer the PBMCs mixed with shRBC onto the Lymphoprep™ layer (as described in Section 3.2.1). Centrifuge at 920 × *g* for 20 min with brake and acceleration set to minimum. $CD2^+$ cells containing T-cells will pellet with the shRBC, while $CD2^-$ cells containing monocytes will remain in the interphase layer between the Lymphoprep™ and the medium (Fig. 5).

d. Collect the $CD2^-$ fraction in a tube pre-coated with FCS and wash the cells twice at 400 × *g* for 5 min. Resuspend the cells in IMDM 10% FCS and plate them in a culture flask at 2×10^6 cells/mL/cm^2. Incubate the flask for 1 h at 37 °C, 8% CO_2 to let the monocytes adhere to the flask. The monocytes will be used to generate autologous mature DC (see Section 3.3).

e. Aspirate the rest of the solution after $CD2^-$ removal.

f. To harvest the $CD2^+$ fraction, lyse shRBC first. Prepare 15 mL of lysis buffer with 12 mL of 160 mM NH_4Cl at pH 7.5, 1.5 mL of FCS and 1.5 mL of RPMI. Gently resuspend the pellet in the lysis buffer for 20 s and quickly centrifuge at 770 × *g* for 1 min. Immediately remove the supernatant and resuspend the cells in fresh RPMI. Wash the cells at 300 × *g* for 5 min.

g. $CD2^+$ cells will be cultured in IMDM complemented with 2% autologous serum and 2 U/mL IL-2 at 37 °C and 8% CO_2 for the 7 days required for the DC differentiation (see Section 3.3). We advise to plate the $CD2^+$ cells at a density of 2×10^6 cells/mL/cm^2 in culture flasks.

3.3 Generation of DCs

The $CD2^-$ fraction obtained after rosetting is used to generate mature DCs, which will be loaded with the candidate peptide and used to stimulate $CD8^+$ T-cells (Fig. 2B). It is essential that autologous cells are used to present the peptide to T-cells; otherwise irrelevant T-cells that recognize allogeneic determinants will expand during the co-culture.

3.3.1 Differentiation and maturation of DCs

a. All IMDM media are supplemented with AAG (0.24 mM L-asparagine, 0.55 mM L-arginine, 1.5 mM L-glutamine) and antibiotics (100 U/mL penicillin, 100 µg/mL streptomycin).

b. Use adherent monocytes (Section 3.2.2.2, step d) to generate DC. Remove non-adherent cells and medium.

c. Add 20 mL of IMDM 10% FCS with 70 ng/mL GM-CSF and 200 U/mL IL-4 per flask. Incubate the cells at 37 °C, 8% CO_2 for 2 days.

d. On day 2, change the medium to replace the consumed IL-4. As some of the cells may have become non-adherent at day 2, collect half of the medium in 50 mL Falcon tube. Centrifuge it at 400 × *g* for 5 min and discard the supernatant. Resuspend the pellet (if present) in a volume of IMDM 10% FCS with 70 ng/mL GM-CSF and 400 U/mL IL-4.

e. Replace medium (as described in c) on day 4.

f. On day 6, collect non-adherent cells in FCS-coated 50-mL Falcon tubes. Pellet cells by centrifugation for 5 min at 400 × *g* and resuspend them at 10^6 cells/mL in IMDM 10% FCS with 70 ng/mL GM-CSF and 200 U/mL IL-4.

g. We used to mature DCs with bacterial extract (Ribomunyl, Pierre Fabre) and 500 U/mL interferon-gamma (IFN-γ). However, as ribomunyl is not available anymore, we use 50 µg/mL PolyIC and 10 ng/mL TNFα as a maturation cocktail instead (Spisek, Bretaudeau, Barbieux, Meflah, & Gregoire, 2001). DCs are matured in the presence of 1 µg/mL of the candidate peptide in 24-well low-adherence tissue culture plates (Costar 3473) at 10^6 cells/mL for 6 h. We reduce the maturation time compare to the classical protocol (48 h) as IL-12 secretion by DCs is very transient (Foster, Weiss, Kaka, Leen, & Rooney, 2008; Ratzinger et al., 2000). Therefore, we cannot check the phenotype of the DC as they still undergo maturation after 6 h.

h. Irradiate cells with 5000 rads (50 Gray).

i. Collect the cells and centrifuge for 5 min at 400 × *g* to wash any unbound peptide and cytokines.

j. Co-culture DCs with T-cells as described in Section 3.5 and freeze the remaining DCs in several vials for further stimulations.

3.3.2 Freezing of DCs

a. Prepare freezing serum, which is composed of FCS 20% DMSO, and store it at 4 °C. Freezing serum has to be used cold.

b. Count DCs and centrifuge cells at 400 × *g* for 7 min.
c. Prepare freezing vials and label them first.
d. Resuspend cells in cold IMDM without serum at 2×10^6 cells/mL.
e. Dilute the cell suspension 1:1 with freezing serum for a final concentration of 10^6 cells/mL.
f. Quickly transfer 1 mL to each cryovial.
g. Transfer the vials to −80 °C. We advise keeping the freezing process as temperature-controlled as possible. To achieve that, we recommend Mr. Frosty™ Freezing Container (Thermo Scientific™ 5100-0001).

3.3.3 Thawing of DC

a. The thawing medium is IMDM complemented with 10% FCS and 5 U/mL DNase I. The medium is supplemented with AAG and antibiotics (see Section 3.3.1 step a).
b. Prepare a Falcon tube with 15 mL of thawing medium warmed at 37 °C before thawing.
c. Gently stir the frozen vial in a water bath heated to 37 °C until almost all the sample has melted.
d. Gently transfer the cells (1 mL) to an empty 15-mL Falcon tube previously coated with FCS. Use 1 mL of warm thawing medium to wash the cells remaining in the vial and transfer them to the Falcon tube.
e. Drop by drop, add 8 mL of warm thawing medium on the cells in the Falcon tube to dilute the DMSO.
f. Centrifuge the cells at 400 × *g* for 5 min.
g. Resuspend them at 2×10^6 cells/mL with thawing medium with 70 ng/mL GM-CSF and 200 U/mL IL-4. Plate the cells in 24-well low-adherence plate (Costar 3473). Incubate them for at least 4 h at 37 °C, 8% CO_2 before use.

3.4 Isolation of $CD8^+$ cells and pHLA-multimer enrichment

$CD8^+$ T-cells are isolated from the $CD2^+$ fraction (Section 3.2.1) by negative magnetic bead selection. Peptide-specific T-cells are subsequently enriched from the $CD8^+$ population by magnetic sorting of cells stained with fluorescent pHLA multimers (Fig. 2C). It is possible to incubate T-cells with up to three different pHLA multimers simultaneously. To maximize the enrichment, the magnetic sorting can be repeated.

3.4.1 Sorting of $CD8^+$ cells from the $CD2^+$ fraction

a. Count $CD2^+$ cells (Section 3.2.2). Centrifuge them at $400 \times g$ for 5 min. Resuspend the cells at 2×10^8 cells/mL in cold PBS with 1% human serum (HS).

b. $CD8^+$ T-cells are isolated by negative selection using the $CD8^+$ T-cell Isolation Kit (Miltenyi, 130-096-495).

c. Use the provided biotin-antibody cocktail that labels all PBMCs except $CD8^+$ T-cells. Add 10 μL of biotin-antibody cocktail per 10^7 cells, mix gently and incubate at 4 °C for 5 min.

d. Add 30 μL of cold PBS 1% HS and 20 μL of the provided anti-biotin "Microbead cocktail" per 10^7 cells and mix gently.

e. Incubate this mix for 10 min at 4 °C. Shake every 5 min to enhance labeling.

f. Dilute to a minimal volume of 600 μL with PBS with 1% HS if required. Pass the mix through a 40-μm filter.

g. We use an AutoMACS™ (Miltenyi) sorter with the deplet05 program. Keep all buffers and samples at 4 °C.

h. Recover the $CD8^+$ cells in the negative fraction and count them. Expect around 30% of the $CD2^+$ cells to be $CD8^+$.

3.4.2 pHLA-multimer enrichment by magnetic sorting

a. Count $CD8^+$ cells and collect them by centrifugation at $400 \times g$ for 7 min.

b. Resuspend cells at 2.5×10^7 cells/mL in PBS 1% HS at RT.

c. Stain the cells with 20 nM of pHLA-multimer conjugated with PE and incubate for 20 min at RT in the dark.

d. Wash the cells with PBS 1% HS and centrifuge at $400 \times g$ for 7 min.

e. For the multimer enrichment, we use the anti-PE Microbead kit (Miltenyi, 130-105-639). Resuspend the cells at 2.5×10^7 cells/80 μL in cold PBS 1% HS. Add 20 μL of anti-PE beads per 80 μL of cell suspension and mix gently.

f. Incubate for 15 min at 4 °C in the dark.

g. Pass the mix through a 40-μm filter.

h. We use AutoMACS with the possel_s program to sort the cells.

i. Recover the positive fraction, which is enriched for pHLA-multimer$^+$ cells. Keep the negative fraction for further flow cytometry staining controls.

3.5 Amplification of peptide-specific $CD8^+$ cells by co-culture

As only few T-cells recognizing the peptide of interest are present in the pHLA-multimer enriched $CD8^+$ T-cell fraction, it is important to first amplify the peptide-specific T-cells with autologous DCs. This step is performed in multiple microwells, which are screened with fluorescent pHLA-multimers after several rounds of specific stimulation with the candidate peptide (Fig. 2D).

a. All IMDM media are supplemented with AAG and antibiotics (see Section 3.3.1 step a).

b. Centrifuge the pHLA-multimer enriched $CD8^+$ T-cells (Section 3.4) and resuspend them in IMDM 10% HS with 12 ng/mL IL-12, 1000 U/mL IL-6, 50 U/mL IL-2 and 10 ng/mL IL-7.

c. Plate the cells in 96-well round-bottomed, low-adherence tissue-culture plates (Costar, 7007). Plate cells into as many wells as possible with a concentration comprised between 2500 and 10,000 cells/well.

d. Plate the autologous DCs loaded with the peptide (Section 3.3) in the wells containing the pHLA-multimer enriched T-cells. We usually use a $CD8^+$:DC ratio of 1:1.

e. Incubate for 7 days at 37 °C, 8% CO_2. If the color of the culture medium turns orange, replace half of it with fresh medium as described in step a.

f. At day 7 post-stimulation, thaw frozen irradiated mature DCs 4 h before use in low-adherence 24-well tissue-culture plates as described (Section 3.3).

g. Pulse the DCs with the peptide by incubating the cells for 1 h with 1 μg/mL peptide at 37 °C, 8% CO_2. Wash the cells with IMDM 10% FCS and centrifuge for 5 min at 400 × *g*.

h. Add the same number of DC per well as in step b in IMDM 10% HS with 50 U/mL IL-2 and 10 ng/mL IL-7. Incubate for 7 days at 37 °C, 8% CO_2.

i. At day 14, repeat steps e to f for a third round of stimulation (optional).

j. Screen the microculture (see Section 3.6) approximately 10–12 days after the last stimulation.

3.6 Screening and sorting of pHLA-multimer$^+$ T-cell clones

After amplification, a sample of each microculture is stained with fluorescent pHLA-multimers and screened for the presence of pHLA-multimer$^+$ cells by flow cytometry (Fig. 6). pHLA-multimer$^+$ cells are sorted from identified

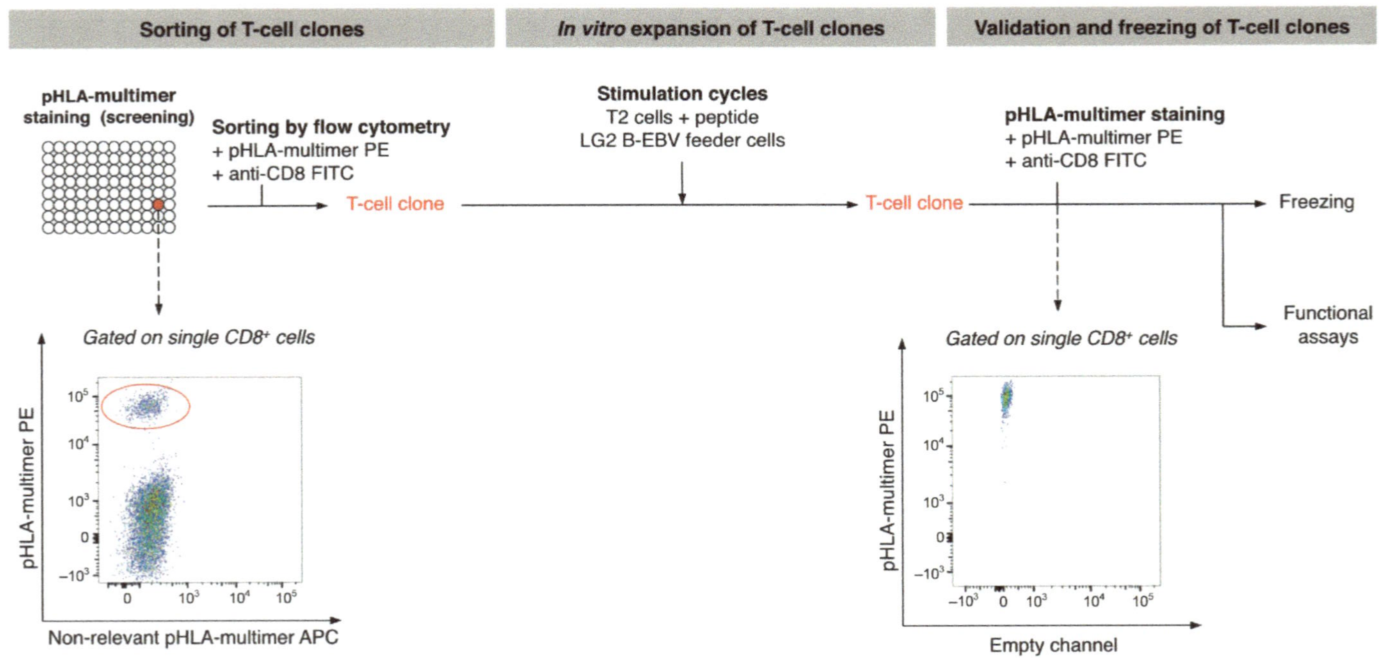

Fig. 6 Amplification of T-cell clones and validation of their ability to recognize fluorescent peptide-HLA-multimers. $CD8^+$ peptide-HLA (pHLA)-multimer$^+$ T-cells were sorted from microcultures by flow cytometry as a clonal T-cell population. T-cells were amplified *in vitro* with T2 cells loaded with 1 μg/mL of candidate peptide and LG2-EBV-B cells as feeder cells for several stimulation rounds. After amplification, purity of pHLA-multimer$^+$ T-cells was assessed by flow cytometry before freezing or proceeding to functional assays.

wells by flow cytometry. Single T-cell clones are separated into individual wells by either limiting dilution or through the flow sorting process. Alternatively, we have sometimes considered that the pHLA-multimer$^+$ cells in one well are derived from the same original pHLA-multimer$^+$ precursor. The sorted cells were amplified as a T-cell line in conditions described in Section 3.7.

3.6.1 Screening of microcultures

a. Screen the microcultures 10–12 days after the last antigen stimulation.

b. Collect 1/3 of the cells (at least 10,000) from each well and transfer them to a FACS-compatible 96-well plate. Remember to keep enough cells for the subsequent sorting of the pHLA-multimer$^+$ cells from the positive wells. If there are insufficient numbers of cells in each microculture, perform one further round of amplification (Section 3.5).

c. We advice to use the CD8$^+$ pHLA-multimer$^-$ fraction (Section 3.4.2) as staining controls. Controls needed are unstained cells, cells stained only for CD8, cells stained only with pHLA-multimer, and cells stained for CD8 and an irrelevant pHLA-multimer.

d. Wash cells with PBS 1% HS.

e. Centrifuge for 7 min at 400 × *g*.

f. Resuspend the pellet in 25 μL PBS 1% HS containing the relevant fluorescent pHLA-multimer at 10 nM, and incubate for 20 min at RT in the dark.

g. Add 25 μL PBS containing anti-CD8α or anti-CD8β antibody conjugated to FITC for 15 min at RT. Note that the optimal concentration of the antibody will depend on the clone, fluorochrome, supplier and machine settings. Titrate the antibodies before use.

h. Wash the cells with 150 μL PBS 1% HS and centrifuge at 400 × *g* for 5 min. Resuspend the cells in 200 μL PBS 1% HS.

i. Perform flow cytometry on your in-house machine and software. Here, we use the BD Fortessa flow cytometer and the DIVA software.

j. Analyze results using DIVA or Flowjo software. Gate on the CD8$^+$ population and analyze pHLA-multimer expression. Define the gating strategy using the staining controls.

3.6.2 Sorting of CD8$^+$ pHLA-multimer$^+$ cells

a. The culture medium is IMDM 10% HS supplemented with AAG and antibiotics as before (Section 3.3.1, step a).

b. 1 or 2 days after the screening step, transfer the remaining cells of a pHLA-multimer$^+$ microculture into a sterile FACS tube or a 15-mL Falcon tube. Wash the cells with 2 mL of sterile PBS 1% HS and centrifuge at 400 × *g* for 5 min.

c. Resuspend the pellet in 25 μL PBS 1% HS containing the relevant fluorescent pHLA-multimer at 20 nM, and incubate for 20 min at RT in the dark.

d. Add 25 μL PBS containing anti-CD8α or anti-CD8β-FITC antibody for 15 min at RT.

e. In the meantime, prepare round-bottom 96-well plates with 100 μL of culture medium with 50 U/mL IL-2 and 15 μg/mL gentamicin per well to collect the sorted cells (step f). Each plate should contain the cells from only one microculture.

f. Wash the cells with 150 μL PBS 1% HS and centrifuge at 400 × *g* for 5 min. Resuspend the cells in 200 μL PBS 1% HS. Pass the cells through a 40-μm filter before sorting.

g. Sort the CD8$^+$ pHLA-multimer$^+$ cells by flow cytometry and collect them in two to three different wells in the plate previously prepared (step e).

h. The number of sorted cells varies from 50 to 10,000.

i. Wrap the plates in aluminum foil to reduce risk of contamination and incubate them for 3 days before stimulating them as described in Section 3.7.

3.7 Amplification and freezing of T-cell clones

T-cell clones have to be amplified to obtain sufficient numbers of cells for validation (Section 4). We use stimulator cells expressing the relevant HLA molecule to present the candidate peptide. In our laboratory, both EBV-B cells pulsed with the peptide and tumor cells naturally presenting the peptide proved to be effective stimulator cells. Using allogeneic stimulator cells is not an issue at this step, as the T-cell clone should only recognize the candidate peptide and not an allogeneic determinant. We also add "feeder cells" to the co-culture whose exact role is ill-defined but whose presence is very important to achieve optimal T-cell proliferation. We always use the LG2-EBV-B cell line obtained in our laboratory as feeder cells, as it does not metabolize the culture medium too rapidly. Alternatively, other EBV-B cell lines, PBMC or a mix of both can also be used as feeder cells. The PBMC are irradiated (50 Gray), frozen, and thawed just before use. Finally, we advise to screen both the stimulator and the feeder cells for mycoplasma contamination on a monthly basis.

3.7.1 Culture and amplification of T-cells

a. All IMDM media are supplemented with AAG and antibiotics (Section 3.1.1.a).

b. Irradiate both stimulator and feeder cells at 100 Gray.

c. For the first amplification round, add 5000 stimulator and 15,000 feeder cells per well. For further stimulation rounds, we tend to add one stimulator cell and three feeder cells per T-cell.

d. Wash the stimulator cells with 10 mL of IMDM without serum. Centrifuge at 400 × *g* for 5 min and resuspend the stimulator cells at 10^6 cells/mL in IMDM without serum. Pulse the cells with 1 μg/mL of peptide of interest. Incubate for 1 h at 37 °C, 8% CO_2.

e. Wash the peptide-pulsed stimulator cells with 10 mL of IMDM without serum. Centrifuge at 400 × *g* for 5 min. Resuspend the stimulator cells in IMDM 10% HS at 10^5 cells/mL.

f. Wash the feeder cells with 10 mL of IMDM without serum. Centrifuge at 400 × *g* for 5 min. Resuspend them at 3×10^5 cells/mL in IMDM 10% HS.

g. Take an appropriate number of stimulator and feeder cells (see step b). Mix them at a 1:1 ratio and add 50 U/mL IL-2.

h. Add 100 μL of this stimulation mix to each well containing the T-cells (Section 3.6.2).

i. Incubate at 37 °C, 8% CO_2.

j. Monitor T-cell proliferation. If the medium's color changes from pink to orange, dilute the culture two- or threefold with fresh medium supplemented with IL-2.

k. Every 7–10 days, repeat steps (a–i) for another round of stimulation. You may have to change the type of microwells depending on degree of T-cell amplification:
- Round-bottom 96-well tissue-culture plates for up to 50,000 T-cells.
- Flat-bottom 96-well tissue-culture plates for 100,000 T-cells.
- 48-well tissue-culture plates for 200,000 T-cells.
- 24-well tissue-culture plates for 400,000 T-cells.

l. After 7–8 rounds of stimulation, T-cell numbers should be sufficient to proceed to functional assays (Section 4). We also advise to freeze at least 2×10^6 T-cells after each round of amplification (Section 3.7.2). These cells can be used for additional functional assays or to amplify the T-cell further.

3.7.2 Freezing of T-cell clones

a. Prepare freezing serum, which is composed of HS with 20% DMSO and store it at 4 °C. Freezing serum has to be used at 4 °C for freezing.
b. Count the T-cells. Do not freeze $<10^6$ per vial. Do not expect to recover >50% of cells after thawing.
c. Resuspend the cells in 500 μL of cold IMDM per aliquot you want to freeze.
d. Add 500 μL of freezing serum per 500 μL of cells suspension and mix gently.
e. Quickly transfer 1 mL of this mix per vial.
f. Transfer the vials at −80 °C. We advise to keep the freezing process as regular as possible. We also recommend Mr. Frosty™ Freezing Container from Thermo Scientific™.

3.7.3 Thawing of T-cell clones

a. The thawing medium is IMDM complemented with 10% HS and 5 U/mL DNase I. The medium is supplemented with AAG and antibiotics as before (Section 3.1.1.a).
b. Warm the thawing medium to 37 °C.
c. Gently stir the frozen cryovial in a water bath heated to 37 °C until most of the contents has melted.
d. Gently transfer the cells (1 mL) to an empty 15-mL Falcon tube. Use 1 mL of warm thawing medium to gently wash the cells remaining in the vial and transfer them to the 15-mL Falcon tube.
e. Drop by drop, add 8 mL of warm thawing medium to the cells in the 15 mL Falcon to dilute the DMSO.
f. Centrifuge the cells at $400 \times g$ for 5 min. Resuspend the cells at 2×10^6 cells/mL with thawing medium with 50 U/mL IL-2. Plate the cells in 24-well plates. Incubate them for at least 24 h at 37 °C, 8% CO_2 before proceeding to functional assays or re-amplification.

4. Validation steps to identify tumor-killing T-cell clones

This protocol's ultimate goal is to obtain T-cells that recognize and mount a reaction to a candidate peptide that is endogenously processed by tumor cells. We propose a pipeline to validate a peptide as a natural tumor antigen that T-cells should recognize:

1. the candidate peptide, using pHLA-multimers and flow cytometry (Section 4.1);

2. peptide-pulsed cells, and secrete cytokines and/or lyse the target cells as sign of activation (Section 4.2);
3. cells presenting an endogenously processed antigen on their surface (Section 4.3). First, T-cells should recognize engineered cells, i.e., cells expressing high levels of the antigenic peptide and HLA molecules. Second, T-cells should recognize tumor cells.

However, if the transfection strategy has not yielded results, we still recommend testing the tumor cells as they might express the peptide anyway. If an allogeneic tumor tests positive, we suggest confirming whether T-cells lose the ability to recognize the tumor following knock-down of the peptide-coding gene by siRNA or Crispr.

We will describe the functional T-cell assays in Section 5.

4.1 Recognition of fluorescent pHLA-multimers

As we suggested to sort the pHLA-multimer$^+$ T-cells by flow cytometry and expand them as an alternative to cloning by limiting dilution, it is essential to regularly verify that any "contaminating" pHLA-multimer$^-$ T-cells do not outgrow the pHLA-multimer$^+$ T-cells (Fig. 6). We recommend testing T-cells 14 days after the last stimulation.

a. Collect 25,000 T-cells of the clonal population and transfer the cells to a conical-bottom 96-well tissue-culture plate. Add 100 μL of PBS 1% HS per well and centrifuge the cells at 400 × *g* for 5 min.
b. Resuspend the pellet in 25 μL PBS 1% HS containing the relevant fluorescent pHLA-multimer at 20 nM, and incubate for 20 min at RT in the dark. Remember to use cells stained with a non-relevant pHLA-multimer as a negative control.
c. Add 25 μL PBS containing 1 μg/mL of anti-CD8α or anti-CD8β FITC antibody for 15 min at RT in the dark. Note that the optimal concentration of the antibody will depend on the clone, fluorochrome, supplier and machine settings. Titrate the antibodies before use.
d. Wash the cells with 150 μL PBS 1% HS and centrifuge at 400 × *g* for 5 min. Resuspend the cells in 200 μL PBS 1% HS. Transfer the cells either to a FACS-compatible 96-well plate or to FACS tubes, depending on the specifications of the flow cytometer available.
e. Analyze binding of the T-cell clones to the pHLA-multimers by flow cytometry. This laboratory uses a BD Fortessa with DIVA software.
f. This laboratory analyses the flow cytometry data with either DIVA or Flowjo software. Use control cells stained with a non-relevant

pHLA-multimer to set the gating strategy. Analyze pHLA-multimer binding to the $CD8^+$ population. After amplification, >95% of the T-cell clones of one population should positively bind to the appropriate pHLA-multimer.

4.2 Recognition of peptide-pulsed cells

Any type of cell expressing the relevant HLA molecule can be used as a potential stimulator. We prefer to use EBV-B cells because they express high levels of HLA molecules. T2 cells are an alternative option to test HLA-A2-binding peptides.

a. All IMDM media are supplemented with AAG and antibiotics (Section 3.1.1.a).
b. Pulse the cells with 1 μg/mL peptide for 30–60 min at RT in IMDM containing 1% HS. This low amount of serum reduces synthetic peptide digestion by peptidases. The peptide will bind unstable empty HLA molecules expressed at the cell surface and stabilize the pHLA complexes.
c. Wash the cells extensively with plain IMDM after the peptide pulse (twice with large volumes) to prevent any residual peptide from binding to the HLA molecules of the T-cells.
d. Cells expressing a different HLA type can be loaded with the corresponding peptide in parallel as negative controls.
e. Compare the avidity of different T-cell clones by testing their ability to lyse target cells pulsed with decreasing amounts of peptide (Section 5.2). We test concentrations ranging from 1 μg/mL to 1 ng/mL using three-fold serial dilutions.

4.3 Recognition of naturally processed antigens

Ultimately, to represent an effective therapeutic approach, the identified T-cell clones have to recognize candidate antigenic peptides presented by HLA molecules at the surface of human tumor cells and exert tumor-killing functions. However, this outcome may not be achieved for various reasons, including:

1. The peptide recognized by the T-cell is the result of an impurity in the synthetic peptide batch, e.g., a truncated peptide or a peptide containing protecting groups (resulting from peptide synthesis). Thus, it does not correspond to peptides produced by tumor cells. We therefore strongly recommend using HPLC-purified peptides (Section 3.1).

2. The proteasome complex expressed by the tumor cell line does not generate the candidate peptide. Consider that tumor cells can express different types of proteasomes (i.e., standard, intermediate and immune-proteasome), each of them having different proteolytic properties (Vigneron & Van Den Eynde, 2014). If the candidate peptide is not produced or if it is destroyed by the proteasome, it cannot be loaded onto newly synthesized HLA molecules.
3. The gene of interest is not expressed at sufficient levels by the tumor cells. We therefore recommend to first testing engineered stimulator cells that express sufficient levels of the transgene of interest (Section 4.3.1).

We therefore propose a first validation step using stimulator cells engineered to express the relevant protein-coding cDNA containing the peptide-coding sequence. Transfection or transduction of stimulator cells should result in high expression of the transgene (Section 4.3.1). If the first step is successful, proceed to the second step: testing T-cell recognition of tumor cells that naturally express the antigen (Section 4.3.2). However, if the first step is unsuccessful, we still recommend testing tumor cells, and in the case of allogeneic tumors, confirm whether T-cells lose the ability to recognize the tumor following knock-down of the peptide-coding gene by siRNA or Crispr.

4.3.1 Engineered stimulator cells: EBV-B cells transduced with a recombinant virus

We previously published protocols using either transiently transfected COS-7 cells or transduction with a recombinant virus to induce the expression of a particular protein (Chaux et al., 1999; De Plaen et al., 1997; Ottaviani et al., 2005). Alternatively, cells engineered to express the different proteasomes, e.g., transfected HEK293 cells, could be used (Guillaume et al., 2010). However, here we describe the use of Modified Vaccinia Ankara (MVA) viruses to transduce EBV-B cells, which can then be used both as stimulator and target cells. EBV-B cells are advantageous because they express high levels of the HLA molecule and are easily infected with MVA. However, a limitation to keep in mind is that EBV-B cells only contain immunoproteasomes, while cultured tumor cell lines mostly express standard or intermediate proteasomes.

a. We perform the infection with MVA in a biosafety level-3 laboratory 1 day before the T-cell recognition assay. Note that, for most transgenes, a biosafety level-2 laboratory should suffice, depending of the biosecurity rules of your institution and the introduced transgene.

b. Use RPMI +2% FCS as infection medium. Low serum concentrations facilitate virus entry into target cells.
c. Wash the target cells with 10 mL of infection medium in a 15-mL Falcon tube and centrifuge at $400 \times g$ for 5 min.
d. Resuspend the pellet in 250 μL of infection medium containing the viral particles. We generally use a Multiplicity of Infection (MOI) of 5 (corresponding to 5 infectious particles per cell), but recommend titrating the virus to your own experimental conditions to achieve optimal infection rates, while avoiding excessive mortality of your recipient cells.
e. Incubate the cells for 1 h at 37 °C, 5% CO_2. Gently shake the tube every 15 min to prevent pelleting.
f. Wash the cells with 10 mL of infection medium and centrifuge for 7 min at $400 \times g$.
g. Resuspend the cells at 10^6 cells/mL in infection medium. Plate the infected cells in 24-well tissue-culture plates at 2×10^6 cells/well, or at 2×10^5 cells/well in 96-well tissue-culture plates if the number of infected cells is less than 2×10^6.
h. Incubate the cells overnight at 37 °C, 5% CO_2.
i. Use the infected cells as soon as possible the next day to avoid high mortality of infected cells. We recommend to not use your cells for a chromium release assay if the mortality exceeds 15%.

4.3.2 Tumor cells that naturally express the gene of interest

Finding a tumor that naturally expresses both the relevant HLA molecules and the peptide-coding sequence can be challenging. If these cells are not readily available, a reasonable alternative is to use tumor cells that naturally express the peptide-coding sequence and transfect these cells with the relevant HLA-coding cDNA. We do not consider transfecting tumor cells with the peptide-coding sequence as valid, as this condition is redundant with engineered target cells. Consider that the goal of this validation step is to demonstrate that: (i) tumor cells can express the peptide-containing protein naturally in sufficient amounts and (ii) the processing machinery produces sufficient amounts of pHLA complexes at the cell surface of the tumor cells to elicit a T-cell response. Additionally, the avidity of the expanded T-cell clones for the pHLA is sufficient to get triggered and react against the tumor cells.

Most of the time, relevant tumor cells available will be allogeneic to the T-cells. This is the reason why it is essential to work with T-cell clones and not polyclonal T-cell populations that might be activated by presence of allogeneic determinants on the tumor cells rather than by the expression of the candidate peptide.

As mentioned earlier, cultured tumor cell lines mostly express standard or intermediate proteasomes, whereas EBV-B cells only contain immunoproteasomes. Nevertheless, treating tumor cells with 50–100 U/mL of IFN-γ for 24–48 h induces expression of the immunoproteasome and up-regulates HLA-expression. This could therefore enhance the presentation of the candidate peptide at the tumor surface.

5. T-cell functional assays

T-cell activation can be analyzed in different ways: up-regulation of activation markers (e.g., CD40L) and intracellular production of cytokines can be analyzed by flow cytometry, and the secretion of cytokines can be measured by ELISA or multiplexed cytokine assays (e.g., the Luminex technology). In Section 5.1, we describe our IFN-γ secretion assay. To measure the lytic activity of T-cells we use the reliable chromium-51 (^{51}Cr) release assay (Section 5.2). We strongly recommend performing these functional assays 6–8 days after the last antigenic stimulation of the T-cell clone. If the T-cells are frozen, thaw them at least 1 day before the test.

5.1 IFN-γ secretion assay

Cytokine release assays are fast and simple methods to evaluate the activation of T-cells stimulated with cells presenting the candidate peptide (target cells). Refer to previous paragraphs in Section 4 for target cell options (engineered cells or tumor cells).

a. Complete medium consists of IMDM supplemented with 10% HS and AAG and antibiotics as previously described (Section 3.1.1.a).
b. Collect, centrifuge and resuspend the target cells at 10^5 cells/mL in complete medium with 25 U/mL IL-2.
c. Seed 10,000 target cells per well in a round-bottom 96-well tissue-culture plate. We perform technical repeats in triplicate.
d. Collect the effector T-cells (Section 3.7), wash them in 10 mL of plain IMDM to eliminate the cytokines contained in the culture medium, centrifuge at 400 × *g* for 5 min and resuspend them at 10^5 cells/mL in complete medium and 25 U/mL IL-2.
e. Add 10,000 effector T-cells per well to the target cells of step c.
f. The total volume of the co-culture is thus 200 μL per well.
g. Incubate for 20 h at 37 °C, 8% CO_2.
h. Carefully harvest 160 μL of supernatant without aspirating cells from the bottom of the wells.

i. Perform an IFN-γ ELISA on 50 μL of supernatant.
j. Freeze the remaining 110 μL of supernatant in conical-bottom 96-well tissue-culture plate at −80 °C to avoid IFN-γ degradation. The frozen supernatant can be used in further analyses, e.g., to test the release of other cytokines.
k. We estimate the amount of IFN-γ contained in the supernatants by ELISA using an anti-IFN-γ capture antibody (clone 350B10G6) and a biotin-conjugated anti-IFN-γ monoclonal antibody (clone 67F12A8).

5.2 Lytic activity assay

Obtaining cytotoxic $CD8^+$ T-cells that kill tumor cells is a highly desired outcome of cancer vaccines. Thus, lytic assays play an important and insightful role in assessing the quality of vaccine-induced immune responses. Lytic assays are often considered to be more sensitive than cytokine release assays, and thus they are extremely useful, particularly when only low amounts of the antigenic peptide are presented at the tumor cell surface. Different approaches to lytic assays exist. Our colleagues, for example, successfully use a degranulation assay based on the up-regulation of CD107 (Dalet et al., 2011). We, however, prefer chromium-release assays, which are more reliable and contain less clone-dependent bias. Live target cells incorporate ^{51}Cr that is released upon T-cell-mediated killing. The resulting release of radioactivity can be quantified with β-radiation scintillation counters. Refer to previous paragraphs in Section 4 for target cell options (engineered cells or tumor cells).

a. All IMDM media are supplemented with AAG and antibiotics (Section 3.1.1.a).
b. Resuspend the target cells in IMDM complemented with 10% FCS and transfer them to a 15-mL Falcon tube coated with FCS. Wash the cells and resuspend them in 5 mL of IMDM 10% FCS.
c. Count cells and determine their viability by using trypan blue exclusion dye. Preferentially, target cells should have a high viability (we recommend >85%).
d. Take 10^6 cells and wash them twice by centrifugation at $400 \times g$ for 5 min at RT. Remove the supernatant entirely after the second wash.
e. Gently resuspend the cells in 10–20 μL of FCS. Add 20 μL of 20 mCi/mL of ^{51}Cr solution (Perkin Elmer Life Science). Mix gently and incubate for 1 h at 37 °C. Gently shake the cells every 10 min during the incubation.

f. Prepare T-cells by washing them with IMDM 10% HS. Centrifuge them for 5 min at 400 × *g*. Resuspend the cells at 3×10^5 cells/mL in IMDM 10% HS.

g. Plate 150 μL T-cell suspension per well in a conical-bottom 96-well tissue-culture plate in the first row. Plate 100 μL of IMDM 10% HS in the next three rows (Fig. 7).

h. Perform a serial 1:3 dilution of the T-cell suspension. T-cell numbers per well should be 30,000, 1000, 3000 and 1000 for the first, second, third and fourth row, respectively.

i. Wash the ^{51}Cr-labeled target cells twice with 14 mL of IMDM 10% FCS. Centrifuge at 300 × *g* for 8 min and discard the supernatant appropriately according to the laboratory's guidelines for the disposal of radioactive solutions. Resuspend the labeled cells at 10^4 cells/mL in IMDM 10% HS. If required, pulse the target cells with the peptide of interest at 1 μg/mL for 30 min at RT, and perform an additional washing step.

j. Add 100 μL of target cells to each well containing T-cells. Each well should contain 1000 target cells, and T-cells at a ratio of 1:30, 10, 3 or 1, depending on the row, in a final volume of 200 μL of IMDM complemented with 10% HS.

k. In 12 empty wells, plate 100 μL of IMDM complemented with 10% HS and 100 μL of target cells. Half of these wells will be used as *minimum ^{51}Cr release* and half as max*imum ^{51}Cr release* (min/max) controls (Fig. 7).

l. Centrifuge the plate at 400 × *g* for 2 min to enhance contact between effector and target cells.

m. To the *max ^{51}Cr release* control wells, add 1% of Triton X-100 in IMDM 10% HS (lysis agent).

n. Incubate the plates at 37 °C, 8% CO_2 for 4 h. It is possible to incubate the assay for 6 h or more, as long as the min/max ratio remains below 20%.

o. Centrifuge the plates for 5 min at 330 × *g*. Harvest 100 μL of supernatant from the samples and the min/max controls. Mix them with 100 μL scintillation fluid (Ultima Gold XR, Perkin Elmer), which converts radioactive signals into luminescent signals.

p. Count ^{51}Cr release with a β-radiation scintillation counter. We use the Microbeta Trilux device.

q. Normalize results to the min/max controls and express results as % lysis of maximum.

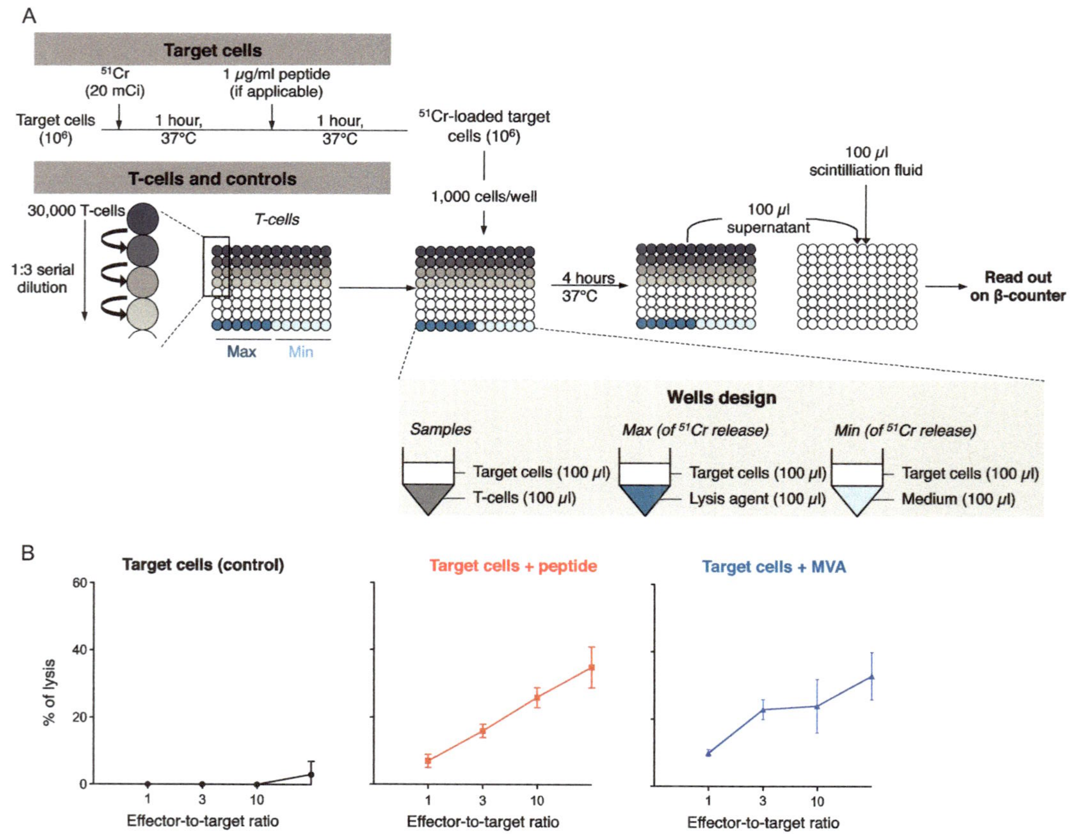

Fig. 7 Overview of chromium-release assay and example of results. (A) Diagram showing the different steps of the chromium-51 (^{51}Cr)-release assay. Target-cells are loaded with 20 mCi ^{51}Cr. Target cells and T-cells are co-cultured in a room dedicated for radioactive work with appropriate safety provisions. Scintillation fluid used is Ultima Gold XR (Perkin Elmer) with a Microbeta Trilux Device. (B) Examples of lytic assay performed with T-cell clones and target cells. Target cells can be cells that do not express the peptide coding sequence. These cells can be either peptide-free control cells, loaded with appropriate the candidate peptide or infected with MVA particles to express the candidate peptide.

6. Concluding remarks

Isolating T-cells specific for a given pHLA complex is challenging due to the low frequency of such $CD8^+$ T-cell precursors (about 3×10^{-7}) in peripheral blood. Here, we presented an approach based on using pHLA-multimers to enrich a starting T-cell population with peptide-specific T-cells. This is followed by an amplification step in microcultures containing peptide-pulsed stimulating cells and growth factors. The pre-enrichment step decreases the workload and increases the probability of identifying microcultures containing relevant peptide-specific T-cells. Researchers should be aware that peptide impurities are unavoidable during synthesis, and use only HPLC-purified synthetic peptides to avoid directing T-cell responses toward impurities, i.e., mainly peptides containing protecting groups added during peptide synthesis. These may bind to some TCR but will never be expressed naturally by tumors. Many publications report functional assays using polyclonal T-cells. We consider these validation assays void, since polyclonal T-cells most likely contain T-cell clones activated by allogeneic determinants, and the tumor cells used in these types of assay are often allogeneic. Thus, T-cell activation might not be peptide-specific. The main drawback of this approach is that it relies on algorithms with significant false positive discovery rates, most of which do not take into account the actual proteasome cleavage sites. One should be aware that the absence of cleavage sites at the C-terminus of the candidate peptide prevents its efficient presentation at the cell surface, whereas the presence of an internal cleavage site prevents its further processing. An *in vitro* screening of the candidate peptides with proteasome extracts is therefore strongly recommended.

Acknowledgments

Pierre van der Bruggen was supported by research program PDR T.0232.19 from the "Fonds de la Recherche Scientifique," Belgium. Christophe Vanhaver was supported by a PhD fellowship grant from FRIA/FNRS (Belgium).

References

Bachmann, H. R. J., Alexander, O., & Stefan, B. (1999). SYFPEITHI: Database for MHC ligands and peptide motifs. *Immunogenetics*, *50*, 213–219.

Bassani-Sternberg, M., Bräunlein, E., Klar, R., Engleitner, T., Sinitcyn, P., Audehm, S., et al. (2016). Direct identification of clinically relevant neoepitopes presented on native human melanoma tissue by mass spectrometry. *Nature Communications*, 7, 13404. https://doi.org/10.1038/ncomms13404. May.

Boon, T., & van der Bruggen, P. (1996). Human tumor antigens recognized by T lymphocytes. *Journal of Experimental Medicine*, *183*, 725–729.

Buchli, R., Vangundy, R. S., Hickman-miller, H. D., Giberson, C. F., Bardet, W., & Hildebrand, W. H. (2005). Development and validation of a fluorescence polarization-based competitive peptide-binding assay for HLA-A * 0201 s a new tool for epitope discovery. *Biochemistry*, *44*, 12491–12507. https://doi.org/10.1021/bi050255v.

Calis, J. J. A., Reinink, P., Keller, C., Kloetzel, P. M., & Kesmir, C. (2015). Role of peptide processing predictions in T cell epitope identification: Contribution of different prediction programs. *Immunogenetics*, *5*, 85–93. https://doi.org/10.1007/s00251-014-0815-0.

Chaux, P., Luiten, R., Demotte, N., Vantomme, V., Stroobant, V., Traversari, C., et al. (1999). Identification of five MAGE-A1 epitopes recognized by cytolytic T lymphocytes obtained by in vitro stimulation with dendritic cells transduced with MAGE-A1. *The Journal of Immunology*, *163*, 2928–2936.

Chaux, P., Vantomme, V., Coulie, P., Boon, T., & van der Bruggen, P. (1998). Estimation of the frequencies of anti-mage-3 cytolytic T-lymphocyte precursors in blood from individuals without cancer. *International Journal of Cancer*, 77(4), 538–542. https://doi.org/10.1002/(SICI)1097-0215(19980812)77:4<538::AID-IJC11>3.0.CO;2-2.

Coulie, P. G., Van den Eynde, B. J., van der Bruggen, P., & Boon, T. (2014). Tumour antigens recognized by T lymphocytes: At the core of cancer immunotherapy. *Nature Reviews. Cancer*, *14*(2), 135–146. https://doi.org/10.1038/nrc3670.

Cox, A. L., Skipper, J., Chen, Y., Henderson, R. A., Darrow, T. L., Shabanowitz, J., et al. (1994). Identification of a peptide recognized by five melanoma-specific human cytotoxic T cell lines. *Science*, *264*(5159), 716–719.

Dalet, A., Robbins, P. F., Stroobant, V., Vigneron, N., Li, Y. F., & El-gamil, M. (2011). An antigenic peptide produced by reverse splicing and double asparagine deamidation. *PNAS*, *108*(29), 323–331. https://doi.org/10.1073/pnas.1101892108.

De Plaen, E., Lurquin, C., Lethe, B., Bruggen, P., Van Der Brichard, V., Renauld, J., et al. (1997). Identification of genes coding for tumor antigens recognized by cytolytic T lymphocytes. *Methods*, *12*, 125–142.

Elvin, J., Potter, C., Elliott, T., Cerundolo, V., & Townsend, A. (1993). A method to quantify binding of unlabeled peptides to class I MHC molecules and detect their allele specificity. *Journal of Immunological Methods*, *158*, 161–171.

Foster, A. E., Weiss, H. L., Kaka, A. S., Leen, A. M., & Rooney, C. M. (2008). Using dendritic cell maturation and IL-12 producing capacity as markers of function: A cautionary tale. *Journal of Immunotherapy*, *31*(4), 359–369. https://doi.org/10.1097/cji.0b013e318165f5d2.

Gmelig-myeling, F., & Waldmann, T. A. (1980). Separation of human blood monocytes and lymphocytes on a continuous percoll gradient. *Journal of Immunological Methods*, *33*, 1–9.

Guillaume, B., Chapiro, J., Stroobant, V., Colau, D., Van Holle, B., & Parvizi, G. (2010). Two abundant proteasome subtypes that uniquely process some antigens presented by HLA class I molecules. *PNAS*, *107*(43), 18599–18604. https://doi.org/10.1073/pnas.1009778107.

Hunt, D. F., Henderson, R. A., Shabanowitz, J., Sakaguchi, K., Michel, H., Sevilir, N., et al. (1992). Characterization of peptides bound to the class I MHC molecule HLA-A2. 1 by mass spectrometry. *Science*, *255*, 1261–1263.

Jojic, N., Reyes-gomez, M., Heckerman, D., Kadie, C., & Schueler-furman, O. (2006). Learning MHC I—Peptide binding. *Bioinformatics*, *22*(14), 227–235. https://doi.org/10.1093/bioinformatics/btl255.

Kalaora, S., Barnea, E., Merhavi-Shoham, E., Qutob, N., Teer, J. K., Shimony, N., et al. (2016). Use of HLA peptidomics and whole exome sequencing to identify human immunogenic neo-antigens. *Oncotarget*, 7(5), 5110–5117. https://doi.org/10.18632/oncotarget.6960.

Karosiene, E., Lundegaard, C., Lund, O., & Nielsen, M. (2012). NetMHCcons: A consensus method for the major histocompatibility complex class I predictions. *Immunogenetics*, *64*, 177–186. https://doi.org/10.1007/s00251-011-0579-8.

Kelly, A., Powis, S. H., Kerr, L.-A., Mockridge, I., Elliott, T., Bastin, J., et al. (1992). Assembly and function of the two ABC transporter proteins encoded in the human major histocompatibility complex. *Nature*, *355*, 641–644.

Kessler, J. H., Mommaas, B., Mutis, T., Huijbers, I., Vissers, D., Benckhuijsen, W. E., et al. (2003). Competition-based cellular peptide binding assays for 13 prevalent HLA class I alleles using fluorescein-labeled synthetic peptides. *Human Immunology*, *64*, 245–255. https://doi.org/10.1016/S0198-8859(02)00787-5.

Li, B., Wang, Y., Chen, J., Wu, H., & Chen, W. (2005). Identification of a new HLA-A * 0201 -restricted CD 8 + T cell epitope from hepatocellular carcinoma-associated antigen HCA 587. *Clinical and Experimental Immunology*, *140*, 310–319. https://doi.org/10.1111/j.1365-2249.2005.02786.x.

Ljunggren, H.-G., Stam, N. J., Ohlen, C., Neefjes, J. J., Höglund, P., Heemels, M.-T., et al. (1990). Empty MHC class I molecules come out in the cold. *Nature*, *346*, 476–480.

Luo, H., Ye, H., Ng, H. W., Shi, L., Tong, W., Mendrick, D. L., et al. (2015). Machine learning methods for predicting HLA—Peptide binding activity. *Bioinformatics and Biology Insights*, *9*, 21–29. https://doi.org/10.4137/BBI.S29466.TYPE.

Mlecnik, B., Bindea, G., Angell, H. K., Maby, P., Angelova, M., Tougeron, D., et al. (2016). Integrative analyses of colorectal cancer show immunoscore is a stronger predictor of patient survival than microsatellite instability. *Immunity*, *44*(3), 698–711. https://doi.org/10.1016/j.immuni.2016.02.025.

Nielsen, M., Lundegaard, C., Worning, P., Lise, S., Øller, L., Lamberth, K., et al. (2003). Reliable prediction of T-cell epitopes using neural networks with novel sequence representations. *Protein Science*, *12*, 1007–1017. https://doi.org/10.1110/ps.0239403.view.

Ocklind, G. (1988). All CD2-positive human lymphocytes rosette with sheep erythrocytes in the presence of polyethylene glycol. *Journal of Immunological Methods*, *112*, 169–172.

Ottaviani, S., Zhang, Y., Boon, T., & Van Der Bruggen, P. (2005). A MAGE-1 antigenic peptide recognized by human cytolytic T lymphocytes on HLA-A2 tumor cells. *Cancer Immunology, Immunotherapy*, *54*(12), 1214–1220. https://doi.org/10.1007/s00262-005-0705-2.

Parker, K. C., Bednarek, M. A., & Coligan, J. E. (1994). Scheme for ranking potential HLA-A2 binding peptides based on independent binding of individual peptide side-chains. *The Journal of Immunology*, *152*, 163–175.

Pascolo, S., Schirle, M., Gu, B., Dumrese, T., Stumm, S., Kayser, S., et al. (2001). A MAGE-A1 HLA-A * 0201 epitope identified by mass spectrometry 1. *Cancer Research*, *61*, 4072–4077.

Paul, S., Kolla, R. V., Sidney, J., Weiskopf, D., Fleri, W., Kim, Y., et al. (2013). Evaluating the immunogenicity of protein drugs by applying in vitro MHC binding data and the immune epitope database and analysis resource. *Clinical and Developmental Immunology*, *2013*, 467852 (7 pages).

Pogue, R. R., Eron, J., Frelinger, J. A., & Matsui, M. (1995). Amino-terminal alteration of the HLA-A * 0201-restricted human immunodeficiency virus pol peptide increases complex stability and in vitro immunogenicity. *Proceedings of the National Academy of Sciences of the United States of America*, *92*, 8166–8170.

Ratzinger, G., Herold, M., Romani, N., Fritsch, P., Kroczek, R. A., Reider, D., et al. (2000). Production of IL-12 by human monocyte-derived dendritic cells is optimal when the stimulus is given at the onset of maturation, and is further enhanced by - IL-4. *The Journal of Immunology*, *166*(1), 633–641. https://doi.org/10.4049/jimmunol.166.1.633.

Skipper, B. J. C. A., Hendrickson, R. C., Gulden, P. H., Brichard, V., Pel, A., Shabanowitz, J., et al. (1996). An HLA-A2-restricted tyrosinase antigen on melanoma cells results from posttranslational modification and suggests a novel pathway for processing of membrane proteins. *The Journal of Experimental Medicine, 183*, 527–534.

Spisek, R., Bretaudeau, L., Barbieux, I., Meflah, K., & Gregoire, M. (2001). Standardized generation of fully mature p70 IL-12 secreting monocyte-derived dendritic cells for clinical use. *Cancer Immunology, Immunotherapy, 50*, 417–427. https://doi.org/10.1007/s002620100215.

Tan, T. L. R., Geluk, A., Toebes, M., Ottenhoff, T. H. M., & Drijfhout, J. W. (1997). A novel, highly efficient peptide-HLA class I binding assay using unfolded heavy chain molecules: Identification of HIV-1 derived peptides that bind to HLA-A*0201 and HLA-A*0301. *Journal of Immunological Methods, 205*, 201–209.

Tong, J. C., Tan, T. W., & Ranganathan, S. (2006). Methods and protocols for prediction of immunogenic epitopes. *Briefings in Bioinformatics, 8*(2), 96–108. https://doi.org/10.1093/bib/bbl038.

Topalian, S. L., Gonzales, M. I., Parkhust, M., Li, Y. F., Southwood, S., Sette, A., et al. (1996). Melanoma-specific CD4+ T cells recognize nonmutated HLA-DR-restricted tyrosinase epitopes. *Journal of Experimental Medicine, 183*(5), 1965–1971. https://doi.org/10.1084/jem.183.5.1965.

van der Bruggen, P., Traversari, C., Chomez, P., Lurquin, C., De Plaen, E., Eynde, D., et al. (1991). A gene encoding an antigen recognized by cytolytic T lymphocytes on a human melanoma. *Science, 254*(5038), 1643–1647.

van der Bruggen, P., Zhang, Y., Chaux, P., Stroobant, V., Panichelli, C., Schultz, E. S., et al. (2002). Tumor-specific shared antigenic peptides recognized by human T cells. *Immunological Reviews, 188*, 51–64. https://doi.org/10.1034/j.1600-065X.2002.18806.x.

Viatte, S., Alves, P. M., & Romero, P. (2006). Reverse immunology approach for the identification of CD8 T-cell-defined antigens: Advantages and hurdles. *Immunology and Cell Biology, 84*(3), 318–330. https://doi.org/10.1111/j.1440-1711.2006.01447.x.

Vigneron, N., & Abi, J. (2015). The capture proteasome assay (CAPA) to evaluate subtype-specific proteasome inhibitors. *Data in Brief, 4*, 146–151. https://doi.org/10.1016/j.dib.2015.04.019.

Vigneron, N., & Van Den Eynde, B. J. (2014). Proteasome subtypes and regulators in the processing of antigenic peptides presented by Class I molecules of the major histocompatibility complex. *Biomolecules, 4*, 994–1025. https://doi.org/10.3390/biom4040994.

Williams, D. S., Bird, M. J., Jorissen, R. N., Yu, Y. L., Walker, F., Zhang, H. H., et al. (2010). Nonsense mediated decay resistant mutations are a source of expressed mutant proteins in colon cancer cell lines with microsatellite instability. *PLoS One, 5*(12), e16012. https://doi.org/10.1371/journal.pone.0016012.

Zhang, H., Wang, P., Papangelopoulos, N., Xu, Y., Sette, A., Bourne, P. E., et al. (2010). Limitations of Ab initio predictions of peptide binding to MHC class II molecules. *PLoS One, 5*(2), e9272. https://doi.org/10.1371/journal.pone.0009272.

Further reading

van der Bruggen, P., Stroobant, V., Vigneron, N., & Van den Eynde, B. J. (2001). *Cancer antigenic peptide database*. Retrieved March 29, 2019, from https://caped.icp.ucl.ac.be/.

CHAPTER TEN

Generation of TCR-engineered reference cell samples to control T-cell assay performance

Nicole Bidmon[a,†], Cécile Gouttefangeas[b,c], Sjoerd H. van der Burg[d,*]

[a]Translational Oncology at the University Medical Center of the Johannes-Gutenberg University Mainz (TRON gGmbH), Mainz, Germany
[b]Department of Immunology, Interfaculty Institute for Cell Biology, University of Tuebingen, Tuebingen, Germany
[c]German Cancer Consortium (DKTK), German Cancer Research Center (DKFZ) Partner Site Tübingen, Tübingen, Germany
[d]Department of Medical Oncology, Leiden University Medical Center, Leiden, The Netherlands
*Corresponding author: e-mail address: shvdburg@lumc.nl

Contents

Abstract

In vitro cellular assays analyzing antigen-specific T cells are characterized by their high complexity and require controlled conditions to lower experimental variations. Without standard cellular reagents, it is difficult to compare results over time and across institutions. To overcome this problem, a simple and robust technology was developed to generate TCR-engineered reference samples (TERS) containing defined numbers of antigen-specific T cells. Utilization of TERS enables performance control of three main T-cell assays: MHC-peptide multimer staining, IFN-γ ELISpot and cytokine flow cytometry. TERS continuously deliver stable results and can be stored for longer periods of time. Here, an optimized manufacturing protocol, based on the electroporation of stable T-cell

† Current address: BioNTech RNA Pharmaceuticals GmbH, An der Goldgrube 12, 55131 Mainz, Germany.

Methods in Enzymology, Volume 631
ISSN 0076-6879
https://doi.org/10.1016/bs.mie.2019.05.010

receptor in vitro-transcribed mRNA, is provided for versatile in-house production of TERS. Included are a guideline to optimize the electroporation settings on locally available electroporation devices and a step-by-step protocol for the production process.

1. Introduction

Multiple assays to monitor the quantity, phenotype, and function of antigen (Ag)-specific T cells are applied in basic research and clinical trials in infectious diseases, autoimmunity, gene therapy and oncology. Since the outcomes of these assays are increasingly used for decision-making, particular attention needs to be paid to assay robustness, quality assessment, and control of assay performance over time (Britten, van der Burg, & Gouttefangeas, 2015). In this context, the inherent complexity of cellular immune assays constitutes a significant challenge. This was highlighted in a range of studies and proficiency panels for most of the commonly used T-cell assays (Britten et al., 2009; Chudley et al., 2014; Jaimes et al., 2011; Janetzki & Britten, 2012; Welters et al., 2012).

One key solution may arise from the use of reference samples that contain a defined and stable number of functional Ag-specific T cells and which may serve the need to control immune assay performance over time and across institutions (Maecker et al., 2010; van der Burg et al., 2011). Typical reference cell samples are (i) cryopreserved PBMCs from healthy donors with confirmed presence of Ag-specific responses against known microbial recall Ags (for example, CMV, EBV, Influenza) for $CD8^+$ and/or $CD4^+$ T cells and established T cell clones (Alanio, Lemaitre, Law, Hasan, & Albert, 2010; Scriba et al., 2005), (ii) specially processed cells (Garcia et al., 2014; Stebbings et al., 2015; Wang et al., 2015), or (iii) cryopreserved PBMCs or TILs from cancer patients with pre-existing T cell responses against tumor-associated Ags (TAAs) (Xu et al., 2008). Notably, all of these samples are available in limited amount, and are cost and/or time intensive to buy, select or establish. Moreover, the specificity, affinity, restriction and frequency of Ag-specific T cells contained in the samples cannot be tailored to the needs of individual laboratories. The generation of reference cells equipped with a chosen T-cell receptor (TCR) and that can be diluted into autologous, non-transduced T cells would bypass most of these problems. In a first step toward such samples, we had demonstrated that retrovirally TCR-transduced T cells spiked into autologous PBMCs would constitute a suitable tool to control immune assay performance. The TCR-transduced cells could be accurately detected down to frequencies of 0.02%, and the inter-laboratory applicability of such samples was confirmed in a small-scale proficiency panel (Cron et al., 2013; Singh et al., 2013). However, manufacturing of standard controls using retroviral vectors requires a specific in-house expertise and qualification, is rather laborious, making upscaling difficult.

Our novel, affordable, easy, robust, versatile and scalable technology for the generation of TCR-engineered reference samples (TERS) is based on the transfection of T cells via electroporation (EP) using T cell receptor (TCR) in vitro transcribed (IVT) mRNA (alpha and beta chains) (Bidmon et al., 2015). After EP, TCR-positive cells (Positive fraction) can be spiked to naïve PBMCs (Negative fraction) at desired final frequencies, aliquoted, frozen and stored in liquid nitrogen (Fig. 1).

In principle, TERS can be engineered to express any chosen $CD4^+$ or $CD8^+$ T cell TCR, provided that the two complementary α and β TCR chains are available (we were able to generate both viral (e.g., CMV, Influenza) and TAAs (e.g., tyrosinase, NY-ESO-1)-specific TERS). These standard samples deliver stable signals in repetitively performed cellular immune monitoring assays such as MHC-peptide multimer (MULTIMER), cytokine flow cytometry (CFC) and IFN-γ ELISpot. Our studies showed that TERS (i) lead to reproducible signals that are comparable to that of endogenous Ag-specific T cells; (ii) are stable over a period of at least 12 months; (iii) work in the hands of different investigators using locally available assay protocols; (iv) can be thawed and applied repetitively to control the performance of a series of experiments; and (v) sensitively capture deviations in assays due to low-quality cell materials, use of reagents with expired shelf life, unintended change of cytometer settings, and inappropriate data analysis (Bidmon et al., 2015). Furthermore, (vi) based on the scalability of the technology, it can be easily produced at various laboratories using a kit-based approach (Bidmon et al., 2018).

Here, we provide a detailed protocol for the optimized in-house production of MHC-restricted TERS containing a defined number of Ag-specific T cells with maximum functionality and high viability.

2. The generation of MHC-restricted TERS via TCR mRNA electroporation

The generation of TERS consists of two parts (Fig. 2). The first part describes the procedure for the optimization of EP settings using GFP and

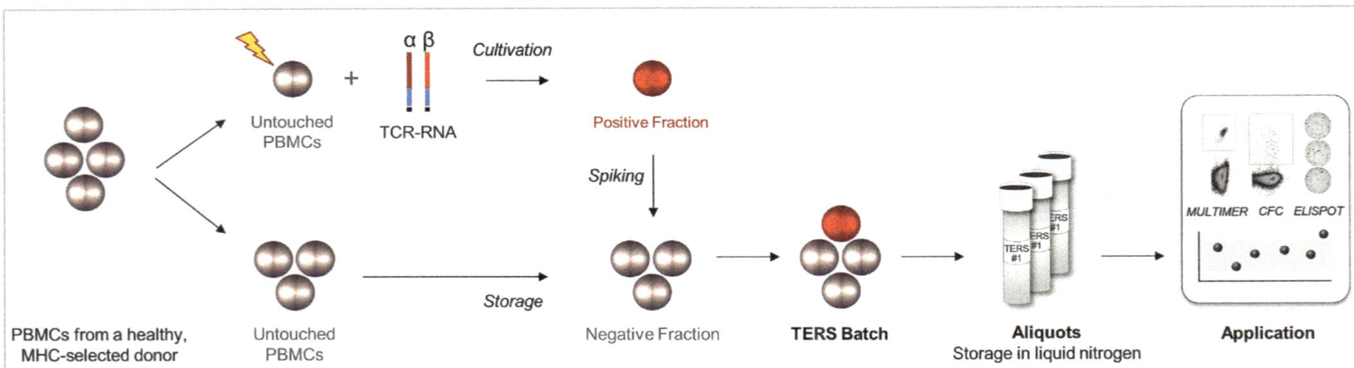

Fig. 1 Schematic principle of the generation of TERS.

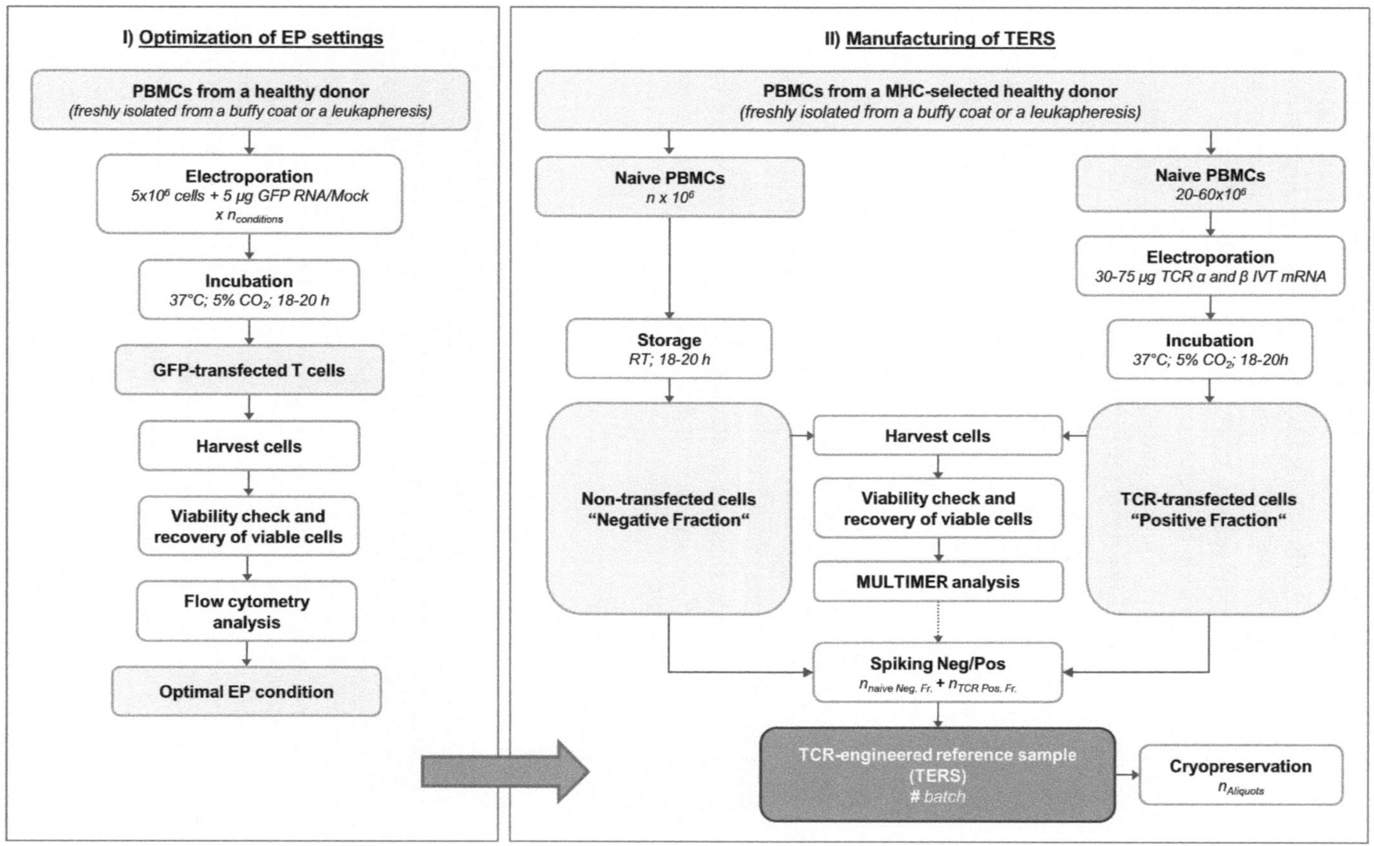

Fig. 2 See figure legend on next page.

has to be performed once for the locally available EP device prior to TERS generation. The second part describes an optimized process for the manufacturing of TERS with maximum functionality and high viability; as example, TCR-transfected $CD8^+$ T cells are shown. Within the production process, the number of spiked TCR^+ T cells can be adjusted over a range of frequencies to tailor the intensity of the Ag-specific signal. The number of frozen cells per aliquot, and the number of aliquots per batch can also be adapted. Each investigator may adapt parts of the production process to in-house preferences using their own protocols for PBMC isolation, freezing and thawing of cells, and T-cell assay.

Tip: When using a particular TCR for the first time, it is recommended to generate TERS containing high, medium and low numbers of Ag-specific cells and to test the correlating signal strength in MULTIMER staining and/or in a functional assay. It will give the opportunity to select the best features for the locally manufacturing process and the appropriateness of the Ag-specific signal of the used TCR.

The protocol below is explained for MHC class I-restricted TCRs in $CD8^+$ T cells; the process would be similar for generating MHC class II-restricted $CD4^+$ TERS (Bidmon et al., 2015); it is however more complex, due to the limited availability of MHC class II-restricted healthy donors and MULTIMERS.

Fig. 2 Overview of the process for TERS generation. (I) Optimization of EP settings (left panel): A range of EP settings needs to be tested at least in one experiment in order to identify the optimal conditions specific for the locally available EP device. For each EP reaction, freshly isolated PBMCs are electroporated with GFP RNA. After overnight culture, the cells of each transfection reaction are harvested and tested for viability and GFP expression. The EP condition showing maximum viability and maximum GFP expression is selected for further manufacturing of TERS. (II) Manufacturing of TERS (right panel): Primarily, for the generation of a MHC-class-I TERS, fresh MHC-matched PBMCs can be transduced with MHC-specific TCR-alpha and TCR-beta mRNA in one shot. The TCR transfected cells (Positive Fraction) are incubated 18–20 h at 37 °C to reach the highest TCR surface expression. The non-transfected, naïve PBMCs (Negative Fraction) are stored 18–20 h at RT. The next day, the cells of both fractions are harvested, counted, tested for viability and stained using the cognate MULTIMER and a CD8 antibody to determine the exact number of TCR-positive T cells. Depending on the detected transfection efficiency (% $MULTIMER^+$ of $CD8^+$), a defined number of the positive fraction cells is spiked to a defined number of naïve cells. Several dilutions can be performed if required. The cells of the generated TERS dilution are then frozen and stored in liquid nitrogen.

2.1 Equipment

2.1.1 Electroporation device

The equipment that is used depends on the locally available EP device. Select the specific EP device and the suitable EP cuvette and EP buffer (Table 1). If using an unlisted EP device, follow manufacturer's instructions of your local EP device or your established EP protocol.

2.1.2 General material

1. six-well tissue culture plates
2. 48-Well tissue culture plates (only for the use of Lonza Amaxa 4D)
3. RNase-free pipette tips (0.1–10 μL; 20–100 μL; 100–1000 μL)
4. 15 and 50 mL reaction tubes
5. 5 mL reaction tubes or 96-well plate (round bottom) for flow cytometry stainings
6. 1.5 mL Cryopreservation tubes
7. Mr. Frosty (or freezing machine)
8. Flow cytometer

Table 1 Overview of common EP devices, convenient EP cuvettes and optimal EP buffers, which use is mandatory depending on the manufacturer.

EP device	EP cuvette	EP buffer
BTX Harvard Apparatus **BTX ECM 830 EP System** **BTX Gemini X2**	BTX 4 mm	X-vivo or BTXpress™ High Performance Electroporation Solution (BTX)
Eppendorf **Multiporator**	Eppendorf 4 mm	Hypoosmolar or Isoosmolar Buffer (Eppendorf)
Bio-Rad **Genepulser® II EP System**	Bio-Rad 4 mm	X-vivo
Bio-Rad **Genepulser Xcell™ EP System**	Bio-Rad 4 mm	OptiMEM (Life-Technologies)
Lonza **Nucleofector II**	Lonza Nucleocuvette™ Vessel	Kit-based Nucleofector solution (Lonza)
Lonza **Amaxa™ 4D**	Lonza Nucleocuvette™ Strip	Kit-based Nucleofector solution (Lonza)

2.1.3 Cell material

Freshly isolated PBMCs from a buffy coat donor or preferentially a leukapheresis are required. High quality PBMCs characterized by a viability >90% and low content in erythrocytes and granulocytes will lead to best performance (cells tend to die and to aggregate after each step within the process, resulting in a low yield of TERS characterized by reduced or even failed Ag-specific signals in functional assays). The PBMCs that are used for TERS manufacturing must display the MHC molecules forming the restriction element for the TCR.

Tip: It is not recommended to use culture-derived or frozen/thawed PBMCs due to the reduced amount of viable cells serving as yield limiting factor for the final TERS batch. The transfection efficiency (TE) will be dramatically reduced if using frozen/thawed PBMCs.

2.2 Reagents

2.2.1 In vitro transcribed mRNA

1. GFP IVT mRNA (stock concentration: 1–5 μg/μL, 5 μg per test transfection)
2. TCR IVT mRNA alpha and beta chain (stock concentration: 3–5 μg/μL, quantity will depend on the number of PBMCs treated, Table 2): the TCR mRNA should be highly concentrated to decrease the amount of RNA buffer within the EP cuvette. Typically, the mRNA is dissolved in RNase-free sterile water or special elution buffers. High amounts of these buffers would alter the EP step, leading to decreased viability of the cells.

Notably, to reach maximum transfection efficiency (GFP or TCR surface expression >90% of T cells), we recommend the use of high quality TCR mRNA supported by the following recommended conditions:

(i) high integrity (>90%) of the mRNA (qualified by Bioanalyzer)
(ii) low RNase content in the mRNA solution
(iii) molecular engineered TCR constructs like chimeric TCRs, consisting of a murine constant domain and human variable domains to prevent mispairing of endogenous and exogenous TCR chains
(iv) chemically engineered TCR mRNA based on an improved mRNA Cap structure such as beta-S-ARCA (D1 or D2)-cap (Kuhn et al., 2010), to prevent fast degradation by RNases

2.2.2 Buffer and media

Manufacturers are indicative unless otherwise stated.

Table 2 Overview of the device-specific amounts of RNA (µg), PBMCs (10^6) and buffer (µL) to be used per test (= per EP reaction or per EP cuvette) for the part (I) Optimization of EP settings or the part (II) Manufacturing of TERS.

		(I) Optimization		(II) Manufacturing	
EP device	**Final volume/ test**	**PBMCs/ test**	**GFP RNA/test**	**PBMCs/test**	**TCR RNA alpha and beta chains/ test**
Harvard Apparatus **BTX ECM 830 EP System BTX Gemini 2×**	250 µL	5×10^6	5.0 µg	(a) 20×10^6 (b) 40×10^6 or (c) 60×10^6	(a + b) 35 µg per chain or (c) 75 µg per chain
Eppendorf **Multiporator**	250 µL				
BioRad **Genepulser® II EP System**	250 µL				
BioRad **Genepulser Xcell™ EP System**	250 µL				
Lonza **Nucleofector II**	100 µL	5×10^6	5.0 µg	5×10^6	5.0 µg per chain
Lonza **Amaxa™ 4D**	20 µL	1×10^6	2.5 µg	1×10^6	2.5 µg per chain

1. Culture medium (mandatory): RPMI 1640 1× + GlutaMAX (Gibco), 5% human AB serum (Invitrogen), 1% MEM Non-Essential Amino Acids (Gibco), 1% $Na_4P_2O_7$ (Gibco), 0.5% Pen/Strep (Sigma-Aldrich)
2. Freezing medium (preferred): RPMI 1640 1× + GlutaMAX (Gibco), 10% human albumin (CSL Behring), 10% DMSO (AppliChem)
3. Thawing medium (preferred): RPMI 1640 1× + GlutaMAX (Gibco), 10% human albumin (CSL Behring)
4. 1× DPBS
5. Flow staining Buffer (preferred): 1× DPBS, 2 mM EDTA, 2% FCS

2.2.3 Additional reagents

1. Ambion® RNaseZap®
2. 2 µg/µL DNase I
3. Cognate fluorescent MULTIMER
4. Flow cytometry: Live/Dead cell dye, anti-CD4 and anti-CD8 antibodies

5. Optionally a synthetic peptide representing the cognate T-cell epitope of the transduced TCR for functional testing in IFN-y ELISpot or flow cytometry
6. Depending on the locally established protocols, additional equipment is needed for the isolation and culture of PBMCs, counting, freezing and thawing of cells, MULTIMER staining, or functional read-outs.

2.3 Protocol

The entire GFP optimization and the TERS production process has to be performed under sterile conditions. For the transfection step, working under RNase-free or -reduced conditions is essential. Therefore, use RNase-free pipette tips and consistently decontaminate the surface area of the clean bench, pipettes and gloves using RNase decontamination solution (RNase Zap®).

If EP devices from Lonza are used, you will need to upscale the number of EP reactions for one specific TCR for the generation of one TERS batch. These systems do not support an increase of the cell concentration within one EP reaction.

If using high amounts of mRNA (>35 μg per TCR chain) for the EP performed with devices from BTX, it is recommended to use the BTX EP buffer instead of X-vivo. This will support high cell viability.

2.3.1 Part I: Optimization of EP settings for the transfection of human T cells with GFP RNA

In the Section 1, a process is described for optimization and setup of EP settings that are valid for the locally available EP device using the transporter RNA GFP. The optimal EP conditions are characterized by maximum GFP expression on the surface of T cells and high viability of the transfected cells.

2.3.1.1 Preparatory work

1. Isolate PBMCs from a buffy coat or a leukapheresis (any MHC-restriction) according to your local protocol.
2. For one EP reaction 5×10^6 or 1×10^6 for Lonza Amaxa 4D are required (Table 1). Transfer the appropriate volume of the cell suspension to a 15 mL reaction tube adapted to the number of tests (e.g., 7 tests = 35×10^6 (7 mL) or 7×10^6 (7 mL) cells that have to be transferred). Include two additional reactions as negative controls (Mock = EP of cells without RNA and untreated (non-electroporated) cells that are directly seeded to the culture plate).

3. Prepare the wells of a six-well culture plate (adapted to number tests). Each well should contain 5 mL culture medium and 1 μg/mL DNase I. Only for Lonza Amaxa 4D, prepare the wells of a 48-well culture plate. Each well should contain 1 mL culture medium and 1 μg/mL DNase I.
4. Store the prepared plates temporarily in the incubator.
5. Prepare the EP device and adjust the parameters to be tested (Table 3).

Table 3 Overview of device-specific EP conditions that have to be tested for the optimization of settings on the locally available EP device.

Harvard apparatus—BTX ECM 830 or BTX Gemini 2× *(square pulse)*										
Voltage (V)	400	400	400	500	500	**500**	500	–	–	–
Pulse length (ms)	6	8	10	1	2	**3**	5	–	–	–
Pulse number	1	1	1	1	1	**1**	1	–	–	–
Eppendorf—Multiporator *(exponential pulse)*										
Voltage (V)	600	**700**	800	700	800	600	700	**800**	700	800
Pulse length (μs)	500	**500**	500	250	250	500	500	**500**	250	250
Pulse number	1	**1**	1	1	1	1	1	**1**	1	1
Buffer	Hypoosmolar					Isoosmolar				
BioRad—Genepulser II *(exponential pulse)*										
Voltage (V)	400	400	300	300	**400**	450	500	–	–	–
Capacitance (μF)	350	150	400	250	**250**	250	250	–	–	–
Pulse number	1	1	1	1	**1**	1	1	–	–	–
BioRad—Genepulser Xcell *(square pulse)*										
Voltage (V)	400	400	**400**	450	**450**	450	450	500	500	**1000**
Pulse length (ms)	5	8	**10**	5	**3**	5	8	3	5	**0.5**
Pulse number	1	1	**1**	2	**2**	1	1	1	1	**1**
Lonza—Nucleofector II										
Program	**U-014**		V-024			T-020		T-023		
Lonza—Amaxa 4D										
Program	FI-115		**EO-115**			FL-113		FF-113		

The EP conditions that showed maximal TCR expression and viability in our hands are highlighted in bold in the gray columns.

2.3.1.2 Electroporation

1. Centrifuge cells at 300*g*, 8 min, RT and discard the supernatants. Resuspend each cell pellet in the appropriate volume using the device-specific EP buffer (Table 2).
2. Transfer the cell suspension (avoid air bubbles) to the device-specific EP cuvette (Tables 1 and 2). Prepare all cuvettes (1/test) and store them on ice until use.
3. For each test, add the device-specific amount of GFP RNA (Table 2) directly to the cell suspension in the EP cuvette and mix by pipetting up and down (avoid air bubbles). Close the cap, place the cuvette into the cuvette holder of the EP device and start the pulsing. The time span from adding the RNA until the pulse occurs should not take longer than 1 min! Working times >1 min causes RNA degradation leading to reduced transfection efficiencies!
4. After pulsing, add 150 μL of culture medium to the EP reaction and mix by pipetting up and down. Transfer the cell suspension to one well of the six-well culture plate and mix carefully to ensure a homogenous cell suspension. Wash the cuvette with 200 μL of medium and transfer it to the same well.
5. The EP reactions of Lonza Nucleofector II and Amaxa 4D are directly transferred to the plates. Wash the cuvette with 100 μL culture medium and transfer to the corresponding well.
6. Additionally, seed 5×10^6 cells/well or 1×10^6 cells/well for Lonza Amaxa 4D of untreated cells as negative control.
7. Incubate the cells at 37 °C, 5% CO_2, for 18–20 h.

2.3.1.3 Harvesting cells and quality control

1. Transfer the cell suspension of each reaction to a separate reaction tube. Rinse each well with 1 mL 1 × DPBS and transfer to each reaction tube.
2. Wash the cells by adding 10 mL 1 × DPBS and centrifuge cells at 300*g*, 8 min, RT. Then discard the supernatants and resuspend the cells in an appropriate counting volume with 1 × DPBS.
3. Count the cells and check the viability according to your lab-specific counting methodology.

2.3.1.4 Flow cytometry analysis to determine GFP expression

1. 1×10^6 cells are used for one staining/test. We recommend staining with at least a Live/Dead cell dye, anti-CD8 and anti-CD4 antibodies, according to the locally established staining protocol with a

recommended gating strategy: Singlets → Living cells → Lymphocytes → CD4 vs. CD8: gate on $CD8^+$ or $CD4^+$ → $CD8^+$ or $CD4^+$ vs. GFP^+: gate on $CD8^+$ or $CD4^+$ GFP^+. The gating strategy is exemplary depicted in Fig. 3A based on the detection of TCR-transfected $CD8^+$ T cells.

2. Set the GFP gate by using the negative control/non-transfected cells and use it for all tests.
3. Select the optimal, device-specific EP settings showing maximal GFP expression (max. Transfection efficiency >80%) and high viability (>80%) of transfected T cells.

These optimized EP settings will be used for the manufacturing of the Ag-specific TERS using TCR mRNA.

2.3.2 Part II: Manufacturing of TERS

2.3.2.1 Preparatory work

1. Isolate fresh PBMCs from a MHC-selected buffy coat donor or leukapheresis according to your local protocol.
2. Transfer for one EP reaction 20×10^6 or 40×10^6 or 60×10^6 or specifically $1–5 \times 10^6$ for Lonza EP devices (Table 2) to a 15 mL reaction tube. Additionally transfer as negative control $1 \times 5 \times 10^6$ cells as Mock (EP without RNA) and transfer $1 \times 5 \times 10^6$ cells as untreated cells directly to one well of the culture plate.
3. Centrifuge the cells at 300*g*, for 8 min, at RT and completely remove the supernatants.
4. The cell pellet for TCR RNA EP is resuspended in 220 μL (for $20–40 \times 10^6$ cells) or 200 μL (60×10^6) BTX Press High Performance EP solution or X-vivo and directly transferred (final: 240 μL) to the EP cuvette. The cell pellet for the Mock EP is resuspended in 240 μL X-vivo and directly transferred (final: 240 μL) to the EP cuvette. The volume of the EP buffer that is added here to the cell pellets differs in order to reach the final volume as described in Table 2 (finally, including the amount of the cell pellet and the addition of the RNA solution). Transfer the cells with care to avoid air bubbles. Store the cuvettes with closed lid on ice until use for EP.
5. Prepare the wells of a six-well culture plate. Each well should contain 5 mL culture medium (37 °C) and 1 μg/mL DNase I. Adapt the number of wells to the number of transfected cells: 1 × well = 10×10^6 of transfected cells, and two wells with 5×10^6 cells of Mock or untreated cells.
6. Store the prepared plates temporarily in the incubator.

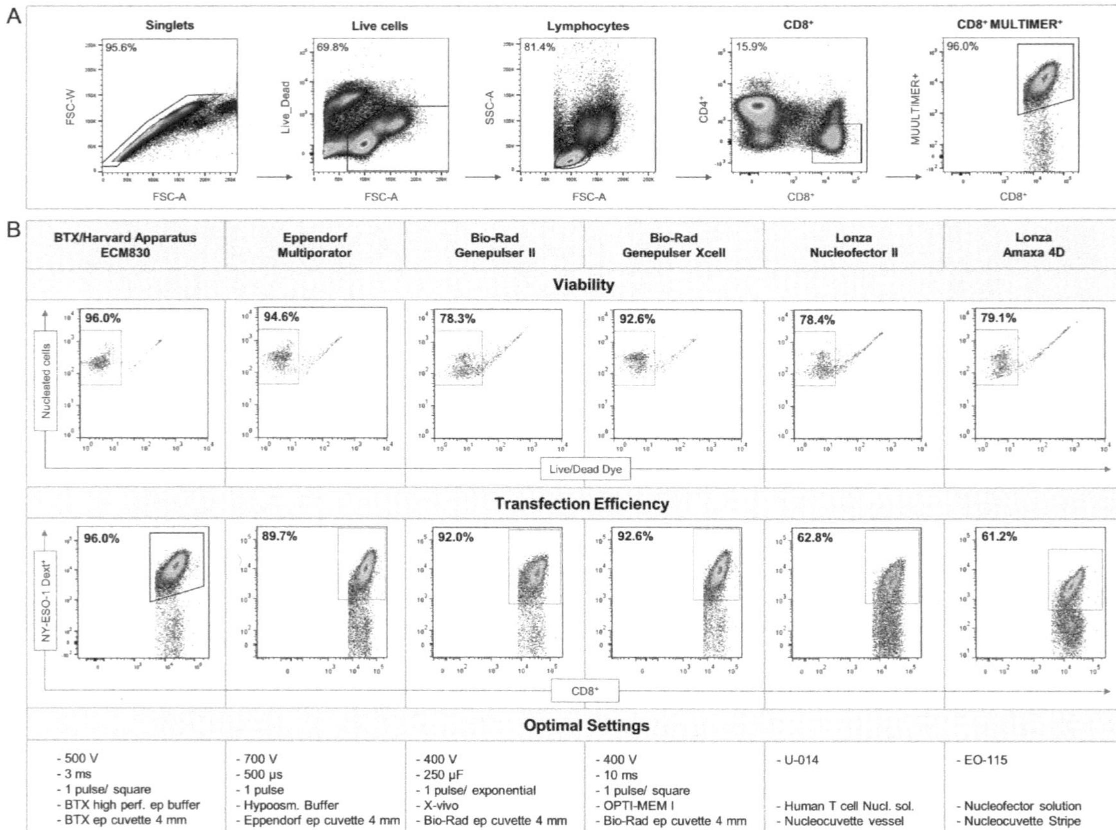

Fig. 3 (A) Example of the gating strategy that was used for the analysis of TCR-transfected $CD8^+$ T cells in MULTIMER staining experiments. Analysis was performed as follows: Singlets, Living cells, Lymphocytes, CD4 vs. CD8: gate on $CD8^+$, $CD8^+$ vs. $MULTIMER^+$, $CD8^+$ $MULTIMER^+$. (B) Optimized EP settings for each EP device used for the generation of TERS specific for NY-ESO-$1_{157–165}$ (SLLMWITQC). 20 h after EP and cultivation, PBMCs were tested in MULTIMER (dextramer: HLA-A2, SLLMWITQV) staining to detect TCR expression on $CD8^+$ T cells = transfection efficiency (TE), and the viability (Via) of transfected cells (Bidmon et al., 2018).

2.3.2.2 Electroporation and culture

1. Prepare the EP device and adjust to optimal parameters (determined in Part I): (Voltage [V], pulse length [ms], pulse number, capacitance [μF] (only for Genepulser II) or select the specific program for Lonza Nucleofector or Amaxa 4D.
2. Add the TCR RNA α and β (amount according to the cell number, Table 2) directly to the cells in the EP cuvette (final volume ~250–260 μL). Mix the RNA with the cell suspension by pipetting 15 times up and down (avoid air bubbles!) Close the lid, place the cuvette in the EP device, and start the pulsing. The time between adding the RNA and the pulsing should not take longer than 1 min. (Working time should be <1 min to prevent RNA degradation and reduced transfection efficiencies.) Add 350 μL of culture medium to the EP reaction and mix by pipetting up and down.
3. When using (I) 20×10^6 or (II) 40×10^6 or (III) 60×10^6 cells for EP, transfer (I) 2 × 300 μL or (II) 4 × 150 μL or (III) or 6 × 1000 μL of the EP reaction to (I) 2× wells or (II) 4× wells or (III) six wells of the six-well culture plate. When using the Lonza Nucleofector or Amaxa 4D combine and seed the cells at 10×10^6 cells in one six-well.
4. Wash the cuvette with 200 μL culture medium taken from one six-well, mix by pipetting up and down and transfer the suspension back to the same well. Finally, mix carefully by slowly pipetting up and down to ensure a homogenous cell suspension in the wells. Each of the six-well contains $\sim 10 \times 10^6$ cells in 5 mL culture medium (final cell conc.: 2×10^6/mL).
5. For the Mock reaction, add an equivalent amount of RNA elution buffer/RNase-free buffer to the cell suspension in the EP cuvette and treat cells as above. Add 250 μL of culture medium to the EP reaction and transfer the cells to one six-well ($\sim 5 \times 10^6$ cells in 5 m culture medium = final cell concentration: 1×10^6/mL) and proceed as described above.
6. Incubate the cells at 37 °C, 5% CO_2, for 18–20 h.

2.3.2.3 Storage of non-transfected PBMCs

1. The non-transfected PBMCs are divided into 50 mL reaction tubes with $200–280 \times 10^6$ PBMCs in 35 mL culture medium per tube ($6–8 \times 10^6$/mL).
2. Store the tubes at RT, for 18–20 h, in a slanting position.

2.3.2.4 Harvesting cells

1. TCR RNA-transfected T cells (Positive fraction): Harvest wells by rinsing the bottom of the plate with the containing cell suspension. Transfer and if needed, combine the suspensions of the TCR-transfected cells to one 50 mL reaction tube. Rinse each well 2 × with 1 mL 1 × DPBS and add to the reaction tube.
2. Non-treated cells (Negative fraction): Resuspend the cells in each reaction tube by pipetting up and down. Centrifuge the cells at 300*g*, for 8 min, at RT and discard most of the supernatant, leave ~3 mL of medium above the cell pellets. Resuspend the pellets in the remaining 3 mL and pool them in one tube. Rinse and wash the emptied tubes with 1 mL and then with 5 mL 1 × DPBS.
3. Wash each tube by adding 1 × DPBS up to 50 mL and centrifuge at 300*g*, 8 min, at RT. Remove the supernatants and resuspend the pellets in an adequate volume for counting.

2.3.2.5 Quality check

Count and check the viability of the cells. For the Negative fraction a viability >90% is expected, whereas the viability of the Positive fraction is expected to be between 80% and 90% (according to pretesting performed in Part I). The recovery is donor-dependent. We recommend not to use cells showing a viability <70% or strong clustering for further TERS production, as functionality might be reduced.

2.3.2.6 Electroporation check: MULTIMER staining to determine the number of TCR-transfected T cells

1. The MULTIMER staining and analysis is performed as per established protocol. We recommend to use at least a Live/Dead cell dye, a CD8 and CD4 antibodies and the gating: Singlets → Living cells → Lymphocytes → CD4 vs. CD8: gate on $CD8^+$ → $CD8^+$ vs. $MULTIMER^+$: gate on $MULTIMER^+$ of $CD8^+$ (Fig. 3A).
2. 1×10^6 cells of the Positive fraction, Mock electroporated cells as well as cells of the Negative fraction are stained with the specific MULTIMER.

Fig. 3B shows the use of perfectly optimized EP settings (on each commonly used EP device) for the EP of PBMCs with a chimeric TCR RNA (D1-capped) specific for NY-ESO-$1_{(157-165)}$ (SLLMWITQC), resulting in maximum TE (>90% $MULTIMER^+$ of $CD8^+$) and high viability >80% of transfected $CD8^+$ T cells. In our hands, the two devices from Lonza do not support best results, probably due to the limited setting possibilities.

2.3.2.7 TERS dilution for batch (TERS#) generation: "Spiking"

First, define the Ag-specific signal required (% MULTIMER$^+$ of CD8$^+$) considering the assay in which the TERS aliquots will be applied. The calculation is based on, and limited by, the current cell number of the Negative and Positive fraction and consists of spiking the cells of the Positive fraction into the Negative fraction in a defined ratio aiming the required signal in the final TERS dilution. Note that a true Negative fraction (w/o addition of TCR-transduced cells) should be frozen as well (min. 18×10^6 for three aliquots to be frozen), as it will be required later as a negative control for the initial TERS quality check after freezing/thawing.

The following information is required for the spiking calculation:

- Total number of living cells of the Negative fraction recovered after ON culture minus 18×10^6 cells (three aliquots for freezing = **Neg**) → $\mathbf{Neg_{Cells}}$
- Total number of living cells of the Positive fraction recovered after ON culture → $\mathbf{Pos_{Cells}}$
- Transfection efficiency (TE) of the Positive fraction % MULTIMER$^+$ of CD8$^+$ → $\mathbf{Pos_{TE}}$ **[%]**
- Required signal (% MULTIMER$^+$ of CD8$^+$) for the final TERS batch → $\mathbf{TERS\#_{MULTIMER}}$ **[%]**

Now calculate the dilution for the TERS batch ($\mathbf{TERS\#_{Dil}}$) based on the factor of the given transfection efficiency of the Positive fraction ($\mathbf{Pos_{TE}}$ **[%]**) and the required signal for the final TERS batch ($\mathbf{TERS\#_{MULTIMER}}$ **[%]**).

$$\mathbf{TERS\#_{Dil}} = \mathbf{Pos_{TE}}\,[\%] / \mathbf{TERS\#_{MULTIMER}}\,[\%] \quad (\mathbf{F1})$$

For the calculation of the number of cells of the Positive fraction that are used for spiking ($\mathbf{Pos_{CellsSpike}}$), divide the number of living cells of the Negative fraction ($\mathbf{Neg_{Cells}}$) by the dilution factor ($\mathbf{TERS\#_{Dil}}$).

$$\mathbf{Pos_{CellsSpike}} = \mathbf{Neg_{Cells}} / \mathbf{TERS\#_{Dil}} \quad (\mathbf{F2})$$

Then the number of cells of the Negative fraction that are used for spiking ($\mathbf{Neg_{CellsSpike}}$) are defined by subtracting the number of cells of the Positive fraction that are used for spiking ($\mathbf{Pos_{CellsSpike}}$) from the number of living cells of the Negative fraction ($\mathbf{Neg_{Cells}}$)

$$\mathbf{Neg_{CellsSpike}} = \mathbf{Neg_{Cells}} - \mathbf{Pos_{CellsSpike}} \quad (\mathbf{F3})$$

Now define the volumes of the cell suspension of the Negative fraction ($\mathbf{Neg_{VolSpike}}$) and the Positive fraction ($\mathbf{Pos_{VolSpike}}$) that have to be combined analogue to the calculated cell numbers in F2 and F3. Use the total volume of the cell suspension ($\mathbf{Neg_{CellsTotVol}}$ or $\mathbf{Pos_{CellsTotVol}}$), the total cell number ($\mathbf{Neg_{Cells}}$ or $\mathbf{Pos_{Cells}}$) and the cell number of the fraction used for spiking ($\mathbf{Neg_{CellsSpike}}$ or $\mathbf{Pos_{CellsSpike}}$).

$$\mathbf{Neg_{VolSpike}\,[mL] = Neg_{CellsTotVol}\,[mL]/Neg_{Cells} * Neg_{CellsSpike}\ (F4)}$$

$$\mathbf{Pos_{VolSpike}\,[mL] = Pos_{CellsTotVol}\,[mL]/Pos_{Cells} * Pos_{CellsSpike}\ (F4)}$$

Transfer the defined volumes of the Negative fraction and the Positive fraction in a new reaction tube and thoroughly mix the combined cell suspensions.

The spiking process to generate the TERS batch (**TERS#**) is summarized as follows:

$$\mathbf{TERS\# = Neg_{CellsSpike} + Pos_{CellsSpike} \rightarrow Neg_{VolSpike}\,[mL] + Pos_{VolSpike}\,[mL]}$$

Example for the spiking and generation of a defined dilution for one TERS batch:

Given Data:

$$\mathbf{Neg_{Cells}} = 290 \times 10^6 - 20 \times 10^6 = 270 \times 10^6;\ \mathbf{Neg_{CellsTotVol}} = 40\,\text{mL};\ \mathbf{Pos_{Cells}} = 10 \times 10^6;\ \mathbf{Pos_{CellsTotVol}} = 15\,\text{mL};\ \mathbf{Pos_{TE}} = 92.0\%;\ \mathbf{TERS\#_{MULTIMER}} = \sim 2.0\%$$

Calculation:

$$\mathbf{TERS\#_{Dil}} = 92\%/2\% = \mathbf{1:45\ (F1)}$$

$$\mathbf{Pos_{CellsSpike}} = 270 \times 10^6/45 = \mathbf{6 \times 10^6\ (F2)}$$

$$\mathbf{Neg_{CellsSpike}} = 270 \times 10^6 - 6 \times 10^6 = \mathbf{264 \times 10^6\ (F3)}$$

$$\mathbf{Neg_{VolSpike}} = 40\,\text{mL}/290 \times 10^6 * 264 \times 10^6 = \mathbf{36.4\,mL\ (F4)}$$

$$\mathbf{Pos_{VolSpike}} = 15\,\text{mL}/10 \times 10^6 * 6 \times 10^6 = \mathbf{9.0\,mL\ (F4)}$$

$$\mathbf{TERS\# = 270 \times 10^6 = 264 \times 10^6 + 6 \times 10^6 \rightarrow 36.4\,mL + 9.0\,mL}$$

This example TERS batch of 270×10^6 cells is generated in a dilution of 1:45 with an expected signal of 2% MULTIMER^+ of CD8^+.

Tip: For the generation of a serial dilution (Fig. 4) with one Ag-specificity, high numbers of PBMCs (min. 60×10^6) need to be electroporated with high amounts (Table 2) of one selected TCR mRNA.

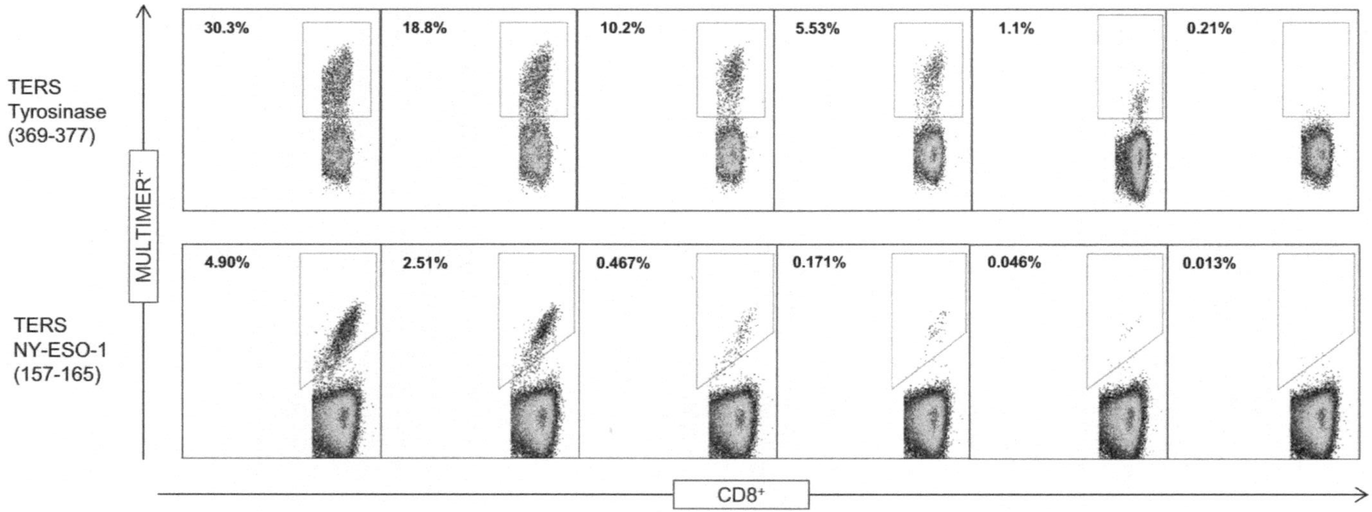

Fig. 4 Example for the generation of various serial diluted TERS derived from one PBMC donor and specific for tyrosinase$_{369–377}$ (upper part) or NY-ESO-1$_{157–165}$ (lower part). Single batches were tested for the number of Ag-specific CD8^{+} T cells in MULTIMER staining (MUL^{+}).

According to the calculation of one TERS batch (corresponds to one dilution), define the required Ag-specific signal and the final cell number for each dilution. We recommend to dilute the highly concentrated batch to the low one by successive adding cells of the Negative fraction with respect to the required signal and cell number for each dilution.

Tip: Generating complex TERS, containing more than one specificity (Bidmon et al., 2015), requires the EP of low to medium numbers of PBMCs ($20–40 \times 10^6$) with low amounts (Table 2) of different, selected TCR mRNAs in independent EP reactions. The cells are cultured independently whereas one Negative fraction is rested overnight. Finally, calculate the amount cells of each independent Ag-specific Positive fraction to be spiked in cells of one Negative fraction, considering the different TEs and the required signal of each Ag-specificity in the final TERS batch.

2.3.2.8 Cryopreservation

Freezing of the cells of the Negative fraction and of the generated TERS batch is performed following the lab established protocol. The number of cells frozen per vial (max. 30×10^6) should be tailored to the lab needs. Prior to freezing, each cell suspension is centrifuged at 300*g*, for 8 min, at 4 °C. The cell pellet is resuspended in freezing medium to a final concentration of, e.g., 5×10^6/1 mL.

Tip: The freezing of 5×10^6 per aliquot will allow a recovery of sufficient TERS cells for performing in average at least three tests in MULTIMER or CFC staining experiments, and at least seven wells in IFN-γ ELISpot.

Then, 1 mL aliquots are transferred for initial freezing either to a Mr. Frosty or to another available freezing system. Finally, the cells can be stored up to 12 months in liquid nitrogen. In our studies, we have shown that TERS deliver stable signals in repetitively performed MULTIMER staining experiments and functional tests after 12 months of storage (Bidmon et al., 2015).

Tip: Some TERS can even be stored up to 24 months in liquid nitrogen depending on the quality of the cells. For storage over 1 year, we recommend to control the batch stability frequently and to compare it to a freshly prepared control batch.

3. Routine application of TERS controls

The use of TERS to routinely control the performance of various cellular assays (MULTIMER staining, but also functional assessment of Ag-specific T cells by ELISpot or CFC) is a three-part concept (Fig. 5) that

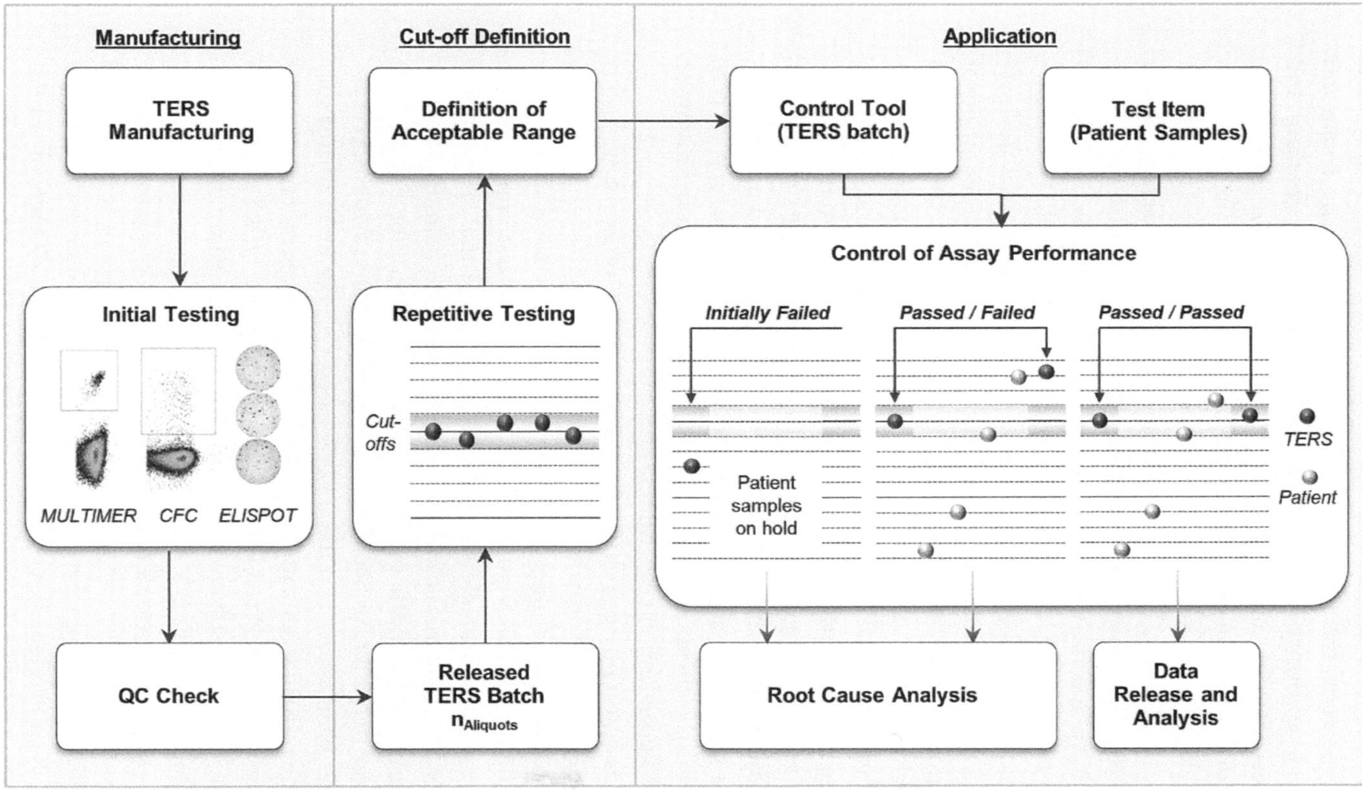

Fig. 5 The three steps of TERS batch QC control and application: manufacturing, cut-off definition, and application. After release of a TERS batch, batch-specific signal size and the acceptable signal range (cut-off values) need to be defined for the lab-specific assay protocols and the experimental setup. In the application phase, the pre-tested TERS aliquots can be subsequently used to control assay performance in a variety of ways, such as prior to assessing a series of test samples (e.g., patient samples) as well as after the last test sample has been run (Bidmon et al., 2015).

has been described in detail (Bidmon et al., 2015). Each newly TERS batch is tested (QC) to verify the actual existing number of Ag-specific T cells and its quality and performance in the user-defined assay after freezing and thawing. The viability and recovery of the cells should be >70%. A batch-to-batch variation of the Ag-specific signal is acceptable, as the variation across aliquots of each batch is within limits. We showed intra-batch variations of <10% in MULTIMER staining and <20% in functional assays using a range of TERS batches containing low to high numbers of T cells with multiple Ag-specificities (Bidmon et al., 2015). The positively tested batch is then released for long-term storage (Fig. 5, left panel).

In the second step, the TERS batch is validated for its assay-format-specific signal intensity and assay-dependent signal variation (definition of assay- and batch-specific cut-off values). For this purpose, we recommend to thaw five aliquots and to test them in independent experiments to define the expected median signal size and acceptable range (to be set at 20% above and below the median) for the specific assay protocol (Fig. 5, middle panel). Notably, the resulting ranges for acceptable assay performance strictly apply to the defined assay protocol and the experimental setup (including the hardware for data acquisition and strategy for analysis). Therefore, any type of changes within the protocol will require a batch-specific re-definition of assay-specific signals and cut-off values.

In the application phase, TERS can be employed at various time points for controlling assay performance: we recommend their testing prior to assessing the test samples (initial TERS-guided QC) as well as after the last test sample (finishing TERS-guided QC), whereas the test sample will in most cases be a patient sample specimen from a clinical trial. If the initial TERS sample leads to a result out of the accepted range (failure), the subsequent experiments with the precious test samples should be put on hold until the source of failure has been identified in a root cause analysis, corrective action has been taken, and the TERS-based QC has been re-run successfully. If the initial TERS sample passes the QC, the experiments with the test samples are initiated. Provided that the finishing last TERS sample also passes the QC test, all results from the test can be released for further use. Whenever the finishing TERS sample fails, this should again lead to a root cause analysis and resolution of problems prior to considering releasing the experimental data from the test samples (Fig. 5, right panel). If the clinical study or project is running for a longer period of time, intermediate TERS controls can be introduced at regular intervals. For a change of TERS batches within one test campaign (e.g., a clinical trial) we propose to implement a bridging test. Testing of five aliquots and the simultaneous cut-off definition of the newly prepared TERS batch (which is already qualified) should be performed in parallel with the last five aliquots of the expiring batch that is currently used within the routine application as assay control.

An alternative concept would be to run one TERS control sample in parallel to each test sample(s) in independent experiments (e.g., in MULTIMER staining or CFC). Especially for the application in IFN-y ELISpot assay, it is advisable to use one TERS control sample on each plate.

The application of TERS enables to sensitively detect the impact of common sources of variation on the results of T cell assays caused by the quality of the cells, reagent stability, cytometer settings and data analysis (Fig. 6).

Additionally, the use of different types of T-cell receptors allows to track the reactions of T cells with different TCR-affinity levels as low-affinity TCRs might provide more sensitive detection of failure. Generating different TERS batches, using low- to high-affinity TAA-specific TCRs (Fig. 7) and containing different numbers of Ag-specific T cells, e.g., serial dilution of TERS (Fig. 4), would allow to mimic different scenarios of signal detection that might occur testing patient samples, for instance when testing low-frequent tumor-specific responses in cancer patients. These TERS samples would help validating a particular assay for its sensitivity and the determination of the detection limits.

For the broad adoption of the TERS technology by the immune monitoring community, TCR mRNA is commercially available in the form of a kit based-approach (Bidmon et al., 2018), currently supported by JPT Peptide Technologies GmbH (https://shop.jpt.com/80-Immunology-Standards-Controls/141-TERS-Reference-Samples/1000739-Universal-T-Cell-Assay-Standard-TERS-trade-Kit-1.html).

4. Summary

The TERS technology offers a robust and versatile tool to overcome a major hurdle in the broad adoption of commonly used T-cell assays in the clinic. It allows tailored generation of reference samples equipped with a defined Ag-specificity that ideally match the demands in local assay applications. These stable, cell-based standard samples, may serve best to sensitively control common sources of variations in T-cell assays, supporting the validation and harmonization of results generated in correlative biomarker studies and for clinical decision-making.

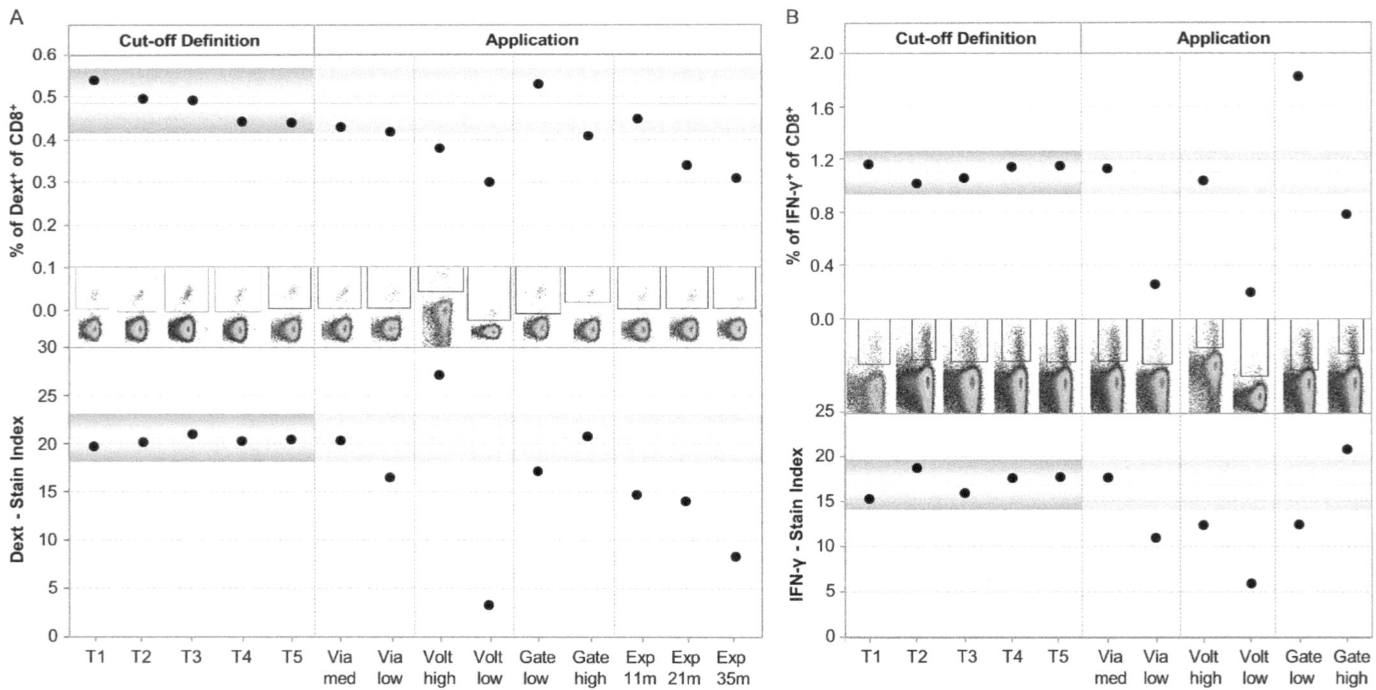

Fig. 6 Implementation of TERS to detect common sources of variation in T cell assays. (A) The figure shows the results of a TERS batch specific for NY-ESO-$1_{157–165}$ (SLLMWITQC) tested in MULTIMER staining: % MULTIMER$^+$ of CD8$^+$ (*y*-axis; upper part), representative dot plots (middle part), and the stain index (lower part) are shown. For cut-off definition (*x*-axis; T1–T5), five TERS aliquots were independently tested under standardized conditions. The median is marked as solid line and the acceptable ranges as gray area. In the application phase, TERS were tested under suboptimal conditions of the following: (1) cell quality (Via: viability); (2) cytometer settings (Volt: voltages); (3) gating (Gate: inaccurate gating); and (4) reagent stability (expired dextramer). (B) One TERS batch specific for $\text{tyrosinase}_{368–376}$ (YMDGTMSQV) was tested in CFC after peptide-specific stimulation. The *y*-axis shows the IFN-γ signals of CD8$^+$ cells (upper part), the corresponding dot plots (middle part), and the stain index (lower part). Standardized testing and cut-off definition was performed as in (A). Applying TERS under suboptimal conditions led to out of range results (Bidmon et al., 2015).

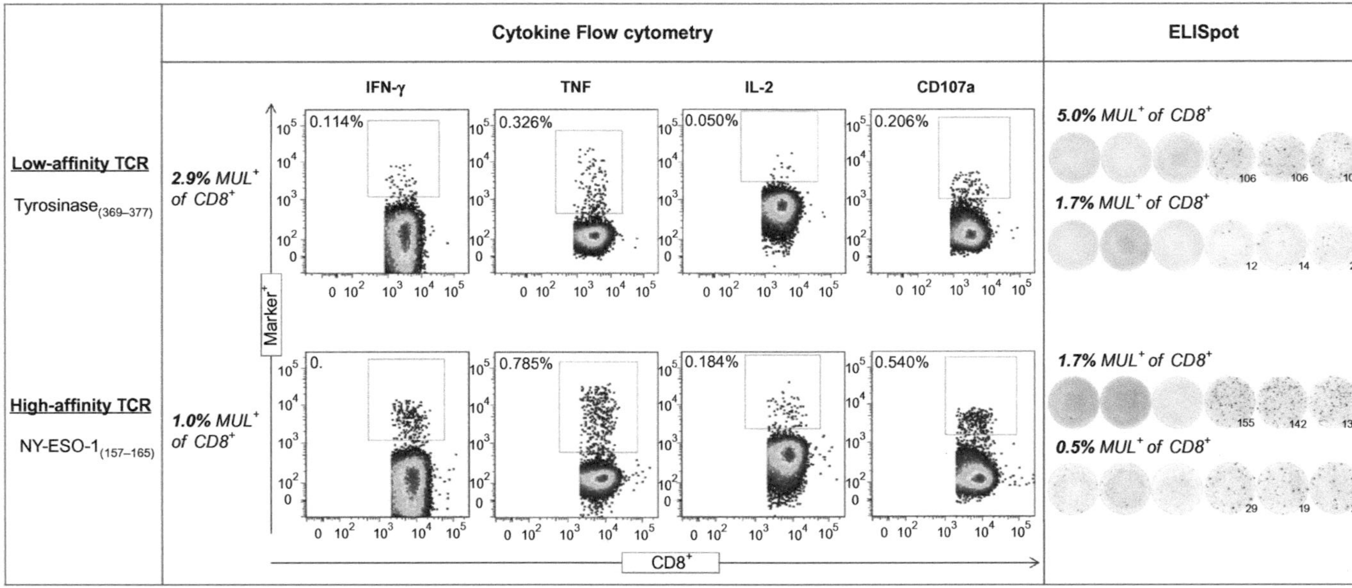

Fig. 7 TCR-affinity-dependent TERS performance in functional assays. Two $CD8^+$ TERS are depicted, one specific for tyrosinase$_{369–377}$ (upper part, low-affinity TCR, 2.9% tyrosinase-MULTIMER$^+$ of $CD8^+$), and the other specific for NY-ESO-1$_{157–165}$ (lower part, high-affinity TCR, 1% NY-ESO-1-MULTIMER$^+$ of $CD8^+$). Left panel: IFN-γ, TNF, IL-2 and the expression of CD107a on $CD8^+$ T cells are shown, after cognate peptide-induced activation (overnight) in CFC. The right panel compares TERS batches specific for tyrosinase$_{369–377}$ (upper part) or NY-ESO-1$_{157–165}$ (lower part), each generated in two different dilutions and tested for IFN-γ production in ELISpot. Both panels show that the tyrosinase (low-affinity)-specific TERS needs to be manufactured containing higher number of TCR^+ $CD8^+$ cells to show clearly detectable signals in functional assays as compared to the NY-ESO-1-specific TERS containing $CD8^+$ T cells expressing a high-affinity TCR.

Acknowledgments

The project was funded by the Wallace Coulter Foundation (WCF), Florida, USA. Many thanks to the supervisor Cedrik M. Britten for his passionate engagement in promoting the successful progress of this project. We thank the whole Association for Cancer Immunotherapy Immunoguiding Program (CIP) team for its continuous support. The experiments were performed in the labs kindly provided by Prof. Dr. Ugur Sahin at the University Medical Center of the Johannes Gutenberg University and the Translational Oncology at the Medical Center of the University Mainz (TRON gGmbH), Mainz, Germany.

References

Alanio, C., Lemaitre, F., Law, H. K., Hasan, M., & Albert, M. L. (2010). Enumeration of human antigen-specific naive CD8+ T cells reveals conserved precursor frequencies. *Blood*, *115*(18), 3718–3725.

Bidmon, N., Attig, S., Rae, R., Schröder, H., Omokoko, T. A., Simon, P., et al. (2015). Generation of TCR-engineered T cells and their use to control the performance of T cell assays. *Journal of Immunology*, *194*(12), 6177–6189.

Bidmon, N., Kind, S., Welters, M. J. P., Joseph-Pietras, D., Laske, K., Maurer, D., et al. (2018). Development of an RNA-based kit for easy generation of TCR-engineered lymphocytes to control T-cell assay performance. *Journal of Immunological Methods*, *458*, 74–82.

Britten, C. M., Janetzki, S., Ben-Porat, L., Clay, T. M., Kalos, M., Maecker, H. T., et al. (2009). Harmonization guidelines for HLA-peptide multimer assays derived from results of a large scale international proficiency panel of the cancer Vaccine Consortium. *Cancer Immunology, Immunotherapy*, *58*(10), 1701–1713.

Britten, C. M., van der Burg, S. H., & Gouttefangeas, C. (2015). A framework for T cell assays. *Oncotarget*, *6*(34), 35143–35144.

Chudley, L., McCann, K. J., Coleman, A., Cazaly, A. M., Bidmon, N., Britten, C. M., et al. (2014). Harmonisation of short-term in vitro culture for the expansion of antigen-specific CD8(+) T cells with detection by ELISPOT and HLA-multimer staining. *Cancer Immunology, Immunotherapy*, *63*(11), 1199–1211.

Cron, A., Gouttefangeas, C., Frelinger, J., Lin, L., Singh, S. K., Britten, C. M., et al. (2013). Hierarchical modeling for rare event detection and cell subset alignment across flow cytometry samples. *PLoS Computational Biology*, *9*(7), e1003130.

Garcia, A., Keinonen, S., Sanchez, A. M., Ferrari, G., Denny, T. N., & Moody, M. A. (2014). Leukopak PBMC sample processing for preparing quality control material to support proficiency testing programs. *Journal of Immunological Methods*, *409*, 99–106.

Jaimes, M. C., Maecker, H. T., Yan, M., Maino, V. C., Hanley, M. B., Greer, A., et al. (2011). Quality assurance of intracellular cytokine staining assays: Analysis of multiple rounds of proficiency testing. *Journal of Immunological Methods*, *363*(2), 143–157.

Janetzki, S., & Britten, C. M. (2012). The impact of harmonization on ELISPOT assay performance. *Methods in Molecular Biology*, *792*, 25–36.

Kuhn, A. N., Diken, M., Kreiter, S., Selmi, A., Kowalska, J., Jemielity, J., et al. (2010). Phosphorothioate cap analogs increase stability and translational efficiency of RNA vaccines in immature dendritic cells and induce superior immune responses in vivo. *Gene Therapy*, *17*(8), 961–971.

Maecker, H. T., McCoy, J. P., Amos, M., Elliott, J., Gaigalas, A., Wang, L., et al. (2010). A model for harmonizing flow cytometry in clinical trials. *Nature Immunology*, *11*(11), 975–978.

Scriba, T. J., Purbhoo, M., Day, C. L., Robinson, N., Fidler, S., Fox, J., et al. (2005). Ultrasensitive detection and phenotyping of CD4+ T cells with optimized HLA class II tetramer staining. *Journal of Immunology*, *175*(10), 6334–6343.

Singh, S. K., Tummers, B., Schumacher, T. N., Gomez, R., Franken, K. L., Verdegaal, E. M., et al. (2013). The development of standard samples with a defined number of antigen-specific T cells to harmonize T cell assays: A proof-of-principle study. *Cancer Immunology, Immunotherapy*, *62*(3), 489–501.

Stebbings, R., Wang, L., Sutherland, J., Kammel, M., Gaigalas, A. K., John, M., et al. (2015). Quantification of cells with specific phenotypes I: Determination of CD4+ cell count per microliter in reconstituted lyophilized human PBMC prelabeled with anti-CD4 FITC antibody. *Cytometry. Part A: The Journal of the International Society for Analytical Cytology*, *87*(3), 244–253.

van der Burg, S. H., Kalos, M., Gouttefangeas, C., Janetzki, S., Ottensmeier, C., Welters, M. J., et al. (2011). Harmonization of immune biomarker assays for clinical studies. *Science Translational Medicine*, *3*(108), 108–144.

Wang, L., Stebbings, R., Gaigalas, A. K., Sutherland, J., Kammel, M., John, M., et al. (2015). Quantification of cells with specific phenotypes II: Determination of CD4 expression level on reconstituted lyophilized human PBMC labelled with anti-CD4 FITC antibody. *Cytometry. Part A: The Journal of the International Society for Analytical Cytology*, *87*(3), 254–261.

Welters, M. J., Gouttefangeas, C., Ramwadhdoebe, T. H., Letsch, A., Ottensmeier, C. H., Britten, C. M., et al. (2012). Harmonization of the intracellular cytokine staining assay. *Cancer Immunology, Immunotherapy*, *61*(7), 967–978.

Xu, Y., Theobald, V., Sung, C., DePalma, K., Atwater, L., Seiger, K., et al. (2008). Validation of a HLA-A2 tetramer flow cytometric method, IFNgamma real time RT-PCR, and IFNgamma ELISPOT for detection of immunologic response to gp100 and MelanA/MART-1 in melanoma patients. *Journal of Translational Medicine*, *6*, 61.

Further reading

Britten, C. M., Janetzki, S., Butterfield, L. H., Ferrari, G., Gouttefangeas, C., Huber, C., et al. (2012). T cell assays and MIATA: The essential minimum for maximum impact. *Immunity*, *37*(1), 1–2.

CHAPTER ELEVEN

In vitro expansion of Vγ9Vδ2 T cells for immunotherapy

Christian Peters[a], Léonce Kouakanou[b], Hans-Heinrich Oberg[a], Daniela Wesch[a], Dieter Kabelitz[a,*]

[a]Institute of Immunology, University Hospital Schleswig-Holstein and Christian-Albrechts University Kiel, Kiel, Germany

[b]Institute of Immunology, Christian-Albrechts-University Kiel, Kiel, Germany

*Corresponding author: e-mail address: dietrich.kabelitz@uksh.de

Contents

Abstract

Vγ9Vδ2 T cells can exhibit potent anti-tumor activity and have gained much attention as an interesting tool for cancer immunotherapy by adoptive cell transfer. In this chapter, we explain the different approaches for the *in vitro* expansion of Vγ9Vδ2 T cells for their subsequent adoptive transfer and present a detailed protocol for the successful *in vitro* expansion at the laboratory scale. Finally, we discuss strategies for optimization and parameters necessary for the adaption to a clinical scale protocol.

Abbreviations

2M3B1-PP	2-methyl-3-butenyl-1-pyrophosphate
BrHPP	bromohydrin pyrophosphate
GMP	good manufacturing practice
HMBPP	(*E*)-4-hydroxy-3-methyl-but-2-enyl pyrophosphate
IPP	isopentenyl pyrophosphate
mAb	monoclonal antibody
nBP	aminobisphosphonates
pAg	phosphoantigens
pVC	L-ascorbic acid 2-phosphate

Methods in Enzymology, Volume 631
ISSN 0076-6879
https://doi.org/10.1016/bs.mie.2019.07.019

1. Introduction

1.1 γδ T cells in immunotherapy

γδ T cells combine certain features of the innate and the adaptive immune system. They express a clonally variable T-cell receptor (TCR) but recognize antigens independently of HLA molecules. Moreover, γδ T cells share the functional expression of certain pattern recognition receptors with innate immune cells. At the transcriptional level γδ T cells share patterns of both NK cells and αβ T cells (Pont et al., 2012). γδ T cells can be differentiated into two major populations according to their Vδ-element usage. γδ T cells expressing a TCR comprising a Vδ1 element are more prevalent in tissues, whereas Vδ2 expressing T cells are the most abundant γδ T-cell population in the peripheral blood. Vδ2 T cells almost exclusively co-express the Vγ9 element (further on referred to as Vγ9Vδ2) and can be specifically activated by small pyrophosphate antigens (pAg), such as intermediates of the mevalonate or non-mevalonate pathway of isoprenoid biosynthesis. Synthetic or naturally occurring pyrophosphates, like bromohydrin pyrophosphate (BrHPP), 2-methyl-3-butenyl-1-pyrophosphate (2M3B1-PP) or (*E*)-4-hydroxy-3-methyl-but-2-enyl pyrophosphate (HMBPP) and isopentenyl pyrophosphate (IPP) activate Vγ9Vδ2 T cells in a similar, BTNA3A1 (CD277)-dependent manner, yet with different potency (Gu, Nawrocka, & Adams, 2014). The mevalonate pathway is often dysregulated in tumor cells leading to the endogenous accumulation of the pAg IPP, which then induces the activation of Vδ2 T cells (Gober et al., 2003; Uchida et al., 2007). The pAg-triggered cytotoxic activity of Vγ9Vδ2 T cells against a broad range of tumor cells classifies them as potent effector cells for tumor immunotherapy. Furthermore, the activation of Vγ9Vδ2 T cells can be pharmaceutically induced by nitrogen-containing aminobisphosphonates (nBP) such as Zoledronate. nBP interfere with the mevalonate pathway by inhibiting an enzyme downstream of IPP (farnesyl pyrophosphate synthase), thereby causing the accumulation of upstream IPP, which in turn leads to the activation of Vγ9Vδ2 T cells. Because of their HLA-independent innate-like features Vγ9Vδ2 T cells have the capacity to recognize and kill many different tumor entities. A meta-analysis in 39 different tumor types recently identified the abundance of tumor-associated γδ T cells as the most important prognostic factor correlated with a favorable outcome (Gentles et al., 2015). Accordingly, Vγ9Vδ2 T cells have been used in clinical studies for the adoptive transfer into tumor patients, unfortunately as yet with limited success (Buccheri, Guggino, Caccamo, Li, & Dieli, 2014).

In general, there are two possible approaches to harness γδ T cells for immunotherapeutic approaches, and both have already been assessed in clinical trials. γδ T cells can be stimulated directly *in vivo* or they can be expanded *in vitro* before adoptive transfer into the patient. The *in vivo* activation can be achieved by the application of γδ T-cell activating drugs and low dose interleukin-2 (IL-2), whereas the *in vitro* expansion implicates a greater workload. Nonetheless, the *ex vivo* expansion of γδ T cells also holds clear advantages.

The *in vivo* response of Vγ9Vδ2 T cells to systemic pAg-stimulation is only transient, most likely because the repetitive stimulation leads to activation-induced cell death and exhaustion of the Vγ9Vδ2 T-cell population (Sicard et al., 2005). In addition, patients who had been treated with Zoledronate before (for bone diseases), showed less Vγ9Vδ2 T-cell numbers within the $CD3^+$ T cells and had a reduced proliferative capacity of Vγ9Vδ2 T cells upon *ex vivo* stimulation (Kalyan, Chandrasekaran, Quabius, Lindhorst, & Kabelitz, 2014; Nicol et al., 2011). In line with this, cancer patients also often have reduced numbers of Vγ9Vδ2 T cells within the $CD3^+$ T cell compartment, which resulted in lower Vγ9Vδ2 T-cell numbers in *ex vivo* generated therapeutic products upon pAg- or Zoledronate-stimulation of tumor patient PBMC. Together with highly variable expansion rates, this limits the efficacy of tumor patient-derived Vγ9Vδ2 T cells for the autologous adoptive transfer (Nicol et al., 2011).

Because γδ T cells are not HLA-restricted, they can in principle also be used in an allogeneic setting with a minimal risk for graft *versus* host reaction. In fact, allogeneic Vγ9Vδ2 T-cells established from healthy donors have been already applied and seem to be well tolerated, without signs of graft *versus* host disease induction (Alnaggar et al., 2019). Therefore, allogeneic Vγ9Vδ2 T cells have the potential to be applied as an easy to generate off-the-shelf cellular therapeutic product.

1.1.1 In vitro *expansion of Vγ9Vδ2 γδ T cells*

Vγ9Vδ2 T cells account for 50–95% of the γδ T cells within the 2–5% γδ T cells contained in PBMC. Since these cells are the only ones responding to pAg, they can be easily stimulated and selectively expanded *ex vivo* to high cell numbers (up to 800-fold expansion) using pAg or nBP plus appropriate growth factors.

Various naturally occurring pAg, such as IPP or the more efficient microbial HMBPP can be used for Vγ9Vδ2 T-cell stimulation and their efficient *in vitro* expansion. However, in clinical studies performing adoptive transfer of Vγ9Vδ2 T cells, so far only the synthetic pAg BrHPP

(IPH1101, Picostim®) at 3 μM (Bennouna et al., 2008) and 2M3B1-PP at 100 μM (Kobayashi, Tanaka, Yagi, Minato, & Tanabe, 2011; Kobayashi et al., 2007) have been used for the Vγ9Vδ2 T cell *ex vivo* expansion.

As mentioned above, nBP can be also used for efficient expansion of Vγ9Vδ2 T cells. In general, for the generation of Vγ9Vδ2 T-cell lines for the adoptive transfer, Zoledronate was used in a range from 1 to 5 μM (Hoeres, Smetak, Pretscher, & Wilhelm, 2018).

An interesting variation of this protocol was recently published by Nada et al., who used much higher concentrations of Zoledronate (100 μM) to pulse PBMC for only a short time period (4 h). Subsequently, PBMC containing the Vγ9Vδ2 T cells were washed and transferred into fresh medium for further expansion in the presence of IL-2 (1000 IU) or IL-15, which were not added to the cell culture before day 3. These cells expanded to higher numbers (~2-fold increase) compared to the cell culture in the continuous presence of lower concentrations of Zoledronate (10 μM) (Nada, Wang, Workalemahu, Tanaka, & Morita, 2017).

1.1.2 In vitro *expansion of Vδ1 T cells*

Within the γδ T cell population in peripheral blood, Vδ1 T cells (expressing various Vγ chains) are the second most frequent subset next to the Vγ9Vδ2 T cells. Vδ1 T cells are more abundant in epithelial tissues and also display potent anti-tumor activity. For this population, a pharmacological activation comparable to the pAg-induced activation of Vγ9Vδ2 T cells is not available. Even though this overview focusses on the expansion of Vγ9Vδ2 T cells, it should be mentioned that Almeida et al. have established a two-step protocol for the expansion of Vδ1 T cells. As a starting product they used γδ T cells positively isolated by magnetic cell sorting, which were expanded for 14 days by anti-CD3 antibody stimulation in the presence of IL-1β, IL-4, IL-21 and IFN-γ. Subsequently, the cells were transferred into fresh cell culture medium supplemented with IL-15 and IFN-γ, were re-stimulated by anti-CD3 antibody and cultured for one more week. This expansion protocol led to an accumulation of ~70% Vδ1 T cells with a high expression of NK-receptors and high activity against chronic lymphocytic leukemia (CLL) cells (Almeida et al., 2016).

1.1.3 *Substances used to modulate Vγ9Vδ2 γδ T-cell differentiation and to augment their anti-tumor reactivity*

Vγ9Vδ2 T cells stimulated with pAg display a cytokine profile, which resembles a Th1 profile, with a characteristic production of IFN-γ and TNF-α

(Dunne, Mangan, Madrigal-Estebas, & Doherty, 2010). The stimulation with nBP also induces IFN-γ release and induces a phenotype similar to direct pAg-stimulation (Kunzmann, Bauer, & Wilhelm, 1999). Such a Th1-like phenotype can be further shaped by additional factors and fits the requirements of immunotherapy. The Vγ9Vδ2 T-cell-based immunotherapy by adoptive transfer so far has been demonstrated to be safe in the application, but the anti-tumor efficacy is rather limited (Buccheri et al., 2014; Hoeres et al., 2018). Therefore, possibilities to improve the therapy have to be elucidated and introduced into the current protocols.

1.1.3.1 Vitamin C

Vitamin C (L-ascorbic acid) is a water-soluble micronutrient with known activity as antioxidant and epigenetic modifier. In a recent study, we found a positive impact of vitamin C on the *in vitro* proliferation and activation of Vγ9Vδ2 T cells. We used the synthetic phospho-modified vitamin C analog L-ascorbic acid 2-phosphate (pVC), which is more stable and less toxic compared to vitamin C when supplemented to the cell culture. When pVC was present at 50 μg/mL during the Vγ9Vδ2 T-cell activation and subsequent cell culture, the proliferative potential of these cells was clearly increased, as was the cytokine production (*e.g.*, IFN-γ) and the cytotoxic effector activity (Kouakanou et al., 2019).

1.1.3.2 IL-2 and IL-15

For the expansion of Vγ9Vδ2 T cells usually a repetitive administration of exogenous IL-2 is used to support/enhance the proliferation and final yield of Vγ9Vδ2 T cells (Capietto et al., 2010). To expand Vγ9Vδ2 T cells for the adoptive transfer, all studies conducted so far only used IL-2, in concentrations ranging from 100 to 320 IU, when used in combination with pAg and from 400 to 1000 IU IL-2 when used in combination with Zoledronate. In most studies IL-2 was replenished by exogenous supplementation in cycles ranging from 1 to 3 days, as summarized in following reviews (Fisher, Heuijerjans, Yan, Gustafsson, & Anderson, 2014; Hoeres et al., 2018).

IL-15 is thought to be more efficient compared to IL-2 in expanding cells with an effector phenotype. When the supplementation of IL-2 and IL-15 were compared for the Zoledronate-induced expansion of Vγ9Vδ2 T cells, it was found that IL-15-expanded Vγ9Vδ2 T cells displayed an early/central memory phenotype ($CD27^{+}CD28^{+/-}$), whereas IL-2-expanded cells displayed an effector phenotype ($CD27^{-}CD28^{-}$) (Nada et al., 2017). Moreover, Vδ2 T cells within tumor-infiltrating lymphocytes of renal cell carcinoma patients displayed mainly a T effector memory (T_{em})-phenotype ($CD27^{-}CD45RA^{-}$) and only proliferated weakly upon pAg-stimulation with BrHPP when combined with IL-2 (100 IU), but responded better to IL-15 (Viey et al., 2008). In a recent study, we found that BrHPP-activated Vγ9Vδ2 T cells, expanded in the presence of IL-15 exhibited a higher cytotoxic activity against different adherent tumor cell lines, when compared to Vγ9Vδ2 T cells expanded in presence of IL-2. Most interestingly, we observed that CD56 surface expression, which is often interpreted as a marker for cytotoxic activity, is specifically induced by IL-15 (Peters et al., 2019). In addition, $CD56^{+}$ γδ T cells were shown to be the population within the γδ T cells, which exerted anti-tumor activity toward squamous cell carcinoma (Alexander et al., 2008). In Zoledronate-expanded Vγ9Vδ2 T cells, IL-15 in combination with IL-2 augmented the proliferation as well as the *in vitro* anti-tumor activity (Van Acker et al., 2016).

Taken together, data from these studies support the use of IL-15 instead or in combination with IL-2 for the generation of Vγ9Vδ2 T cells with an augmented effector potential, which might improve the efficacy of the Vγ9Vδ2 T-cell based immunotherapy.

1.1.3.3 TGF-β

When studying the effects of TGF-β on the differentiation of Vδ2 T cells, we found that purified Vδ2 T cells cultured in the presence of transforming growth factor-β (TGF-β), expanded to high cell numbers and displayed enhanced cytotoxic potential. In line with this, we observed that the adhesion molecule profile was substantially different in these TGF-β-exposed Vδ2 T cells. Moreover, the conjugate formation with tumor cells was enhanced, for which we identified the E-cadherin-binding molecule CD103 to play an essential role (Peters et al., 2019). The higher proliferative potential and the altered adhesion molecule profile point to a possibly increased migratory capacity into tissues and a longer *in vivo* lifespan of these cells, which would both be positive features for the immunotherapy of solid tumors.

1.1.3.4 Antibodies and antibody constructs

The specificity of adoptively transferred Vγ9Vδ2 T cells can be enhanced by directly targeting them to tumor cell surface antigens. This can be achieved by antibody-mediated cellular cytotoxicity (ADCC), taking advantage of the activation of Vγ9Vδ2 T cells *via* antibody binding to their FcR. In a mouse model of xenografted human mammary carcinoma, it was found that the adoptive transfer of BrHPP + IL-2-expanded Vγ9Vδ2 T cells on its own had only very little impact on the tumor growth. However, the adoptive transfer strongly enhanced the efficacy of the trastuzumab (anti-Her2) antibody-treatment, when both were given in combination (Capietto, Martinet, & Fournie, 2011). In a Phase I/II clinical study the application of rituximab (anti-CD20) in low grade follicular lymphoma was combined with *in vivo* BrHPP-activation of Vγ9Vδ2 T cells, which led to enhanced efficacy of the treatment (Laurent et al., 2009). Another promising antibody-based approach to specifically enhance γδ T-cell cytotoxicity is the usage of bispecific antibodies such as [(HER2)$_2$xVγ9] or [(Her2)$_2$xCD16] to target Vγ9- or CD16-expressing γδ T cells to HER2/neu-expressing tumor cells. In a pancreatic ductal adeno carcinoma xenograft mouse model it was further demonstrated that adoptive transfer of Vγ9Vδ2 T cells in combination with the bispecific antibody [(HER2)$_2$xVγ9] inhibited the tumor growth *in vivo*, whereas Vγ9Vδ2 T cells alone were ineffective (Oberg et al., 2018, 2014).

2. Protocol for the Vγ9Vδ2 T-cell expansion at laboratory scale

2.1 Material and substances

1. Ficoll-Hypaque solution (Biocoll; Biochrom AG, Berlin, Germany, #L6115)
2. 6-Well cell culture plates (*e.g.*, Greiner Bio-One, Frickenhausen, Germany, #6557160)
3. LS-columns (Miltenyi Biotec, Bergisch Gladbach, Germany, #130-042-401)
4. MB (MACS-buffer; PBS + 0.5% BSA + 2 mM EDTA)
5. Fc-Blocking reagent (Miltenyi Biotec, #130-059-901)
6. Anti-mouse IgG micro beads (Miltenyi Biotec, #130-048-401)
7. BrHPP (Innate Pharma, Marseille, France), alternatively HMBPP (Echelon Biosciences, Salt Lake City, UT, USA, #I-M055) or Zoledronate (Zometa®, Novartis, Basel, Switzerland)
8. r-hu-IL-2 (recombinant human IL-2: Aldesleukin; Proleukin®, Novartis)
9. pVC (L-ascorbic acid 2-phosphate, Sigma-Aldrich/Merck, Darmstadt, Germany, #A8960)
10. RPMI-1640 cell culture medium, 25 mM Hepes, 2 mM L-glutamine, phenol red (*e.g.*, Thermo Fisher Scientific/Gibco™, Waltham, MA,

USA, #52400-025) with 10% heat-inactivated FCS (*e.g.*, Life Technologies/Gibco™, Waltham, Massachusetts, #10270-106) and antibiotics (penicillin/streptomycin); CST Optimizer™ T-Cell expansion serum-free medium with phenol red, no glutamine (Thermo Fisher Scientific/Gibco™, #A10485-01); TexMACS with glutamine and phenol red (Miltenyi, #130-097-196); X-Vivo™ 15 with L-glutamine, gentamicin and phenol red (Lonza, Walkersvill, MD, USA, #BE02-060F)

11. Monoclonal antibodies for flow cytometry-based assessment of the purity of Vγ9Vδ2 T cells. Anti-Vγ9-Fitc (clone: B3, BD Biosciences, Franklin Lakes, NJ, USA, #555732) or anti-Vδ2-FITC (clone B6, BD Scientific, #555739) and anti-CD3-PE (clone: OKT-3, BD Scientific, #345765).

2.2 Protocol

Using the protocol listed below (Fig. 1A) purity of Vγ9Vδ2 T cells of above 95% can be achieved starting from total PBMC, without further depletion of unwanted cells (Fig. 1B).

1. Ficoll separation of PBMC from the peripheral blood or leukocyte concentrate.
   1.1 Bring Ficoll-Hypaque solution to room temperature before use.
   1.2 Overlay a 15 mL Ficoll cushion with 30 mL diluted peripheral blood. For a total of 60 mL peripheral blood, 60 mL PBS (1 ×) is added before the mixture is carefully overlaid on the Ficoll cushions of four tubes. Centrifugate the tubes at 693 *g* for 20 min and collect the PBMC from the interphase on top of the Ficoll cushion into a separate tube. Fill up with PBS and centrifugate at 444 *g* for 5 min, discard the supernatant and wash the pelleted PBMC twice with PBS at 174 *g*, 10 min.
2. Prepare the appropriate volume of cell culture medium + substances. Usually we use RPMI-1640 cell culture medium with 25 mM Hepes, 2 mM L-glutamine, antibiotics and phenol red supplemented with 10% heat-inactivated FCS for the generation of Vγ9Vδ2 T-cell lines.
   2.1 Add Zoledronate at 2.5 μM and IL-2 at 50 IU/mL to the cell culture medium.
   2.2 Or alternatively, add HMBPP at 1–10 nM (or BrHPP at 300 nM) and IL-2 at 50 IU/mL.
   2.3 In addition to above 2.1 or 2.2, pVC (50 μg/mL) can be added.

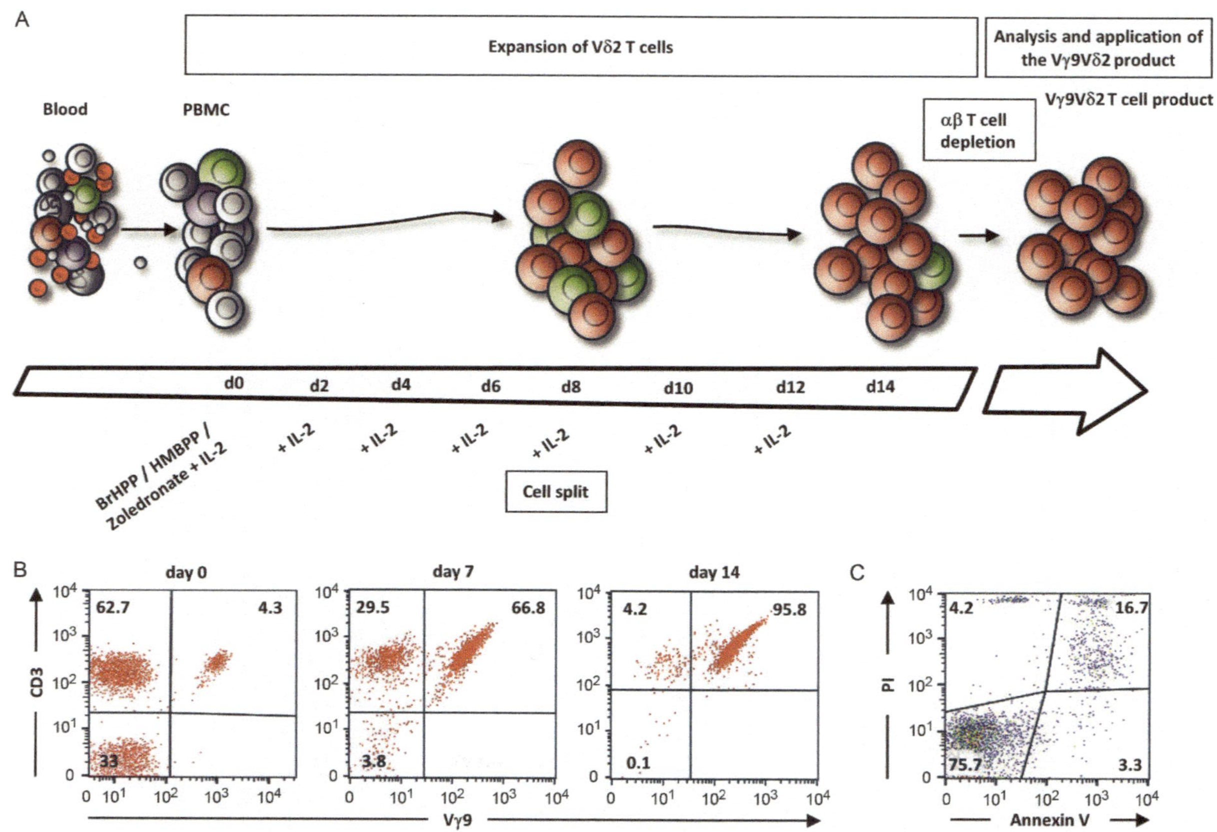

Fig. 1 (A) Scheme of the Vγ9Vδ2 T-cell expansion by direct stimulation of PBMC. (B) The purity of Zoledronate-expanded Vγ9Vδ2 T-cells (2.5 μM Zoledronate and 50 IU IL-2 in RPMI + 10% FCS) was determined on days 0, 7, and 14 by flow cytometry using anti-CD3 (clone: OKT-3) and anti-Vγ9 (clone: 7A5) monoclonal antibodies. (C) The percentage of viable cells was determined by flow cytometry using annexin V/propidium iodide staining, identifying viable cells (Annexin V^-/PI^-), early apoptotic cells (Annexin V^+/PI^-) and late apoptotic cells (Annexin V^+/PI^+).

3. *Re*-suspend PBMC in cell culture medium + substances at 1×10^6 cells/mL in a 6-well flat bottom plate, at 5 mL per well (5×10^6 cells/well).
4. Supplement the Vγ9Vδ2 T cell culture with IL-2 (1:2000, Stock concentration 100.000 IU/mL) to a final concentration of 50 IU/mL every second day during the whole course of the cell culture period.
5. Cell split: Whenever the cell culture medium color changes to yellow (usually in between day 6–8) collect the whole cell culture (all cells and medium) into a 50 mL tube, add 0.5 × volume of fresh cell culture medium, add IL-2 to a final concentration of 50 IU/mL and re-seed the cells in 5 mL per well.
6. The expansion of the Vγ9Vδ2 T-cell product will take approximately 12–14 days. Assess the purity by flow cytometry on day 10, 12, 14. A purity of >95% is desirable. This is routinely achieved in our laboratory if the starting PBMC contain at least 2–3% γδ T cells. The Vγ9Vδ2 T-cell lines can be usually kept in culture for up to ~21 days without further re-stimulation upon constant IL-2 supplementation and can be used for functional assays. The viability of the Vγ9Vδ2 T-cell lines can be assessed by flow cytometry using annexin V/propidium iodide staining (Vermes, Haanen, Steffens-Nakken, & Reutelingsperger, 1995), and usually should exceed 80% (Fig. 1C).
7. If Vγ9Vδ2 T-cell purity higher than 95% is required or such a high purity was not achieved but is needed, αβ T cells can be depleted from the Vγ9Vδ2 T-cell product.
   - **7.1** Vγ9Vδ2 T-cell containing cell cultures are collected into 50 mL Falcon tubes and are pelleted by centrifugation (300 *g*, 5 min).
   - **7.2** 100×10^6 cells are re-suspended in 600 μL MB + 100 μL Fc-blocking reagent and incubated for 15 min at 4 °C. Subsequently, the cells are washed with 10 mL MB and pelleted by centrifugation (5 min, 300 *g*).
   - **7.3** Discard the supernatant and re-suspend the cells in 850 μL MB +anti-TCRαβ (clone: WT31, BD Biosciences, #347770) at a final concentration of 10 μg/mL. Incubate 25 min at 4 °C and wash by adding 10 mL MB and pellet by centrifugation (5 min, 300 *g*).
   - **7.4** Re-suspend the cell pellet in 800 μL MB plus 200 μL anti-mouse IgG micro beads. Incubate 15 min at 4 °C and wash the cells by adding 10 mL MB and subsequent centrifugation (5 min, 300 *g*).
   - **7.5** Re-suspend the cell pellet in 3 mL MB and apply to a LD-column equilibrated with 3 mL MB. Wash the column twice with 1.5 mL MB. Collect the flow through and analyze the Vγ9Vδ2 purity by flow cytometry.

3. Adaptations for the clinical application

For the adaptation of a Vγ9Vδ2 T-cell expansion protocol for clinical application several requirements have to be met. Any such clinical therapeutic approach requires the approval of the respective administrative authorities. In the EU an immunotherapy by *in vitro* expanded Vγ9Vδ2 T cells is presumably classified as an advanced therapy medicinal product (ATMP) or advanced therapy investigational medicinal product (ATIMP) according to guidelines by the European Medicines Agency (EMA). These products are grouped into three classes, gene therapeutics, tissue-engineering-products and somatic cell therapeutics. A Vγ9Vδ2 T-cell-based immunotherapy would be considered a somatic cell therapy, as it employs *in vitro* expansion conditions including the use of certain cytokines to expand and differentiate somatic cells so that they exert the desired immunologic function *in vivo*. The generation of such a product is controlled by the EU guideline[1] according to the EU Regulation (EC) N° 1394/2007. In the United States, an immunotherapy by adoptive transfer of Vγ9Vδ2 T cells could be classified by the FDA (U.S. Food & Drug Administration) as regenerative medicine advanced therapy (RMAT), which includes similar categories as ATMP and is regulated according to Section 3033 of the 21st Century Cures Act.

The most important point for the development of an *ex vivo* expansion protocol according to EMA/FDA regulations is that the product for the adoptive transfer has to be manufactured under good manufacturing practice (GMP) conditions and in appropriately classified laboratories. Therefore, all materials and cytokines used have to be GMP-grade. In general, there are two possible ways for the generation of such a Vγ9Vδ2 T-cell product. The expansion can be done in a completely closed system (such as the Miltenyi Biotech CliniMACS Prodigy) or in a semi-closed system, culturing the cells in air-permeable Teflon cell culture bags under GMP conditions.

Under standard research laboratory conditions, Vγ9Vδ2 T cells are usually cultured in RPMI-1640 cell culture medium supplemented with 10% FCS. For clinical application, components of animal origin are usually not approved. Therefore, serum-free culture conditions are preferred, which in turn can affect the *ex vivo* pAg-induced activation and proliferation of the Vγ9Vδ2 T cells. To use the optimal concentration of the respective

[1] https://ec.europa.eu/health/sites/health/files/files/eudralex/vol-4/2017_11_22_guidelines_gmp_for_atmps.pdf.

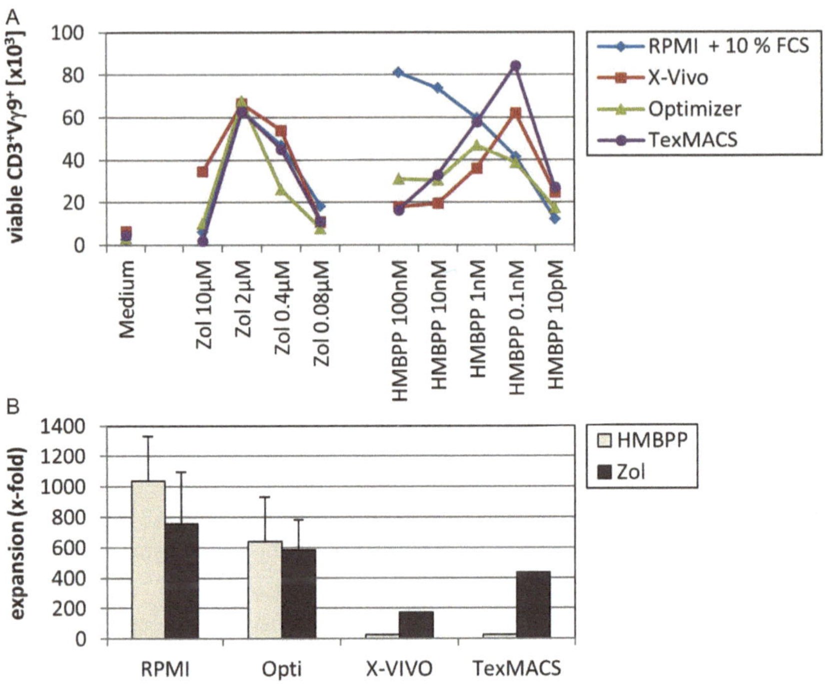

Fig. 2 The influence of the cell culture medium on the Vγ9Vδ2 T-cell expansion. (A) Vγ9Vδ2 T cells within PBMC were stimulated by different concentrations of Zoledronate (Zol) or HMBPP in different cell culture media (RPMI + 10% FCS, Optimizer, X-Vivo, TexMACS) in the presence of 50 IU/mL IL-2. The number of viable $V\gamma9^+$ T cells within the $CD3^+$ cells (total T cells) was determined by flow cytometry on day 7 after initial activation. Depicted are mean values of four independent experiments using PBMC from different donors. (B) Vγ9Vδ2 T-cell lines were expanded in different cell culture media containing 50 IU/mL IL-2 using 1 nM HMBPP or 2 μM Zoledronate. The expansion ratio was calculated according to the starting cell number of Vδ2 T cells within the PBMC and the flow cytometrically determined number of viable $V\gamma9^+$ T cells on day 13 (RPMI, Optimizer: $n = 2–3$; X-VIVO, TexMACS: $n = 1$).

Vγ9Vδ2 T-cell stimulus it is necessary to titrate these substances for the selected cell culture medium, as we did for RPMI + 10% FCS and three commercially available serum-free cell culture media (Fig. 2). As can be seen, serum-free media enabled comparable γδ T-cell expansion in response to Zoledronate whereas the dose-response patterns in response to pAg HMBPP differed significantly between RPMI + 10% FCS and serum-free culture media. In cases in which the addition of serum is necessary human AB serum can be used, which is available as clinical grade product.

In case of an autologous adoptive transfer the purity of expanded Vγ9Vδ2 T cells does not necessarily need to be very high as contaminating NK cells and αβ T cells have no negative influence (Kobayashi et al., 2011). To avoid the risk of a graft *versus* host disease in the case of an allogeneic therapy, it is necessary to transfer a cell product, which is free of contaminating αβ T cells (as well as B cells). Therefore, in most instances αβ T cells (and B cells) will have to be depleted at some point, either from the PBMC before the initial expansion or from the expanded Vγ9Vδ2 T-cell line. The depletion of residual NK cells is not necessary, since these cells will not induce a graft *versus* host reaction, but themselves exert anti-tumor functions. Furthermore, depletion of NK cells by *e.g.* anti-CD56 mAb would also eliminate a considerable proportion of the γδ T cells and therefore is not recommended.

To augment the efficiency of the γδ T-cell transfer, in some studies the patients were treated with nBP after the adoptive transfer of Vδ2 T cells to re-activate the transferred γδ T cells (Kobayashi et al., 2011; Nicol et al., 2011). Moreover, *in vitro* experiments suggest that tumor cells can be sensitized to γδ T-cell mediated killing by treatment with nBP and subsequent accumulation of pAg in the tumor cells (D'Asaro et al., 2010; Dhar & Chiplunkar, 2010; Mattarollo, Kenna, Nieda, & Nicol, 2007). Therefore, the combination of adoptive Vγ9Vδ2 T-cell transfer with Zoledronate treatment might enhance the efficacy of the cellular immunotherapy, as might additional strategies like enhancing the accessibility of the tumor microenvironment by pro-inflammatory stimuli (Hoeres et al., 2018).

References

Alexander, A. A., Maniar, A., Cummings, J. S., Hebbeler, A. M., Schulze, D. H., Gastman, B. R., et al. (2008). Isopentenyl pyrophosphate-activated CD56+ {gamma}{delta} T lymphocytes display potent antitumor activity toward human squamous cell carcinoma. *Clinical Cancer Research*, *14*, 4232–4240.

Almeida, A. R., Correia, D. V., Fernandes-Platzgummer, A., da Silva, C. L., da Silva, M. G., Anjos, D. R., et al. (2016). Delta one T cells for immunotherapy of chronic lymphocytic leukemia: Clinical-grade expansion/differentiation and preclinical proof of concept. *Clinical Cancer Research*, *22*, 5795–5804.

Alnaggar, M., Xu, Y., Li, J., He, J., Chen, J., Li, M., et al. (2019). Allogenic Vgamma9Vdelta2 T cell as new potential immunotherapy drug for solid tumor: A case study for cholangiocarcinoma. *Journal for Immunotherapy of Cancer*, *7*, 36.

Bennouna, J., Bompas, E., Neidhardt, E. M., Rolland, F., Philip, I., Galea, C., et al. (2008). Phase-I study of Innacell gammadelta, an autologous cell-therapy product highly enriched in gamma9delta2 T lymphocytes, in combination with IL-2, in patients with metastatic renal cell carcinoma. *Cancer Immunology, Immunotherapy*, *57*, 1599–1609.

Buccheri, S., Guggino, G., Caccamo, N., Li, D. P., & Dieli, F. (2014). Efficacy and safety of gammadeltaT cell-based tumor immunotherapy: A meta-analysis. *Journal of Biological Regulators and Homeostatic Agents*, *28*, 81–90.

Capietto, A. H., Martinet, L., Cendron, D., Fruchon, S., Pont, F., & Fournie, J. J. (2010). Phosphoantigens overcome human TCRVgamma9 + gammadelta cell immunosuppression by TGF-beta: Relevance for cancer immunotherapy. *Journal of Immunology*, *184*, 6680–6687.

Capietto, A. H., Martinet, L., & Fournie, J. J. (2011). Stimulated gammadelta T cells increase the in vivo efficacy of trastuzumab in HER-2 + breast cancer. *Journal of Immunology*, *187*, 1031–1038.

D'Asaro, M., La, M. C., Di, L. D., Orlando, V., Todaro, M., Spina, M., et al. (2010). V gamma 9V delta 2 T lymphocytes efficiently recognize and kill zoledronate-sensitized, imatinib-sensitive, and imatinib-resistant chronic myelogenous leukemia cells. *Journal of Immunology*, *184*, 3260–3268.

Dhar, S., & Chiplunkar, S. V. (2010). Lysis of aminobisphosphonate-sensitized MCF-7 breast tumor cells by Vgamma9Vdelta2 T cells. *Cancer Immunity*, *10*, 10.

Dunne, M. R., Mangan, B. A., Madrigal-Estebas, L., & Doherty, D. G. (2010). Preferential Th1 cytokine profile of phosphoantigen-stimulated human Vgamma9Vdelta2 T cells. *Mediators of Inflammation*, *2010*, 704941.

Fisher, J. P., Heuijerjans, J., Yan, M., Gustafsson, K., & Anderson, J. (2014). Gammadelta T cells for cancer immunotherapy: A systematic review of clinical trials. *Oncoimmunology*, *3*, e27572.

Gentles, A. J., Newman, A. M., Liu, C. L., Bratman, S. V., Feng, W., Kim, D., et al. (2015). The prognostic landscape of genes and infiltrating immune cells across human cancers. *Nature Medicine*, *21*, 938–945.

Gober, H. J., Kistowska, M., Angman, L., Jeno, P., Mori, L., & De, L. G. (2003). Human T cell receptor gammadelta cells recognize endogenous mevalonate metabolites in tumor cells. *The Journal of Experimental Medicine*, *197*, 163–168.

Gu, S., Nawrocka, W., & Adams, E. J. (2014). Sensing of pyrophosphate metabolites by Vgamma9Vdelta2 T cells. *Frontiers in Immunology*, *5*, 688.

Hoeres, T., Smetak, M., Pretscher, D., & Wilhelm, M. (2018). Improving the efficiency of Vgamma9Vdelta2 T-cell immunotherapy in cancer. *Frontiers in Immunology*, *9*, 800.

Kalyan, S., Chandrasekaran, V., Quabius, E. S., Lindhorst, T. K., & Kabelitz, D. (2014). Neutrophil uptake of nitrogen-bisphosphonates leads to the suppression of human peripheral blood gammadelta T cells. *Cellular and Molecular Life Sciences*, *71*, 2335–2346.

Kobayashi, H., Tanaka, Y., Yagi, J., Minato, N., & Tanabe, K. (2011). Phase I/II study of adoptive transfer of gammadelta T cells in combination with zoledronic acid and IL-2 to patients with advanced renal cell carcinoma. *Cancer Immunology, Immunotherapy*, *60*, 1075–1084.

Kobayashi, H., Tanaka, Y., Yagi, J., Osaka, Y., Nakazawa, H., Uchiyama, T., et al. (2007). Safety profile and anti-tumor effects of adoptive immunotherapy using gamma-delta T cells against advanced renal cell carcinoma: A pilot study. *Cancer Immunology, Immunotherapy*, *56*, 469–476.

Kouakanou, L., Xu, Y., Peters, C., He, J., Wu, Y., Yin, Z., et al. (2019). Vitamin C promotes the proliferation and effector functions of human gammadelta T cells. *Cellular & Molecular Immunology*. https://doi.org/10.1038/s41423-019-0247-8.

Kunzmann, V., Bauer, E., & Wilhelm, M. (1999). Gamma/delta T-cell stimulation by pamidronate. *The New England Journal of Medicine*, *340*, 737–738.

Laurent, G., Lafaye de Micheaux, S., Solal-Celigny, P., Soubeyran, P., Delwail, V., Ghesquières, H., et al. (2009). Phase I/II study of IPH1101, gammadelta T cell agonist, combined with rituximab, in low grade follicular lymphoma patients. *Blood*, *114*(22), 1649. Abstract only.

Mattarollo, S. R., Kenna, T., Nieda, M., & Nicol, A. J. (2007). Chemotherapy and zoledronate sensitize solid tumour cells to Vgamma9Vdelta2 T cell cytotoxicity. *Cancer Immunology, Immunotherapy*, *56*, 1285–1297.

Nada, M. H., Wang, H., Workalemahu, G., Tanaka, Y., & Morita, C. T. (2017). Enhancing adoptive cancer immunotherapy with Vgamma2Vdelta2 T cells through pulse zoledronate stimulation. *Journal for Immunotherapy of Cancer*, *5*, 9.

Nicol, A. J., Tokuyama, H., Mattarollo, S. R., Hagi, T., Suzuki, K., Yokokawa, K., et al. (2011). Clinical evaluation of autologous gamma delta T cell-based immunotherapy for metastatic solid tumours. *British Journal of Cancer*, *105*, 778–786.

Oberg, H. H., Kellner, C., Gonnermann, D., Sebens, S., Bauerschlag, D., Gramatzki, M., et al. (2018). Tribody [(HER2)2xCD16] is more effective than trastuzumab in enhancing gammadelta T cell and natural killer cell cytotoxicity against HER2-expressing cancer cells. *Frontiers in Immunology*, *9*, 814.

Oberg, H. H., Peipp, M., Kellner, C., Sebens, S., Krause, S., Petrick, D., et al. (2014). Novel bispecific antibodies increase gammadelta T-cell cytotoxicity against pancreatic cancer cells. *Cancer Research*, *74*, 1349–1360.

Peters, C., Meyer, A., Kouakanou, L., Feder, J., Schricker, T., Lettau, M., et al. (2019). TGF-beta enhances the cytotoxic activity of Vdelta2 T cells. *OncoImmunology*, *8*, e1522471.

Pont, F., Familiades, J., Dejean, S., Fruchon, S., Cendron, D., Poupot, M., et al. (2012). The gene expression profile of phosphoantigen-specific human gammadelta T lymphocytes is a blend of alphabeta T-cell and NK-cell signatures. *European Journal of Immunology*, *42*, 228–240.

Sicard, H., Ingoure, S., Luciani, B., Serraz, C., Fournie, J. J., Bonneville, M., et al. (2005). In vivo immunomanipulation of V gamma 9V delta 2 T cells with a synthetic phosphoantigen in a preclinical nonhuman primate model. *Journal of Immunology*, *175*, 5471–5480.

Uchida, R., Ashihara, E., Sato, K., Kimura, S., Kuroda, J., Takeuchi, M., et al. (2007). Gamma delta T cells kill myeloma cells by sensing mevalonate metabolites and ICAM-1 molecules on cell surface. *Biochemical and Biophysical Research Communications*, *354*, 613–618.

Van Acker, H. H., Anguille, S., Willemen, Y., Van den Bergh, J. M., Berneman, Z. N., Lion, E., et al. (2016). Interleukin-15 enhances the proliferation, stimulatory phenotype, and antitumor effector functions of human gamma delta T cells. *Journal of Hematology & Oncology*, *9*, 101.

Vermes, I., Haanen, C., Steffens-Nakken, H., & Reutelingsperger, C. (1995). A novel assay for apoptosis. Flow cytometric detection of phosphatidylserine expression on early apoptotic cells using fluorescein labelled annexin V. *Journal of Immunological Methods*, *184*, 39–51.

Viey, E., Lucas, C., Romagne, F., Escudier, B., Chouaib, S., & Caignard, A. (2008). Chemokine receptors expression and migration potential of tumor-infiltrating and peripheral-expanded Vgamma9Vdelta2 T cells from renal cell carcinoma patients. *Journal of Immunotherapy*, *31*, 313–323.

CHAPTER TWELVE

CFSE dilution to study human T and NK cell proliferation in vitro

Iñigo Terrén[a], Ane Orrantia[a], Joana Vitallé[a], Olatz Zenarruzabeitia[a], Francisco Borrego[a,b,*]

[a]Biocruces Bizkaia Health Research Institute, Immunopathology Group, Barakaldo, Spain
[b]Ikerbasque, Basque Foundation for Science, Bilbao, Spain
*Corresponding author: e-mail address: francisco.borregorabasco@osakidetza.eus

Contents

Abstract

Lymphocytes proliferate in response to several stimuli. In many situations, a rapid lymphocyte expansion, or the identification of a slow dividing cell subpopulation may be of great interest. Thus, it is necessary to perform reliable assays to study and compare lymphocyte subsets proliferation. For this purpose, carboxifluorescein diacetate succinimidyl ester (CFSE) dilution assay has been stablished as a very useful tool that provides cumulative information about cell proliferation. Unlike other techniques that measure a static parameter of a specific time-point, CFSE staining allows to distinguish between subsequent cell divisions. Here, we show a simple protocol to study human T and NK cell proliferation with CFSE dilution assay by flow cytometry.

Methods in Enzymology, Volume 631
ISSN 0076-6879
https://doi.org/10.1016/bs.mie.2019.05.020

1. Introduction

Lymphocytes proliferate following contact with antigens, cytokines, and other stimuli. However, under different conditions, this proliferation could happen at different rates. A crucial step in most adoptive cellular therapies is to achieve a sufficient cell number to be infused into cancer patients, or to maintain high proliferation rates once cells are infused. Distinct strategies have been employed for this purpose, and proliferation studies are essential to determine the optimal proliferative conditions. Some of these techniques could be useful to identify cell subsets that differentially respond to a specific stimulus, or to evaluate the efficacy of antiproliferative drugs, among others.

Several methodologies have been developed to study and compare lymphocyte proliferation rates. The simplest method may be determining cell number with a Neubauer chamber before, during and after the stimulation period. However, this method does not provide information in a heterogeneous cell population about which cell subset is responding to the stimuli or the phenotype of responding cells. Also, it should be considered the time required for cell division cycle and the limitations in the data interpretation. For example, a specific stimulus could induce proliferation only in certain cell subsets while non-responding cells may die, so the total cell number could be the same at the beginning and at the end of the stimulation period. Even if the experiments are performed with purified cell types, there are many T and NK cell subsets that could differentially respond to the same stimulus. For example, and to consider the relevance and magnitude of cell diversity, Horowitz et al. have analyzed the phenotype of human NK cells and estimated that there are from 6,000 to 30,000 subpopulations within an individual (Horowitz et al., 2013). Thus, the interpretation of data obtained by measuring total cell number could be misleading in several circumstances.

Metabolic activity has been also used as a parameter for measuring proliferation, since it is increased during this process. Typical assays include tetrazolium salts that are reduced to the colored product formazan in viable cells. The first described proliferation assay with tetrazolium salts was based on 3-(4,5-dimethylthiazol-2-yl)-2,5-diphenyltetrazolium bromide (MTT), a positive charged and permeable tetrazolium reagent that forms solid formazan crystals upon reduction (Mosmann, 1983). Formazan can be measured with a spectrophotometer, but it must be solubilized prior to absorbance reading. Of note, MTT derived formazan could form crystals that destroy cell membranes and thus increase cell death (Riss et al., 2004). Later, some negatively-charged tetrazolium derivatives were developed, including 3-(4,5-dimethylthiazol-2-yl)-5-(3-carboxymethoxyphenyl)-2-(4-sulfophenyl)-2H-tetrazolium (MTS), 2,3-bis-(2-methoxy-4-nitro-5-sulfophenyl)-2H-tetrazolium-5-carboxanilide (XTT) and water soluble tetrazolium salts (WST). These salts cannot penetrate into cells, so they are used in combination with intermediate electron acceptors, such as phenazine methyl sulfate (PMS) or phenazine ethyl sulfate (PES). These mediators are reduced in the cytoplasm and exit the cells to convert tetrazolium salts in formazan. The advantage of MTS, XTT and WST, compared with MTT, is that the produced formazan is soluble and thus solubilization step can be avoided. Importantly, intermediate electron acceptors may be toxic, so cell viability could be affected by their concentration (Präbst, Engelhardt, Ringgeler, & Hübner, 2017; Riss et al., 2004). Similar to tetrazolium reduction assays, metabolic activity of viable cells could be measured, with higher sensitivity, with the resazurin reduction assay. This permeable dye is reduced to soluble resorufin into viable cells, which can also be measured with a spectrophotometer. It should be considered that resorufin could be converted to dihydroresorufin in some cells, which is toxic and thus may affect to cell viability (Präbst et al., 2017). Although these assays are widely used and can be worthwhile to estimate cell viability, they are very sensitive to small changes of pH or temperature, and could be influenced by other factors such as the concentration of superoxide and superoxide dismutase (Berridge, Herst, & Tan, 2005; Wang, Yu, & Wickliffe, 2011). Despite the advantages of these techniques for several experimental settings, they might not be the best option to study lymphocyte proliferation. Metabolic activity of cells can be modulated under different conditions, and it might not directly reflect the proliferative status of the cells, e.g., activated lymphocytes may be producing cytokines but not proliferating. Moreover, like total cell count, metabolic activity assays also lack information about responding and non-responding cell subsets, and it could be a limitation when working with such heterogeneous cells as lymphocytes.

An alternative, and more direct, way to measure cell proliferation is to analyze DNA synthesis that occurs during the S phase (synthesis phase) of cell division cycle (Cavanagh, Walker, Norazit, & Meedeniya, 2011). This technique is based on the incorporation of labeled nucleosides or nucleoside analogues into newly synthetized DNA strands. First protocols included tritium labeled thymidine (^{3}H-thymidine), which could be measured with a liquid scintillation counter. Later, radioactive probes were replaced by the thymidine analogue 5-bromodeoxyuridine (BrdU). This analogue can be detected with specific anti-BrdU antibodies. However, DNA denaturation treatment is required to make BrdU accessible for the antibodies, and during this process many protein epitopes could be affected (Cappella, Gasparri, Pulici, & Moll, 2015). Alterations in the epitopes may imply that specific antibodies against those proteins could not bind them, thus limiting flow cytometry multiparametric analysis. More recently, the thymidine analogue 5-ethynyl-2′-deoxyuridine (EdU) has emerged as a substitute for BrdU-based techniques. EdU can be detected with a small, permeable and fluorescent dye-conjugated azide, and unlike BrdU, it does not require neither DNA denaturation nor permeabilization of cell and nuclear membranes for its detection. Moreover, EdU labeling shows more reproducible results and better signal-to-noise ratio than BrdU assay (Flomerfelt & Gress, 2016). A great advantage of assays based on DNA synthesis, in comparison with the above mentioned techniques, is that BrdU/EdU-labeled cells can be analyzed by flow cytometry simultaneously with other markers, thus allowing a more comprehensive analysis of the cells. However, data obtained with labeled nucleosides and analogues is restricted to identify proliferating cells, but it fails to distinguish between cells that have undergone multiple divisions. Alternatively, proliferation of different samples can be compared by measuring proliferation-associated markers (e.g., Ki-67) or cell-cycle markers (e.g., PCNA) (Shipkova & Wieland, 2012), although they show the same limitation than BrdU/EdU-labeling being unable to differentiate the number of divisions cells may go through.

Previously described procedures provide useful information about cell proliferation, although they rely on static measures. Other alternative technique is based on the intracellular fluorescent dye carboxifluorescein diacetate succinimidyl ester (CFSE). Originally described in 1990 for lymphocyte migration studies (Weston & Parish, 1990), CFSE was also used to study cell cytotoxicity (Jin et al., 2017; Lecoeur, Février, Garcia, Rivière, & Gougeon, 2001), although it is more commonly used to determine lymphocyte division since the first description of this application in 1994 (Lyons & Parish, 1994). This versatile dye is membrane permeant, so it enters into the cytoplasm where its acetate groups are removed by cellular esterases. Then, CFSE stably binds to the abundant amine groups present in cytoplasmic molecules, conferring a stable fluorescence intensity to cells which is equally divided between daughter cells after each division (Parish, 1999). CFSE labeling allows tracking cells both *in vivo* and *in vitro*. However, although CFSE fluorescence can be detected for months in non-proliferating cells, current flow cytometry technology is able to distinguish between no more than eight cell divisions of proliferating cells, due to the subsequent reduction of fluorescence intensity (Quah, Warren, & Parish, 2007). Here, we describe a simple protocol to CFSE-label human T and/or NK cells and analyze their proliferation *in vitro* by flow cytometry.

2. CFSE staining and cell culture

2.1. Prepare single cell suspension of peripheral blood mononuclear cells (PBMC) or purified T/NK cells. If cells are obtained from a buffy coat or whole blood, proceed as follows:

- **2.1.1.** Add 10 mL Ficoll® Paque Plus (GE Healthcare, ref. 17-1440-02) to two 50 mL conical tubes. Add 30 mL sterile PBS (Note 6.1) to 10 mL buffy coat (1:4 dilution) and mix gently. If starting from whole blood, prepare a 1:2 dilution in sterile PBS. Then, carefully add 20 mL of the mixture on top of the Ficoll to each conical tube. Avoid mixing Ficoll and diluted buffy coat (Note 6.2).
- **2.1.2.** Centrifuge tubes (750*g* for 25–30 min, at room temperature) and collect PBMC layer with a Pasteur pipette in a 50 mL conical tube.
- **2.1.3.** Add 30 mL sterile PBS and centrifuge cells (150*g* for 10 min, at room temperature).
- **2.1.4.** Discard the supernatant, resuspend pellet in 30 mL sterile PBS. Centrifuge cells (300*g* for 10 min, at room temperature). Repeat this step.
- **2.1.5.** Resuspend pellet in 20 mL sterile PBS and filter cells with 70 μm cell strainers (Falcon, ref. 352350) to avoid cell clumps (optional). Then, enrich for T or NK cells (optional). There are many commercial kits for T and NK cells purification.

2.2. Prepare CFSE stock solution (5 mM) by adding 18 μL dimethyl sulfoxide (DMSO) (included in the CellTrace™ CFSE kit, Invitrogen, ref. C34554) to the vial, and keep always protected from light.

2.3. Adjust cell concentration to 4×10^6 cells/mL (in sterile serum-free PBS) in a 50 mL conical tube (Note 6.3).

2.4. Add the appropriate volume of CFSE (working concentration: 0.5 μM), mix gently and incubate for 20 min, protected from light, at room temperature (Notes 6.4 and 6.5). For example, for 80×10^6 PBMC in a 50 mL conical tube, add 20 mL sterile PBS (for a final cell concentration of 4×10^6 cells/mL) and 2 μL of CFSE stock solution.

2.5. Add Iscove's Modified Dulbecco's Medium (IMDM) (Gibco, ref. 12440053) supplemented with 10% human AB serum, 1% GlutaMAX™ (Gibco, ref. 35050038) and 1% penicillin-streptomycin (Gibco, ref. 15140122), hereinafter cIMDM. For the example in the previous step, after the 20 min incubation, add 30 mL cIMDM (Note 6.6).

2.6. Mix gently and incubate for additional 5 min, protected from light, at room temperature.

2.7. Centrifuge cells (300*g* for 5 min, at room temperature) and discard the supernatant (Note 6.7).

2.8. Resuspend pellet and wash cells with 15 mL cIMDM (centrifuge at 300*g* for 10 min at room temperature) (Note 6.6). Discard the supernatant.

2.9. Resuspend pellet and culture cells under the desired conditions (Notes 6.8, 6.9 and 6.10). As an example, purified CD4+ T cells can be cultured in plates coated with anti-CD3 and anti-CD28 monoclonal antibodies, in the presence of IL-2 for 3 days (Narayanan et al., 2010). Also, in our laboratory, we have labeled PBMC with CFSE and cultured them for 4 days with different cytokines combination (Terrén et al., 2018). In any case, we recommend to separately culture non-CFSE-labeled cells to use them as an unstained control.

3. Flow cytometry staining

3.1. Viability staining is optional but highly recommended due to CFSE toxicity. In our laboratory, we have used the LIVE/DEAD™ Fixable Near-IR Dead Cell Stain Kit (Invitrogen, ref. L34976):

- **3.1.1.** Collect CFSE labeled 10^6 cells in flow cytometry tubes and wash twice with 2 mL PBS (300*g* for 5 min, at 4 °C).
- **3.1.2.** Resuspend pellet in 1 mL PBS.
- **3.1.3.** Incubate cells with the viability dye (mentioned in step 3.1) for 30 min at 4 °C, protected from light (Note 6.11).
- **3.1.4.** Wash cells (300*g* for 5 min, at 4 °C) with 2 mL PBS supplemented with 2.5% bovine serum albumin (BSA) (Sigma-Aldrich, ref. 82100-M). Discard supernatant and resuspend pellet. Repeat this step.

3.2. Extracellular staining (Note 6.12):

- **3.2.1.** Incubate cells with fluorochrome-conjugated antibodies for 30 min at 4 °C, protected from light (Note 6.13). For example, staining with PerCP/Cyanine5.5 anti-human CD3 (clone SK7) (BioLegend, ref. 344808) and APC anti-human

CD56 (clone MEM-188) (BioLegend, ref. 304610) may allow to distinguish between NK (CD3−CD56+) and T (CD3+CD56+ and CD3+CD56−) cell subsets, which will be used as an example in the next section of "proliferation analysis."

3.2.2. Wash cells with 2mL PBS supplemented with 2.5% BSA (300*g* for 5min, at 4°C), discard supernatant and resuspend pellet. Repeat this step.

3.3. Fixation (optional) (Notes 6.14 and 6.15):

3.3.1. Fix cells following BD Cytofix/Cytoperm™ kit (BD Biosciences, ref. 554714) instructions. Briefly, resuspend cells in flow cytometry tubes in 100μL PBS with 4% formaldehyde (Sigma-Aldrich, ref. P6148), and incubate for 15min on ice (Note 6.16).

3.3.2. Add 2mL PBS supplemented with 2.5% BSA and centrifuge them at 300*g* for 5min at 4°C. Discard the supernatant. Repeat this step.

3.3.3. Add 1mL PBS supplemented with 2.5% BSA and store cells at 4°C until the next day (Note 6.17).

3.4. Permeabilization and intracellular staining (optional) (Note 6.15):

3.4.1. Permeabilize cells following BD Cytofix/Cytoperm™ kit instructions. Briefly, centrifuge fixed cells at 300*g* for 5min at 4°C. Discard the supernatant and resuspend cells in 1mL PermWash Buffer 1× (BD Biosciences, ref. 554723, it is included in the BD Cytofix/Cytoperm™ kit), diluted with distilled water. Incubate for 15min on ice (Note 6.16).

3.4.2. Centrifuge cells at 300*g* for 5min at 4°C and discard the supernatant.

3.4.3. Incubate cells with the desired fluorochrome-conjugated antibodies for 30min at 4°C, protected from light.

3.4.4. Add 1mL PermWash Buffer 1× and centrifuge cells at 300*g* for 5min at 4°C. Discard the supernatant and repeat this step.

3.5. Resuspend cells in PBS, vortex them and acquire in a flow cytometer (Notes 6.18 and 6.19).

4. Proliferation analysis

Here, we present two procedures to analyze cell divisions. First, we recommend excluding dead cells before performing proliferation analysis. Also, we have separately analyzed NK cells (CD3−CD56+), CD3+CD56+ and CD3+CD56− T cell subpopulations (Fig. 1).

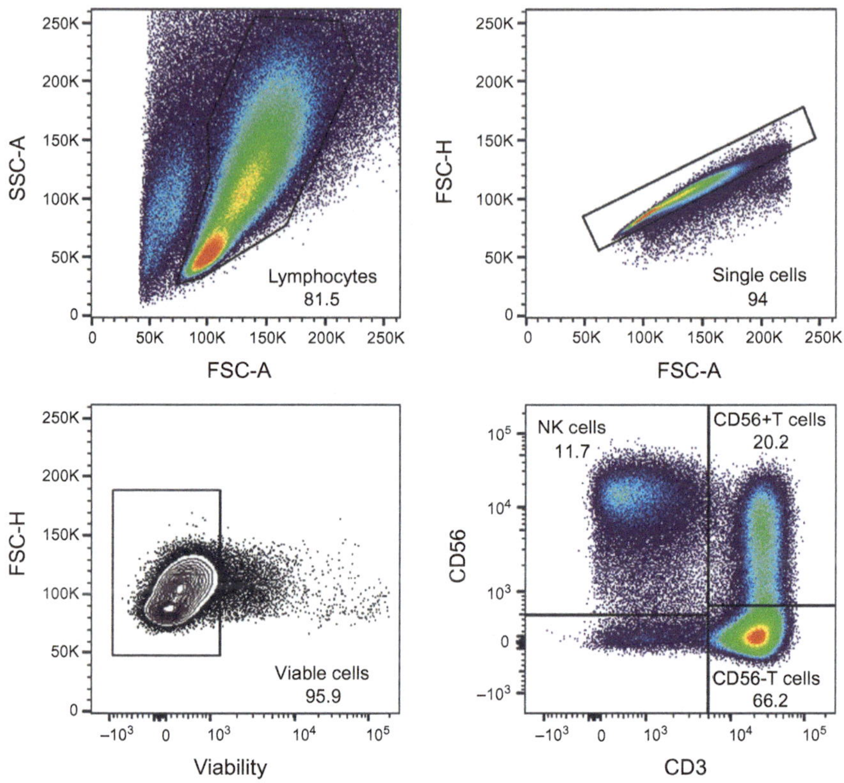

Fig. 1 Gating strategy. First, lymphocytes and single cells are gated from PBMC that have been stimulated for 16–18h with IL-12, IL-15 and IL-18 (10, 100 and 50ng/mL, respectively), and then cultured for 4 days with 20U/mL IL-2. Next, dead cells are excluded using a viability dye. Finally, based on CD3 and CD56 expression, cell subsets are gated. NK cells are defined by the phenotype CD3−CD56+. T cells are CD3+ and they can be divided in two subsets according to CD56 expression, CD3+CD56− and CD3+CD56+ T cells.

4.1 Manual gating

The simplest way to identify several divisions is to manually gate each division peak, starting from the one with the highest fluorescence (division 0 or undivided), then the peak with the second highest fluorescence (division 1), etc. (Fig. 2A). As mentioned above, CFSE fails to distinguish more than eight divisions, so the peak that would correspond to the division 8 will actually include cells that have undergone eight or more divisions. Assuming that the fluorescence intensity is exactly equally divided between daughter cells, those that have divided seven times would have lost more than 99% of the fluorescence of the original population. Therefore, in a situation where

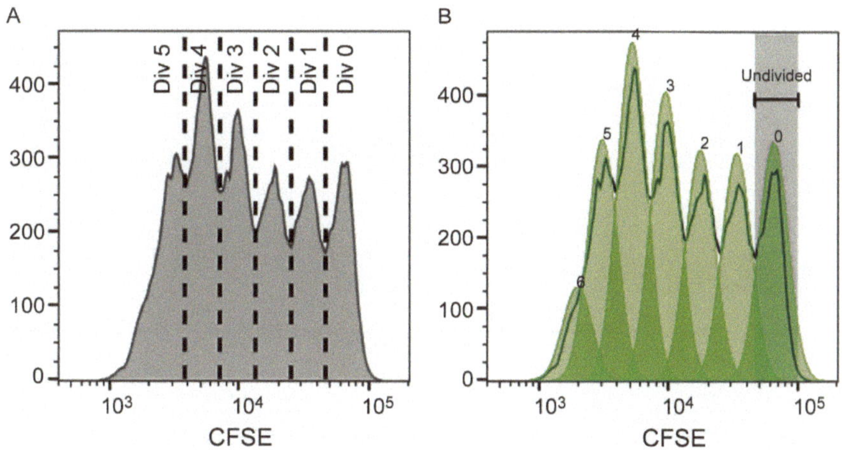

Fig. 2 CFSE histograms with (A) the manual gating of each division peak, including undivided cells (Div 0), and (B) the model (green peaks) of the same sample (black histogram) adjusted with the proliferation tool from FlowJo software (v10.4.1).

CFSE-labeled cells exhibit low fluorescence intensity before proliferation, it could be difficult to distinguish between their intrinsic autofluorescence and the fluorescence of CFSE from those cells that have divided a significant number of times. These problems can be avoided by optimizing CFSE labeling (Note 6.4), performing an appropriate compensation for the multicolor flow cytometry panel (Note 6.19), and taking into consideration the extent of the culture period (Note 6.8). Once ensured the correct experimental settings, manual gating could be very useful to quickly show which cell subset is responding to a determined stimulus (Fig. 3A), to compare the percentage of cells that have undergone more or less than a determined number of divisions (e.g., the percentage of cells that have divided at least once), or to compare the expression of a marker in cells that have gone through different divisions (Fig. 3B).

4.2 Proliferation analysis tools

Many flow cytometry analysis software packages include tools to analyze cell proliferation. Here, we have used the proliferation platform from FlowJo (v10.4.1) software, although this tool is not available in all the versions of this software. The procedure is as follows: once gated the population of interest within viable cells, open the proliferation tool and adjust the model to fit with the data (Fig. 2B). First, gate the undivided cells and set the division 0 (Peak 0). Then, set the number of peaks (Note 6.20), correct the

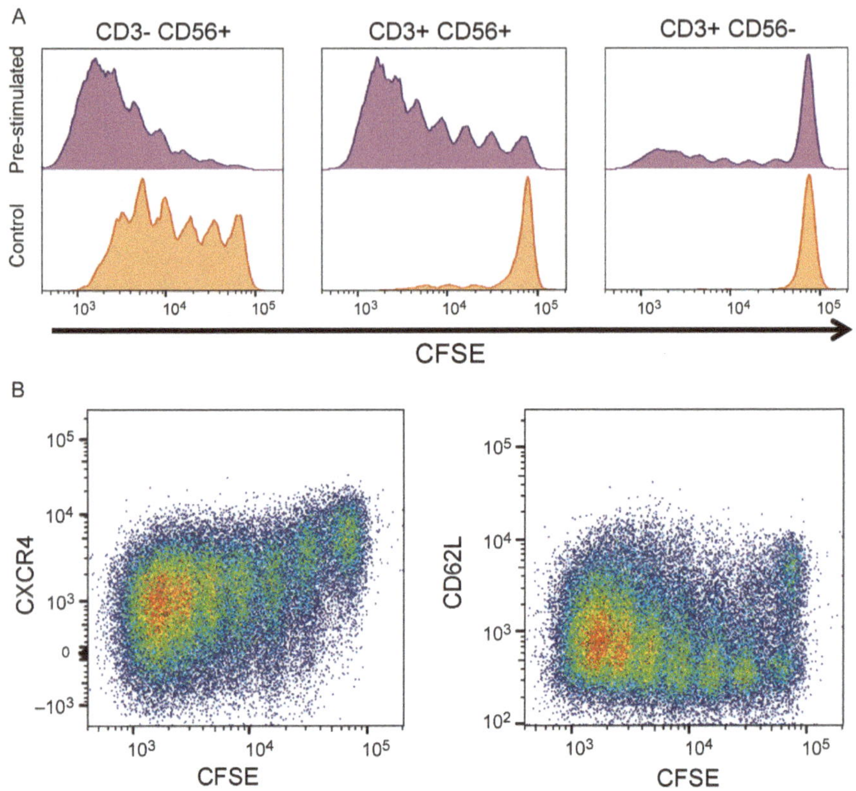

Fig. 3 (A) CFSE histograms of CD3−CD56+ NK cells and both CD3+CD56+ and CD3+CD56− T cell subsets cultured for 4 days with 20U/mL IL-2 (Control) or stimulated with IL-12, IL-15 and IL-18 (10, 100 and 50ng/mL, respectively) for 16–18h and then cultured for 4 days with 20U/mL IL-2 (Pre-stimulated). (B) Pseudocolor plot representing the expression of lymphocyte trafficking markers CXCR4 and CD62L in subsequent cell divisions.

background (Note 6.21), and fix the coefficient of variation (CV) and the ratio of fluorescence between subsequent peaks (Note 6.22). Some values like the number of peaks may need to be adjusted for each sample. Therefore, it is recommendable to check if the model fits the data in all the samples. The output of this tool are several parameters that can be used to compare cell proliferation, including the percentage of cells from the original population that have divided at least once (Percent Divided), the average number of divisions excluding (Proliferation Index) or including (Division Index) undivided cells, and fold-expansion of the cells excluding (Replication Index) or including (Expansion Index) undivided cells (Table 1). Of note,

Table 1 Proliferation parameters obtained with proliferation analysis tool of CD3− CD56+ NK cells and both CD3+CD56+ and CD3+CD56− T cell subsets cultured for 4 days with 20 U/mL IL-2 (Control) or stimulated with IL-12, IL-15 and IL-18 (10, 100 and 50 ng/mL, respectively) for 16–18 h and then cultured for 4 days with 20 U/mL IL-2 (Pre-stimulated).

	NK cells		T cells			
	CD3−CD56+		CD3+CD56+		CD3+CD56−	
	Control	Pre-stim	Control	Pre-stim	Control	Pre-stim
Proliferation Index	1.97	3.05	1.56	2.35	1	2.18
Division Index	0.99	2.55	0.11	1.31	0.011	0.19
Percent Divided	50	83.4	7.37	55.9	1.06	8.86
Expansion Index	3.49	14	1.22	6.21	1.01	1.73
Replication Index	5.98	16.6	4.01	10.3	2	9.27

the same designation has been used to name different parameters in different analysis software packages, so the designation of the parameter should be clearly defined to avoid misunderstandings (Roederer, 2011). Additionally, this software allows gating and separately studying each division, a function that can be helpful to compare the expression of a specific marker in distinct cell divisions. Proliferation analysis tools can facilitate the analysis when working with a large number of samples and/or experimental conditions, and they also may help to avoid data misinterpretation. For instance, the CFSE histograms of pre-stimulated CD3−CD56+ NK cells and CD3+CD56+ T cells have similar aspect (Fig. 3A), but the fold-expansion (Expansion and Replication Indexes), the average number of divisions (Proliferation and Division Indexes) and the fraction of responding parental cells (Percent Divided) are higher in NK cells (Table 1). Thus, the correct interpretation would be that the overall proliferation rate, in this experimental setup, is higher in pre-stimulated CD3−CD56+ NK cell subset than in CD3+CD56+ T cells.

5. Concluding remarks

CFSE staining has a particular advantage for measuring cell proliferation compared with other techniques. This assay provides cumulative information about cells during the culture period, rather than just information about proliferative status of the cells in a determined time-point.

This characteristic can be exploited to track the expression of specific markers or cytokine production within different cell subpopulations along subsequent divisions (Narayanan et al., 2010) (Fig. 3). CFSE can be used both *in vitro* and *in vivo*, so CFSE-labeled cells can also be adoptively transferred into host animal and track their proliferation (Quah et al., 2007). Furthermore, CFSE has been used for the identification of slow-dividing cancer cell subpopulations. These self-renewing and infrequently cycling cells, present in leukemia and solid tumors, has been identified as tumor-initiating cells and could be a relevant target to focus on (Deleyrolle et al., 2011). Also, CFSE can be particularly helpful to separately study different cells in co-culture experiments. For instance, in an experimental setting where regulatory T (Treg) cell suppression capacity is measured by their potential to inhibit target cell proliferation, it is crucial to discriminate between the proliferation of Treg and target cells, which would be impossible to do with ^{3}H-thymidine assay (Venken et al., 2007).

Interestingly, there are other fluorescent dyes similar to CFSE that can be used for the same purpose. However, some of these CFSE-like dyes may have disadvantages such as lower capacity to be retained by proliferating cells, thus limiting its application in proliferation studies (Quah & Parish, 2012). More recently, it has been developed a nanotechnology-based method to study cell proliferation. Like CFSE labeling, these non-toxic nanoparticles are internalized by cells (including PBMC) and are divided between daughter cells, although equal distribution of nanoparticles has not been demonstrated yet and division peaks have low resolution. Nevertheless, a great advantage is that fluorescent nanoparticles can be tagged with a repertoire of fluorophores, which facilitate multicolor panel configuration and may increase peak resolution (Altea-Manzano et al., 2017).

Although CFSE dilution assay is a very useful technique, it also presents some weaknesses. As previously mentioned, this dye stably labels cells by binding to amine groups present in several molecules. Thus, it is reasonable to assume that CFSE staining may interfere with cellular processes. Indeed, CFSE has dose-dependent toxicity that should be considered when defining a working concentration. In PBMC, it has been shown that concentrations ranging from 2.5 to 10 μM of CFSE affect cell viability (Lašt'ovička, Budinský, Špíšek, & Bartůňková, 2009). Some cell subsets (e.g., CD8 +CD57+ T cells) may exhibit higher sensitivity to even lower concentrations of CFSE (Chong et al., 2008; Strioga, Pasukoniene, & Characiejus, 2011). Therefore, this assay may not be entirely suitable to perform CFSE proliferation studies in these specific situations. For this reason, we strongly recommend including viability staining in the protocol. Also, CFSE could interfere with cell proliferation, as it has been shown by ^{3}H-thymidine incorporation and Ki-67 expression assays, using PBMC labeled with an increasing concentration of CFSE. However, lower concentrations (1.25 μM) of CFSE show minimal effect on cell proliferation (Laštovička, Rataj, & Bartůňková, 2016). Additionally, it has to be considered that ^{3}H-thymidine assay also slows lymphocyte cycling and induces a dose-dependent inhibition of the DNA synthesis rate (Laštovička et al., 2016). Lastly, CFSE staining protocol requires cells to be stained before cell division, so this assay may not be the most suitable to study the proliferation of tissue resident lymphocytes *in vivo*. Nonetheless, CFSE dilution assay has been stablished as a very powerful technique to study cell proliferation. Most of its disadvantages can be avoided by using a low concentration of this fluorescent dye. In our protocol, we propose to use a concentration of 0.5 μM, which has been useful and enough for our experimental conditions, although it can be increased if it is necessary since the usual working concentration is 5 μM. As mentioned before, in contrast to other techniques that measure a static value, CFSE dilution assay provides cumulative information about cell proliferation, and it can be used both *in vivo* and *in vitro*. These characteristics, the simple and rapid staining protocol, and the reliability of the technique have made CFSE to become a highly valuable tool for immunologists and cell biologists.

6. Notes

6.1. Use PBS without calcium or magnesium (Gibco, ref. 20012068; Lonza, ref. 17-516Q) to avoid cell aggregation.

6.2. Tilt the conical tube with Ficoll and slowly pour the diluted buffy coat or whole blood in the tube by sliding it through the vessel wall.

6.3. Although we recommend using 4×10^6 cell/mL for CFSE labeling, the concentration can be increased. In our experiments, we have used cell concentrations ranging from 4×10^6 to 5×10^6 cell/mL.

6.4. Commonly, a working concentration of 5 μM is used, but the toxicity of CFSE has to be considered. Thus, it is preferable to use a lower concentration (0.5 μM in this protocol), but ensuring the correct labeling and appropriate fluorescence intensity.

6.5. Vortex the samples several times during the incubation to avoid cell sedimentation in the tube and ensure the CFSE to be homogeneously distributed.

6.6. Different media can be used, but they must be supplemented with serum/proteins to remove the CFSE that has not been incorporated into cells.

6.7. After centrifugation, cell pellet would acquire a yellow color.

6.8. To stablish the extent of the culture, it has to be considered that CFSE staining will not be useful to study more than eight cell divisions.

6.9. Avoid the exposure of the cells to light sources during their manipulation.

6.10. If cell degranulation or cytokines production are going to be analyzed, monensin and/or brefeldin A should be added during the stimulation. In our laboratory, we have used BD GolgiStop™ (monensin, BD Biosciences, ref. 554724) and BD GolgiPlug™ (brefeldin A, BD Biosciences, ref. 555029), following manufacturer's instructions.

6.11. If cells are fixed in step 3.3, ensure to use a fixable viability dye.

6.12. Other authors have optimized flow cytometry staining protocols to simultaneously perform viability and extracellular staining, which can be useful to save time.

6.13. The number of fluorochrome-conjugated antibodies that can be used depends on the flow cytometer. If this is a limiting factor, working with purified cells may help to use fewer antibodies to identify the cell subsets of interest.

6.14. CFSE is a fixable dye, but other CFSE-like dyes may not be fixable, so we recommend checking it and to contact the supplier. Also, check if fixation affects all the fluorochromes used for the staining, such as the viability dye.

6.15. The fixation and permeabilization protocol has been adapted for fixing cells following the culture period and to continue the staining the next day. It has been adapted from the "Alternative Fixation and Permeabilization Protocol" of BD Cytofix/Cytoperm™ kit. However, this protocol can be modified to perform the whole experiment in the same day.

6.16. Vortex tube before the incubation to have a homogeneous cell suspension.

6.17. If tubes will be stored for a long period, such as overnight, cover tubes to avoid dust entering into them.

6.18. Cells should be resuspended in a PBS volume adequate for cell acquisition, which depends on the flow cytometer. For example, in our hands, we resuspend 1×10^6 non-fixed or fixed cells in 350 μL or 1 mL PBS, respectively, to acquire them with a BD FACSCanto™ II (BD Biosciences). Fixed cells that have been stored overnight tend to form aggregates, thus it is important to dilute cells (or acquire them slowly) to avoid any obstruction in the flow cytometer.

6.19. Proper compensation must be performed prior to data analysis. The flow cytometer model will determine if compensation must be done before cells acquisition or it can be performed after it. However, in both cases, compensation matrix should be examined in the analysis software.

6.20. If using FlowJo software, and following the indications of software developers, it should be set a number of peaks higher than visible. For instance, in a situation where 6 peaks can be visually distinguished (including the undivided peak), adjust the number of peaks of the model to 7, at least (Fig. 2B). This adjustment would help to fit the model to the data. However, this correction is only for FlowJo software and may not be applied to other flow cytometry analysis software packages.

6.21. The autofluorescence of the unstained sample can be used to set the background value.

6.22. The ratio of fluorescence between subsequent peaks will be usually 0.5 or lower, and small changes in this value will significantly modify the model.

Acknowledgments

This study was supported by grants from Health Department, Basque Government (2018222038), Basque Foundation for Research and Innovation-EiTB Maratoia (BIO14/TP/003) and AECC-Spanish Association Against Cancer (PROYE16074BORR). I.T. is recipient of a fellowship from the Jesús de Gangoiti Barrera Foundation (FJGB18/002) and a predoctoral contract funded by the Department of Education, Basque Government (PRE_2018_1_0032). J.V. is recipient of a predoctoral contract funded by the Department of Education, Basque Government (PRE_2017_2_0242). O.Z. is recipient of a postdoctoral contract funded by "Instituto de Salud Carlos III-*Contratos Sara Borrell 2017* (CD17/00128)" and the European Social Fund (ESF)-*The ESF invests in your future.* F.B. is an Ikerbasque Research Professor, Ikerbasque, Basque Foundation for Science.

Conflict of interests

The authors declare that the research was conducted in the absence of any commercial or financial relationships that could be construed as a potential conflict of interest.

References

Altea-Manzano, P., Unciti-Broceta, J. D., Cano-Cortes, V., Ruiz-Blas, M. P., Valero-Griñan, T., Diaz-Mochon, J. J., et al. (2017). Tracking cell proliferation using a nanotechnology-based approach. *Nanomedicine, 12*(13), 1591–1605. https://doi.org/10.2217/nnm-2017-0118.

Berridge, M. V., Herst, P. M., & Tan, A. S. (2005). Tetrazolium dyes as tools in cell biology: New insights into their cellular reduction. In M. Raafat Rl-Gewely (Ed.), *Vol. 11. Biotechnology annual review* (pp. 127–152). Elsevier. https://doi.org/10.1016/S1387-2656(05)11004-7.

Cappella, P., Gasparri, F., Pulici, M., & Moll, J. (2015). Cell proliferation method: Click chemistry based on BrdU coupling for multiplex antibody staining. In J. Paul Robinson, Z. Darzynkiewicz, P. Dean, A. Orfao, P. Rabinovitch, C. C. Stewart, H. Tanke, & L. Wheeless (Eds.), *Current protocols in cytometry* (pp. 7.34.1–7.34.17). Hoboken, NJ: John Wiley & Sons, Inc. https://doi.org/10.1002/0471142956.cy0734s72.

Cavanagh, B. L., Walker, T., Norazit, A., & Meedeniya, A. C. B. (2011). Thymidine analogues for tracking DNA synthesis. *Molecules*, *16*(9), 7980–7993. https://doi.org/10.3390/molecules16097980.

Chong, L. K., Aicheler, R. J., Llewellyn-Lacey, S., Tomasec, P., Brennan, P., & Wang, E. C. Y. (2008). Proliferation and interleukin 5 production by CD8hiCD57+ T cells. *European Journal of Immunology*, *38*(4), 995–1000. https://doi.org/10.1002/eji.200737687.

Deleyrolle, L. P., Harding, A., Cato, K., Siebzehnrubl, F. A., Rahman, M., Azari, H., et al. (2011). Evidence for label-retaining tumour-initiating cells in human glioblastoma. *Brain*, *134*(5), 1331–1343. https://doi.org/10.1093/brain/awr081.

Flomerfelt, F. A., & Gress, R. E. (2016). Analysis of cell proliferation and homeostasis using EdU labeling. In R. Bosselut & M. S. Vacchio (Eds.), *Vol. 1323. Methods in molecular biology* (pp. 211–220). Clifton, NJ. https://doi.org/10.1007/978-1-4939-2809-5_18.

Horowitz, A., Strauss-Albee, D. M., Leipold, M., Kubo, J., Nemat-Gorgani, N., Dogan, O. C., et al. (2013). Genetic and environmental determinants of human NK cell diversity revealed by mass cytometry. *Science Translational Medicine*, *5*(208), 208ra145. https://doi.org/10.1126/scitranslmed.3006702.

Jin, Q., Jiang, L., Chen, Q., Li, X., Xu, Y., Sun, X., et al. (2017). Rapid flow cytometry-based assay for the evaluation of γδ T cell-mediated cytotoxicity. *Molecular Medicine Reports*, 3555–3562. https://doi.org/10.3892/mmr.2017.8281.

Lašt'ovička, J., Budinský, V., Špíšek, R., & Bartůňková, J. (2009). Assessment of lymphocyte proliferation: CFSE kills dividing cells and modulates expression of activation markers. *Cellular Immunology*, *256*(1–2), 79–85. https://doi.org/10.1016/j.cellimm.2009.01.007.

Lašťovička, J., Rataj, M., & Bartůňková, J. (2016). Assessment of lymphocyte proliferation for diagnostic purpose: Comparison of CFSE staining, Ki-67 expression and 3H-thymidine incorporation. *Human Immunology*, 77(12), 1215–1222. https://doi.org/10.1016/j.humimm.2016.08.012.

Lecoeur, H., Février, M., Garcia, S., Rivière, Y., & Gougeon, M. L. (2001). A novel flow cytometric assay for quantitation and multiparametric characterization of cell-mediated cytotoxicity. *Journal of Immunological Methods*, *253*(1–2), 177–187. Retrieved from http://www.ncbi.nlm.nih.gov/pubmed/11384679.

Lyons, A. B., & Parish, C. R. (1994). Determination of lymphocyte division by flow cytometry. *Journal of Immunological Methods*, *171*(1), 131–137. https://doi.org/10.1002/andp.18501560508.

Mosmann, T. (1983). Rapid colorimetric assay for cellular growth and survival: Application to proliferation and cytotoxicity assays. *Journal of Immunological Methods*, *65*(1–2), 55–63. Retrieved from http://www.ncbi.nlm.nih.gov/pubmed/6606682.

Narayanan, S., Silva, R., Peruzzi, G., Alvarez, Y., Simhadri, V. R., Debell, K., et al. (2010). Human Th1 cells that express CD300a are polyfunctional and after stimulation up-regulate the T-box transcription factor eomesodermin. *PLoS One*, *5*(5), e10636. https://doi.org/10.1371/journal.pone.0010636.

Parish, C. R. (1999). Fluorescent dyes for lymphocyte migration and proliferation studies. *Immunology and Cell Biology*, 77(6), 499–508. https://doi.org/10.1046/j.1440-1711.1999.00877.x.

Präbst, K., Engelhardt, H., Ringgeler, S., & Hübner, H. (2017). Basic colorimetric proliferation assays: MTT, WST, and resazurin. *Methods in Molecular Biology (Clifton, N.J.)*, *1601*, 1–17. https://doi.org/10.1007/978-1-4939-6960-9_1.

Quah, B. J. C., & Parish, C. R. (2012). New and improved methods for measuring lymphocyte proliferation in vitro and in vivo using CFSE-like fluorescent dyes. *Journal of Immunological Methods*, *379*(1–2), 1–14. https://doi.org/10.1016/j.jim.2012.02.012.

Quah, B. J. C., Warren, H. S., & Parish, C. R. (2007). Monitoring lymphocyte proliferation in vitro and in vivo with the intracellular fluorescent dye carboxyfluorescein diacetate succinimidyl ester. *Nature Protocols*, *2*(9), 2049–2056. https://doi.org/10.1038/nprot.2007.296.

Riss, T. L., Moravec, R. A., Niles, A. L., Duellman, S., Benink, H. A., Worzella, T. J., et al. (2004). *Cell viability assays. In Assay guidance manual.* Retrieved from http://www.ncbi.nlm.nih.gov/pubmed/23805433.

Roederer, M. (2011). Interpretation of cellular proliferation data: Avoid the panglossian. *Cytometry. Part A*, *79 A*(2), 95–101. https://doi.org/10.1002/cyto.a.21010.

Shipkova, M., & Wieland, E. (2012). Surface markers of lymphocyte activation and markers of cell proliferation. *Clinica Chimica Acta*, *413*(17–18), 1338–1349. https://doi.org/10.1016/j.cca.2011.11.006.

Strioga, M., Pasukoniene, V., & Characiejus, D. (2011). CD8 + CD28 − and CD8 + CD57 + T cells and their role in health and disease. *Immunology*, *134*(1), 17–32. https://doi.org/10.1111/j.1365-2567.2011.03470.x.

Terrén, I., Mikelez, I., Odriozola, I., Gredilla, A., González, J., Orrantia, A., et al. (2018). Implication of interleukin (IL)-12/15/18 and ruxolitinib in the phenotype, proliferation and polyfunctionality of human cytokine-preactivated NK cells. *Frontiers in Immunology*, *9*, 737. https://doi.org/10.3389/FIMMU.2018.00737.

Venken, K., Thewissen, M., Hellings, N., Somers, V., Hensen, K., Rummens, J., et al. (2007). A CFSE based assay for measuring CD4 + CD25 + regulatory T cell mediated suppression of auto-antigen specific and polyclonal T cell responses. *Journal of Immunological Methods*, *322*(1–2), 1–11. https://doi.org/10.1016/j.jim.2007.01.025.

Wang, S., Yu, H., & Wickliffe, J. K. (2011). Limitation of the MTT and XTT assays for measuring cell viability due to superoxide formation induced by nano-scale TiO2. *Toxicology In Vitro*, *25*(8), 2147–2151. https://doi.org/10.1016/j.tiv.2011.07.007.

Weston, S. A., & Parish, C. R. (1990). New fluorescent dyes for lymphocyte migration studies. *Journal of Immunological Methods*, *133*(1), 87–97. https://doi.org/10.1016/0022-1759(90)90322-M.

CHAPTER THIRTEEN

Rapid isolation and enrichment of mouse NK cells for experimental purposes

Maite Alvarez[a,b,c,†], Maria C. Ochoa[a,b,c,d,†], Luna Minute[a,b], Ignacio Melero[a,b,c,d,*], Pedro Berraondo[a,b,c]

[a]Program of Immunology and Immunotherapy, CIMA Universidad de Navarra, Pamplona, Spain
[b]Navarra Institute for Health Research (IDISNA), Pamplona, Spain
[c]Centro de Investigación Biomédica en Red de Cáncer (CIBERONC), Madrid, Spain
[d]Department of Immunology and Immunotherapy, Clinica Universidad de Navarra, Pamplona, Spain
*Corresponding author: e-mail address: imelero@unav.es

Contents

Abstract

Natural killer (NK) cells have shown to play a critical, but as yet poorly defined, role in the process by which the immune system controls tumor progression. Indeed, NK cell-based immunotherapy, particularly NK cell adoptive transfer therapy, has become a very attractive cancer weapon against multiple types of cancers such as metastatic and hematological cancers. Unfortunately, the implementation of these therapies has been challenged by the existence of immunosuppression mechanisms that have prevented NK cell functionality. Additionally, the development of protocols to obtain purified and functional NK cells has faced some difficulties due to the limitations in the numbers of

† These authors equally share authorship.

Methods in Enzymology, Volume 631
ISSN 0076-6879
https://doi.org/10.1016/bs.mie.2019.10.025

cells that can be obtained and the development of an exhaustion phenotype with impaired proliferative and functional capabilities during lengthy ex vivo NK cell expansion protocols. Thus, the development of new strategies to obtain a rapid expansion of highly functional NK cells without the appearance of exhaustion is still much needed. This is particularly true in the case of mouse NK cells, a surrogate commonly used to evaluate NK cell biology and human NK cell-based immunotherapeutic alternatives. Here, we describe a feasible and rapid protocol to produce strongly activated mouse NK cells in vivo taking advantage of the hydrodynamic delivery of a plasmid that contains interleukin-15, a cytokine known to cause NK cell expansion and activation, fused with the binding domain of the IL-15Rα ("sushi" domain) and apolipoprotein A-I.

1. Introduction

NK cells are lymphocytes that form part of the innate immunity which, unlike T cell lymphocytes, are capable of recognizing target cells in an MHC-I unrestricted manner (Karre, Ljunggren, Piontek, & Kiessling, 1986; Pierini, Alvarez, & Negrin, 2016). They act as part of the first barrier against microbial and malignant cells, directly eliminating infected and transformed cells through a variety of mechanisms, of which the release of inflammatory cytokines and chemokines as well as the mediation of antibody-dependent cellular cytotoxicity (ADCC) of opsonized cells are some examples (Morvan & Lanier, 2016; Pierini et al., 2016). In recent years, some cancer immunotherapies have focused on increasing NK cell activity in tumors, with one of the main strategies being NK cell-based adoptive transfer therapy for hematological cancers (Burns et al., 2003; Cooley, Parham, & Miller, 2018; Miller et al., 2005; Ruggeri et al., 2002). For this purpose, different protocols to isolate, expand, and activate human NK cells have been developed. Some of the protocols include long-term culture of NK cells with stimulating cytokines such as interleukin-2 (IL-2), IL-12, IL-18 or IL-15 or with irradiated feeder cells (Alvarez, Sungur, Perez, Sckisel, & Murphy, 2012; Tognarelli et al., 2018). IL-2 and IL-15 are perhaps the two cytokines most commonly used for ex vivo NK cell expansion protocols (Cooley et al., 2018; Robinson & Schluns, 2017). Both are known to induce the proliferation and activation of cultured NK cells with a marked increase in the expression of typical activating receptors and markers as well as an improvement in their function (Alvarez et al., 2019).

However, human NK cells in cultures using only cytokines do not reach very high levels of expansion (less than a fivefold increase), and the cultures cannot be prolonged more than 2–3 weeks because cell viability decreases. Additionally, recent studies have shown that long-term in vitro culture of NK cells results in the induction of exhaustion in both human and mouse models characterized by impaired proliferative and functional capabilities (Alvarez et al., 2019; Felices et al., 2018; Gill et al., 2012). The combination of cytokines with irradiated feeder cells to obtain NK cell expanded populations has been more successful with up to a 500-fold increase in viable and activated NK cells (Kweon et al., 2019). The administration of NK cells expanded with irradiated EBV-transformed lymphoma or K562 clone 9.mbIL-21 feeder cells supplemented with low doses of IL-2 has shown strong cytotoxic potential against orthotopic pancreatic and glioblastoma tumor-bearing NSG mouse models (Ames et al., 2015; Luna et al., 2019). Similarly, NKs expanded with irradiated K562-mb15-41BBL cells, or freshly isolated and irradiated autologous peripheral blood mononuclear cells have shown potential against different cancer models (Lee et al., 2017).

Despite the differences that exist between mouse and human NK cells, mouse models have been extensively used to expand our knowledge of NK cell biology and, more importantly, to evaluate the clinical translation of NK cell-based therapies (Murphy, Parham, & Miller, 2012). As in human models, to study NK cells and to model NK cell adoptive cell transfer therapy in mice, it is necessary to obtain pure populations. In most experiments, murine NK cells are purified from the spleen using commercially available NK selection kits that can then be used either fresh or further expanded with stimulating cytokines, mainly IL-2 and IL-15. These protocols, as in human NK cells, tend to have a limited expansion rate and cause pronounced exhaustion if cultures are maintained for over a week (Alvarez et al., 2019). The alternative of adding irradiated feeder cells to the murine NK cell culture has not been developed.

In this study, we propose an efficient, fast, easy, and inexpensive protocol to obtain high numbers of murine NK cells that can be utilized during ex-vivo experiments or to model NK cell adoptive transfer therapy. Briefly, this method is based on the transference of an IL-15 construction by hydrodynamic injection into *B6.129S7-Rag1tm1Mom/J* (Rag1$^{-/-}$) mice to increase NK cell production and subsequent isolation from the mouse spleen 3 days after plasmid administration.

2. IL-15

IL-15 is a cytokine that induces activation and proliferation of NK and $CD8^+$ T cells. This cytokine is expressed on the membrane of dendritic cells (DC) and other antigen-presenting cells (APC) tethered to its receptor subunit α and exerts its action on $CD8^+$ T cells and NK cells by triggering the heterodimeric receptor formed by the subunits IL-2Rβ and the common γ chain in a process referred to as trans-presentation (Leonard, Lin, & O'Shea, 2019; Stoklasek, Schluns, & Lefrancois, 2006).

Unlike IL-2, IL-15 has shown to be essential for NK cell development and homeostasis, as proven by the absence of mature NK cells in IL-15 deficient mice and the inability of wild type NK cells to survive when transfer into mice lacking this cytokine (Cooper et al., 2002; Kennedy et al., 2000; Koka et al., 2003; Prlic, Blazar, Farrar, & Jameson, 2003). In particular, IL-15 seems to be involved during the later stages of NK cell maturation due to the presence of a normal NK cell precursor population in $IL\text{-}15^{-/-}$ mice (Castillo, Stonier, Frasca, & Schluns, 2009). Supporting this notion, IL-15 administration to accelerate NK cell reconstitution after bone marrow transplantation causes an increase in mature donor-derived NK cells, but not of immature NK cells (Barao et al., 2011; Huntington et al., 2009). Additionally, it has been shown that human $CD34^+$ progenitor cells can be differentiated toward precursor NK cells in an IL-15 independent-manner (McCullar et al., 2008). However, the need for IL-15 can be overridden by the presence of other growth factors. During mouse cytomegalovirus (MCMV) infection, a strong proliferation of the Ly49H NK cell subset, which recognizes the MCMV encoded product m157, in mice deficient for IL-15 or IL-15Rα has been reported (Sun, Ma, & Lanier, 2009). The presence of high levels of IL-12 was likely responsible for this effect (Sun et al., 2009). Another characteristic that makes IL-15 attractive for immunotherapy is that IL-15 is not involved in the expansion of regulatory T cells and the activation-induced cell death of effector cells typically associated with IL-2 (Jakobisiak, Golab, & Lasek, 2011).

Systemic delivery of IL-15 causes an increase in effector cells with a higher activation and cytotoxic profile that leads to stronger anti-tumor responses (Conlon et al., 2015; Eisenman et al., 2002; Mortier et al., 2006; Robinson & Schluns, 2017). Unfortunately, the short half-life of this cytokine has rendered the need for repetitive IL-15 dosage to accomplish an effective anti-tumor response that is frequently associated with severely toxic side effects (Berger et al., 2009; Conlon et al., 2015; Fehniger et al., 2001; Waldmann et al., 2011). Alternatively, approaches that aim to increase the stability of the cytokine have also been explored. Following this direction, the therapeutic potential in cancer of 15N72D/IL-15Rα-Fc superagonist complexes (N-803, formerly designated as ALT-803) (Kim et al., 2016; Xu et al., 2013) and heterodimeric IL-15/IL-15Rα (Felices et al., 2017; Liu et al., 2016) are currently being studied in more than 15 clinic trials. However, these approaches can also cause toxicity, and thus, subcutaneous delivery seems to be more protective of the treatment-related toxic side effects (Liu et al., 2016; Xu et al., 2013). Additionally, it is important to note that recent mouse and human studies have demonstrated that sustained IL-15 treatment results in NK cell impairment (Alvarez et al., 2019; Elpek, Rubinstein, Bellemare-Pelletier, Goldrath, & Turley, 2010; Felices et al., 2018). Therefore, the development of new approaches to better exploit the anti-tumor properties of IL-15 in NK cell-based therapies are still needed.

3. Sushi-IL15-Apo construction

One of the mechanism that increases the efficacy of IL-15 when used as an enhancer of NK cell proliferation or activation is the mimicking of the IL-15Rα trans-presentation constructing fusion proteins encompassing IL-15 and the binding domain of the IL-15Rα (called "sushi" domain), which bypasses the need for DC or APC (Mortier et al., 2006; Stoklasek et al., 2006). Additionally, several modifications can be added in order to improve the pharmacokinetic profile of IL-15, increasing its half-life in plasma (Berraondo et al., 2019).

In our group, we have developed a construction (pSushi-IL15-Apo) encompassing murine IL-15, the "sushi" domain, and the apolipoprotein A-I (Ochoa et al., 2013). Apolipoprotein A-I is the main protein component of the high-density lipoproteins (HDL) that circulate collecting cholesterol from tissues and delivering it to the liver (Francis, 2010). The fusion of apolipoprotein A-I to the Sushi-IL15 complex increases the plasma half-life and stabilizes the protein, that is integrated into circulating HDLs. This plasmid has been proved to increase the number of $CD8^+$ T and NK cells in the spleen and liver when transferred by hydrodynamic injection into mice with

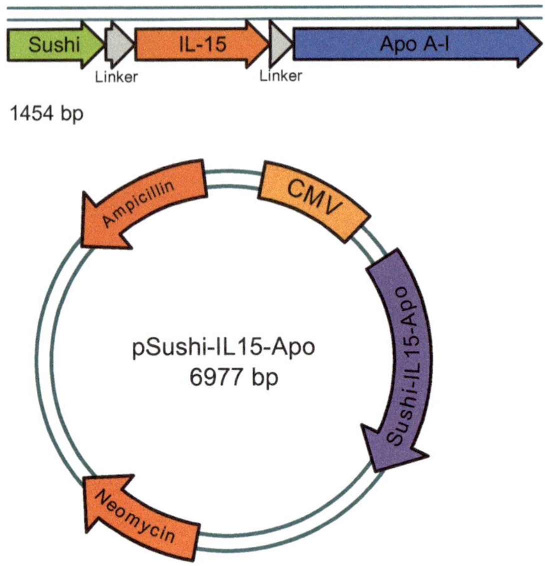

Fig. 1 Schematic representation of pSushi-IL15-Apo construct.

higher efficacy than other constructions with IL-15 (Ochoa et al., 2013). Moreover, the CD8$^+$ T and NK cells resulting from this treatment were activated and presented anti-tumor efficacy (Ochoa et al., 2013).

Details of plasmid cloning were previously described (Ochoa et al., 2013; Ochoa et al., 2018). Briefly, the construction pSushi-IL15-Apo contains the sushi receptor sequence in the N-terminal site together with the IL-15 sequence (UniProt P48346), both linked with the GGSGGGSGGG SGGGGSLQ linker, and the Apolipoprotein A-I (UniProt Q00623) sequence in the C-terminal end. IL-15 and Apolipoprotein A-I sequences are separated by a GAP linker (Fig. 1) The sushi receptor sequence was obtained from the previously described plasmid IL-15RαΔMD by Duitman, Orinska, Bulanova, Paus, and Bulfone-Paus (2008). This construction contains a sequence of 21 amino acids coding for the signal peptide of the immunoglobulin kappa variable chain (Uniprot P01658). The Sushi-IL15-Apo sequence contains 1407 nucleotides, coding for a 468 amino acid protein (Table 1). This sequence was cloned into the pcDNA™3.1/V5-His TOPO® TA commercial plasmid, which includes a CMV promoter and a resistance gene for ampicillin (for production in prokaryotic cells) and for neomycin for transfection in eukaryotic cells. The whole construction contains 6977 base pairs. (Fig. 1) Plasmid cloning

Table 1 The sequence of Sushi-IL15-Apo.

METDTLLLWVLLLWVPGSTGDGTTCPPPVSIEHADIRVKNYSVNSRERY VCNSGFKRKA
GTSTLIECVINKNTNVAHWTTPSLKCIRDPSGGSGGGGSGGGSGGGGSL QEANWIDVRY
DLEKIESLIQSIHIDTTLYTDSDFHPSCKVTAMNCFLLELQVILHEYSNMT LNETVRNVLY
LANSTLSSNKNVAESGCKECEELEEKTFTEFLQSFIRIVQMFINTSGAPDE PQSQWDKVK
DFANVYVDAVKDSGRDYVSQFESSSLGQQLNLNLLENWDTLGSTVSQL QERLGPLTRD
FWDNLEKETDWVRQEMNKDLEEVKQKVQPYLDEFQKKWKEDVELYR QKVAPLGAEL
QESARQKLQELQGRLSPVAEEFRDRMRTHVDSLRTQLAPHSEQMRES LAQRLAELKSNP
TLNEYHTRAKTHLKTLGEKARPALEDLRHSLMPMLETLKTKAQSVIDK ASETLTAQ

was performed with a pcDNA™3.1/V5-His TOPO™ TA Expression Kit (Thermo Fisher Scientific), which includes the *E. coli* cells to produce the plasmid.

4. Hydrodynamic delivery

Hydrodynamic injection is an experimental technique to transfer naked plasmids into the liver of mice (Barao, Alvarez, Redelman, et al., 2011; Liu, Song, & Liu, 1999; Zhang, Budker, & Wolff, 1999). The plasmid is injected intravenously (iv) by the tail-vein in a volume of around 10% of the rodent weight in a short period (3–5 s). The large volume of plasmid solution rapidly injected through the tail vein goes by the inferior cava vein and induces cardiac congestion followed by a temporal retrograde flow into the hepatic veins that induces an increase in pressure. This produces an enlargement of the endothelial fenestrae, disrupts liver sinusoids, and the fluid enters through the damaged membrane of the hepatocytes where the insert of the vector is expressed (Suda & Liu, 2007). This process induces a transitory enlargement of liver volume (lasting around 30 min). Although the procedure appears to be very dramatic, survival is almost 100% in mice, and the animals recover their normal behavior in a brief period (10–20 min). The pores in the plasma membrane are resealed, and the endothelium regains its original structure in approximately 24 h (Suda & Liu, 2007).

Because the apolipoprotein A-I is a protein expressed mainly in the liver, its transference to the hepatocytes favors the integration of the fusion protein Sushi-IL15-Apo into the HDLs, chaperoning the trans-presented form of IL-15 to different organs, including the lungs and spleen. Although the increase in the NK cell proportion in mice transferred with pSushi-IL15-Apo can be observed in those organs, it is easier to isolate NK cells from the spleen, given the higher percentage of NK cells present in this organ.

5. NK cell expansion and isolation protocol

5.1 Hydrodynamic delivery of pSushi-IL15-Apo plasmid into Rag1$^{-/-}$ mice by intravenous injection

1. Use of the Rag1$^{-/-}$ mouse strain is recommended so as to obtain a more efficient NK cell expansion given the lack of mature T cells and B cells present in this strain, limiting the competition for IL-15, and thus facilitating and favoring NK cell expansion.
2. Because of the nature of the tail vein injection, a minimal mouse weight of 18 g (7–8 weeks old) is desirable to facilitate intravenous injection and prevent the detrimental effects of hydrodynamic delivery.
3. The plasmid solution is made in a physiological saline solution (NaCl 0.9%) at room temperature. The saline solution must not be cold to prevent hypothermia. The volume of solution is calculated as 10% of the mouse weight, and 10 μg of the plasmid is transferred per mouse.
4. The mouse is immobilized in an appropriate trap with the tail vein accessible. To make tail visualization easier, the mice can be pre-warmed in a warm chamber at 37 °C.
5. The tail is disinfected with 70% ethanol, and then the solution with the plasmid is injected iv. The solution must be administered with a 27-gauge needle, and the complete dose must be given in less than 10 s. At that moment, the mouse is paralyzed for a few seconds, and the heartbeat stops. Shortly after the injection, the mouse starts breathing at a high frequency for a couple of minutes. The normal respiratory rate is recovered within 5–10 min, and the mouse behavior goes back to normal in around 10–20 min.

5.2 Spleen extraction and NK cell isolation

1. Three days after the hydrodynamic plasmid delivery, mice are sacrificed and spleens are collected in a 50 mL Falcon™ conical tube containing 2–3 mL of 1 × phosphate-buffered saline (PBS) by making a short incision (around 1 cm) on the left side of the disinfected (70% ethanol) abdomen (inferior to the stomach). Spleens can be collected earlier than 3 days post-hydrodynamic plasmid injections; however, a reduction in NK cell recovery is to be expected. In contrast, longer waiting times result in the death of the animal due to NK cell-related toxicities in the lungs (Ochoa et al., 2013).
2. Spleens are mechanically disrupted in a 6 mm Petri-dish (Corning®) containing 2–3 mL of RPMI medium (RPMI 1640 medium supplemented with 10% fetal bovine serum, 1% penicillin-streptomycin, 1 mM Sodium Pyruvate, 2 mM L-Glutamine, 10 mM HEPES, 50 μM 2-mercaptoethanol) using a scalpel and the piston of a syringe (Becton Dickinson) or with a tissue dissociator such as a gentleMACS™ Dissociator (Miltenyi).
3. The single-cell suspension is then filtered into a Falcon™ 50 mL conical tube using a 70 μm cell strainer (Corning®). Several washes with RPMI medium or 1 × PBS are performed in the petri-dish and cell strainer to ensure maximum cell collection.
4. The Falcon™ 50 mL conical tube is centrifuged at 300 × *g* for 5 min at 4 °C, and the resulting supernatant is discarded.
5. The cell pellet is re-suspended in a buffer solution containing 1 × PBS (pH 7.2), 0.5% of fetal bovine serum and 2 mM ethylenediaminetetraacetic acid (EDTA).
6. Cells are counted using a hemocytometer with a 1:2 dilution with 0.4% trypan blue to evaluate the total number of live cells.
7. Cells are centrifuged, as explained in step 4.
8. From this point, the manufacturer's instructions for the negative selection NK cell isolation kit (Miltenyi, reference 130-115-818) are followed.
   a. Cells are incubated with 10 μL per 10^7 cells of NK Cell Biotin-Antibody Cocktail and 40 μL per 10^7 cells of buffer for 5 min at 4 °C.
   b. Cells are washed with buffer (2 mL per 10^7 cells).
   c. Repeat step 4.

d. Cells are then incubated with 20 μL per 10^7 cells of anti-Biotin Microbeads and 80 μL per 10^7 cells of buffer for 10 min at 4 °C.
e. Repeat steps 8b and c.
f. Cells are passed (2×10^8 cell/mL) through previously risen LS columns while collecting the unlabeled cells that pass through in a 50 or 15 mL Falcon™ conical tube.
g. Perform a wash step by rising the columns with 3 mL of buffer.

9. Unlabeled cells are kept. Once the NK cell isolation kit protocol has finalized, cells are again centrifuged (step 4) and re-suspended in 1 × PBS or RPMI complete medium to proceed with the NK cell in vivo infusion or cell culture, respectively.
10. Standard flow cytometry analysis to evaluate NK cell purity (CD3, CD49b, NK1.1) and activation status (Eomes, T-bet, Thy1.2, NKG2D, CD226, Ly49G2, and NKG2A) can be performed at this point (Alvarez et al., 2019).

Hydrodynamic pSushi-IL15-Apo delivery in Rag1$^{-/-}$ mice results in approximately a fivefold NK cell expansion with 40% of total NK cells in the spleen (Fig. 2A). After negative magnetic selection of NK cells using the mouse NK Cell Isolation Kit (Miltenyi), a purity of between 85% and 90% is accomplished (Fig. 2B) and an average of 5 million NK cells per mouse can be recovered per spleen (Fig. 2C).

6. NK cell characterization

With this protocol, a total number of NK cells ranging between 3×10^6 and 6×10^6 per mouse can be obtained (Fig. 2C) due to strong NK cell proliferation as measured by the expression of the proliferation marker Ki67 (Fig. 2D and E). This represents at least three to five times the total number of NK cells that can be recovered from unstimulated Rag1$^{-/-}$ mice. Importantly, pSushi-IL15-Apo-treated NK cells display a mature, activated, and functional phenotype.

As observed in Fig. 2A, pSushi-IL15-Apo treatment results in the selective expansion of CD3$^-$NK1.1$^+$CD49b$^+$ mature NK cells with barely any impact on the CD3$^-$NK1.1$^+$CD49b$^-$ precursor NK cell compartment. This result correlates with previous studies that have suggested that IL-15 is involved in later stages of NK cell development (Barao, Alvarez, Redelman, et al., 2011; Huntington et al., 2009). Similarly, when the maturation markers CD27 and

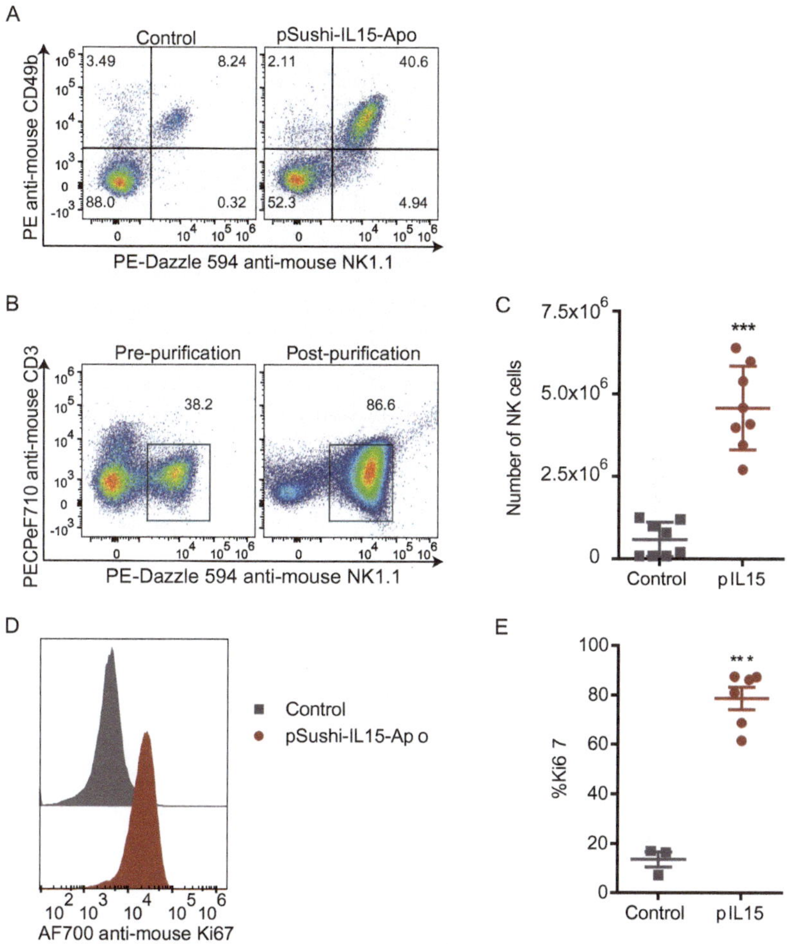

Fig. 2 pSushi-IL15-Apo causes a rapid and strong expansion of NK cells in Rag1$^{-/-}$ mice. Rag1$^{-/-}$ mice received 10 μg of pSushi-IL15-Apo or control plasmid in 2 mL by hydrodynamic injection. Three days after plasmid administration, spleens were recovered, and NK cell phenotype was characterized by flow cytometry. (A) Percentage of total NK cells expressing the NK cell markers CD49b and NK1.1 is shown on CD3^{-} gated cells 3 days after control or pSushi-IL15-Apo treatment. (B) Distribution of CD3 and NK1.1 cells before and after NK cell purification in pSushi-IL15-Apo-treated mice is shown. (C) Total number of NK cells gated as CD3^{-}NK1.1^{+}CD49b^{+} from the spleen of control or pSushi-IL15-Apo treated Rag1$^{-/-}$ mice is shown. (D and E) Representative histogram of the expression (D) or total percentage (E) of the proliferation maker Ki67 on gated NK cells after pSushi-IL15-Apo treatment is shown. Data are representative of two independent experiments with an n of three or six for control- and pSushi-IL15-Apo-treated mice, respectively (mean ± SEM). Student *T* tests were used to assess significance (*$P<0.05$, **$P<0.01$, ***$P<0.001$).

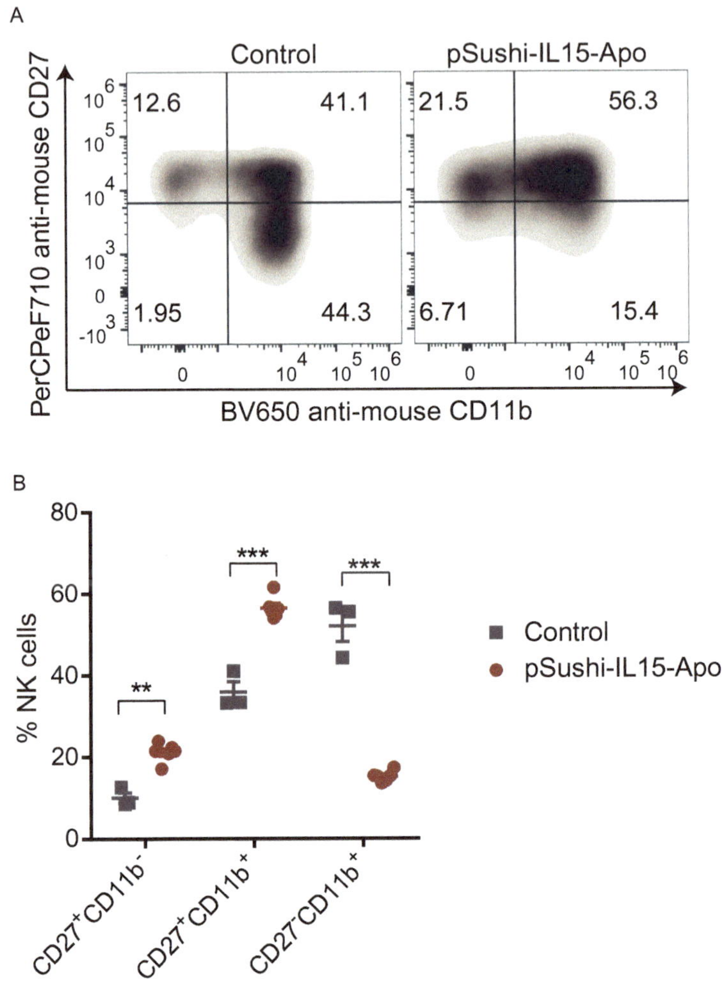

Fig. 3 pSushi-IL15-Apo treatment favors the expansion of the mature $CD27^+CD11b^+$ NK cell subset. (A) Representative histograms of the distribution of CD27 and CD11b on gated NK cells (as $CD3^-NK1.1^+CD49b^+$) is shown. (B) Total percentage of immature- ($CD27^+CD11b^-$) and mature-like NK cell subsets ($CD27^+CD11b^+$ and $CD27^-CD11b^-$) is shown for gated NK cells. Data are representative of two independent experiments with an n of three or six for control- and pSushi-IL15-Apo-treated mice, respectively (mean ± SEM). Student *T* tests were used to assess significance (*$P<0.05$, **$P<0.01$, ***$P<0.001$).

CD11b are analyzed within the mature population ($NK1.1^+CD49b^+$), a preferential increase of the subset that co-expresses the maturation markers CD27 and CD11b is observed after pSushi-IL15-Apo treatment when compared to unstimulated NK cells (Fig. 3A and B). According to Hayakawa and Smyth, this subset is correlated with superior effector functions and higher proliferative potential when compared to the more differentiated NK cell subset ($CD11^{+}CD27^{-}$) or the immature-like NK cell subset ($CD11^{-}CD27^{+}$) (Chiossone et al., 2009; Hayakawa & Smyth, 2006).

Further phenotypic analysis shows that NK cells from mice that receive the hydrodynamic injection of pSushi-IL15-Apo present a significantly higher percentage and surface expression of the mouse activation marker Thy1.2 as well as the activating NK cell receptors NKG2D and CD226 when compared to NK cells from unstimulated mice (Fig. 4A–C). Accordingly, a significant increase in the percentage and surface expression of $Ly49G2^{+}$ NK cells is observed after pSushi-IL15-Apo treatment (Fig. 4D–F). Although Ly49G2 is an inhibitory marker that identifies unlicensed and licensed NK cells in $H2^{k}$ and $H2^{d}$ mice respectively (Kim et al., 2005), an expansion of $Ly49G2^{+}$ NK cells has been previously reported upon activation in several mouse strains independently of their licensing status, which suggested that in a stimulatory environment, the expression of Ly49G2 is associated with activation (Alvarez, Sun, & Murphy, 2016; Barao et al., 2011; Sun et al., 2012; Sungur et al., 2013). Interestingly, Ly49G2 expansion has been recently linked to the expression of the activating receptor Ly49R as NK cells that co-express both receptors have higher proliferative potential and cytokine responsiveness toward MCMV (Brown, Gamache, Nash, & Cronk, 2019). In contrast, the inhibitory receptor NKG2A remains unchanged by pSushi-IL15-Apo treatment (Fig. 4D and F). In line with a higher activation status, hydrodynamic delivery of pSushi-IL15-Apo also causes a significant increase in the percentage and surface expression of the activating transcription markers eomesodermin (Eomes) and T-box transcription factors TBX21 (T-bet) when compared to control NK cells (Fig. 4G–I). Taken together, these data demonstrate that pSushi-IL15-Apo can efficiently induce NK cell activation.

To study the cytotoxic capacity of the NK cells coming from $Rag1^{-/-}$ mice transfected with pSushi-IL15-Apo, NK cells were co-cultured with MC38 tumor cells in an NK:tumor cell ratio of 3:1 and the viability of the target cells were studied at different times by staining with the cell viability dye 7AAD (ThermoFisher Scientific). As can be seen in Fig. 5A, the viability of the target cells decreased with time, indicating the capacity of the NK cells to retain cytotoxicity ex vivo. Additionally, sushi-IL15-Apo activated NK cells expressed more granzyme B in their cytoplasm compared to unstimulated NK cells (Fig. 5B and C).

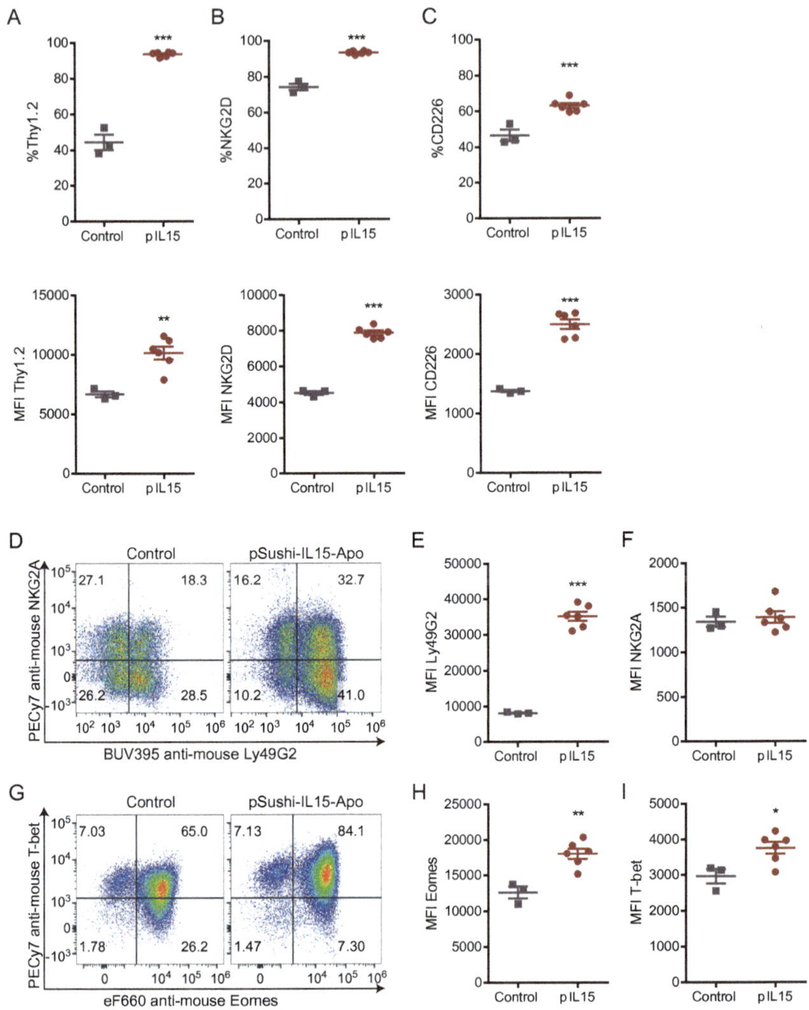

Fig. 4 Hydrodynamic pSushi-IL15-Apo delivery results in an NK cell activation phenotype. (A–C) Total percentage and MFI of the activation marker (Thy1.2) and receptors (NKG2D and CD226) is shown for gated NK cells. (D) Representative dot plots of the expression of the inhibitory receptors Ly49G2 and NKG2A on gated NK cells. (E–F) MFI expression of Ly49G2 (E) and NKG2A (F) is shown on gated NK cells. (G) Representative dot plots of the distribution of the activating transcription factors Eomes and T-bet are shown for gated NK cells. (H–I) MFI expression of Eomes (H) and T-bet (I) is shown. Data is representative of two independent experiments with an n of three or six for control- and pSushi-IL15-Apo-treated mice, respectively (mean ± SEM). Student *T* tests were used to assess significance (*$P < 0.05$, **$P < 0.01$, ***$P < 0.001$).

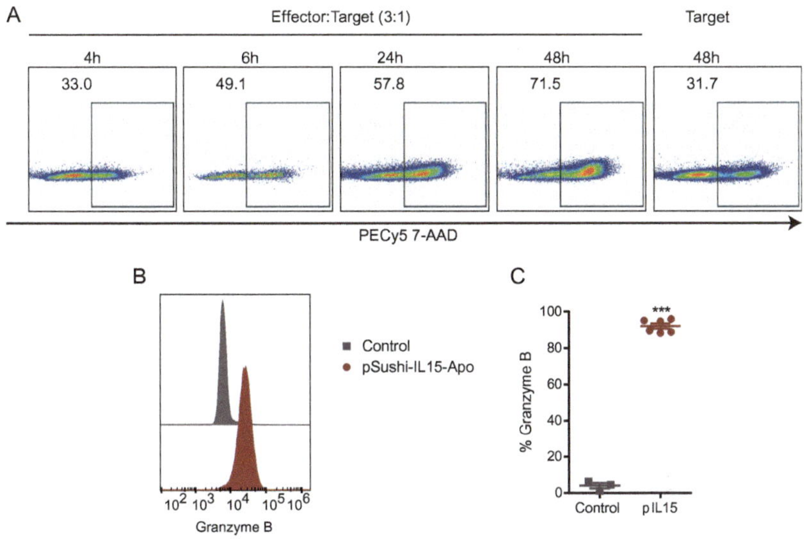

Fig. 5 Enhanced NK cell cytotoxic function observed after hydrodynamic pSushi-IL15-Apo delivery. (A) Representative dot plots of the cell death marker 7-AAD is shown on gated MC38 tumor cells co-cultured with purified NK cells at different time points for a cytotoxicity assay at 3:1 effector: target ratio. (B) Representative histograms of Granzyme B expression on gated NK cells. (C) Total percentage of Granzyme B producing NK cells is shown. Data are representative of two independent experiments with an n of three or six for control- and pSushi-IL15-Apo-treated mice, respectively (mean $\pm$ SEM). Student *T* tests were used to assess significance (*$P < 0.05$, **$P < 0.01$, ***$P < 0.001$).

7. Conclusions

In recent years, the interest in IL-15 for its cancer immunotherapeutic application has grown intensively given its direct influence on effector cells, mainly cytotoxic T cells and NK cells, and the lack of some of the shortcomings associated to IL-2 therapy. IL-15 has proven to not only induce a strong proliferation and activation of these effector cells but also, and more importantly, to produce a robust increase in their cytotoxic functions leading to promising results in cancer treatments either when administered alone or combined with other immunotherapeutic strategies (Cooley et al., 2018; Robinson & Schluns, 2017).

Here we show that the hydrodynamic delivery of the pSushi-IL15-Apo construct into $Rag1^{-/-}$ mice induces a strong activation and proliferation of NK cells in the spleen resulting in a fivefold expansion and enhanced functional properties associated with better anti-tumoral responses (Ochoa et al., 2013). This method, thus, provides an easy and affordable way to expand and activate fully functional murine NK cells to perform ex vivo experiments or to model NK cell adoptive therapies that can potentially be translated into the clinical settings in a timely manner.

Acknowledgments

This work was supported by the Spanish Association against Cancer (Foundation AECC), Foundation for Applied Medical Research (FIMA) Worldwide Cancer Research (AIRC), Joint Translational Call for Proposals 2015 (JTC 2015) TRANSCAN-2 (code: TRS-2016-00000371), and Fondo de Investigación Sanitaria-Fondo Europeo de Desarrollo Regional (FEDER) under Grants PI16/00668 and PI19/01128. This project has received funding from the European Union's Horizon 2020 research and innovation program (grant agreement n° 635122—PROCROP). M.A. was supported by the Marie Skłodowska-Curie fellowship (CINK 746985).

Conflict of interest disclosure

I.M. is a consultant for Bristol-Myers Squibb, Boehringer Ingelheim, AstraZeneca, Roche-Genentech, Merck Serono, Alligator, Tusk, Genmab, F-Star, Molecular Partners, Numab, Catalym, Bioncotech and Bayer. P.B. reports advisory roles with Tusk and Moderna, research funding from Sanofi, and Bavarian Nordic and speaker honoraria from BMS, MSD, Novartis, and AstraZeneca. The other authors declare no conflict of interest.

References

Alvarez, M., Simonetta, F., Baker, J., Pierini, A., Wenokur, A. S., Morrison, A. R., et al. (2019). Regulation of murine NK cell exhaustion through the activation of the DNA damage repair pathway. *JCI Insight*, *5*, e127729.

Alvarez, M., Sun, K., & Murphy, W. J. (2016). Mouse host unlicensed NK cells promote donor allogeneic bone marrow engraftment. *Blood*, *127*, 1202–1205.

Alvarez, M. A. E., Sungur, C. M., Perez, J., Sckisel, G. D., & Murphy, W. J. (2012). Cancer immunotherapy using Natural Killer (NK) cells. *Trends Cancer Research*, *8*, 19.

Ames, E., Canter, R. J., Grossenbacher, S. K., Mac, S., Smith, R. C., Monjazeb, A. M., et al. (2015). Enhanced targeting of stem-like solid tumor cells with radiation and natural killer cells. *Oncoimmunology*, *4*, e1036212.

Barao, I., Alvarez, M., Ames, E., Orr, M. T., Stefanski, H. E., Blazar, B. R., et al. (2011). Mouse Ly49G2+ NK cells dominate early responses during both immune reconstitution and activation independently of MHC. *Blood*, *117*, 7032–7041.

Barao, I., Alvarez, M., Redelman, D., Weiss, J. M., Ortaldo, J. R., Wiltrout, R. H., et al. (2011). Hydrodynamic delivery of human IL-15 cDNA increases murine natural killer cell recovery after syngeneic bone marrow transplantation. *Biology of Blood and Marrow Transplantation*, *17*, 1754–1764.

Berger, C., Berger, M., Hackman, R. C., Gough, M., Elliott, C., Jensen, M. C., et al. (2009). Safety and immunologic effects of IL-15 administration in nonhuman primates. *Blood*, *114*, 2417–2426.

Berraondo, P., Sanmamed, M. F., Ochoa, M. C., Etxeberria, I., Aznar, M. A., Perez-Gracia, J. L., et al. (2019). Cytokines in clinical cancer immunotherapy. *British Journal of Cancer, 120*, 6–15.

Brown, M. G., Gamache, A., Nash, W. T., & Cronk, J. (2019). Natural selection for killer receptors and their MHC class I ligands: In pursuit of gene pairs that fit well in tandem. *Journal of Leukocyte Biology, 105*, 489–495.

Burns, L. J., Weisdorf, D. J., DeFor, T. E., Vesole, D. H., Repka, T. L., Blazar, B. R., et al. (2003). IL-2-based immunotherapy after autologous transplantation for lymphoma and breast cancer induces immune activation and cytokine release: A phase I/II trial. *Bone Marrow Transplantation, 32*, 177–186.

Castillo, E. F., Stonier, S. W., Frasca, L., & Schluns, K. S. (2009). Dendritic cells support the in vivo development and maintenance of NK cells via IL-15 trans-presentation. *Journal of Immunology, 183*, 4948–4956.

Chiossone, L., Chaix, J., Fuseri, N., Roth, C., Vivier, E., & Walzer, T. (2009). Maturation of mouse NK cells is a 4-stage developmental program. *Blood, 113*, 5488–5496.

Conlon, K. C., Lugli, E., Welles, H. C., Rosenberg, S. A., Fojo, A. T., Morris, J. C., et al. (2015). Redistribution, hyperproliferation, activation of natural killer cells and CD8 T cells, and cytokine production during first-in-human clinical trial of recombinant human interleukin-15 in patients with cancer. *Journal of Clinical Oncology, 33*, 74–82.

Cooley, S., Parham, P., & Miller, J. S. (2018). Strategies to activate NK cells to prevent relapse and induce remission following hematopoietic stem cell transplantation. *Blood, 131*, 1053–1062.

Cooper, M. A., Bush, J. E., Fehniger, T. A., VanDeusen, J. B., Waite, R. E., Liu, Y., et al. (2002). In vivo evidence for a dependence on interleukin 15 for survival of natural killer cells. *Blood, 100*, 3633–3638.

Duitman, E. H., Orinska, Z., Bulanova, E., Paus, R., & Bulfone-Paus, S. (2008). How a cytokine is chaperoned through the secretory pathway by complexing with its own receptor: Lessons from interleukin-15 (IL-15)/IL-15 receptor alpha. *Molecular and cellular biology, 28*, 4851–4861.

Eisenman, J., Ahdieh, M., Beers, C., Brasel, K., Kennedy, M. K., Le, T., et al. (2002). Interleukin-15 interactions with interleukin-15 receptor complexes: Characterization and species specificity. *Cytokine, 20*, 121–129.

Elpek, K. G., Rubinstein, M. P., Bellemare-Pelletier, A., Goldrath, A. W., & Turley, S. J. (2010). Mature natural killer cells with phenotypic and functional alterations accumulate upon sustained stimulation with IL-15/IL-15Ralpha complexes. *Proceedings of the National Academy of Sciences of the United States of America, 107*, 21647–21652.

Fehniger, T. A., Suzuki, K., Ponnappan, A., VanDeusen, J. B., Cooper, M. A., Florea, S. M., et al. (2001). Fatal leukemia in interleukin 15 transgenic mice follows early expansions in natural killer and memory phenotype CD8 + T cells. *The Journal of Experimental Medicine, 193*, 219–231.

Felices, M., Chu, S., Kodal, B., Bendzick, L., Ryan, C., Lenvik, A. J., et al. (2017). IL-15 super-agonist (ALT-803) enhances natural killer (NK) cell function against ovarian cancer. *Gynecologic Oncology, 145*, 453–461.

Felices, M., Lenvik, A. J., McElmurry, R., Chu, S., Hinderlie, P., Bendzick, L., et al. (2018). Continuous treatment with IL-15 exhausts human NK cells via a metabolic defect. *JCI Insight, 3*(3), pii: 96219.

Francis, G. A. (2010). The complexity of HDL. *Biochimica et Biophysica Acta, 1801*, 1286–1293.

Gill, S., Vasey, A. E., De Souza, A., Baker, J., Smith, A. T., Kohrt, H. E., et al. (2012). Rapid development of exhaustion and down-regulation of eomesodermin limit the antitumor activity of adoptively transferred murine natural killer cells. *Blood, 119*, 5758–5768.

Hayakawa, Y., & Smyth, M. J. (2006). CD27 dissects mature NK cells into two subsets with distinct responsiveness and migratory capacity. *Journal of Immunology, 176*, 1517–1524.

Huntington, N. D., Legrand, N., Alves, N. L., Jaron, B., Weijer, K., Plet, A., et al. (2009). IL-15 trans-presentation promotes human NK cell development and differentiation in vivo. *The Journal of Experimental Medicine, 206*, 25–34.

Jakobisiak, M., Golab, J., & Lasek, W. (2011). Interleukin 15 as a promising candidate for tumor immunotherapy. *Cytokine & Growth Factor Reviews, 22*, 99–108.

Karre, K., Ljunggren, H. G., Piontek, G., & Kiessling, R. (1986). Selective rejection of H-2-deficient lymphoma variants suggests alternative immune defence strategy. *Nature, 319*, 675–678.

Kennedy, M. K., Glaccum, M., Brown, S. N., Butz, E. A., Viney, J. L., Embers, M., et al. (2000). Reversible defects in natural killer and memory CD8 T cell lineages in interleukin 15-deficient mice. *The Journal of Experimental Medicine, 191*, 771–780.

Kim, P. S., Kwilas, A. R., Xu, W., Alter, S., Jeng, E. K., Wong, H. C., et al. (2016). IL-15 superagonist/IL-15RalphaSushi-Fc fusion complex (IL-15SA/IL-15RalphaSu-Fc; ALT-803) markedly enhances specific subpopulations of NK and memory CD8+ T cells, and mediates potent anti-tumor activity against murine breast and colon carcinomas. *Oncotarget*, 7, 16130–16145.

Kim, S., Poursine-Laurent, J., Truscott, S. M., Lybarger, L., Song, Y. J., Yang, L., et al. (2005). Licensing of natural killer cells by host major histocompatibility complex class I molecules. *Nature, 436*, 709–713.

Koka, R., Burkett, P. R., Chien, M., Chai, S., Chan, F., Lodolce, J. P., et al. (2003). Interleukin (IL)-15R[alpha]-deficient natural killer cells survive in normal but not IL-15R[alpha]-deficient mice. *The Journal of Experimental Medicine, 197*, 977–984.

Kweon, S., Phan, M. T., Chun, S., Yu, H., Kim, J., Kim, S., et al. (2019). Expansion of human NK cells using K562 cells expressing OX40 ligand and short exposure to IL-21. *Frontiers in Immunology, 10*, 879.

Lee, H. R., Son, C. H., Koh, E. K., Bae, J. H., Kang, C. D., Yang, K., et al. (2017). Expansion of cytotoxic natural killer cells using irradiated autologous peripheral blood mononuclear cells and anti-CD16 antibody. *Scientific Reports*, 7, 11075.

Leonard, W. J., Lin, J. X., & O'Shea, J. J. (2019). The gammac Family of cytokines: Basic biology to therapeutic ramifications. *Immunity, 50*, 832–850.

Liu, B., Kong, L., Han, K., Hong, H., Marcus, W. D., Chen, X., et al. (2016). A novel fusion of ALT-803 (Interleukin (IL)-15 superagonist) with an antibody demonstrates antigen-specific antitumor responses. *The Journal of Biological Chemistry, 291*, 23869–23881.

Liu, F., Song, Y., & Liu, D. (1999). Hydrodynamics-based transfection in animals by systemic administration of plasmid DNA. *Gene Therapy, 6*, 1258–1266.

Luna, J. I., Grossenbacher, S. K., Sturgill, I. R., Ames, E., Judge, S. J., Bouzid, L. A., et al. (2019). Bortezomib augments natural killer cell targeting of stem-like tumor cells. *Cancers (Basel), 11*(1), pii: E85.

McCullar, V., Oostendorp, R., Panoskaltsis-Mortari, A., Yun, G., Lutz, C. T., Wagner, J. E., et al. (2008). Mouse fetal and embryonic liver cells differentiate human umbilical cord blood progenitors into CD56-negative natural killer cell precursors in the absence of interleukin-15. *Experimental Hematology, 36*, 598–608.

Miller, J. S., Soignier, Y., Panoskaltsis-Mortari, A., McNearney, S. A., Yun, G. H., Fautsch, S. K., et al. (2005). Successful adoptive transfer and in vivo expansion of human haploidentical NK cells in patients with cancer. *Blood, 105*, 3051–3057.

Mortier, E., Quemener, A., Vusio, P., Lorenzen, I., Boublik, Y., Grotzinger, J., et al. (2006). Soluble interleukin-15 receptor alpha (IL-15R alpha)-sushi as a selective and potent agonist of IL-15 action through IL-15R beta/gamma. Hyperagonist IL-15 x IL-15R alpha fusion proteins. *The Journal of Biological Chemistry, 281*, 1612–1619.

Morvan, M. G., & Lanier, L. L. (2016). NK cells and cancer: You can teach innate cells new tricks. *Nature Reviews. Cancer, 16*, 7–19.

Murphy, W. J., Parham, P., & Miller, J. S. (2012). NK cells—from bench to clinic. *Biology of Blood and Marrow Transplantation, 18*, S2–S7.

Ochoa, M. C., Fioravanti, J., Rodriguez, I., Hervas-Stubbs, S., Azpilikueta, A., Mazzolini, G., et al. (2013). Antitumor immunotherapeutic and toxic properties of an HDL-conjugated chimeric IL-15 fusion protein. *Cancer Research, 73*, 139–149.

Ochoa, M. C., Minute, L., Lopez, A., Perez-Ruiz, E., Gomar, C., Vasquez, M., et al. (2018). Enhancement of antibody-dependent cellular cytotoxicity of cetuximab by a chimeric protein encompassing interleukin-15. *Oncoimmunology, 7*, e1393597.

Pierini, A., Alvarez, M., & Negrin, R. S. (2016). NK Cell and CD4+FoxP3+ regulatory T cell based therapies for hematopoietic stem cell engraftment. *Stem Cells International, 2016*, 9025835.

Prlic, M., Blazar, B. R., Farrar, M. A., & Jameson, S. C. (2003). In vivo survival and homeostatic proliferation of natural killer cells. *The Journal of Experimental Medicine, 197*, 967–976.

Robinson, T. O., & Schluns, K. S. (2017). The potential and promise of IL-15 in immuno-oncogenic therapies. *Immunology Letters, 190*, 159–168.

Ruggeri, L., Capanni, M., Urbani, E., Perruccio, K., Shlomchik, W. D., Tosti, A., et al. (2002). Effectiveness of donor natural killer cell alloreactivity in mismatched hematopoietic transplants. *Science, 295*, 2097–2100.

Stoklasek, T. A., Schluns, K. S., & Lefrancois, L. (2006). Combined IL-15/IL-15Ralpha immunotherapy maximizes IL-15 activity in vivo. *Journal of Immunology, 177*, 6072–6080.

Suda, T., & Liu, D. (2007). Hydrodynamic gene delivery: Its principles and applications. *Molecular Therapy, 15*, 2063–2069.

Sun, K., Alvarez, M., Ames, E., Barao, I., Chen, M., Longo, D. L., et al. (2012). Mouse NK cell-mediated rejection of bone marrow allografts exhibits patterns consistent with Ly49 subset licensing. *Blood, 119*, 1590–1598.

Sun, J. C., Ma, A., & Lanier, L. L. (2009). Cutting edge: IL-15-independent NK cell response to mouse cytomegalovirus infection. *Journal of Immunology, 183*, 2911–2914.

Sungur, C. M., Tang-Feldman, Y. J., Ames, E., Alvarez, M., Chen, M., Longo, D. L., et al. (2013). Murine natural killer cell licensing and regulation by T regulatory cells in viral responses. *Proceedings of the National Academy of Sciences of the United States of America, 110*, 7401–7406.

Tognarelli, S., Wirsching, S., von Metzler, I., Rais, B., Jacobs, B., Serve, H., et al. (2018). Enhancing the activation and releasing the brakes: A double hit strategy to improve NK cell cytotoxicity against multiple myeloma. *Frontiers in Immunology, 9*, 2743.

Waldmann, T. A., Lugli, E., Roederer, M., Perera, L. P., Smedley, J. V., Macallister, R. P., et al. (2011). Safety (toxicity), pharmacokinetics, immunogenicity, and impact on elements of the normal immune system of recombinant human IL-15 in rhesus macaques. *Blood, 117*, 4787–4795.

Xu, W., Jones, M., Liu, B., Zhu, X., Johnson, C. B., Edwards, A. C., et al. (2013). Efficacy and mechanism-of-action of a novel superagonist interleukin-15: Interleukin-15 receptor alphaSu/Fc fusion complex in syngeneic murine models of multiple myeloma. *Cancer Research, 73*, 3075–3086.

Zhang, G., Budker, V., & Wolff, J. A. (1999). High levels of foreign gene expression in hepatocytes after tail vein injections of naked plasmid DNA. *Human Gene Therapy, 10*, 1735–1737.

CHAPTER FOURTEEN

Assessment of NK cell-mediated cytotoxicity by flow cytometry after rapid, high-yield isolation from peripheral blood

Peter Holicek[a,b], Iva Truxova[b], Lenka Kasikova[a,b], Sarka Vosahlikova[b], Cyril Salek[c,d], Jana Rakova[b], Monika Holubova[e], Daniel Lysak[f], Isabelle Cremer[g,h], Radek Spisek[a,b], Jitka Fucikova[a,b,*]

[a]Department of Immunology, Charles University, 2nd Faculty of Medicine and University Hospital Motol, Prague, Czech Republic
[b]Sotio, Prague, Czech Republic
[c]Institute of Hematology and Blood Transfusion, Prague, Czech Republic
[d]Institute of Clinical and Experimental Hematology, 1st Faculty of Medicine, Charles University, Prague, Czech Republic
[e]Biomedical Center, Medical Faculty in Pilsen, Charles University, Pilsen, Czech Republic
[f]Department of Hematology and Oncology, Hospital in Pilsen, Pilsen, Czech Republic
[g]Team Inflammation, Complement and Cancer, INSERM, U1138, Centre de Recherche des Cordeliers, Paris, France
[h]Sorbonne Universite de Paris, Paris, France
*Corresponding author: e-mail address: fucikova@sotio.com

Contents

Abstract

Natural killer (NK) cells constitute the predominant innate lymphocyte subset that mediates the anti-viral and anti-tumor immune responses. NK cells use an array of innate receptors to sense their environment and to respond to infections, cellular stress and transformation. The resulting NK cell activation, including cytotoxicity and cytokine production, is a fundamental component of the early immune response. The most recent

Methods in Enzymology, Volume 631
ISSN 0076-6879
https://doi.org/10.1016/bs.mie.2019.05.034

discoveries in NK cell biology have stimulated the translational research that has led to remarkable results for the treatment of human malignancies. Therefore, the rapid isolation of NK cells from the peripheral blood or tumor microenvironment and the subsequent assessment of cytolytic function are crucial to the study of their potency and NK cell-mediated immunosurveillance. Here, we provide protocols for NK cell isolation and the assessment of NK cell cytotoxicity using flow cytometry.

1. Introduction

Natural killer (NK) cells are lymphocytes that function at the interface between innate and adaptive immunity (Vivier, Tomasello, Baratin, Walzer, & Ugolini, 2008). There are distinct phenotypic and functional similarities between NK cells and T cells, especially with $CD8^+$ T cells (Freud, Mundy-Bosse, Yu, & Caligiuri, 2017). However, in contrast to $CD8^+$ T cells, NK cells develop outside the thymus in various other tissues, and they do not express a rearranged T-cell receptor (TCR) (Freud et al., 2017). Instead, their final activity is regulated by the different types of germline-encoded, non-rearranged activating and inhibitory receptors (Chiesa et al., 2006; Elliott & Yokoyama, 2011). They play a crucial role in immunosurveillance and host defense against virus-infected cells and transformed tumor cells. NK cells contribute (1) directly to immune defense through their effector functions, including cytotoxicity and cytokine production, and (2) indirectly by regulating the population of antigen-presenting cells (APCs) and T lymphocytes by cytokines (Artis & Spits, 2015; Long, Kim, Liu, Peterson, & Rajagopalan, 2013). In humans, NK cells can be divided into $CD56^{dim}$ and $CD56^{bright}$ subsets, which differ in their homing properties and immunoregulatory functions (Cooper, Fehniger, & Caligiuri, 2001). As indicated in numerous preclinical and clinical studies, NK cells play an important role in cancer immunosurveillance and may also represent a powerful prognostic biomarker for distinct cancer malignancies (Coca et al., 1997; Ishigami et al., 2000; Lowry & Zehring, 2017; Platonova et al., 2011). Here, we describe the methods for NK cell isolation from peripheral blood and assays for the assessment of NK cell cytotoxicity and effector functions using flow cytometry.

2. NK cell isolation: Principle

Human NK cells comprise 10–15% of circulating lymphocytes, and they are present in both lymphoid organs and nonlymphoid peripheral tissues (Artis & Spits, 2015; Cooper et al., 2001; Freud et al., 2017). They are commonly defined as a population of CD3-negative and CD56-positive cells (Spits et al., 2013; Walker, Barlow, & McKenzie, 2013). Two decades ago, two distinct populations of human NK cells have been determined, based on their cell surface density of CD56 and homing properties (Chiesa et al., 2006; Vivier et al., 2008). The vast majority of NK cells in the peripheral blood and spleen have a low expression of CD56 ($CD56^{dim}$) and, in contrast, they express high levels of Fcγ receptor III (FcγIIIA, CD16), whereas approximately 10% of NK cells are $CD56^{bright}CD16^{dim}$ or $CD56^{bright}CD16^{-}$ (Freud et al., 2017). Functional studies of these subsets revealed, that the $CD56^{dim}$ population represents the fully mature NK cells with high cytotoxic and secretory functions (they produce IFNγ upon interaction with tumor cells in vitro) compared to the those of the $CD56^{bright}$ subset (De Maria, Bozzano, Cantoni, & Moretta, 2011; Vivier et al., 2008). In contrast, most NK cells in lymph nodes and tonsils are $CD56^{bright}CD16^{-}$ and lack perforin. However, these cells produce cytokines, such as IFNγ in response to stimulation with different cytokines (IL-12, IL-15, and IL-18) (Cooper et al., 2001; Freud et al., 2017).

3. NK cell isolation: Protocol

(1) To isolate mononuclear cells (PBMCs), dilute fresh peripheral blood 1:1 with PBS + EDTA (*Lonza BioScience, Basel, Switzerland*) (see Note 1).

(2) Carefully layer 35 mL of diluted blood onto 15 mL of Ficoll Paque Plus (*GE HealthCare, Stockholm, Sweden*) in a 50 mL conical tube.

(3) Centrifuge the diluted peripheral blood at 1800 rpm for 30 min at 20 °C under the settings acceleration 1 and deceleration 0 (see Note 2).

(4) Carefully transfer the PBMC layer, located at the interphase, into a new conical 50 mL tube.

(5) Wash the isolated PBMCs in 50 mL of PBS + EDTA.

(6) Centrifuge the PBMCs at 1200 rpm for 10 min at 4 °C.

(7) To remove the platelets, discard the supernatant, resuspend the cells again in 50 mL PBS + EDTA and centrifuge them twice at 1000 and 800 rpm, respectively, for 10 min each at 4 °C.

(8) Discard the supernatant and resuspend the cells in appropriate volume of PBS (see Note 3).

(9) Determine the number of isolated PBMCs.

(10) Transfer the cells into 50 mL conical tubes and centrifuge the cells at 1200 rpm for 10 min.

(11) Proceed to the negative selection of NK cells using the NK Cell Isolation Kit (*MACS, Miltenyi Biotec, Bergish Gladbach, Germany*; See Note 4 for full kit protocol)

(12) Add 10 μL of NK Cell Biotin-Antibody Cocktail (*MACS, Miltenyi Biotec, Bergish Gladbach, Germany*) per 1×10^7 cells in 40 μL isolation buffer for 5 min at 4 °C.

(13) Add 30 μL of isolation buffer (see Note 5).

(14) Add 20 μL NK Cell MicroBead Cocktail (*MACS, Miltenyi Biotec, Bergish Gladbach, Germany*) per 1×10^7 cells, mix properly and incubate the cell suspensions for 10 min at 4 °C.

(15) For further manual NK cell isolation, use separation columns (*MACS, Miltenyi Biotec, Bergish Gladbach, Germany*). Apply the suspension onto the column and collect the flow-through containing the unlabelled cells (NK cells).

4. NK cell cytotoxicity: Principles

The activation of NK cell cytotoxicity is thought to be mediated by a balance of inhibitory and activating NK cell receptors, as well as by other adhesion and costimulatory molecules, as reviewed previously in detail elsewhere (Fauriat, Long, Ljunggren, & Bryceson, 2010; Krzewski & Coligan, 2012). The final process of killing their target cells is a complex, multi-stage process that concludes with the secretion of lytic granules at the immunological synapse (Krzewski & Coligan, 2012; Long et al., 2013). The contents of the granules include a variety of enzymes typical for lysosomes as well as proteins unique to lytic granules, including perforin, granzyme, Fas ligand, TNF-related apoptosis-inducing ligand (TRAIL), granulysin, and small antimicrobial peptides (Long et al., 2013). These components are able to generate pores into the targeted cell's membrane and initiate apoptosis after exocytosis (Krzewski & Coligan, 2012). Therefore, the assessment of NK cell cytolytic functions is critical to the study of their potency and NK cell-mediated immunosurveillance. Various cytotoxicity assays measuring target cell death have been reported, including cytotoxicity and degranulation assays (Somanchi, McCulley, Somanchi, Chan, & Lee, 2015). During the degranulation step, cytolytic granules in NK cells are released, and the lysosome-associated membrane protein-1 (LAMP-1; CD107a), which is present on the cytolytic granule surface, is transported to the cell surface, which allows the identification of activated NK cells by flow cytometry as reported here (Long et al., 2013; Vivier et al., 2008). Moreover, quantitative assays of NK cell killing traditionally measure the release of chromium or a fluorescent dye from labeled target cells, which improve our understanding of the cytolytic impact of NK cells. Here, we report a rapid flow cytometry assay for quantifying target cell killing after the single-color labeling of target cells prior to co-incubation with NK cells. In our protocol, target cells are first labelled with Vybrant™ Cell-Labeling Solution. Using this assay, we observed consistent cytotoxicity results demonstrated by NK cell dose-dependent responses.

5. NK Cell cytotoxicity: Protocols

5.1 Cytotoxicity assay: Protocol

(1) Stain 1×10^6 of NK cells using 2 μL Invitrogen™ Vybrant™ DiI Cell-Labeling Solution (*Invitrogen™, Carlsbad, CA, USA*) in 1 mL of RPMI medium for 20 min at 37 °C and 5% CO_2.

(2) Stain 1×10^6 of K562 cells using 2 μL Invitrogen™ Vybrant™ DiD Cell-Labeling Solution (*Invitrogen™, Carlsbad, CA, USA*) in 1 mL of RPMI medium for 20 min at 37 °C and 5% CO_2.

(3) Wash both NK and K562 cells in 5 mL of RPMI and centrifuge them at 1200 rpm for 10 min at RT.

(4) Discard the supernatant and resuspend the NK cells and K562 cells in lymphocyte medium (See Note 6) at final concentrations of 4×10^6 cells per 1 mL and 1×10^6 cells per 1 mL, respectively.

(5) Place the cells into 96 U-shaped well plates and cultivate them as indicated in Table 1 for 4 h at 37 °C and 5% CO_2.

(6) Add 200 UI/mL of IL-2 into each well containing a mixture of NK cells and K562 cells.

Table 1 Experimental setup for the cytotoxicity assay.

Condition	Volume of NK cell suspension (μL)	Volume of K562 cell suspension (μL)	Volume of medium (μL)	Incubation time (h)
NK: K562 cells ~10:1	50	20	130	4
NK: K562 cells ~5:1	25	20	155	4
K562 cells	–	20	180	4

(7) After the incubation, transfer the cells from a 96 U-shaped well plate into a 96 V-shaped well plate and centrifuge them at 2200 rpm for 2 min at 4 °C.
(8) The Annexin V/DAPI Staining Kit was used according to the manufacturer's instructions (*eBioscience, San Diego, CA, USA*) to determine the population of apoptotic/necrotic cells.
(9) Analyse the cell suspension by flow cytometry.
(10) Determine the percentage of early apoptotic (Annexin V^+/$DAPI^-$), late apoptotic (AnnexinV^+/$DAPI^+$) and necrotic (Annexin V^-/$DAPI^+$) K562 cells using the gating strategy indicated in Fig. 1.

5.2 Degranulation assay: Protocol

(1) Prepare 1×10^7 PBMCs in 1 mL of RPMI 1640 medium supplemented with 10% FBS, 2 mmol/L L-glutamine and 100 U/mL penicillin (*Lonza, BioScience, Basel, Switzerland*).
(2) Prepare 5×10^5 K562 cells in 1 mL of RPMI medium.
(3) Cultivate PBMCs and K562 cells in 96 U-shaped well plates as indicated in Table 2 for 4 h at 37 °C and 5% CO_2.
(4) Incubate the cells from all conditions with 4 μL of anti-CD107a-FITC primary antibody for 1 h at 37 °C and 5% CO_2 (Table 3).
(5) Add phorbol myristate acetate (PMA) and ionomycine at final concentrations of 50 ng/mL and 1000 ng/mL, respectively, only to the wells, as indicated in Table 2.
(6) Subsequently, incubate the cells from all conditions with 3.0 μg/mL of Brefeldin A Solution (*eBioscience, San Diego, CA, USA*) for 3 h at 37 °C and 5% CO_2.
(7) Transfer the cells into 96 V-shaped well plates and wash the cells with 200 μL of PBS.
(8) Stain the cell suspension with a panel of primary antibodies: anti-CD3 AF700, anti-CD56 ECD and anti-CD45 HV500 (Table 3) for 20 min at 4 °C.
(9) Wash the cells twice in PBS.
(10) Add 100 μL of fixation buffer, prepared as follows: Fixation/Permeabilization Concentrate and Fixation/Permeabilization Diluent in a 1:3 ratio (*eBioscience, San Diego, CA, USA*).
(11) Add 100 μL permeabilization buffer, diluted in distilled water at a 1:9 ratio.

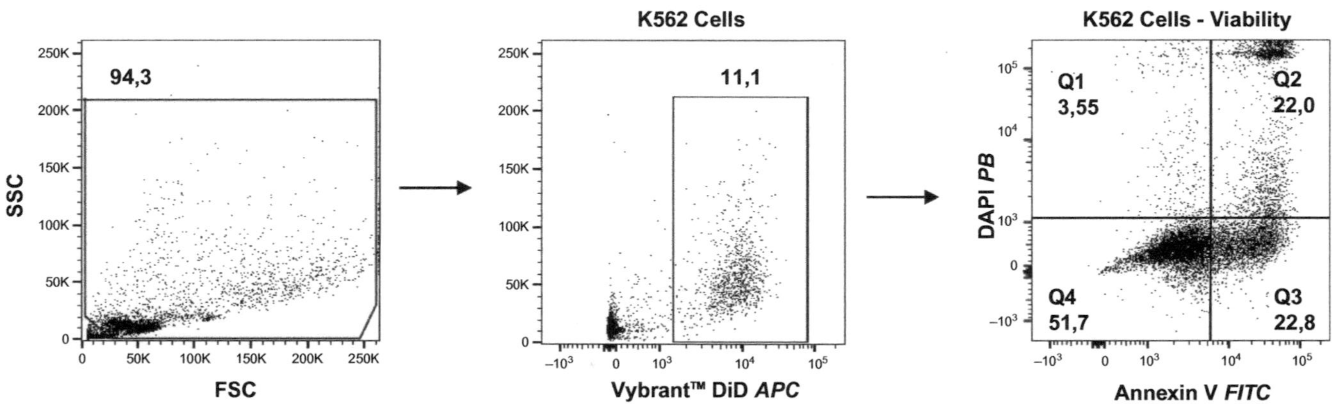

Fig. 1 Cytotoxicity assay. Gating strategy for the detection of the percentage of Annexin$^+$ and DAPI$^+$ K562 cells.

Table 2 Experimental setup for degranulation assay.

Condition	Volume of PBMC suspension (μL)	Volume of K562 cell suspension (μL)
Non-stimulated PBMC	200	–
PBMC: K562 cells ~10:1	200	10
PBMC + PMA/Iono	200	–

Table 3 Antibodies used for flow cytometry.

Parameter	Source	Producer	Clone	Fluorochrome	Dilution
CD3	Mouse	EXBIO	MEM-57	Alexa Fluor 700	5:100
CD45	Mouse	BD Biosciences	HI30	HV500	6:100
CD56	Mouse	Beckman Coulter		ECD	6:100
CD107a	Mouse	BioLegend	H4A3	FITC	8:100
Granzyme B	Mouse	BD Biosciences	GB11	Brilliant Violet 421	4:100
Perforin	Mouse	BioLegend	dG9	APC	4:100
IFN-γ	Mouse	eBioscience	4S·B3	PE-Cy7	1:100

(12) Add a mixture of intracellular antibodies: anti-IFNγ PE-Cy7, anti-granzyme B BV421 and anti-perforin APC for 20 min at 4 °C, as indicated in detail in Table 3.

(13) The cells were incubated in a refrigerator (2–8 °C) for 20 min and were then resuspended in 150 μL of 1 × permeabilization buffer.

(14) Wash the cells twice with PBS.

(15) The cell suspension was analyzed by flow cytometry (Fig. 2).

6. Concluding remarks

NK cells represent a subset of lymphocytes that play a fundamental role in immunosurveillance and host defense against cancer and viral infections (Krzewski & Coligan, 2012). The killing of target cells is a complex process that involves the phenotypical and functional activation of NK cells,

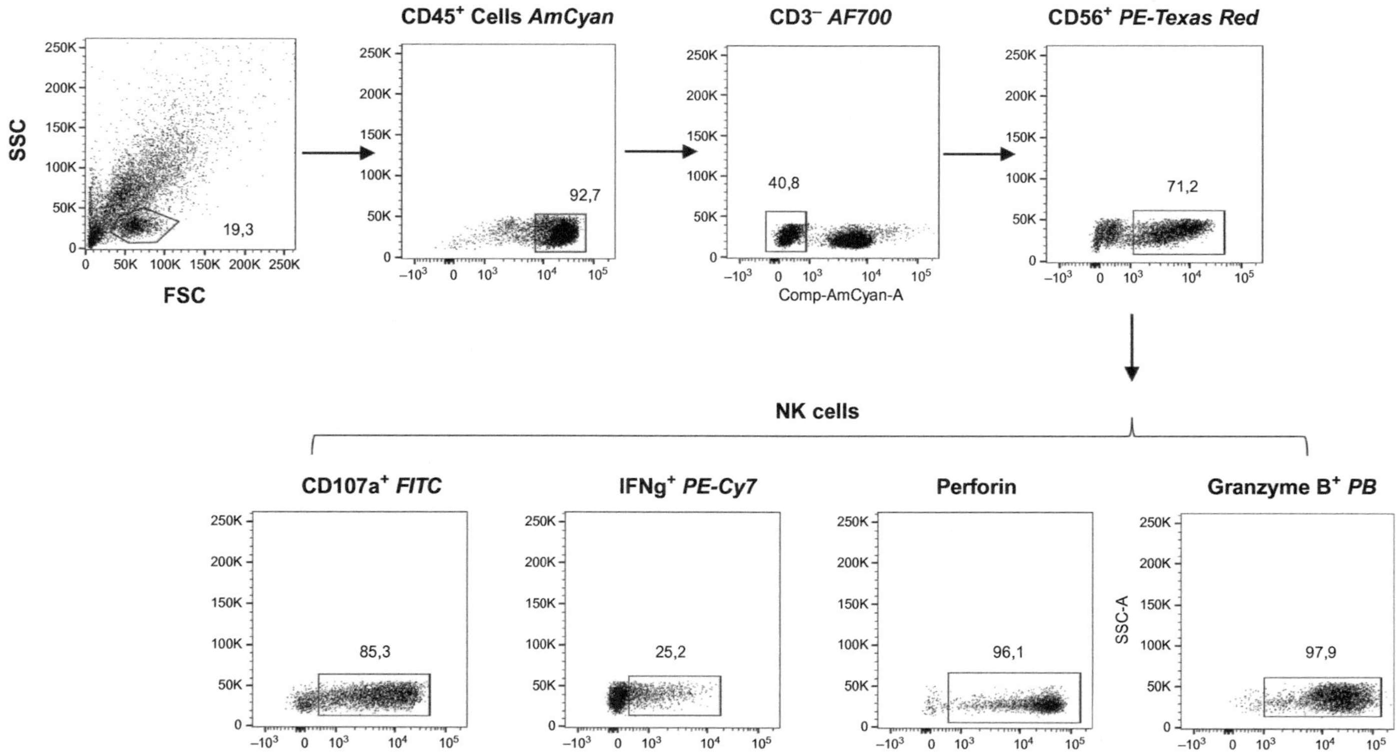

Fig. 2 Gating strategy for the detection of NK cells and their surface receptors.

leading to the secretion of lytic granules containing perforin and granzymes at the immunological synapse (Krzewski & Coligan, 2012). Hence, the evaluation of NK cell cytotoxicity represents a crucial step in understanding the role of NK cell-mediated immunosurveillance. Numerous methods have been described for the assessment of NK cell cytotoxicity, the most prominent method being the ^{51}chromium release assay, with toxicity limitations and the necessity to dispose of radioactive material (Shabrish, Gupta, & Madkaikar, 2016; Somanchi et al., 2015). Here, we provide two protocols for rapid NK cell isolation from peripheral blood and the assessment of NK cell-mediated cytotoxicity using flow cytometry.

7. Notes

(1) The peripheral blood or buffy coat should not be older than 12h and should be supplemented with anticoagulants (e.g., heparin, EDTA, citrate or citrate phosphate dextrose).

(2) Use a centrifuge with adjustable acceleration and deceleration settings. The rotor brakes must be inactivated entirely to effectively prevent mixing of the phases.

(3) The appropriate PBS volume is estimated by the expected total cell count of isolated PBMCs and by the cell counting device/technique, which might require different cell concentrations in suspension.

(4) A detailed protocol provided by the manufacturer is found: https://www.miltenyibiotec.com/UN-en/products/macs-cell-separation/cell-separation-reagents/microbeads-and-isolation-kits/nk-cells/nk-cell-isolation-kit-human.htmL

(5) Isolation buffer composition: PBS supplemented with 0.5% BSA and 2mM EDTA. The isolation components included were from the human NK Cell Isolation Kit (*MACS, Miltenyi Biotec, Bergish Gladbach, Germany*).

(6) Lymphocyte medium composition: RPMI Medium 1640 ((−)L-glutamine) supplemented with 10% AB serum, penicillin 1 U/mL, streptomycin 1 μg/mL and Glutamax 20 μM, sodium pyruvate 1 mM, MEM NEAA 1 × Gibco, β-mercaptoethanol 50 μM.

Acknowledgments

This study was exclusively sponsored by Sotio, Prague, Czech republic.

References

Artis, D., & Spits, H. (2015). The biology of innate lymphoid cells. *Nature*, *517*, 293–301.

Chiesa, S., Mingueneau, M., Fuseri, N., Malissen, B., Raulet, D. H., Malissen, M., et al. (2006). Multiplicity and plasticity of natural killer cell signaling pathways. *Blood*, *107*, 2364–2372.

Coca, S., Perez-Piqueras, J., Martinez, D., Colmenarejo, A., Saez, M. A., Vallejo, C., et al. (1997). The prognostic significance of intratumoral natural killer cells in patients with colorectal carcinoma. *Cancer*, *79*, 2320–2328.

Cooper, M. A., Fehniger, T. A., & Caligiuri, M. A. (2001). The biology of human natural killer-cell subsets. *Trends in Immunology*, *22*, 633–640.

De Maria, A., Bozzano, F., Cantoni, C., & Moretta, L. (2011). Revisiting human natural killer cell subset function revealed cytolytic CD56(dim)CD16+ NK cells as rapid producers of abundant IFN-gamma on activation. *Proceedings of the National Academy of Sciences of the United States of America*, *108*, 728–732.

Elliott, J. M., & Yokoyama, W. M. (2011). Unifying concepts of MHC-dependent natural killer cell education. *Trends in Immunology*, *32*, 364–372.

Fauriat, C., Long, E. O., Ljunggren, H. G., & Bryceson, Y. T. (2010). Regulation of human NK-cell cytokine and chemokine production by target cell recognition. *Blood*, *115*, 2167–2176.

Freud, A. G., Mundy-Bosse, B. L., Yu, J., & Caligiuri, M. A. (2017). The broad spectrum of human natural killer cell diversity. *Immunity*, *47*, 820–833.

Ishigami, S., Natsugoe, S., Tokuda, K., Nakajo, A., Che, X., Iwashige, H., et al. (2000). Prognostic value of intratumoral natural killer cells in gastric carcinoma. *Cancer*, *88*, 577–583.

Krzewski, K., & Coligan, J. E. (2012). Human NK cell lytic granules and regulation of their exocytosis. *Frontiers in Immunology*, *3*, 335.

Long, E. O., Kim, H. S., Liu, D., Peterson, M. E., & Rajagopalan, S. (2013). Controlling natural killer cell responses: Integration of signals for activation and inhibition. *Annual Review of Immunology*, *31*, 227–258.

Lowry, L. E., & Zehring, W. A. (2017). Potentiation of natural killer cells for cancer immunotherapy: A review of literature. *Frontiers in Immunology*, *8*, 1061.

Platonova, S., Cherfils-Vicini, J., Damotte, D., Crozet, L., Vieillard, V., Validire, P., et al. (2011). Profound coordinated alterations of intratumoral NK cell phenotype and function in lung carcinoma. *Cancer Research*, *71*, 5412–5422.

Shabrish, S., Gupta, M., & Madkaikar, M. (2016). A modified NK cell degranulation assay applicable for routine evaluation of NK cell function. *Journal of Immunology Research*, *2016*, 3769590.

Somanchi, S. S., McCulley, K. J., Somanchi, A., Chan, L. L., & Lee, D. A. (2015). A novel method for assessment of natural killer cell cytotoxicity using image cytometry. *PLoS One*, *10*, e0141074.

Spits, H., Artis, D., Colonna, M., Diefenbach, A., Di Santo, J. P., Eberl, G., et al. (2013). Innate lymphoid cells—A proposal for uniform nomenclature. *Nature Reviews. Immunology*, *13*, 145–149.

Vivier, E., Tomasello, E., Baratin, M., Walzer, T., & Ugolini, S. (2008). Functions of natural killer cells. *Nature Immunology*, *9*, 503–510.

Walker, J. A., Barlow, J. L., & McKenzie, A. N. (2013). Innate lymphoid cells—How did we miss them? *Nature Reviews. Immunology*, *13*, 75–87.

CHAPTER FIFTEEN

Two-dimensional dynamic evaluation of natural killer cell-mediated lysis of adherent target cells

Alexandra Frazao[†], **Louise Rethacker**[†], **Anne Caignard***
INSERM, U1160, Institut de Recherche Saint-Louis, Hôpital Saint-Louis, Paris, France
*Corresponding author e-mail address: anne.caignard@inserm.fr

Contents

Abstract

Cell-mediated cytotoxicity is a major function of cytotoxic lymphocytes. The cytotoxic function of immune effectors requires accurate and sensitive quantification as tumor immunotherapies are actively developed to treat growing types of tumors. Various methods have been developed to quantify this function. A first approach consists in measuring the externalization of LAMP-1 (lysosomal associated membrane 1), CD107a molecules transitory expressed at the cell surface by degranulating cytotoxic cells and is determined by flow cytometry. The focus of the chapter concerns the second approach that quantifies target cell lysis resulting from the close contact interaction with cytotoxic Natural Killer (NK) lymphocytes. For long time, target cell lysis was evaluated by chromium release assay, requiring use of radioactive labeled salt (^{51}Cr) and specific devices not compatible with repeated tests performed for immunomonitoring of patients. Other methods include fluorimetric and bioluminescence assays. Monitoring immune cell-mediated lysis of adherent targets by the dynamic measure of cell proliferation using the Real time cell analyzer (RTCA) is a good alternative. The test relies on

† These authors are equally contributed to this chapter.

Methods in Enzymology, Volume 631
ISSN 0076-6879
https://doi.org/10.1016/bs.mie.2019.07.033

sensitive dynamic measure of cell index values during their interactions with cytotoxic immune effectors. The cell index increasing with cell proliferation rapidly drops in presence of immune cells. The lysis is quantified in reference with targets alone and expressed as percentages of lysis curves. Moreover, the dynamic monitoring of target cell index allows the evaluation of drugs, cytokines, mAbs on target cell lysis. Here we describe a robust and sensitive method for quantification of immune cell-mediated lysis of adherent targets.

1. Introduction

Cytotoxicity is a key function of immune lymphoid cell subsets such as $CD8^+$ T lymphocytes and NK cells implicated in tumor immunosurveillance. Cell cytotoxicity is a sensitive read out to monitor the impact of immunotherapies that are now reference treatments for cancer patients. Actual tumor immunotherapies involve on one hand adoptive transfer of engineered chimeric antigen receptor CAR-T cells or CAR-NK cells that show efficiency in hematologic cancers (Sadelain, Brentjens, & Riviere, 2013; Salter, Pont, & Riddell, 2018) and are evaluated in solid tumors (Bollino & Webb, 2017; Watanabe, Kuramitsu, Posey, & June, 2018). On the other hand, patients are treated with immune checkpoint blockers (ICB), mAbs that block immune inhibitory receptors to restore a high lytic function by cytotoxic T lymphocytes (CTL) or NK cells infiltrating the tumor (Andre et al., 2018; Ribas & Wolchok, 2018). To follow the impact of these immunotherapies in patients, it is important to develop and use accurate and standardized methods to measure the cytotoxic capacity of transferred engineered CAR-T cells, CTL or NK cells in ICB treated patients, starting from ex vivo samples immune cells. Lymphoid cell-mediated cytolysis is dependent of close contact between effectors and targets and involves the formation of an immune synapse that gather receptor ligands interactions inducing activation of cell signaling in immune effectors and leading to apoptosis of the target (Dieckmann, Frazer, Asano, Stinchcombe, & Griffiths, 2016; Montoya, Sancho, Vicente-Manzanares, & Sanchez-Madrid, 2002). NK cell-mediated cytotoxicity is also governed by the formation of a lytic immune synapse leading to the directed secretion of cytolytic granules and subsequent target cell death (Mace et al., 2014).

Different approaches for evaluating effector cell lytic capacity or target cell lysis were developed (Fig. 1). Lytic potential of effector cells is assessed by flow cytometry by the measure of LAMP-1 externalization (CD107a assay) that comes out as a surrogate marker of cytotoxic potential of immune

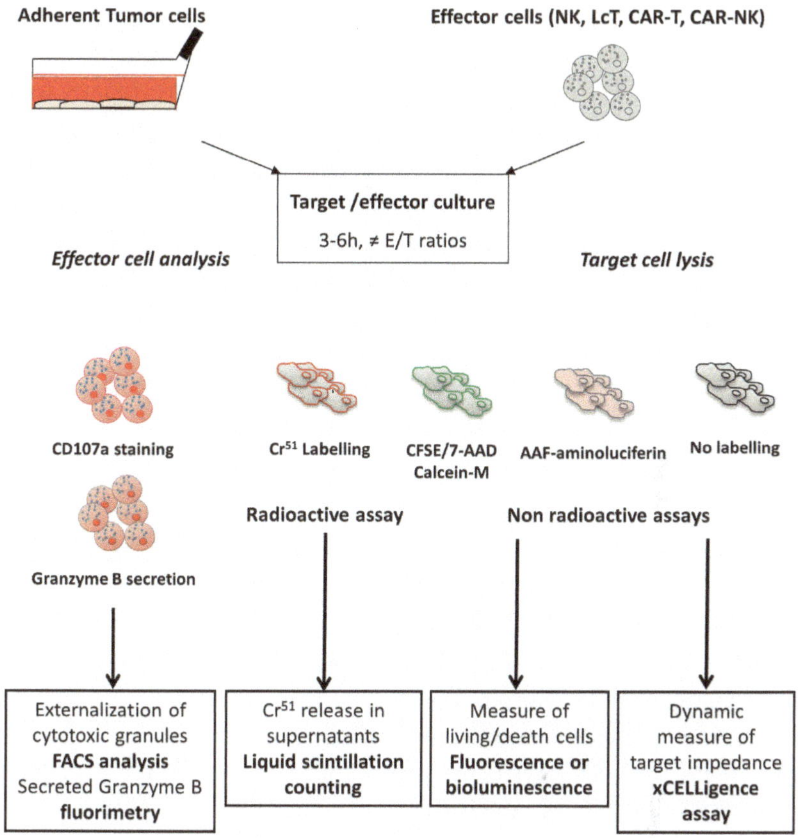

Fig. 1 Measure of cytotoxic effector cell activation and cell-mediated lysis assay of adherent tumor cells. Effector and target cells are mixed together at different E/T ratio for short term cocultures (2–4 h). Activation of effector cell cytotoxic capacity can be evaluated by externalization of CD107a (LAMP-1) detected by flow cytometry in cell suspensions. Membrane staining of immune cells with specific mAbs allows the detection of CD107a in distinct cytotoxic cell subsets in cell suspensions. Concentration of granzyme B in supernatants can be monitored. Tumor target cell lysis can be monitored by different methods. The ^{51}Cr release assay, the gold standard assay for decades is progressively replaced by nonradioactive assays. These tests allow detection of dead and living cell ratios detected by fluorescence or bioluminescence. RTCA xCELLigence assay allows dynamic measure of cell proliferation of adherent cells and has been adapted to determine immune effector cell-mediated death.

cells (Zaritskaya, Shurin, Sayers, & Malyguine, 2010). The quantification of Granzyme B by ELISPOT assay also refers to lytic potential of NK cells and is used for monitoring cell-mediated cytotoxicity (Shafer-Weaver et al., 2003) and can be used to follow cell-mediated cytotoxicity in cancer vaccine trials (Shafer-Weaver et al., 2006). However, these assays do not inform on the fate of the target cells.

The alternative is to measure target cell lysis by assessing target cell death. Target lysis occurs rapidly after formation of the synapse and leads to apoptosis of the target cell. Immune cell-mediated cytotoxicity focused on the lysis of the target can be determined by different techniques that display various sensitivity and reproducibility (Fig. 1). For decades the gold standard technique has been the use of chromium release assay developed in 1970 (Brunner, Mauel, Rudolf, & Chapuis, 1970), based on insertion of chromium salts in the membrane of the target during a short incubation. Then the effector cells at different E/T ratios are added and cocultures are maintained for 2–6 h. The measure of the radioactive counts in supernatants of cocultures are determined at given time points of cocultures. Lysis is estimated by the formula:

$$[(\text{experimental release} - \text{minimum release})/(\text{maximum release} - \text{minimum release}) \times 100],$$

where minimum release is the supernatant from labeled targets incubated with medium, maximum release the supernatant from detergent treated labeled targets. This technique is reliable and sensitive but highly depends on the first step of insertion of chromium salts in the membrane (Fig. 1).

Determination of cytotoxic immune cell-mediated lysis is a useful read out to follow along immunotherapy treatments by adoptive cell transfer of engineered T and NK cells expressing a chimeric antigen receptor (CAR). The precise and standardized evaluation of tumor cell survival after coculture with cytotoxic immune cells is a prerequisite to calculate the percentages of tumor cell lysis.

In that context, the ^{51}Cr release assay is not appropriate as manipulation of radioactive elements is restricted, requiring safe disposal of radioactive waste and cannot be used for routine assay by immunomonitoring laboratories. RTCA with xCELLigence has been used in preclinic trials (Cerignoli et al., 2018; Quereda et al., 2010) and to follow manufacture of CAR-T cells (Gargett, Truong, Ebert, Yu, & Brown, 2019).

Alternative methods to measure target cell death include assays with vital dyes that are incorporated in target before addition of cytotoxic effectors and assessment of dilution of the dye. These nonradioactive assays have been developed in which target cells are loaded with fluorescent dyes. Different Fluorometric Cell Viability Kits have been designed to quantify live cell numbers based on the presence of their cytoplasmic membrane integrity, as end-point assay for cell viability. The fluorescent signal generated from the assay is proportional to the number of living cells in the sample.

Calcein-AM is a nonfluorescent, hydrophobic compound that easily permeates intact, living cells. The hydrolysis of Calcein-AM by intracellular esterases produces Calcein, a hydrophilic strongly fluorescent compound that is retained in the cell cytoplasm. Cells grown in black-walled plates can be quantified in less than 2h using a fluorescence plate reader with excitation at 490nm and emission at 520nm. Calcein-AM provides a simple, rapid and accurate method to measure cell viability and or cytotoxicity (Roden, Lee, Panelli, & Marincola, 1999). It has been adapted for measuring immune cell-mediated cytotoxicity (Somanchi, McCulley, Somanchi, Chan, & Lee, 2015).

Fluorescent probes such as 5-(6)-carboxyfluorescein diacetate succinimidyl ester (CFSE) to label target cells, in combination with red fluorescent probes such as 7-AAD or propidium iodide to label dead effector and dead target cells can be used. CFSE passively diffuses into cells and its acetate groups are cleaved by intracellular esterases to yield highly fluorescent carboxyfluorescein succinimidyl ester. 7-AAD and propidium iodide are fluorescent dyes excluded from live cells but penetrating dead or damaged cells to label DNA. The labeling distinguishes living target cells in green, dead target cells in green and red, dead effector cells in red, and live effector cells, which remain unstained. Thus, effector cells cytotoxicity can be evaluated quantitatively at a single cell level (Kim, Donnenberg, Donnenberg, Gooding, & Whiteside, 2007).

Other assays use luminescence for assessing relative dead cell number in cell populations. The CytoTox-Glo™ Assay is a luminescent cytotoxicity assay that measures the extracellular activity of a distinct intracellular protease activity released from membrane-compromised cells. A luminogenic cell-impermeant peptide substrate (AAF-aminoluciferin) is used to measure dead-cell protease activity. The liberated aminoluciferin product is measured as "glow type" luminescence generated by Ultra-Glo™ Recombinant Luciferase provided in the assay reagent. The AAF-aminoluciferin substrate cannot cross the intact membrane of viable cells and does not generate any appreciable signal from the live-cell population. The amount of luminescence directly correlates with the percentage of cells undergoing cytotoxic stress. With the addition of a lysis reagent, the CytoTox-Glo™ Assay also can deliver the luminescent signal associated with the total number of cells in each assay well. Viability is calculated by subtracting the luminescent dead-cell signal from the total luminescent value, thus allowing normalization of the data to cell number. The cytotoxicity protease biomarker is constitutive and conserved across cell lines, and the CytoTox-Glo™ Assay demonstrates excellent correlation with other methods of assessing cell viability. Bioluminescence based cytotoxicity assay with target cells stably transduced with luciferase was used to measure NK cell activity (Karimi et al., 2014).

However, these tests lack sensitivity, are not easily standardized and are time consuming when numerous conditions are evaluated (E/T ratio, duration of coculture, different target ...).

The RTCA are suitable for addressing cell responses in different applications. It was first used for assessing cell proliferation, cell adhesion and spreading, and cytotoxicity (Atienza et al., 2006; Atienza, Zhu, Wang, Xu, & Abassi, 2005; Solly, Wang, Xu, Strulovici, & Zheng, 2004). Our team has developed for several years a sensitive assay to measure the dynamic lysis of adherent tumor cells by NK cells. We used the RTCA, xCELLigence system (ACEA Biosciences) an impedance based label free cell analysis system that has been adapted to immune cell-mediated cytolysis of adherent tumor cells. RTCA is used to follow cell impedance. Our data demonstrate that RTCA can be used to detect tumor cells lysis by immune NK cells in a label free and real time fashion.

2. Assessment of NK cell-mediated lysis using cell impedance dynamic measure

2.1 Principles

The methodology relies on the monitoring by RTCA, xCELLigence System (ACEA Biosciences) of the cell adhesion and proliferation of adherent cells. The RTCA xCELLigence system is a microelectronic biosensor system for dynamic, real-time, label free cellular analysis for a variety of applications such as drug discovery and screening, cell migration and invasion (Atienzar et al., 2011; Ke, Wang, Xu, & Abassi, 2011; Roshan Moniri et al., 2015). This technology allows accurate dynamic measure by capturing data throughout the entire time course of an experiment and obtaining more physiologically time lapse feature of cell-mediated cytolysis.

It measures electric impedance across gold plated interdigitated microelectrodes on specialized 96 well E-Plate (s). Impedance is reported as a cell index (CI) value. The CI value at each time point is calculated as

$$\mathrm{CI} = (\mathrm{Zi} - \mathrm{Z0})/15\,\Omega,$$

where Zi is the impedance at an individual time point during the experiment and Z0 is the impedance at the beginning of the experiment. The CI is thus self-calibrated value derived from the ratio of measured impedances. CI is a unitless parameter.

The impedance measurement provides quantitative information about the status of adherent cells, including cell number, viability and morphology. The RTCA station located in the incubator transmits signals from the E-Plate to the RTCA unit. The RTCA unit controls cells, selects wells to automatically measure impedance data to the computer. The control Unit is a laptop computer with a mobile port replicator device connected to it. The software tools are installed and CI derived from impedance measures are continuously displayed on the software user interface. The average measurement rate is from one each 20 s to one each 15 min for a 96-well E-Plate. Proliferation of adherent cells will increase the numbers of adherent cells and thus the cell index.

We have adapted the RTCA for immune cell-mediated lysis of adherent tumor targets. NK cells induce the apoptosis of the target following interactions between the two cells. Target lysis leads to cell detachment, cell index decrease observed from 10 to 20 min after NK cells addition and reaches a plateau at 120–180 min. The inhibition of cell index is proportional to effector/target ratio (Fig. 2). The assay has been used to successfully quantify the cell-mediated lysis of NK cells, TCRγ/δ T cells (Schiller et al., 2016; Seidel et al., 2014) and $CD8^+$ T cells (Erskine, Henle, & Knutson, 2012; Guan et al., 2013; Morisada et al., 2017; Peper et al., 2014).

Additional possible applications include the determination of the impact of drug treatment of target on NK cell-mediated lysis, modulation of NK cell-mediated cytotoxicity by blocking mAbs (anti-NKG2D, NKp30, NKp46, anti-HLA-I) (Frazao et al., 2019; Fregni et al., 2013, 2011; Messaoudene et al., 2014). Antibody dependent cell cytotoxicity can also be studied with this method (Glamann & Hansen, 2006).

2.2 Protocols

2.2.1 Cell culture of adherent tumor targets

Various types of human tumor cell lines, renal cell carcinoma RCC5, RCC6, melanoma SK28, M14, A375, MelC, MelS, breast cancer MCF-7, MDA-MB-321, colon cancer HT29, DLD1 are routinely maintained at 37 °C (5% CO_2) in Dulbecco's Modified Eagle Medium (DMEM) or RPMI 1640 Glutamax supplemented with 10% fetal calf serum (FCS) and 1% penicillin–streptomycin. Cells are passed 24–48 h before the experiment to allow appropriate confluence in the flask and assure exponential growth phase of cells in the xCElligence assay.

When cultures reach 80% confluence (corresponding to different cell numbers according to the size and shape of cells), exhausted culture medium is discarded and cells are washed with prewarmed PBS. Then

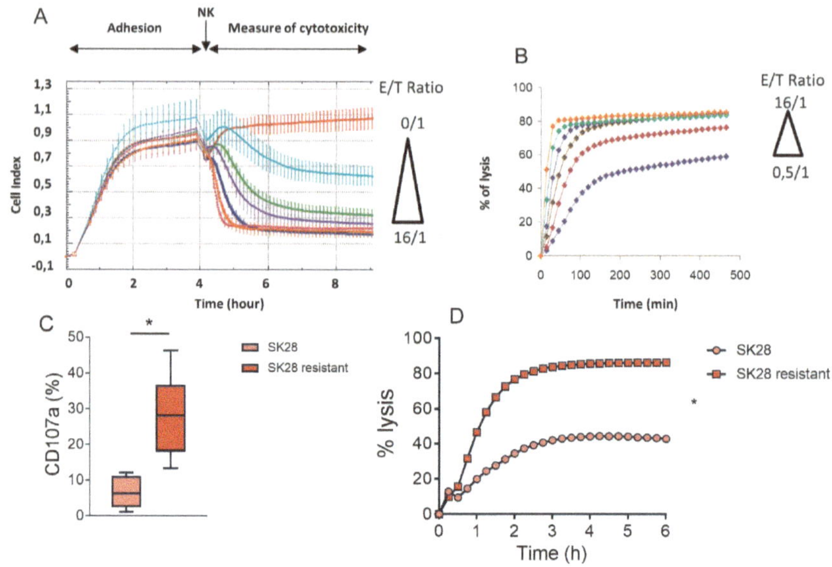

Fig. 2 Evaluation of the dynamic NK cell-mediated lysis of adherent tumor target cells. Tumor targets seeded in the E-Plate adhere, spread and grow for 5–6 h before the addition of NK cells. NK cells induce a rapid drop of CI values in an effector/target ratio dependent manner (A). CI data from the addition of NK cells allow the calculation of percentages of target lysis. Parallel NK cell degranulation (C) and curves of percentages of lysis from RTCA data (D) derived from the same experiment are shown.

the cell monolayer is detached with 0.52% Trypsin/EDTA, washed, diluted and cells are seeded 1 in 5. This procedure is used for maintenance of cell culture and provides cells for experiments. Before performing an assay, cell viability is assessed by Trypan blue exclusion labeling to ensure a viability >90%.

2.2.2 NK cell selection and culture

Blood samples from healthy donors are obtained from Etablissement Français du Sang (EFS) or from patients. Peripheral blood mononuclear cells (PBMC) are isolated by Ficoll-Paque PLUS (GE Healthcare) density gradient centrifugation. NK cells are purified by negative immunoselection using the NK Cell Isolation Kit and MS Columns (Miltenyi Biotec), leading to >95% of $CD3^-CD56^+$ cells.

Alternatively NK cells are selected from lymph node cell suspensions using the same procedure. When isolated from tumors, NK cells are immunoselected from $EpCAM^-$ or $CD45^+$ subset of tumor cell suspensions.

Purified NK cells (1×10^6/mL) are cultured in RPMI medium (Life Technologies) supplemented with 10% human serum AB (Biowest) and IL-2 (10 ng/mL, Miltenyi Biotec) or IL-15 (10 ng/mL, Miltenyi Biotec) for 4–6 days. Half of the medium is changed every 2 days with cytokine renewal.

2.2.3 Monitoring target cell proliferation

The process to monitor cell growth and proliferation is the following. First, for calibration of the E Plate "measure of the background noise," each wells of the plate are filled with 50 µL of medium and 150 µL of PBS are put between the wells. The plate is put in the system and background is measured. PBS between the wells of the E plate allows homogeneous humidity and avoids edge effects altering cell spreading and growth in these E-Plate wells.

Then, 1.3×10^4 (M14, A375) or 1×10^4 (SK28) cells/well are seeded in duplicates in a volume of 100 µL/well and their adhesion monitored for 5 h. Cell index values depend on the size, shape and spreading of target cells and must be adjusted for each cell line. Preliminary assay of proliferation of each target cell type is required to determine the optimal numbers of cells to be seeded. Indeed a cell index >2 is appropriate at 4–6 h after adhesion of the cells to monitor changes induced by drugs as minimum CI values above background signal is 0.2.

After spreading, targets can be treated with different concentrations of drugs or cytokines to evaluate the impact of treatments on cell growth dynamics. The method sensitivity allows following different phases of the response such as initial growth acceleration before decline of cell index. Results are expressed as Cell index (CI) curves (Fig. 3A).

2.2.4 NK cell-mediated lysis assay

For cell-mediated lysis assays, adherent targets (10,000–13,000) are seeded into 96 E-Plate wells in 100 µL of medium. Cell adhesion and spreading are monitored for 4–6 h, the optimal time for adhesion and spreading of most tumor cell lines.

After this phase, the E-Plate is taken out of the station and IL-2 or IL-15-activated NK cells are added at different Effector/Target (E/T) ratios (0.5/1, 1/1, 2/1, 5/1...), in a volume of 50 µL/well. The E-Plate is put back in the station and Cell index (CI) is measured every 15 min for up to 6 h. Addition of NK cells induces a rapid and dramatic decline of target cell index (Fig. 2A).

Lysis of various types of tumor cells (melanoma, breast cancer, colon cancer, renal cell carcinoma) by NK cells was assessed giving sensitive and reproducible results (Frazao et al., 2019; Fregni et al., 2011; Messaoudene et al., 2014; Perier et al., 2011; Rocca et al., 2016).

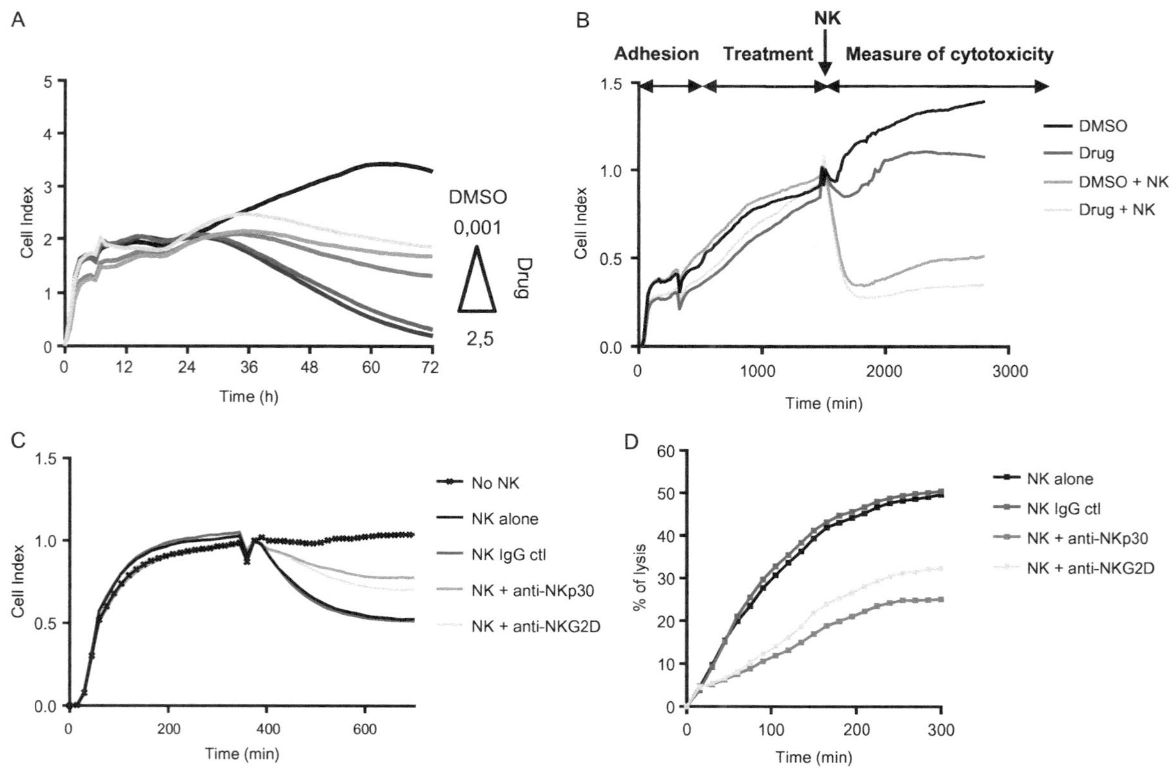

Fig. 3 Additional applications of the RTCA XCELLigence assay for evaluation of NK cell fitness. The treatment of targets before the addition of NK cells allows the evaluation of immunogenicity of this treatment (A and B): cell proliferation of treated tumor targets (A), CI data curves showing the different steps: adhesion, cell growth in response to treatment and NK cell-mediated lysis (B). The modulation of NK cell-mediated lysis by anti-NK receptors mAbs can be monitored: CI data curves (C), percentages of NK mediated lysis (D).

2.2.5 Calculation and statistical analysis

The immune cell-mediated cytolytic function is expressed as percentages of lysis during the first 6 h after the addition of immune effectors. CI Curves are normalized at the last point before addition of NK cells (Fig. 2A). Results are expressed as percentage of lysis determined from normalized CI curves (normalized CI, nCI) with the following equation:

$$\text{Percentage of lysis} = [\text{nCI}\ (\text{no effector}) - \text{nCI}\ (\text{effector})]/\text{nCI}\ (\text{no effector}) \times 100.$$

Percentages of lysis are calculated taking as reference CI curves from targets without NK cells. Lysis percentage curves depicted dynamic percentages of target cell lysis for the first 4–6 h of NK/Target cocultures (Fig. 2A and B). Lysis percentages curves allow comparing NK cell-mediated lysis of target cells displaying different CI curves before normalization (Fig. 2A).

Duplicates are used as dynamic CI data generate multiple time points allowing statistical comparison of CI curves.

Nonparametric paired Wilcoxon test is used to compare lysis percentage curves, and to compare CI data curves (treatment of targets with drugs).

2.2.6 Treatment of targets before addition of NK cells

In such assays, target cells are seeded, let to adhere and spread for 4–6 h, and treated with drugs (BRAF, MEK inhibitors, DITC…) or cytokines (IFNα, TGFβ..) or DMSO in control conditions, for various times (6–20 h) (in a volume of 25 μL/well) before the addition of cytotoxic immune effectors. The treatment duration with the drugs is chosen based on tumor cells impedance curves to avoid overgrowth of nontreated tumors cells in control wells. NK cells are added when control curves are still increasing (Fig. 3B). Normalization at the time of drug addition allows determining the effect of the drug on target cell survival/proliferation in comparison to wells plated with targets incubated with drug diluent (usually DMSO) corresponding to the highest drug concentration condition. Normalization at the time of NK cell addition is done to compare the effect of immune cells on targets pretreated with different concentrations of the drug in reference to targets incubated in presence of diluent.

Percentages of lysis by NK cells are calculated in reference to treated and nontreated targets. Variations in percentages of lysis toward treated cells indicate that the agent used modulates the recognition and lysis of targets by NK cells. Using this setting, we showed that BRAF or MEK inhibitors decreased the lysis of melanoma cells (Frazao et al., 2017).

2.2.7 Modulation of NK cell-mediated lysis by blocking mAbs

To determine the involvement of NK receptors in NK cell-mediated lysis, anti-NKR specific mAbs are used to block the interaction with the cognate ligand on the target. NK cells are incubated for 15–20 min with 40 μg/mL of the following mAbs: NKG2D, NKG2A, NKp30, NKp46, DNAM1 (R&D Systems) or control Igs before their addition to targets in the E-Plate (Fregni et al., 2013, 2011; Perier et al., 2011). The E-Plate is put back in the station and the Cell index (CI) is measured every 15 min for up to 6 h. CI curves are normalized at the last point before NK cell addition. Percentages of lysis are calculated as above and lysis percentage curves in presence of control Igs or anti-NKR mAbs are depicted (Fig. 3C). Blocking with mAbs can be expressed as percentages of inhibition or increase compared to the control condition (Igs, or no mAbs). Comparisons of mean values are performed using Wilcoxon matched paired test and $P < 0.05$ is regarded as significant.

2.2.8 Limits of the method

The dynamic measure of NK cell-mediated lysis is restricted to adherent tumor target cells.

CI values refer to cell detachment and/or cell death. Although CI curves can be monitored for hours, it is important to assess NK-mediated lysis during 2–6 h and to ensure that in control wells (without NK cells) cells are still growing. Immunoselection of effectors of interest (T versus NK cells) is required as there is no possibility to differentiate specific effectors present in heterogeneous cell suspensions.

3. Concluding remarks

First developed for accurate, sensitive and dynamic monitoring of adherent cell spreading and cell growth, the xCELLigence system has become a useful and reliable tool to measure the dynamic of adherent cell lysis induced by T and NK immune cells. xCELLigence system presents advantages compared to ^{51}Cr release assay (Erskine et al., 2012; Peper et al., 2014) and correlates with CD107a (Fregni et al., 2013, 2011), (Fasbender & Watzl, 2018), (Friedman et al., 2018).

Impedance based technology is efficient to quantify Antibody dependent cell cytotoxicity (ADCC) against adherent cells (Toth, Szollosi, & Vereb, 2017).

As immunotherapies with Chimeric Antigen receptor (CAR)-T cells (CAR-NKs) display growing interest for solid tumors, the method can be extended to monitor the antitumor activities of engineered T and NK cells.

4. Notes

1. Based on recommendations of the American Type Cell Culture Collection (ATCC, Manassas, VA). Optimal culture conditions for cell lines established in the lab were determined in the lab.
2. Cell lines are maintained in small or larger (25 or 75 cm^2 flasks). For melanoma cell lines, a 75 cm^2 flask at 75%–80% confluence contain $0.8 \times 10^6 - 2 \times 10^6$ cells.

   Gentle washing of cells with PBS to remove serum (known to inhibit trypsin) facilitates and speeds the detachment of the cells. The time required for Trypsin/EDTA to detach a monolayer varies with the cell type and degree of confluence from 1 to 4 min incubation at 37 °C. Detachment is confirmed by inspection on light microscopy. Clumps are dissolved by repeated pipetting assuring good counting of cell numbers.
3. Note that each cell line has its own pattern of CI curves depending of the cell size and morphology. Preliminary experiments are therefore required to optimize the kinetics of adhesion/spreading time to exponential growth phase.
4. The RTCA station is located in a regular incubator to ensure a controlled temperature, humidity and CO_2 atmosphere required for reproducible cell line growth the E-Plate.

Acknowledgments

This work was supported by Institut National du Cancer (INCa), by Cancérôpole Ile de France (PhD grant A.F.).

References

Andre, P., Denis, C., Soulas, C., Bourbon-Caillet, C., Lopez, J., Arnoux, T., et al. (2018). Anti-NKG2A mAb is a checkpoint inhibitor that promotes anti-tumor immunity by unleashing both T and NK cells. *Cell, 175*, 1731-1743.e1713.

Atienza, J. M., Yu, N., Kirstein, S. L., Xi, B., Wang, X., Xu, X., et al. (2006). Dynamic and label-free cell-based assays using the real-time cell electronic sensing system. *Assay and Drug Development Technologies, 4*, 597–607.

Atienza, J. M., Zhu, J., Wang, X., Xu, X., & Abassi, Y. (2005). Dynamic monitoring of cell adhesion and spreading on microelectronic sensor arrays. *Journal of Biomolecular Screening, 10*, 795–805.

Atienzar, F. A., Tilmant, K., Gerets, H. H., Toussaint, G., Speeckaert, S., Hanon, E., et al. (2011). The use of real-time cell analyzer technology in drug discovery: defining optimal cell culture conditions and assay reproducibility with different adherent cellular models. *Journal of Biomolecular Screening, 16*, 575–587.

Bollino, D., & Webb, T. J. (2017). Chimeric antigen receptor-engineered natural killer and natural killer T cells for cancer immunotherapy. *Translational Research: The Journal of Laboratory and Clinical Medicine, 187*, 32–43.

Brunner, K. T., Mauel, J., Rudolf, H., & Chapuis, B. (1970). Studies of allograft immunity in mice. I. Induction, development and in vitro assay of cellular immunity. *Immunology, 18*, 501–515.

Cerignoli, F., Abassi, Y. A., Lamarche, B. J., Guenther, G., Santa Ana, D., Guimet, D., et al. (2018). In vitro immunotherapy potency assays using real-time cell analysis. *PLoS One, 13*, e0193498.

Dieckmann, N. M., Frazer, G. L., Asano, Y., Stinchcombe, J. C., & Griffiths, G. M. (2016). The cytotoxic T lymphocyte immune synapse at a glance. *Journal of Cell Science, 129*, 2881–2886.

Erskine, C. L., Henle, A. M., & Knutson, K. L. (2012). Determining optimal cytotoxic activity of human Her2neu specific CD8 T cells by comparing the Cr51 release assay to the xCELLigence system. *Journal of Visualized Experiments, 66*, e3683.

Fasbender, F., & Watzl, C. (2018). Impedance-based analysis of Natural Killer cell stimulation. *Scientific Reports, 8*, 4938.

Frazao, A., Colombo, M., Fourmentraux-Neves, E., Messaoudene, M., Rusakiewicz, S., Zitvogel, L., et al. (2017). Shifting the balance of activating and inhibitory natural killer receptor ligands on BRAF(V600E) melanoma lines with vemurafenib. *Cancer Immunology Research, 5*, 582–593.

Frazao, A., Messaoudene, M., Nunez, N., Dulphy, N., Roussin, F., Sedlik, C., et al. (2019). CD16(+)NKG2A(high) Natural Killer cells infiltrate breast cancer-draining lymph nodes. *Cancer Immunology Research, 7*, 208–218.

Fregni, G., Messaoudene, M., Fourmentraux-Neves, E., Mazouz-Dorval, S., Chanal, J., Maubec, E., et al. (2013). Phenotypic and functional characteristics of blood natural killer cells from melanoma patients at different clinical stages. *PLoS One, 8*, e76928.

Fregni, G., Perier, A., Pittari, G., Jacobelli, S., Sastre, X., Gervois, N., et al. (2011). Unique functional status of natural killer cells in metastatic stage IV melanoma patients and its modulation by chemotherapy. *Clinical Cancer Research: An Official Journal of the American Association For Cancer Research, 17*, 2628–2637.

Friedman, J., Morisada, M., Sun, L., Moore, E. C., Padget, M., Hodge, J. W., et al. (2018). Inhibition of WEE1 kinase and cell cycle checkpoint activation sensitizes head and neck cancers to natural killer cell therapies. *Journal for Immunotherapy of Cancer, 6*, 59.

Gargett, T., Truong, N., Ebert, L. M., Yu, W., & Brown, M. P. (2019). Optimization of manufacturing conditions for chimeric antigen receptor T cells to favor cells with a central memory phenotype. *Cytotherapy, 6*, 593–602.

Glamann, J., & Hansen, A. J. (2006). Dynamic detection of natural killer cell-mediated cytotoxicity and cell adhesion by electrical impedance measurements. *Assay and Drug Development Technologies, 4*, 555–563.

Guan, N., Deng, J., Li, T., Xu, X., Irelan, J. T., & Wang, M. W. (2013). Label-free monitoring of T cell activation by the impedance-based xCELLigence system. *Molecular BioSystems, 9*, 1035–1043.

Karimi, M. A., Lee, E., Bachmann, M. H., Salicioni, A. M., Behrens, E. M., Kambayashi, T., et al. (2014). Measuring cytotoxicity by bioluminescence imaging outperforms the standard chromium-51 release assay. *PLoS One, 9*, e89357.

Ke, N., Wang, X., Xu, X., & Abassi, Y. A. (2011). The xCELLigence system for real-time and label-free monitoring of cell viability. *Methods in Molecular Biology, 740*, 33–43.

Kim, G. G., Donnenberg, V. S., Donnenberg, A. D., Gooding, W., & Whiteside, T. L. (2007). A novel multiparametric flow cytometry-based cytotoxicity assay simultaneously immunophenotypes effector cells: Comparisons to a 4 h 51Cr-release assay. *Journal of Immunological Methods, 325*, 51–66.

Mace, E. M., Dongre, P., Hsu, H. T., Sinha, P., James, A. M., Mann, S. S., et al. (2014). Cell biological steps and checkpoints in accessing NK cell cytotoxicity. *Immunology and Cell Biology*, *92*, 245–255.

Messaoudene, M., Fregni, G., Fourmentraux-Neves, E., Chanal, J., Maubec, E., Mazouz-Dorval, S., et al. (2014). Mature cytotoxic CD56(bright)/CD16(+) natural killer cells can infiltrate lymph nodes adjacent to metastatic melanoma. *Cancer Research*, *74*, 81–92.

Montoya, M. C., Sancho, D., Vicente-Manzanares, M., & Sanchez-Madrid, F. (2002). Cell adhesion and polarity during immune interactions. *Immunological Reviews*, *186*, 68–82.

Morisada, M., Moore, E. C., Hodge, R., Friedman, J., Cash, H. A., Hodge, J. W., et al. (2017). Dose-dependent enhancement of T-lymphocyte priming and CTL lysis following ionizing radiation in an engineered model of oral cancer. *Oral Oncology*, *71*, 87–94.

Peper, J. K., Schuster, H., Loffler, M. W., Schmid-Horch, B., Rammensee, H. G., & Stevanovic, S. (2014). An impedance-based cytotoxicity assay for real-time and label-free assessment of T-cell-mediated killing of adherent cells. *Journal of Immunological Methods*, *405*, 192–198.

Perier, A., Fregni, G., Wittnebel, S., Gad, S., Allard, M., Gervois, N., et al. (2011). Mutations of the von Hippel-Lindau gene confer increased susceptibility to natural killer cells of clear-cell renal cell carcinoma. *Oncogene*, *30*, 2622–2632.

Quereda, J. J., Martinez-Alarcon, L., Mendoca, L., Majado, M. J., Herrero-Medrano, J. M., Pallares, F. J., et al. (2010). Validation of xCELLigence real-time cell analyzer to assess compatibility in xenotransplantation with pig-to-baboon model. *Transplantation Proceedings*, *42*, 3239–3243.

Ribas, A., & Wolchok, J. D. (2018). Cancer immunotherapy using checkpoint blockade. *Science*, *359*, 1350–1355.

Rocca, Y. S., Roberti, M. P., Julia, E. P., Pampena, M. B., Bruno, L., Rivero, S., et al. (2016). Phenotypic and functional dysregulated blood NK Cells in colorectal cancer patients can be activated by Cetuximab Plus IL-2 or IL-15. *Frontiers in Immunology*, 7, 413.

Roden, M. M., Lee, K. H., Panelli, M. C., & Marincola, F. M. (1999). A novel cytolysis assay using fluorescent labeling and quantitative fluorescent scanning technology. *Journal of Immunological Methods*, *226*, 29–41.

Roshan Moniri, M., Young, A., Reinheimer, K., Rayat, J., Dai, L. J., & Warnock, G. L. (2015). Dynamic assessment of cell viability, proliferation and migration using real time cell analyzer system (RTCA). *Cytotechnology*, *67*, 379–386.

Sadelain, M., Brentjens, R., & Riviere, I. (2013). The basic principles of chimeric antigen receptor design. *Cancer Discovery*, *3*, 388–398.

Salter, A. I., Pont, M. J., & Riddell, S. R. (2018). Chimeric antigen receptor-modified T cells: CD19 and the road beyond. *Blood*, *131*, 2621–2629.

Schiller, C. B., Braciak, T. A., Fenn, N. C., Seidel, U. J., Roskopf, C. C., Wildenhain, S., et al. (2016). CD19-specific triplebody SPM-1 engages NK and gammadelta T cells for rapid and efficient lysis of malignant B-lymphoid cells. *Oncotarget*, 7, 83392–83408.

Seidel, U. J., Vogt, F., Grosse-Hovest, L., Jung, G., Handgretinger, R., & Lang, P. (2014). Gammadelta T cell-mediated antibody-dependent cellular cytotoxicity with CD19 antibodies assessed by an impedance-based label-free real-time cytotoxicity assay. *Frontiers in Immunology*, *5*, 618.

Shafer-Weaver, K., Rosenberg, S., Strobl, S., Gregory Alvord, W., Baseler, M., & Malyguine, A. (2006). Application of the granzyme B ELISPOT assay for monitoring cancer vaccine trials. *Journal of Immunotherapy*, *29*, 328–335.

Shafer-Weaver, K., Sayers, T., Strobl, S., Derby, E., Ulderich, T., Baseler, M., et al. (2003). The Granzyme B ELISPOT assay: An alternative to the 51Cr-release assay for monitoring cell-mediated cytotoxicity. *Journal of Translational Medicine*, *1*, 14.

Solly, K., Wang, X., Xu, X., Strulovici, B., & Zheng, W. (2004). Application of real-time cell electronic sensing (RT-CES) technology to cell-based assays. *Assay and Drug Development Technologies*, *2*, 363–372.

Somanchi, S. S., McCulley, K. J., Somanchi, A., Chan, L. L., & Lee, D. A. (2015). A novel method for assessment of Natural Killer cell cytotoxicity using image cytometry. *PLoS One*, *10*, e0141074.

Toth, G., Szollosi, J., & Vereb, G. (2017). Quantitating ADCC against adherent cells: Impedance-based detection is superior to release, membrane permeability, or caspase activation assays in resolving antibody dose response. *Cytometry. Part A: The Journal of the International Society for Analytical Cytology*, *91*, 1021–1029.

Watanabe, K., Kuramitsu, S., Posey, A. D., Jr., & June, C. H. (2018). Expanding the therapeutic window for CAR T cell therapy in solid tumors: The knowns and unknowns of CAR T cell biology. *Frontiers in Immunology*, *9*, 2486.

Zaritskaya, L., Shurin, M. R., Sayers, T. J., & Malyguine, A. M. (2010). New flow cytometric assays for monitoring cell-mediated cytotoxicity. *Expert Review of Vaccines*, *9*, 601–616.

CHAPTER SIXTEEN

Rapid isolation of mouse ILCs from murine intestinal tissues

Kyle Burrows, Pailin Chiaranunt, Louis Ngai, Arthur Mortha*
Department of Immunology, University of Toronto, Toronto, ON, Canada
*Corresponding author: e-mail address: arthur.mortha@utoronto.ca

Contents

Abstract

Tissue-resident immune cells like innate lymphoid cells (ILCs) are regulators of local immunity and tissue homeostasis. Similar to Natural Killer (NK) cells, ILCs express germline-encoded natural cytotoxicity receptors (NCRs) that facilitate the rapid execution of effector functions. Recent advances using transgenic animal models have further uncovered the developmental, transcriptional, epigenetic, and functional differences between members of the ILC family. Isolation of ILCs, which are particularly enriched in non-lymphoid tissues, can often be challenging and time consuming. Here, we provide a simple and rapid protocol for the isolation of NK cells and ILCs from murine intestinal tissues. This protocol is suitable for Fluorescence Activated Cell Sorting (FACS) and intracellular analysis of cytokine and transcription factor expression using flow and mass cytometry.

Methods in Enzymology, Volume 631
ISSN 0076-6879
https://doi.org/10.1016/bs.mie.2019.10.001

1. Introduction

1.1 Development of and relationships between ILCs

Recombination activated gene 1 and 2 (*Rag1* and *Rag2*) are a prerequisite for antigen receptor rearrangements and the generation of MHC-restricted, antigen-specific adaptive immunity by T and B lymphocytes (Mombaerts et al., 1992; Shinkai et al., 1992). T and B lymphocytes, as implicated in their name, develop along the lymphoid branch of hematopoiesis and diverge from the common lymphoid precursor cell. Lineage determining transcriptional master regulators of T cell and B cell development were identified within the *E*-box protein and Paired box protein family (Dias, Xu, McGregor, & Kee, 2008). Interestingly, the identification of MHC-independent cellular immunity against tumor cells was the spark that ignited research into the concept of innate germline-encoded self/non-self-recognition by specialized *Rag1*- and *Rag2*-independent lymphocytes. These cells, later termed Natural Killer (NK) cells, are cytotoxic, secrete cytokines comparable to $CD8^+$ cytotoxic T cells, and express activating and inhibiting NCRs that receive signals from target cells independent of MHC (Karre, Ljunggren, Piontek, & Kiessling, 1986; Stern, Gidlund, Orn, & Wigzell, 1980). NK cells are thus considered to be the innate counterpart of cytotoxic CD8+ T cells due to their strong functional overlap. Findings demonstrating that T and B cell-deficient $Rag1/2^{-/-}$ mice are able to mount type 1, 2, and 3 immune responses raised questions as to whether other immune cells may be capable of executing effector functions comparable to helper CD4+ T cells. In fact, analysis of tissue-resident cells revealed that an uncharacterized population of lymphoid cells make up a large proportion of the lineage (CD3, B220) negative cell fraction (Fallon et al., 2006; Sciume et al., 2012; Takatori et al., 2009). These cells, which are particularly found at mucosal surfaces including the intestine, urogenital tract, lung, and skin produce cytokines typically expressed by $CD4^+$ T helper(h) cells in response to intracellular and extracellular pathogens (Cella et al., 2009; Neill et al., 2010; Saenz et al., 2010). Consequently, these innate responses were demonstrated to be required for optimal host defense and tissue homeostasis (Klose et al., 2013; Luci et al., 2009; Neill et al., 2010; Saenz et al., 2010; Sanos et al., 2009; Satoh-Takayama et al., 2008).

Lineage commitment and effector functions of $CD8^+$ T cells and $CD4^+$ Th cells are coupled to tight transcriptional regulation (Rothenberg, 2014). Type 1 immune responses by Th1 cells require the actions of the transcription factor Tbx21 (T-bet), while cytotoxic CD8 T cell responses are impaired in the absence of Eomesodermin (Eomes) (Pearce et al., 2003; Szabo et al., 2000). Similarly, transcriptional regulation of Th2 and Th17 responses are directed by GATA3 and RORγt, respectively (Ivanov et al., 2006; Zheng & Flavell, 1997). ILCs are identified as Lineage^{-} CD127^{+} cells, share the production of T cell-related cytokines, and depend on the expression of Eomes, T-bet, GATA3, and RORγt for lineage commitment and development (Eberl et al., 2004; Hoyler et al., 2012; Klose et al., 2013, 2014). Eomes^{+} cells, co-expressing T-bet while lacking the expression of the lineage marker CD3 and B220, were identified as classical NK cells with potent cytotoxic capability and high levels of interferon(IFN)-gamma(γ) secretion, comparable to CD8^{+} T cells. Lineage^{-} cells expressing only T-bet without additional T cell-associated transcription factors were found to produce IFNγ while lacking cytotoxic activity and were consequently termed ILC1s due to their similarity to Th1 cells (Klose et al., 2014). Lineage^{-} GATA3 and RORa-expressing ILC2s, the innate counterparts of Th2 cells, release predominantly interleukin(IL)-5 and IL-13, as well as IL-4 and IL-9 (Halim et al., 2012; Takatori et al., 2009). Meanwhile, RORγt^{+} Lineage^{-} ILC3s release IL-22, IL-17, and Granulocyte Macrophage-Colony Stimulating Factor (GM-CSF), rendering them the innate counterpart of Th17 cells (Cella et al., 2009; Neill et al., 2010; Saenz et al., 2010). Collectively, these data demonstrate an underlying network of innate lymphoid cells mirroring the network of T lymphocytes on both functional and transcriptional levels (Fang & Zhu, 2017).

T cells develop from bone marrow derived Early Thymic Precursors (ETPs), which arise from the pool of lymphoid-primed multipotent progenitors (LMPP) through the Common Lymphoid Precursor stage (De Obaldia & Bhandoola, 2015). The commitment toward the T cell lineage requires the action of *E*-box proteins like E2A or E2–2 (Bain et al., 1997). The Id protein family are negative regulators of E protein activity, inhibit T cell lineage commitment and were first identified during development of classical NK cells (Yokota et al., 1999). Within NK cells, Id protein activity determines the final steps of maturation, but otherwise appears dispensable for full NK cell development (Boos, Yokota, Eberl, & Kee, 2007). Interestingly, while sparing classical NK cell development, deletion of Id2 results in an almost entire loss of cytokine secreting Th-like ILCs, suggesting divergence of cytotoxic NK cells from cytokine-secreting ILCs at the stage of a lymphoid precursor expressing high levels of Id2 (Xu et al., 2019). In line with these observations, fate labeling studies of cells expressing the transcription factor Zbtb16 identified a lymphoid precursor cell capable of generating all ILCs including NK cells, while the use of Id2 reporter mice allowed the identification of a precursor common to all helper-like ILCs (Constantinides, McDonald, Verhoef, & Bendelac, 2014; Klose et al., 2014). In accordance with these results, a hierarchical commitment towards NK cells and ILCs was described along a Zbtb16^{high} Id2^{low} Common Innate Lymphocyte Precursor (CILP) via the Zbtb16^{-} Id2^{high} Common Helper Innate Lymphocyte Precursor (CHILP) (Constantinides et al., 2014; Klose et al., 2014). Further refinement of the lymphoid precursor pool identified ILC potential within the LMPP population identifying those precursors as Early Innate Lymphocyte Precursor (EILP) (Yang et al., 2015). EILP consequently developed into CILP and CHILP driven by the orchestrated actions of the transcription factors TOX, Nfil3, Id2 and TCF-1 (Harly, Cam, Kaye, & Bhandoola, 2018).

Further specification of fully committed ILC precursors (ILCp) demonstrated the presence of at least four distinct developmental lineages generating NK cells, ILC1s, ILC2s and ILC3s. NK cell lineage commitment requires the transcriptional activity of Tox, Nfil3, T-bet, Eomes and Id2, while ILC1 commitment requires the additional activity of GATA3 in the absence of Eomes. Interestingly, GATA3 levels play a critical role in the discrimination between ILC2 and ILC3. ILC2 commitment was shown to require high levels of GATA3, Id2, RORa, Gfi-1 and Bcl-11b, while ILC3 lineage commitment was driven by intermediate levels of GATA3 and the transcription factors Id2, Ahr, Runx3 and RORγt (Vivier et al., 2018). Notably, the ILC3 lineage contains additional subsets that can be discriminated by the presence of NCRs and the activity of Tbet. NCR-expressing ILC3s are absent in Tbet-deficient mice and require continuous transcriptional activity to retain their identity and function (Klose et al., 2013). Interestingly, NCR^{+} ILC3s were further shown to downregulate the expression of RORγt under circumstances of inflammation or upon changes in the microenvironment, allowing them to adopt an ILC1-like/Ex-ILC3 phenotype (Vonarbourg et al., 2010). Comparable to the plasticity observed in NCR^{+} ILC3s, ILC2s show a down-regulation of their core ILC2 signature profile under strong type 1 and 3 inflammatory conditions, suggesting adaptations of ILC2 to environmental signal through silencing their effector functions (Cai et al., 2019; Duerr et al., 2016; Moro et al., 2016; Ohne et al., 2016). Cytokines released by local tissue and myeloid cells can trigger the activity of fully committed ILCs and have been discussed in detail elsewhere (Mortha & Burrows, 2018). In brief, NK cell, ILC1, and Ex-ILC3 IFNγ-driven activity requires the presence of IL-12 and IL-18. ILC2 IL-5- and IL-13-driven effector functions are regulated by the cytokines IL-25 and IL-33. ILC3 secretion of IL-17, IL-22, and

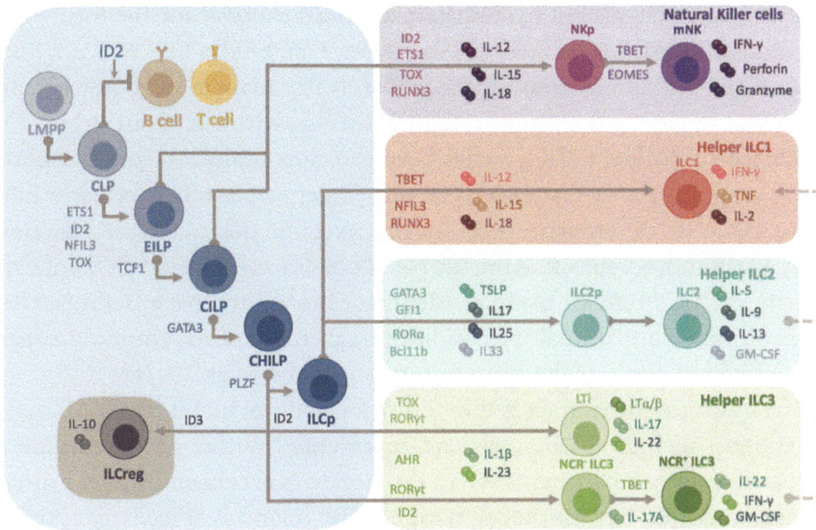

Fig. 1 ILC development. ILCs arise from early lymphoid-primed multipotent progenitors (LMPPs) through the common lymphoid precursor (CLP) in the bone marrow (blue space). The Inhibitor of DNA binding 2 (Id2) blocks *E*-protein activity at this stage and promotes differentiation into the early innate lymphoid precursor (EILP). Its progeny, the common innate lymphocyte precursor (CILP), retains the ability to generate all ILCs including cytotoxic Natural Killer cells (violet lineage). While NK cell specification requires ID2, ETS1, TOX, TBET, EOMES and RUNX3, differentiation and divergence into further precursors require the transcription factors ETS1, ID2, NFIL3, TOX, TFC1 and GATA3. The common helper innate lymphocyte precursor (CHILP) diverges from the CILP and loses the ability to generate NK cells but retains the ability to generate ILC1s, ILC2s, and ILC3s. Lineage-restricted ILC progenitors (ILCps) have been identified in the bone marrow and can generate distinct helper ILC lineages following the expression of ILC1 (red line), ILC2 (turquoise line), and ILC3 (green line) lineage specific transcription factors. The exact origin of presumably regulatory ILC (ILCregs) remains unknown but requires the activity of ID3 (gray line). Helper ILCs require instructive actions of lineage-specific transcription factors for their terminal differentiation. Plasticity of terminally differentiated ILCs has been reported (dashed light gray line).

GM-CSF depends on IL-1 and IL-23 and are controlled by the intestinal microbiota (Cortez, Robinette, & Colonna, 2015). In conclusion, ILCs are lymphoid cells with distinct environmental responsiveness and lineage defining transcriptional programs and effector functions (Fig. 1).

1.2 ILCs as a rheostat for intestinal immunity and tissue homeostasis

ILCs seed peripheral tissues during embryogenesis and early postnatal life (Chea et al., 2015; Mebius, Rennert, & Weissman, 1997; Possot et al., 2011; Schneider et al., 2019). Their tissue-resident nature and ability to proliferate and self-sustain makes them a crucial rheostat for the local cytokine milieu (Dudakov et al., 2012; Gasteiger, Fan, Dikiy, Lee, & Rudensky, 2015). ILC cytokine expression also controls the adaptation of newly infiltrating leukocytes to the immunological settings within a given tissue. The abundance of different ILCs varies between non-lymphoid and lymphoid tissues, suggesting that tissue-derived signals generate niches for early infiltrating ILCs to differentiate and locally execute their effector functions (Kim, Hashimoto-Hill, & Kim, 2016; Yudanin et al., 2019). While the majority of ILCs remain tissue resident, recent reports have demonstrated that ILCs under inflammatory conditions can migrate to lymphoid organs and traffic between non-lymphoid organs (Huang et al., 2018).

The gastrointestinal tract is the body's largest mucosal surface. All major ILC subsets, including NK cells, are represented within the multitude of microanatomic compartments across this organ. NK cells and ILC1s primarily make up the intraepithelial lymphocyte fraction of cells supporting epithelial defense (Fuchs et al., 2013). The lamina propria, Peyer's patches, cryptopatches (CPs), and isolated lymphoid follicles (ILFs) are rich in ILC1s, ex-ILC3s, ILC2s, and ILC3s, with preferential accumulation of $CCR6^{+}$ ILC3s in CPs/ILFs (Fuchs et al., 2013; Hoyler et al., 2012; Klose et al., 2014; Luci et al., 2009; Sanos et al., 2009; Vonarbourg et al., 2010). The intestinal tract, as a passaging route for ingested antigens and possible pathogens, is highly vulnerable to infection and inflammation and thus challenged with sustaining a tightly regulated tissue homeostasis.

Intestinal pathologies ranging from infections to autoimmune diseases pose a threat to this organ such that the entire functional diversity of all ILC subsets is necessary to facilitate optimal host defense and immune homeostasis (Eberl, 2010). The intestines, as home to all ILC subsets, is thus a suitable organ for the identification of ILC-specific functions during homeostasis and inflammation. In particular, NK cells and ILC1s exhibit type 1 immune responses against intracellular pathogens (e.g., *Toxoplasma* spp. and *Salmonella spp.*) and during autoimmune pathologies, including celiac disease, Inflammatory Bowel Disease (IBD), and cancer (Bernink et al., 2015; Dadi et al., 2016; Fuchs et al., 2013; Kupz et al., 2013; Schulthess et al., 2012). Type 2 immune responses against parasitic worm or enteric protozoan infections require ILC2 activation for appropriate defense and tissue repair (Price et al., 2010). Intriguingly, exacerbated type 2 immune response in the intestinal tract is further associated with the development of allergies (Hams et al., 2014; Lee et al., 2016; Noval Rivas, Burton, Oettgen, & Chatila, 2016). Lastly, the execution of type 3 immune responses against extracellular pathogens (i.e., bacteria and fungi) is a key function of ILC3s. In particular, the defense against enteric invasive *Enterobacteriaceae* (*Citrobacter rodentium*), blooming *Clostridia* species (*Clostridium difficile*) post-antibiotic treatment, and oropharyngeal *Candida* infections require the ILC3 -specific cytokine profile (Abt et al., 2015; Kinnebrew et al., 2010). A specialized subset of ILC3s, termed lymphoid tissue inducer (LTi) cells, contributes to the organogenesis of secondary and tertiary lymphoid structures along the intestinal tract and thus significantly impacts adaptive immunity (Kruglov et al., 2013; Tsuji et al., 2008). More strikingly, ILC3s were reported to control Th cell responses either directly by antigen presentation or indirectly through cytokine secretion (Hepworth et al., 2015; Melo-Gonzalez et al., 2019; Mortha et al., 2014). Collectively, these reports assign ILCs with a variety of functions to sustain intestinal immune and organ homeostasis. Strikingly, in instances of imbalanced immune homeostasis, such as that during IBD, ILC3s were shown to contribute to tissue inflammation (Zeng et al., 2019). However, additional reports using ILC3-deficient mice have demonstrated increased susceptibility towards the development of IBD, highlighting that the finetuning of ILC3 functions is key for proper mucosal immune balance (Mortha & Burrows, 2018). Finally, ILC1-, ILC2-, and ILC3-derived cytokines were reported to have both pro- and anti-tumorigenic effects within the intestinal tract (Hernandez, Gronke, & Diefenbach, 2018; Leon-Cabrera et al., 2017; Wang, Wang, Song, Chu, & Qu, 2015). In line with observations made in tumors outside of the intestinal tract, type 1 immune responses and the associated cytokines promote cytotoxicity and anti-tumor immunity.

In essence, ILC biology has progressed tremendously over the last 10 years. However, more investigation is needed to fully understand the role and scope of each individual ILC subset within intestinal biology and beyond. Their diversity, particularly within the intestinal tract, makes this organ a very attractive tissue to conduct studies on ILCs. Here, we provide a fast and cost-efficient protocol for the isolation of all ILC subsets with minimal use of transgenic ILC reporter mice. This protocol will hopefully be useful to investigators interested in purifying and culturing intestinal ILCs. It will hopefully facilitate transcriptional analysis and surface marker phenotyping. We have used this protocol to successfully isolate and sort ILC subsets for adoptive transfer into lymphopenic or irradiated hosts and conduct ex vivo functional studies, including culturing and gene expression analysis. We hope that the outlined procedure of isolating and identifying ILCs within the intestinal tract will motivate researchers to tackle questions regarding the biology of these exciting cells and help move our understanding of ILCs to the next level.

2. Identification of gut innate lymphoid cell populations by flow cytometry

2.1 Equipment

Flow cytometer (BD LSR Fortessa X20), five lasers (355/405/488/561/640).
5 mL Polystyrene Round-Bottom Tubes (Falcon, cat. no. 352052)
Pipettes, 0.5–10 μL(10 μL), 2–20 μL(20 μL), 20–200(200 μL), and 100–1000(1000 μL)
Serological pipettes, 5, 10, 25 and 50 mL
Pipette Controller
Conical tubes, 15 and 50 mL (Falcon)
Microcentrifuge tubes, 1.5 mL (Eppendorf)
Centrifuge (Eppendorf 5810 R)
Inverted Phase Microscope (Fisher Scientific)
Cell strainers, 70 μm (Corning, cat. no. CLS431751)
Fine curved scissors (Almedic)
Fine forceps (Almedic)
Petri Dish 100 × 15 mm (Fisherbrand, cat. no. FB0875712)
6 well plate (Corning, cat. No. 3506)
Incubator Shaker (New Brunswick Scientific, cat. no. I24/I24R)

2.2 Mice

C57BL/6 mice, male or female, 8–10 weeks old (The Jackson Laboratory, cat. no. 000644)
RORγt $^{gfp/+}$ mice (C57BL/6 *Rorc*^{tm2Litt}; The Jackson Laboratory, cat. no. 007572)

2.3 Chemicals/reagents

Deoxyribonuclease I (DNase I) from bovine pancreas (SIGMA Life Science, cat. no. DN25).
Collagenase from Clostridium histolyticum (SIGMA Life Science, cat. no. 5138).
Hank's balanced salt solution without calcium/magnesium (HBSS −/−) (Gibco, cat. no. 14175).
HBSS with calcium and magnesium (HBSS +/+) (Gibco, cat. no. 14025).

Percoll (GE Healthcare, cat. no. 17-0891-01) (see Note).
HEPES (1 M) (Gibco, cat. no. 15630).
Ethylenediaminetetraacetic acid (EDTA) (Bioshop, cat. no. EDT111).
Fetal bovine serum (Gibco, cat. no. 12483), heat inactivated at 56 °C for at least 30 min, aliquoted and stored at −20 °C.
Dulbecco's Phosphate Buffered Saline (Modified, without calcium chloride and magnesium chloride, liquid, sterile-filtered, suitable for cell culture.) (SIGMA Life Science, cat. no. D8537).
Dulbecco's Modified Eagle's Medium (DMEM)—high glucose, with 4500 mg/L glucose, L-glutamine, and sodium bicarbonate, without sodium pyruvate, liquid, sterile-filtered, suitable for cell culture (SIGMA Life Science, cat. no. D5976).
Penicillin-Streptomycin (100 ×) (SIGMA Life Science, cat. no. P4333).
10 × Dulbecco's phosphate buffered saline (DPBS; Invitrogen, cat. no. D1408).
Dissociation buffer: HBSS −/− supplemented with 5 mM EDTA and 10 mM Hepes (see Note).
Digestion buffer: HBSS +/+ containing 4% FBS, 1 mM HEPES, 0.5 mg/mL collagenase, and 0.5 mg/mL DNase I.
D-10 cell culture medium: High-glucose DMEM, supplemented with 10% FBS, 200 mg/L glutamine, 50,000 U/L penicillin. Stored at 4 °C.
FACS buffer: 1 × DPBS supplemented with 2% FBS and 5 mM EDTA. Stored at 4 °C.
eBioscience™ FOXP3/Transcription Factor Staining Buffer Set (ThermoFisher, cat. no. 00-5523-00).
Alexa Fluor 488 Antibody Labeling Kit (ThermoFisher, cat. no. A20181).
LIVE/DEAD™ Fixable Aqua Dead Cell Stain Kit (ThermoFisher, cat. no. L34966).

2.4 Antibodies

CD45.2 (104) mAb APC-Cy7 (BioLegend, cat. no. 109824) (1:200 dilution)
CD3e (145-2C11) mAb PerCP-Cy5.5 (Biolegend, cat. no. 100328) (1:100 dilution)
B220 (RA3-6B2) mAb PerCP-Cy5.5 (Biolegend, cat. no. 103236) (1:200 dilution)

NKp46 (29A1.4) mAb APC (Biolegend, cat. no. 137608) (1:100 dilution)
NK1.1 (PK136) mAb PE (Biolegend, cat. no. 108708) (1:200 dilution)
MHCII (I-A/I-E) (M5/114.15.2) mAb Pacific Blue (BioLegend, cat. no. 107620) (1:1000 dilution)
CD4 (GK1.5) mAb FITC (BioLegend, cat. no. 100406) (1:200 dilution)
RORγt (B2D) mAb Alexa 488* (eBioscience, cat. no. 14-6981-82) (1:100 dilution)
TBET (eBio4B10) mAb PE-Cy7 (eBioscience, cat. no. 25-5825-82) (1:100 dilution)
EOMES (Dan11mag) mAb eFluor450 (eBioscience, cat. no. 50-4875-82) (1:100 dilution)
GATA3 (TWAJ) mAb PE (eBioscience, cat. no. 12-9966-42) (5 μL per test)
KLRG1 (2F1/KLRG1) mAb FITC (Biolegend, cat. no. 138409) (1:200 dilution)
KLRG1 (2F1/KLRG1) mAb PE (Biolegend, cat. no. 138407) (1:200 dilution)
CD25 (PC61.5) mAb eFluor 450 (eBioscience, cat. no. 48-0251-82) (1200 dilution)
CD90.2 (53-2.1) mAb Pacific Blue (Biolegend, cat. no. 140305) (1:200 dilution)
CD27 (LG.7F9) mAb PE-Cy7 (eBioscience, cat. no. 25-0271-82) (1:100 dilution)
CD127 (A7R34) mAB Pacific Blue (Biolenged, cat. no. 135024) (1:100 dilution)
IL17RB (9B10) mAb APC (Biolegend, cat. no. 146304) (1:100 dilution)

*Conjugated to Alexa Fluor 488 using the Alexa Fluor 488 Antibody Labeling Kit (Thermo, cat. no. A20181).

3. Protocol

3.1 Lamina propria lymphocyte isolation

Time is of the essence! Lamina propria cells die rapidly once removed from the mouse; small intestinal tissue is more sensitive and should be processed before the colon.

Prepare all buffers prior to starting the experiment. Solutions containing enzymes should be prepared fresh every time.

1. Euthanize mouse according to institutional guidelines. Open the peritoneal cavity, pull the small intestine below the stomach using forceps and cut at the duodenum/stomach junction. Carefully pull the intestine out to not damage the tissue, sliding it through the scissors to trim out the mesentery, and then cut at the cecum to fully remove the small intestine. Next, grasp the large intestine, cut out the cecum, then pull the intestine and cut as close to the rectum as possible to remove the colon.
2. Place small intestine and colon into separate petri dishes containing ice-cold D-10 media and keep moist at all times.
3. Using forceps and scissors, remove the mesentery, residual fat, and connective tissue. A well-cleaned tissue will lead to a cleaner lymphocyte preparation, which will give better staining.
4. Remove the Peyer's patches of the small intestine by cutting out small sections with curved scissors. Peyer's patches can be recognized by their opaque color and nodal structure; on average there are 7–10 Peyer's patches per small intestine. The Peyer's patches of the distal end are small and more difficult to visualize, but they become larger toward the proximal end of the small intestine. Peyer's patches are not found on the colon.
5. Cut open the intestine longitudinally using scissors while gently holding the tissue with forceps or a gloved hand. Once opened, gently shake the intestine in the petri dish containing ice-cold D-10 in order to remove the fecal material. Next, gently scrape the luminal side of the intestine with curved forceps to remove any remaining fecal matter and excess mucous. For the small intestine, cut the sample into four equal parts.
6. Place intestine into a 50 mL conical tube containing dissociation buffer (10 mL for colon, 15 mL for small intestine), which will remove the epithelial layer containing both epithelial cells and intraepithelial lymphocytes.
7. Incubate at 37 °C for 15 min with slow rotation at 100 rpm in a shaker.
8. Vortex vigorously for 30 s, and discard the supernatant containing epithelial cells and intraepithelial lymphocytes.
9. Collect the tissue, gently scrape both sides with the forceps to remove any remaining epithelium and mucous, and rinse in a petri dish containing HBSS +/+. Then place the tissue in a 1.5 mL microcentrifuge tube and mince into fine pieces using scissors. The minced tissue should be able to pass through a 1000 μL pipette tip with ease.

10. Transfer tissue into a 50 mL conical tube with digestion buffer (10 mL for colon, 15 mL for small intestine) and incubate at 37 °C for 20 min with slow rotation at 100 rpm in a shaker.
11. Vortex well for 30 s, then collect the supernatant by filtering through a 70 μm cell strainer into a 50 mL conical tube.
12. Spin down the cells at 500 × *g* for 10 min.
13. Prepare Percoll solutions. (90% Percoll is prepared by adding 9 vol. of Percoll with 1 vol. of 10 × PBS. 40% and 80% Percoll are prepared by diluting the 90% Percoll with D-10.)
14. Resuspend the cells from step 12 in 5 mL of 40% Percoll.
15. Fill a fresh 15 mL centrifuge tube with 5 mL 80% Percoll and overlay the 40% Percoll containing the cells on top of the 80%. A slow and constant pace while pipetting is key to prevent mixing while overlaying the Percoll.
16. Centrifuge the tubes at 1300 × *g* for 20 min at 20 °C with brakes off.
17. Remove and aspirate out the top layer of epithelial cells.
18. Leukocytes will form an opaque layer at the interface between the 40% and 80% Percoll. Collect these cells using a fresh serum pipette.
19. Transfer the leukocytes to a fresh 15 mL conical tube and fill to 10 mL with FACS buffer. Spin down the cells at 500 × *g* for 7 min, then resuspend in FACS buffer and keep cells on ice. Save a small aliquot for cell counting. Cells are now ready for staining.

Notes

It is very important to use Ca/Mg-free HBSS as Ca/Mg-containing HBSS will not allow for separation of epithelial cells and intraepithelial lymphocytes. Excessive epithelial cell contaminations during digestion results in decreased viability of lamina propria lymphocytes. Percoll is not an isotonic solution. Thus, 1 vol. of 10 × DPBS is added to 9 vol. of Percoll in order to obtain an isotonic 90% Percoll solution. Dilute 90% Percoll with the appropriate amount of cell culture medium in order to obtain 40% and 80% Percoll solutions used for the separation gradient. Typical cell yields of ILC subsets for colons and small intestines isolated from wild-type C57BL/6 male or female, 8–10-week-old mice housed in our facility are listed in Fig. 2 We have previously performed this protocol in animal facilities across three different countries on animals housed under different grades of pathogen exclusion. The cell yields reported here may vary with those observed at different institutions

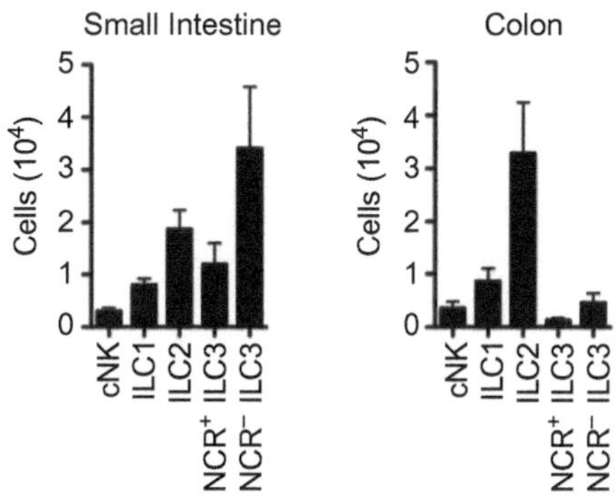

Fig. 2 Quantification of ILC subsets in small and large intestine. Bar graphs show absolute cell numbers of ILC subsets isolated from small intestine and colon of 8 week old C57Bl/6 mice. Data are representative of three independent experiments (with an $n = 3$ mice/experiment).

and may even reach higher numbers when performed on mice housed under reduced pathogen exclusion criteria. Specifically, exclusion of segmented filamentous bacteria (SFB), mouse norovirus (MNV), mouse hepatitis virus (MHV), helicobacter spp. and intestinal protozoa, as is the case in our facility, are important factors affecting cell yields. These pathogens are not commonly excluded across animal facilities and will result in higher yields of gut CD45-expressing cells.

3.2 Staining procedure for innate lymphoid cells

3.2.1 Surface marker staining

1. Wash cells with PBS. Incubate in 100 μL of Live/Dead Aqua (1:2000 of stock diluted in PBS) for 20 min on ice.
2. Wash cells twice with FACS buffer.
3. Incubate cells with Fc block (purified CD16/CD32 (2.4G.2) MAb) for 20 min on ice.
4. Spin down and discard the supernatant.
5. Incubate cells with a 100 μL cocktail of surface staining, fluorescently labeled antibodies for 20 min on ice. For ILCs we have found that by minimally co-staining with PerCP-Cy5.5 labeled anti-CD3 and anti-B220 as our lineage cocktail and APC-Cy7 labeled anti-CD45 as our hematopoietic marker we can utilize combinations of other FITC-, PE-, PE-Cy7-, APC-, and eFluor450- labeled antibodies to assess further surface markers on lamina propria cells.

6. Antibody panels for ILC surface stains:

	ILC1 panels	ILC2 panels	ILC3 panels
GFP/FITC	RORγt(GFP)	RORγt(GFP)	RORγt(GFP)
PE	NK1.1	Klrg1	NK1.1
PerCP-Cy5.5	Lineage cocktail	Lineage cocktail	Lineage cocktail
PE-Cy7	CD27	ICOS	CD4
APC	NKp46	IL17RB	NKp46
APC-Cy7	CD45	CD45	CD45
eFluor450/Pacific Blue	CD127	CD25/CD90	MHCII/CD90
Aqua	Live/Dead	Live/Dead	Live/Dead

7. Wash stained cells twice with FACS buffer. Analyze on a flow cytometer (Figs. 3 and 4) or continue with transcription factor staining.

3.2.2 Transcription factor staining

1. After the last wash post-surface staining, fix and permeabilize cells using the fixation/permeabilization buffer from the eBioscience Foxp3 Staining Kit for 20 min at 4 °C in the dark.
2. Wash 2 × with the kit's wash/permeabilization buffer.
3. Intracellular staining is performed with a 100 μL cocktail of Alexa-488-conjugated anti-RORγt, PE-conjugated anti-GATA3, PE-Cy7-conjugated TBET and/or eFluor 450-conjugated anti-EOMES antibodies diluted in wash/permeabilization buffer solution for 1 h at 4 °C in the dark.
4. Antibody panels for ILC intracellular transcription factor staining:

	ILC1 panels	ILC2 panels	ILC3 panels
Alexa 488/FITC	RORγt(IC)[a]	Klrg1	RORγt(IC)
PE	NK1.1	Gata3(IC)	NK1.1
PerCP-Cy5.5	Lineage cocktail	Lineage cocktail	Lineage cocktail
PE-Cy7	Tbet(IC)	ICOS	CD4
APC	NKp46	IL17RB	NKp46
APC-Cy7	CD45	CD45	CD45
eFluor450/Pacific Blue	Eomes(IC)	CD25/CD90	MHCII/CD90
Aqua	Live/Dead	Live/Dead	Live/Dead

[a](IC) Intracellular transcription factor staining.

5. Wash 2 × with the kit's wash/permeabilization buffer.
6. Resuspend stained cells in FACS buffer. Analyze on a flow cytometer (Figs. 5 and 6).

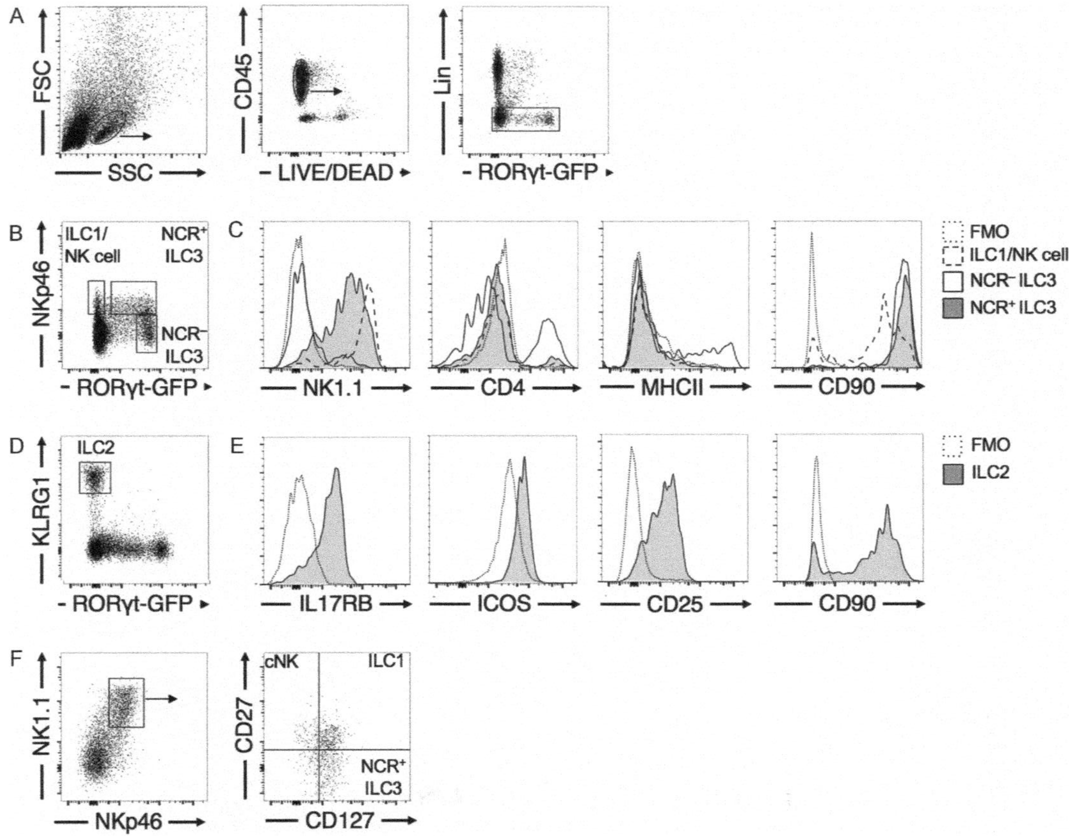

Fig. 3 Gating strategy for small intestinal ILCs via surface staining. (A) Small intestine lamina propria lymphocytes are first gated by forward scatter (FSC) and side scatter (SSC), then positive for hematopoietic cell marker CD45 and negative for Live/Dead stain, and finally negative for adaptive lymphocyte lineage markers (Lin^-: $CD3e^-$ $B220^-$). (B) ILC1/3 subsets can further be differentiated based on RORγt-GFP and NKp46 expression (NK/ILC1: $NKp46^+$ RORγt-GFP^-, NCR+ ILC3: $NKp46^+$ RORγt-GFP^+, NCR^- ILC3: $NKp46^-$ RORγt-GFP^+). (C) ILC1/3 subsets can be characterized with additional surface markers such as: NK1.1, CD4, MHCII and CD90. (D) ILC2s can be identified from the $CD45^+$ Lin^- population (final gate in panel A) as $KLRG1^+$. (E) ILC2s can also be identified with combinations of other surface markers including: IL17RB, ICOS, CD25, and CD90. (F) ILC1/cNK cells can be identified as NK1.1/NKp46 double positive, with cNK cells as $CD27^+CD127^-$ and ILC1s as $CD27^+CD127^+$.

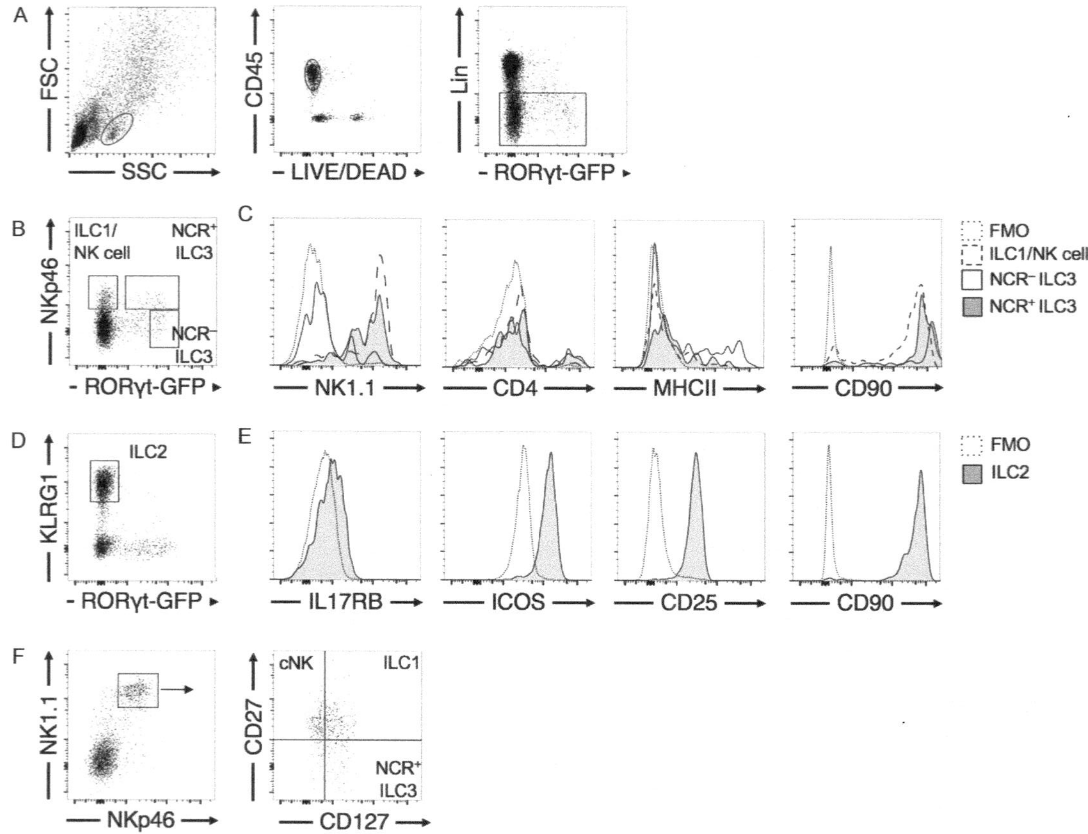

Fig. 4 Gating strategy for colonic ILCs via surface stain. (A) Colon lamina propria lymphocytes are first gated by FSC and SSC, then positive for hematopoietic cell marker CD45 and negative for Live/Dead stain, and finally negative for adaptive lymphocyte lineage markers (Lin^-: $CD3e^-$,$B220^-$). (B) ILC1/3 subsets can further be differentiated based on RORγt-GFP and NKp46 expression (NK/ILC1: $NKp46^+$ RORγt-GFP^-, NCR+ ILC3: $NKp46^+$ RORγt-GFP^+, NCR^- ILC3: $NKp46^-$ RORγt-GFP^+). (C) ILC1/3 subsets can then be characterized with additional surface markers such as: NK1.1, CD4, MHCII and CD90. (D) ILC2 can be identified from the $CD45^+$ Lin^- population as $KLRG1^+$. (E) ILC2 can also be identified with combination of other surface markers including IL17RB, ICOS, CD25, and CD90. (F) ILC1/cNK cells can be identified as NK1.1/NKp46 double positive with cNK cells as CD27+CD127- and ILC1 as CD27+CD127+.

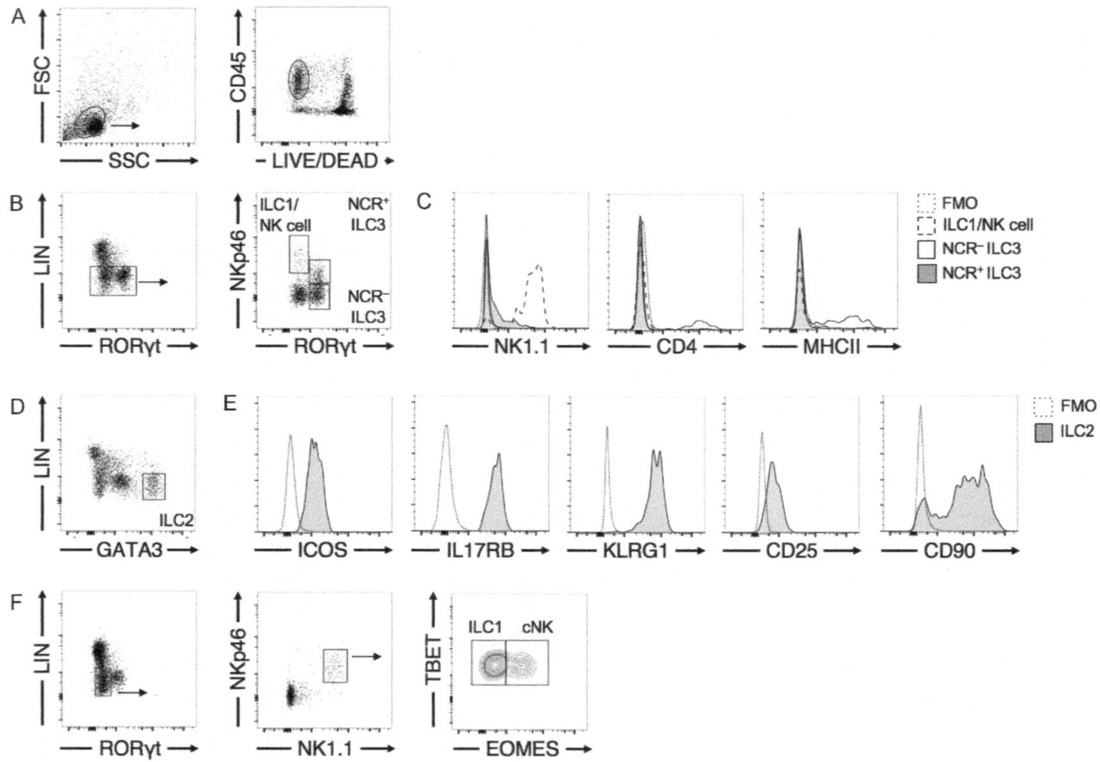

Fig. 5 Gating strategy for small intestinal ILCs via intracellular transcription factor stain. (A) Small intestine lamina propria lymphocytes are first gated by FSC and SSC, and then positive for hematopoietic cell marker CD45 and negative for Live/Dead stain. (B) ILC1/3 subsets are negative for adaptive lymphocyte lineage markers (Lin^{-}: $CD3e^{-}$, $B220^{-}$) and then differentiated based on RORγt and NKp46 expression (NK/ILC1: $NKp46^{+}$ $ROR\gamma t^{-}$, NCR+ ILC3: $NKp46^{+}$ $ROR\gamma t^{+}$, NCR^{-} ILC3: $NKp46^{-}$ $ROR\gamma t^{+}$). (C) ILC1/3 subsets can be characterized with additional surface markers, such as: NK1.1, CD4, MHCII and CD90. (D) ILC2s can be identified as Lin^{-} $GATA3^{High}$. (E) Other ILC2 surface markers include: IL17RB, ICOS, CD25, and CD90. (F) ILC1/cNK cells can be identified as Lin^{-} $ROR\gamma t^{-}$, then NK1.1/NKp46 double positive, with ILC1s as $TBET^{+}$ and cNK cells $TBET^{+}$ $EOMES^{+}$.

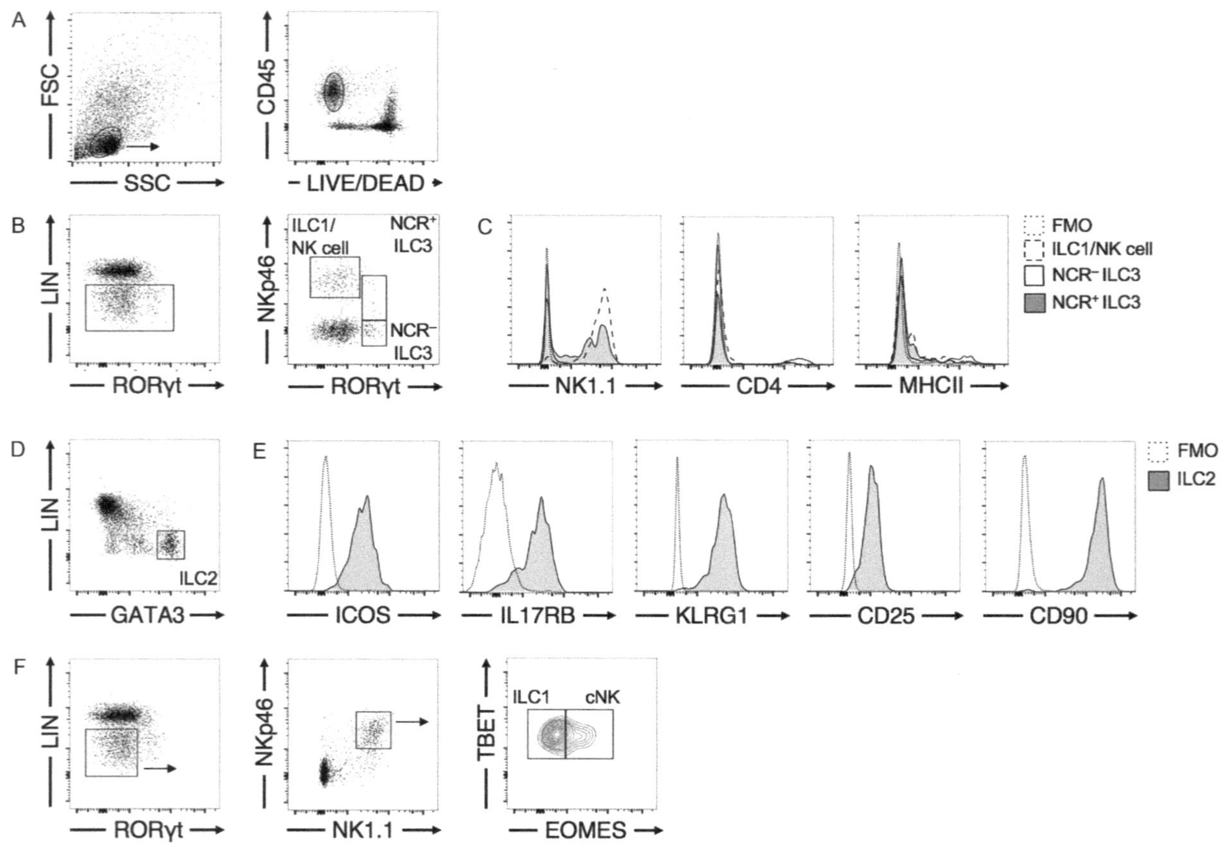

Fig. 6 Gating strategy for colon ILC via intracellular transcription factor stain. (A) Colon lamina propria lymphocytes are first gated by FSC and SSC, then positive for hematopoietic cell marker CD45 and negative for Live/Dead stain. (B) ILC1/3 subsets are negative for adaptive lymphocyte lineage markers (Lin^-: $CD3e^-$, $B220^-$) and then differentiated based on RORγt and NKp46 expression (NK/ILC1: $NKp46^+ ROR\gamma t^-$, NCR+ ILC3: $NKp46^+ ROR\gamma t^+$, NCR^- ILC3 $NKp46^- ROR\gamma t^+$). (C) ILC1/3 subsets can then be characterized with additional surface markers such as: NK1.1, CD4, MHCII and CD90. (D) ILC2 can be identified as Lin^- $GATA3^{High}$. (E) Additional surface markers such as: IL17RB, ICOS, CD25, and CD90 can confirm ILC2. (F) ILC1/cNK cells can be identified as $Lin^- ROR\gamma t^-$, then NK1.1/NKp46 double positive with ILC1 as $TBET^+$ and cNK cells $TBET^+ EOMES^+$.

Acknowledgments

A.M. is supported by CIHR-Project Grant (388337) and NSERC-Discovery Grant (RGPIN-2019-04521). A.M. is the Tier 2 Canadian Research Chair in Mucosal Immunology and supported by the Tier 2 CRC-CIHR program. Fig. 1 was "Created with BioRender."

References

Abt, M. C., Lewis, B. B., Caballero, S., Xiong, H., Carter, R. A., Susac, B., et al. (2015). Innate immune defenses mediated by two ILC subsets are critical for protection against acute Clostridium difficile infection. *Cell Host & Microbe, 18*(1), 27–37. https://doi.org/10.1016/j.chom.2015.06.011.

Bain, G., Engel, I., Robanus Maandag, E. C., te Riele, H. P., Voland, J. R., Sharp, L. L., et al. (1997). E2A deficiency leads to abnormalities in alphabeta T-cell development and to rapid development of T-cell lymphomas. *Molecular and Cellular Biology, 17*(8), 4782–4791. https://doi.org/10.1128/mcb.17.8.4782.

Bernink, J. H., Krabbendam, L., Germar, K., de Jong, E., Gronke, K., Kofoed-Nielsen, M., et al. (2015). Interleukin-12 and -23 control plasticity of CD127(+) group 1 and group 3 innate lymphoid cells in the intestinal Lamina propria. *Immunity, 43*(1), 146–160. https://doi.org/10.1016/j.immuni.2015.06.019.

Boos, M. D., Yokota, Y., Eberl, G., & Kee, B. L. (2007). Mature natural killer cell and lymphoid tissue-inducing cell development requires Id2-mediated suppression of E protein activity. *The Journal of Experimental Medicine, 204*(5), 1119–1130. https://doi.org/10.1084/jem.20061959.

Cai, T., Qiu, J., Ji, Y., Li, W., Ding, Z., Suo, C., et al. (2019). IL-17-producing ST2(+) group 2 innate lymphoid cells play a pathogenic role in lung inflammation. *The Journal of Allergy and Clinical Immunology, 143*(1), 229–244. e229. https://doi.org/10.1016/j.jaci.2018.03.007.

Cella, M., Fuchs, A., Vermi, W., Facchetti, F., Otero, K., Lennerz, J. K., et al. (2009). A human natural killer cell subset provides an innate source of IL-22 for mucosal immunity. *Nature, 457*(7230), 722–725. https://doi.org/10.1038/nature07537.

Chea, S., Possot, C., Perchet, T., Petit, M., Cumano, A., & Golub, R. (2015). CXCR6 expression is important for retention and circulation of ILC precursors. *Mediators of Inflammation, 2015*, 368427. https://doi.org/10.1155/2015/368427.

Constantinides, M. G., McDonald, B. D., Verhoef, P. A., & Bendelac, A. (2014). A committed precursor to innate lymphoid cells. *Nature, 508*(7496), 397–401. https://doi.org/10.1038/nature13047.

Cortez, V. S., Robinette, M. L., & Colonna, M. (2015). Innate lymphoid cells: New insights into function and development. *Current Opinion in Immunology, 32*, 71–77. https://doi.org/10.1016/j.coi.2015.01.004.

Dadi, S., Chhangawala, S., Whitlock, B. M., Franklin, R. A., Luo, C. T., Oh, S. A., et al. (2016). Cancer immunosurveillance by tissue-resident innate lymphoid cells and innate-like T cells. *Cell, 164*(3), 365–377. https://doi.org/10.1016/j.cell.2016.01.002.

De Obaldia, M. E., & Bhandoola, A. (2015). Transcriptional regulation of innate and adaptive lymphocyte lineages. *Annual Review of Immunology, 33*, 607–642. https://doi.org/10.1146/annurev-immunol-032414-112032.

Dias, S., Xu, W., McGregor, S., & Kee, B. (2008). Transcriptional regulation of lymphocyte development. *Current Opinion in Genetics & Development, 18*(5), 441–448. https://doi.org/10.1016/j.gde.2008.07.015.

Dudakov, J. A., Hanash, A. M., Jenq, R. R., Young, L. F., Ghosh, A., Singer, N. V., et al. (2012). Interleukin-22 drives endogenous thymic regeneration in mice. *Science*, *336*(6077), 91–95. https://doi.org/10.1126/science.1218004.

Duerr, C. U., McCarthy, C. D., Mindt, B. C., Rubio, M., Meli, A. P., Pothlichet, J., et al. (2016). Type I interferon restricts type 2 immunopathology through the regulation of group 2 innate lymphoid cells. *Nature Immunology*, *17*(1), 65–75. https://doi.org/10.1038/ni.3308.

Eberl, G. (2010). A new vision of immunity: Homeostasis of the superorganism. *Mucosal Immunology*, *3*(5), 450–460. https://doi.org/10.1038/mi.2010.20.

Eberl, G., Marmon, S., Sunshine, M. J., Rennert, P. D., Choi, Y., & Littman, D. R. (2004). An essential function for the nuclear receptor RORgamma(t) in the generation of fetal lymphoid tissue inducer cells. *Nature Immunology*, *5*(1), 64–73. https://doi.org/10.1038/ni1022.

Fallon, P. G., Ballantyne, S. J., Mangan, N. E., Barlow, J. L., Dasvarma, A., Hewett, D. R., et al. (2006). Identification of an interleukin (IL)-25-dependent cell population that provides IL-4, IL-5, and IL-13 at the onset of helminth expulsion. *The Journal of Experimental Medicine*, *203*(4), 1105–1116. https://doi.org/10.1084/jem.20051615.

Fang, D., & Zhu, J. (2017). Dynamic balance between master transcription factors determines the fates and functions of CD4 T cell and innate lymphoid cell subsets. *The Journal of Experimental Medicine*, *214*(7), 1861–1876. https://doi.org/10.1084/jem.20170494.

Fuchs, A., Vermi, W., Lee, J. S., Lonardi, S., Gilfillan, S., Newberry, R. D., et al. (2013). Intraepithelial type 1 innate lymphoid cells are a unique subset of IL-12- and IL-15-responsive IFN-gamma-producing cells. *Immunity*, *38*(4), 769–781. https://doi.org/10.1016/j.immuni.2013.02.010.

Gasteiger, G., Fan, X., Dikiy, S., Lee, S. Y., & Rudensky, A. Y. (2015). Tissue residency of innate lymphoid cells in lymphoid and nonlymphoid organs. *Science*, *350*(6263), 981–985. https://doi.org/10.1126/science.aac9593.

Halim, T. Y., MacLaren, A., Romanish, M. T., Gold, M. J., McNagny, K. M., & Takei, F. (2012). Retinoic-acid-receptor-related orphan nuclear receptor alpha is required for natural helper cell development and allergic inflammation. *Immunity*, *37*(3), 463–474. https://doi.org/10.1016/j.immuni.2012.06.012.

Hams, E., Armstrong, M. E., Barlow, J. L., Saunders, S. P., Schwartz, C., Cooke, G., et al. (2014). IL-25 and type 2 innate lymphoid cells induce pulmonary fibrosis. *Proceedings of the National Academy of Sciences of the United States of America*, *111*(1), 367–372. https://doi.org/10.1073/pnas.1315854111.

Harly, C., Cam, M., Kaye, J., & Bhandoola, A. (2018). Development and differentiation of early innate lymphoid progenitors. *The Journal of Experimental Medicine*, *215*(1), 249–262. https://doi.org/10.1084/jem.20170832.

Hepworth, M. R., Fung, T. C., Masur, S. H., Kelsen, J. R., McConnell, F. M., Dubrot, J., et al. (2015). Immune tolerance. Group 3 innate lymphoid cells mediate intestinal selection of commensal bacteria-specific CD4(+) T cells. *Science*, *348*(6238), 1031–1035. https://doi.org/10.1126/science.aaa4812.

Hernandez, P., Gronke, K., & Diefenbach, A. (2018). A catch-22: Interleukin-22 and cancer. *European Journal of Immunology*, *48*(1), 15–31. https://doi.org/10.1002/eji.201747183.

Hoyler, T., Klose, C. S., Souabni, A., Turqueti-Neves, A., Pfeifer, D., Rawlins, E. L., et al. (2012). The transcription factor GATA-3 controls cell fate and maintenance of type 2 innate lymphoid cells. *Immunity*, *37*(4), 634–648. https://doi.org/10.1016/j.immuni.2012.06.020.

Huang, Y., Mao, K., Chen, X., Sun, M. A., Kawabe, T., Li, W., et al. (2018). S1P-dependent interorgan trafficking of group 2 innate lymphoid cells supports host defense. *Science*, *359*(6371), 114–119. https://doi.org/10.1126/science.aam5809.

Ivanov, I. I., McKenzie, B. S., Zhou, L., Tadokoro, C. E., Lepelley, A., Lafaille, J. J., et al. (2006). The orphan nuclear receptor RORgammat directs the differentiation program of proinflammatory IL-17+ T helper cells. Cell, *126*(6), 1121–1133. https://doi.org/10.1016/j.cell.2006.07.035.

Karre, K., Ljunggren, H. G., Piontek, G., & Kiessling, R. (1986). Selective rejection of H-2-deficient lymphoma variants suggests alternative immune defence strategy. *Nature*, *319*(6055), 675–678. https://doi.org/10.1038/319675a0.

Kim, C. H., Hashimoto-Hill, S., & Kim, M. (2016). Migration and tissue tropism of innate lymphoid cells. *Trends in Immunology*, *37*(1), 68–79. https://doi.org/10.1016/j.it.2015.11.003.

Kinnebrew, M. A., Ubeda, C., Zenewicz, L. A., Smith, N., Flavell, R. A., & Pamer, E. G. (2010). Bacterial flagellin stimulates toll-like receptor 5-dependent defense against vancomycin-resistant Enterococcus infection. *The Journal of Infectious Diseases*, *201*(4), 534–543. https://doi.org/10.1086/650203.

Klose, C. S. N., Flach, M., Mohle, L., Rogell, L., Hoyler, T., Ebert, K., et al. (2014). Differentiation of type 1 ILCs from a common progenitor to all helper-like innate lymphoid cell lineages. *Cell*, *157*(2), 340–356. https://doi.org/10.1016/j.cell.2014.03.030.

Klose, C. S., Kiss, E. A., Schwierzeck, V., Ebert, K., Hoyler, T., d'Hargues, Y., et al. (2013). A T-bet gradient controls the fate and function of CCR6-RORgammat+ innate lymphoid cells. *Nature*, *494*(7436), 261–265. https://doi.org/10.1038/nature11813.

Kruglov, A. A., Grivennikov, S. I., Kuprash, D. V., Winsauer, C., Prepens, S., Seleznik, G. M., et al. (2013). Nonredundant function of soluble LTalpha3 produced by innate lymphoid cells in intestinal homeostasis. *Science*, *342*(6163), 1243–1246. https://doi.org/10.1126/science.1243364.

Kupz, A., Scott, T. A., Belz, G. T., Andrews, D. M., Greyer, M., Lew, A. M., et al. (2013). Contribution of Thy1+ NK cells to protective IFN-gamma production during Salmonella typhimurium infections. *Proceedings of the National Academy of Sciences of the United States of America*, *110*(6), 2252–2257. https://doi.org/10.1073/pnas.1222047110.

Lee, J. B., Chen, C. Y., Liu, B., Mugge, L., Angkasekwinai, P., Facchinetti, V., et al. (2016). IL-25 and CD4(+) TH2 cells enhance type 2 innate lymphoid cell-derived IL-13 production, which promotes IgE-mediated experimental food allergy. *The Journal of Allergy and Clinical Immunology*, *137*(4), 1216–1225. e1215. https://doi.org/10.1016/j.jaci.2015.09.019.

Leon-Cabrera, S. A., Molina-Guzman, E., Delgado-Ramirez, Y. G., Vazquez-Sandoval, A., Ledesma-Soto, Y., Perez-Plasencia, C. G., et al. (2017). Lack of STAT6 attenuates inflammation and drives protection against early steps of colitis-associated colon cancer. *Cancer Immunology Research*, *5*(5), 385–396. https://doi.org/10.1158/2326-6066.CIR-16-0168.

Luci, C., Reynders, A., Ivanov, I. I., Cognet, C., Chiche, L., Chasson, L., et al. (2009). Influence of the transcription factor RORgammat on the development of NKp46+ cell populations in gut and skin. *Nature Immunology*, *10*(1), 75–82. https://doi.org/10.1038/ni.1681.

Mebius, R. E., Rennert, P., & Weissman, I. L. (1997). Developing lymph nodes collect CD4+CD3- LTbeta+ cells that can differentiate to APC, NK cells, and follicular cells but not T or B cells. *Immunity*, 7(4), 493–504.

Melo-Gonzalez, F., Kammoun, H., Evren, E., Dutton, E. E., Papadopoulou, M., Bradford, B. M., et al. (2019). Antigen-presenting ILC3 regulate T cell-dependent IgA responses to colonic mucosal bacteria. *The Journal of Experimental Medicine*, *216*(4), 728–742. https://doi.org/10.1084/jem.20180871.

Mombaerts, P., Iacomini, J., Johnson, R. S., Herrup, K., Tonegawa, S., & Papaioannou, V. E. (1992). RAG-1-deficient mice have no mature B and T lymphocytes. *Cell*, *68*(5), 869–877. https://doi.org/10.1016/0092-8674(92)90030-g.

Moro, K., Kabata, H., Tanabe, M., Koga, S., Takeno, N., Mochizuki, M., et al. (2016). Interferon and IL-27 antagonize the function of group 2 innate lymphoid cells and type 2 innate immune responses. *Nature Immunology*, *17*(1), 76–86. https://doi.org/10.1038/ni.3309.

Mortha, A., & Burrows, K. (2018). Cytokine networks between innate lymphoid cells and myeloid cells. *Frontiers in Immunology*, *9*, 191. https://doi.org/10.3389/fimmu.2018.00191.

Mortha, A., Chudnovskiy, A., Hashimoto, D., Bogunovic, M., Spencer, S. P., Belkaid, Y., et al. (2014). Microbiota-dependent crosstalk between macrophages and ILC3 promotes intestinal homeostasis. *Science*, *343*(6178), 1249288. https://doi.org/10.1126/science.1249288.

Neill, D. R., Wong, S. H., Bellosi, A., Flynn, R. J., Daly, M., Langford, T. K., et al. (2010). Nuocytes represent a new innate effector leukocyte that mediates type-2 immunity. *Nature*, *464*(7293), 1367–1370. https://doi.org/10.1038/nature08900.

Noval Rivas, M., Burton, O. T., Oettgen, H. C., & Chatila, T. (2016). IL-4 production by group 2 innate lymphoid cells promotes food allergy by blocking regulatory T-cell function. *The Journal of Allergy and Clinical Immunology*, *138*(3), 801–811. e809. https://doi.org/10.1016/j.jaci.2016.02.030.

Ohne, Y., Silver, J. S., Thompson-Snipes, L., Collet, M. A., Blanck, J. P., Cantarel, B. L., et al. (2016). IL-1 is a critical regulator of group 2 innate lymphoid cell function and plasticity. *Nature Immunology*, *17*(6), 646–655. https://doi.org/10.1038/ni.3447.

Pearce, E. L., Mullen, A. C., Martins, G. A., Krawczyk, C. M., Hutchins, A. S., Zediak, V. P., et al. (2003). Control of effector CD8+ T cell function by the transcription factor Eomesodermin. *Science*, *302*(5647), 1041–1043. https://doi.org/10.1126/science.1090148.

Possot, C., Schmutz, S., Chea, S., Boucontet, L., Louise, A., Cumano, A., et al. (2011). Notch signaling is necessary for adult, but not fetal, development of RORgammat(+) innate lymphoid cells. *Nature Immunology*, *12*(10), 949–958. https://doi.org/10.1038/ni.2105.

Price, A. E., Liang, H. E., Sullivan, B. M., Reinhardt, R. L., Eisley, C. J., Erle, D. J., et al. (2010). Systemically dispersed innate IL-13-expressing cells in type 2 immunity. *Proceedings of the National Academy of Sciences of the United States of America*, *107*(25), 11489–11494. https://doi.org/10.1073/pnas.1003988107.

Rothenberg, E. V. (2014). Transcriptional control of early T and B cell developmental choices. *Annual Review of Immunology*, *32*, 283–321. https://doi.org/10.1146/annurev-immunol-032712-100024.

Saenz, S. A., Siracusa, M. C., Perrigoue, J. G., Spencer, S. P., Urban, J. F., Jr., Tocker, J. E., et al. (2010). IL25 elicits a multipotent progenitor cell population that promotes T(H)2 cytokine responses. *Nature*, *464*(7293), 1362–1366. https://doi.org/10.1038/nature08901.

Sanos, S. L., Bui, V. L., Mortha, A., Oberle, K., Heners, C., Johner, C., et al. (2009). RORgammat and commensal microflora are required for the differentiation of mucosal interleukin 22-producing NKp46+ cells. *Nature Immunology*, *10*(1), 83–91. https://doi.org/10.1038/ni.1684.

Satoh-Takayama, N., Vosshenrich, C. A., Lesjean-Pottier, S., Sawa, S., Lochner, M., Rattis, F., et al. (2008). Microbial flora drives interleukin 22 production in intestinal NKp46+ cells that provide innate mucosal immune defense. *Immunity*, *29*(6), 958–970. https://doi.org/10.1016/j.immuni.2008.11.001.

Schneider, C., Lee, J., Koga, S., Ricardo-Gonzalez, R. R., Nussbaum, J. C., Smith, L. K., et al. (2019). Tissue-resident group 2 innate lymphoid cells differentiate by layered ontogeny and in situ perinatal priming. *Immunity*, *50*(6), 1425–1438. e1425. https://doi.org/10.1016/j.immuni.2019.04.019.

Schulthess, J., Meresse, B., Ramiro-Puig, E., Montcuquet, N., Darche, S., Begue, B., et al. (2012). Interleukin-15-dependent NKp46+ innate lymphoid cells control intestinal inflammation by recruiting inflammatory monocytes. *Immunity*, *37*(1), 108–121. https://doi.org/10.1016/j.immuni.2012.05.013.

Sciume, G., Hirahara, K., Takahashi, H., Laurence, A., Villarino, A. V., Singleton, K. L., et al. (2012). Distinct requirements for T-bet in gut innate lymphoid cells. *The Journal of Experimental Medicine*, *209*(13), 2331–2338. https://doi.org/10.1084/jem.20122097.

Shinkai, Y., Rathbun, G., Lam, K. P., Oltz, E. M., Stewart, V., Mendelsohn, M., et al. (1992). RAG-2-deficient mice lack mature lymphocytes owing to inability to initiate V(D)J rearrangement. *Cell*, *68*(5), 855–867. https://doi.org/10.1016/0092-8674(92)90029-c.

Stern, P., Gidlund, M., Orn, A., & Wigzell, H. (1980). Natural killer cells mediate lysis of embryonal carcinoma cells lacking MHC. *Nature*, *285*(5763), 341–342. https://doi.org/10.1038/285341a0.

Szabo, S. J., Kim, S. T., Costa, G. L., Zhang, X., Fathman, C. G., & Glimcher, L. H. (2000). A novel transcription factor, T-bet, directs Th1 lineage commitment. *Cell*, *100*(6), 655–669. https://doi.org/10.1016/s0092-8674(00)80702-3.

Takatori, H., Kanno, Y., Watford, W. T., Tato, C. M., Weiss, G., Ivanov, I. I., et al. (2009). Lymphoid tissue inducer-like cells are an innate source of IL-17 and IL-22. *The Journal of Experimental Medicine*, *206*(1), 35–41. https://doi.org/10.1084/jem.20072713.

Tsuji, M., Suzuki, K., Kitamura, H., Maruya, M., Kinoshita, K., Ivanov, I. I., et al. (2008). Requirement for lymphoid tissue-inducer cells in isolated follicle formation and T cell-independent immunoglobulin a generation in the gut. *Immunity*, *29*(2), 261–271. https://doi.org/10.1016/j.immuni.2008.05.014.

Vivier, E., Artis, D., Colonna, M., Diefenbach, A., Di Santo, J. P., Eberl, G., et al. (2018). Innate lymphoid cells: 10 years on. *Cell*, *174*(5), 1054–1066. https://doi.org/10.1016/j.cell.2018.07.017.

Vonarbourg, C., Mortha, A., Bui, V. L., Hernandez, P. P., Kiss, E. A., Hoyler, T., et al. (2010). Regulated expression of nuclear receptor RORgammat confers distinct functional fates to NK cell receptor-expressing RORgammat(+) innate lymphocytes. *Immunity*, *33*(5), 736–751. https://doi.org/10.1016/j.immuni.2010.10.017.

Wang, L., Wang, Y., Song, Z., Chu, J., & Qu, X. (2015). Deficiency of interferon-gamma or its receptor promotes colorectal cancer development. *Journal of Interferon & Cytokine Research*, *35*(4), 273–280. https://doi.org/10.1089/jir.2014.0132.

Xu, W., Cherrier, D. E., Chea, S., Vosshenrich, C., Serafini, N., Petit, M., et al. (2019). An Id2(RFP)-reporter mouse redefines innate lymphoid cell precursor potentials. *Immunity*, *50*(4), 1054–1068. e1053. https://doi.org/10.1016/j.immuni.2019.02.022.

Yang, Q., Li, F., Harly, C., Xing, S., Ye, L., Xia, X., et al. (2015). TCF-1 upregulation identifies early innate lymphoid progenitors in the bone marrow. *Nature Immunology*, *16*(10), 1044–1050. https://doi.org/10.1038/ni.3248.

Yokota, Y., Mansouri, A., Mori, S., Sugawara, S., Adachi, S., Nishikawa, S., et al. (1999). Development of peripheral lymphoid organs and natural killer cells depends on the helix-loop-helix inhibitor Id2. *Nature*, *397*(6721), 702–706. https://doi.org/10.1038/17812.

Yudanin, N. A., Schmitz, F., Flamar, A. L., Thome, J. J. C., Tait Wojno, E., Moeller, J. B., et al. (2019). Spatial and temporal mapping of human innate lymphoid cells reveals elements of tissue specificity. *Immunity*, *50*(2), 505–519. e504. https://doi.org/10.1016/j.immuni.2019.01.012.

Zeng, B., Shi, S., Ashworth, G., Dong, C., Liu, J., & Xing, F. (2019). ILC3 function as a double-edged sword in inflammatory bowel diseases. *Cell Death & Disease*, *10*(4), 315. https://doi.org/10.1038/s41419-019-1540-2.

Zheng, W., & Flavell, R. A. (1997). The transcription factor GATA-3 is necessary and sufficient for Th2 cytokine gene expression in CD4 T cells. Cell, *89*(4), 587–596. https://doi.org/10.1016/s0092-8674(00)80240-8.

CHAPTER SEVENTEEN

Detecting and analyzing murine innate lymphoid cells

Alejandra Gomez-Cadena[a,b], Pedro Romero[a], Sara Trabanelli[a,b], Camilla Jandus[a,b,]*

[a]Department of Oncology UNIL CHUV, University of Lausanne, Lausanne, Switzerland
[b]Ludwig Institute for Cancer Research, University of Lausanne, Lausanne, Switzerland
*Corresponding author: e-mail address: camilla.jandus-marone@unil.ch

Contents

Abstract

During the last 10 years, different subsets of Innate lymphoid cells (ILCs) have been identified in murine models. ILCs play and important role in maintaining immune barriers, tissue homeostasis, and are able to regulate the immune response in several anatomic sites. They can be found in lymphoid and non-lymphoid organs of adult mice but are mainly tissue-resident cells that can expand locally under physiologic or pathologic conditions (Gasteiger, Fan, Dikiy, Lee, & Rudensky, 2015). Because ILCs need to be identified by a complex combination of several cell-surface and intracellular markers and by their production of specific sets of cytokines, multiparametric flow cytometry remains one of the most efficient methods to analyze and isolate the different ILC sub-populations. This chapter describes how ILCs can be identified in different murine organs and how ILC subsets can be isolated and functionally analyzed.

Methods in Enzymology, Volume 631
ISSN 0076-6879
https://doi.org/10.1016/bs.mie.2019.10.022

1. Introduction

Innate lymphoid cells (ILCs) are a heterogenous population of innate effector lymphocytes. They lack recombined antigen receptors and provide a fast and first line of defense against pathogen invasion, before the adaptive immune response comes into play. Three major ILC groups (ILC1, ILC2 and ILC3) have been described based on the master transcription factors that drive their differentiation, and their cytokine signature upon activation. In terms of function, the ILC1, ILC2 and ILC3 classification mirrors the one of adaptive CD4 Th1, Th2 and Th17 cells, respectively. $Tbet^+$ ILC1, produce IFNγ as Th1, $GATA3^+$ ILC2 produce IL-4, IL-5 and IL-13, as Th2, and finally $RORgt^+$ ILC3 produce IL-17 and IL-22 (Fig. 1; Vivier et al., 2018).

The principal functions of ILCs are presented in Fig. 1.

ILCs also include lymphoid tissue-inducer (LTi) cells, which promote lymphoid organ formation during development. They are often described as belonging to the ILC3 group since they also express the RORγτ and share some extracellular markers. However, some studies suggest that they might belong to a different lineage (Bar-Ephraim & Mebius, 2016), and for that reason, they will not be addressed in this chapter. Concerning the helper ILC subsets, it has been described that ILC1 cells function as first line of defense against bacterial infections such as *T. gondii* (Gigley, 2016),

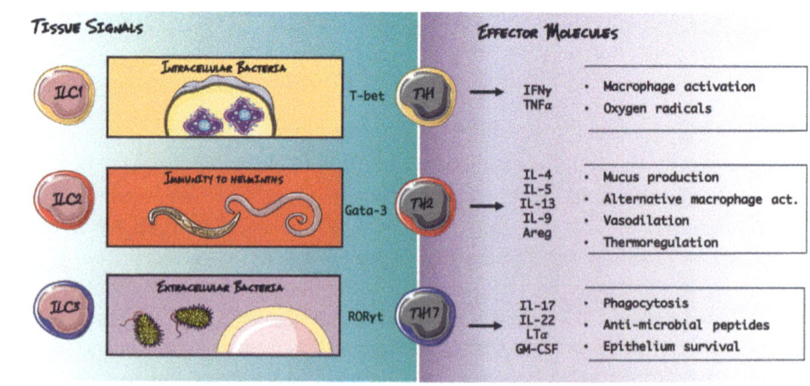

Fig. 1 The main immunological function of ILC subsets (left); the characteristic transcription factors and their adaptive immunity counterparts (middle) and finally, the effector molecules produced by activated ILCs according to the different subsets (right). *Adapted from Eberl, G., Colonna, M., Di Santo, J. P., & McKenzie, A. N. (2015). Innate lymphoid cells. Innate lymphoid cells: A new paradigm in immunology.* Science, *348(6237), aaa6566. https://doi.org/10.1126/science.aaa6566.*

C. difficile or virus infections (Weizman et al., 2017). ILC2 are involved in the immune response to parasites such as helminths and also favor tissue repair (Monticelli et al., 2015). Finally, ILC3 are abundant at mucosal sites and are involved in the selective containment of extracellular bacteria from the intestinal flora (Sonnenberg et al., 2012). In this chapter, we will discuss how to identify, isolate and stimulate the three main helper murine ILC populations.

2. Identification, isolation and stimulation of murine ILCs

2.1 Reagents

Processing media: DMEM GlutaMax, 10% FCS, 50 μg/mL Penicillin-Streptomycin, HEPES 10 mM, 50 μM β-Mercaptoethanol.

Mouse ILC culturing media: RPMI 1640, HEPES 10 mM, 50 μg/mL Penicillin-Streptomycin, 1 mM Na-Pyruvate, 50 μM β-Mercaptoethanol, 10% FCS.

Staining buffer: Sterile PBS 1 ×, 0.2% BSA, 5 mM EDTA, 0.2% Sodium azide.

Sorting buffer: Sterile PBS 1 ×, 0.2% BSA (Highly pure), 5 mM EDTA.

Fix/Perm solutions: eBioscience™ Foxp3/Transcription Factor Staining Buffer Kit.

Digestion media: 5 mL of Processing media/sample, 385 μL of Liberase™ TL Research Grade (Roche) prepared according to manufacturers' instructions, 1 U/mL DNAse I.

Percoll: 90% Percoll, 10% PBS 10 × = Stock solution.

- o 40% Solution = 40 mL Stock solution, 60 mL Processing media
- o 70% Solution = 70 mL Stock solution, 30 mL Processing media

2.2 Isolation of innate lymphoid cells from lymphoid and non-lymphoid organs: General considerations

To isolate and/or analyze innate lymphoid cells, homogenous single cell suspensions from different lymphoid organs can be easily obtained following organ harvesting and processing. However, ILC subset distribution might change depending on the organ, and therefore the organ of interest should be selected depending on the specific questions to be addressed. For instance, in steady-state, all ILC subsets, including ILC1, ILC2, ILC3 and LTi, can be successfully isolated from the small intestine lamina propria, from the lymph nodes (LNs), the mesenteric ones in particular, and from the spleen.

Regarding the liver, ILC1 represent a significant percentage of the tissue-resident lymphocytes, along with NK cells (Liu & Zhang, 2017), while the lungs are characterized by the presence of almost exclusively the ILC2 subset (Gasteiger, Fan, Dikiy, Lee, & Rudensky, 2015). In addition, the bone marrow hosts mainly ILC progenitors and some ILC2 ready to egress, expressing ST2 and/or KLRG1 (Stier et al., 2018; Tait Wojno & Beamer, 2018). In the following protocol we describe how to obtain ILCs from lymph nodes (LN), spleen, bone marrow (BM), lungs and liver. For the isolation of ILCs from the mouse lamina propria, we refer to the excellent method article recently published by Gronke, Kofoed-Nielsen, and Diefenbach (2017).

2.2.1 Organ processing: Protocols

A. Lymph nodes

- LN are collected in 48-well plates containing processing media (*Notes 1 and 2*).
- Prepare 50 mL falcon tubes and one 40 μm cell strainer/tube.
- Transfer the LN into the cell strainer placed on the top of the 50 mL tube and add 500 μL of processing media.
- Use the piston of a plastic syringe to mechanically dissociate the LN over the strainer regularly adding processing media.
- After disaggregation, wash the filter at least 2 times with 1 mL of processing media and complete up to 5 mL total volume.
- Centrifuge the cell suspension at 530*g* 5 min at 4 °C and discard the supernatant.
- Add 1 mL of fresh Mouse ILC culture media and count the cells.

B. Spleen

- Spleens are collected in 48-well plates containing processing media (*Notes 1 and 2*).
- Prepare 50 mL falcon tubes and one 70 μm cell strainer/tube.
- Transfer the spleen into the cell strainer placed on the top of the 50 mL tube and add 500 μL of processing media.
- Use the piston of a plastic syringe to mechanically dissociate the spleen over the strainer regularly adding processing media.
- After disaggregation, wash the filter at least 3 times with 1 mL of processing media and complete up to 10 mL total volume.
- Centrifuge the cell suspension at 530*g* 5 min at 4 °C and discard the supernatant.

- Resuspend the pellet in 3 mL of red blood cell lysis buffer and incubate 3 min at room temperature (RT).
- Add 3 mL of processing media and centrifuge at 530*g* for 5 min at 4 °C. Discard the supernatant.
- Add 1 mL of fresh Mouse ILC culture media and count the cells.

C. Lung

- Lungs are collected in 48-well plates containing processing media (*Notes 1 and 2*).
- Prepare a 1.5 mL Eppendorf tube/sample with 500 μL of digestion media. Add the lung tissues.
- Using scissors cut the lungs in small pieces and add additional 500 μL of digestion media.
- Incubate at 37 °C with agitation for 20 min (*Note 3*).
- Prepare 50 mL falcon tubes and one 70 μm cell strainer/tube.
- Transfer the digested pieces into the cell strainer placed on the top of the 50 mL tube and add 500 μL of processing media.
- Use the piston of a plastic syringe to mechanically dissociate the pieces left after the digestion over the strainer, adding regularly processing media.
- After disaggregation, wash the filter at least 3 times with 1 mL of processing media and complete up to 10 mL total volume.
- Centrifuge the cell suspension at 530*g* 5 min at 4 °C and discard the supernatant.
- Prepare 15 mL falcon tubes with 4 mL of a 70% Percoll solution.
- Resuspend the pellet in 4 mL of 40% Percoll solution and load it carefully on the 70% Percoll layer.
- Centrifugate at RT for 20 min at 940*g* with low increasing speed and without break.
- Recover the white ring containing the lymphocytes and transfer it in tubes containing 1 mL of processing media.
- Centrifuge the cell suspension at 530*g* 5 min at 4 °C and discard the supernatant.
- Add 1 mL of fresh Mouse ILC culture media and count the cells.

D. Liver

- Liver tissue is collected in 6-well plates containing processing media (*Note 2*).
- Prepare a Petri dish/sample.

- Using the piston of a 10 mL syringe smash the liver over a 200 μm stainless steel mesh. Add regularly processing media to the strainer. Once the dissociation is finished recover the solution from the Petri plate to transfer the content into a 15 mL tubes. Wait 1 min while the bigger pieces sediment.
- Recover the supernatant and centrifuge at 530*g* 5 min at 4 °C and discard the supernatant.
- Prepare 15 mL falcon tubes with 4 mL of a 70% Percoll solution.
- Resuspend the pellet in 4 mL of 40% Percoll solution and load it carefully on the 70% Percoll layer.
- Centrifugate at RT for 20 min at 940*g* with slow acceleration and without break during deceleration.
- Recover the white ring containing the lymphocytes and transfer it in tubes containing 1 mL of processing media.
- Centrifuge the cell suspension at 530*g* 5 min at 4 °C and discard the supernatant.
- Add 3 mL of fresh Mouse ILC culture media and count the cells.

E. Bone Marrow

- When dissecting the animal remove the muscles from the lower extremities using scissors, carefully cut the femur at the hip level and separate the femur and the tibia at the knee joint. Place bones in a 6-well plate with processing media.
- In a petri dish cut off the epiphyses of the bone and use a 25-gauge needle and an insulin syringe filled with processing media to flush the bone marrow cells from both ends (*Note 2*).
- Transfer the cell suspension into a 15 mL tube and centrifuge at 530*g* 5 min at 4 °C and discard the supernatant.
- Resuspend the pellet in 1 mL of red blood cell lysis buffer and incubate 5 min at 37 °C. Wash.
- Add 3 mL of fresh Mouse ILC culture media and count the cells.

F. Lamina Propria

For the lamina propria processing, we referred to the protocol published by Gronke et al. (2017), with the following minor modifications:

- The digestion media is prepared as described in the Section 2.1, containing Liberase™ TL Research Grade and DNAse I.
- Add 1 mM DTT in the pre-digestion media.

- Only one digestion is performed and the GentleMacs™ Dissociator is used to perform an extra mechanical disaggregation of the tissue under the program m_inestin_01.

2.3 Innate lymphoid cell identification and flow cytometry staining: General considerations

In recent years, improvements in the identification of ILC subsets have been made and a general consensus on specific markers characterizing murine ILCs has been reached. Murine innate lymphocytes include natural killer cells (NK), ILC1, ILC2, ILC3 and LTi, but in this chapter we will focus on the identification of helper ILCs (ILC1, ILC2 and ILC3). We refer to other excellent articles for the identification of murine NK cells (Walzer et al., 2007) and LTi (Bando, Liang, & Locksley, 2015). In general terms, helper ILCs are relatively rare, and consistently lack classical lineage-specific cell surface markers. For this reason, at first, a careful exclusion of lineage positive cells has to be achieved by the addition of different antibodies, all labeled by the same fluorochrome, recognizing each lineage specific cell surface marker. Next, total ILCs and each distinct ILC subset are further characterized by the expression of specific surface markers and transcription factors described in Table 1. For a more in-depth immune phenotyping of ILC subpopulations, in terms of chemokine receptors, activation status and proliferation, more molecules can be monitored, as recently reported by Vivier et al. (2018).

Table 1 Marker combination for murine ILC subset identification by multicolor flow cytometry.

	Mouse ILCs			
Lineage markers	Lin$^-$ (CD3ε, CD5, CD8α, CD11c, CD11b, CD19, B220, FCRεI, TCRαβ, TCRγδ, DX5, Ter119)			
Total ILCs	CD45$^+$, CD127$^+$, CD90.2$^+$			
ILC subsets: extracellular markers	ILC1	ILC2	ILC3	
	NK1.1$^+$	ST2$^+$ (IL33R)	cKit (CD117)	
	NKp46$^+$	KLRG1$^+$	NKp46$^+$	NKp46$^-$
			CCR6$^+$	CCR6$^-$
ILC subsets: transcription factors	T-bet$^+$	GATA3$^+$	RORγt$^+$	

Notes 4 and 5.

2.3.1 Innate lymphoid cell identification and flow cytometry staining: Protocol

- For a staining using 96-well plates, add 100 μL of your cell suspension/sample and centrifuge at 530*g*, 5 min at 4 °C (*Note 6*).
- Add 100 μL of cell viability dye like LIVE/DEAD™ Fixable Green Dead dye (Life Technologies) prepared according to manufacturer recommendations and incubate for at least 20 min at 4 °C.
- Wash the cells with 100 μL PBS, centrifugate at 530*g* for 5 min at 4 °C and discard supernatant.
- Prepare your antibody mix in staining buffer using the markers reported in Table 1 or additional ones of interest and stain the cells in a 50 μL volume. Incubate for 30 min at 4 °C (*Note 7*).
- Wash the cells with 150 μL staining buffer, centrifuge at 530*g* for 5 min at 4 °C and discard supernatant.
- Fix the cells using a traditional Fix/perm protocol according to manufacturer recommendations (*Note 8*).
- Prepare the intracellular staining mix with the markers or TF proposed in Table 1 in permeabilization buffer and incubate for 30 min at RT.
- Wash the cells with 150 μL staining buffer, centrifugate at 530*g* for 5 min at 4 °C and discard supernatant.
- Resuspend in 200 μL staining buffer and cells are ready to be analyzed by flow cytometry.

The gating strategy to identify the different ILC subsets in the different organs is presented in Fig. 2.

2.4 Innate lymphoid cell fluorescence-activated cell sorting: General considerations

To isolate primary ILCs for subsequent in vitro culture and functional assays, a slightly different protocol should be used. As mentioned before, the extracellular markers used for the staining should be carefully selected depending on the ILC subset of interest. Since no intracellular staining can be performed when viable cells need to be collected, we recommend using a reporter mouse strain for the isolation of ILC3 (e.g., RORγt-fate mapping mouse strain), (Vonarbourg et al., 2010). Indeed, since extracellular markers, such as cKit can be downregulated during cell activation, some ILC3 might be missed if sorting is based on this marker. As an alternative for ILC3 purification, CCR6 can be used, since it has been reported that the vast majority of ILC3 in LN express CCR6 (Dutton et al., 2017); however, this method will not allow to specifically detect 100% of the cells. ILC1 and ILC2 subsets can be detected more accurately through the combination of the different extracellular markers mentioned in Table 1.

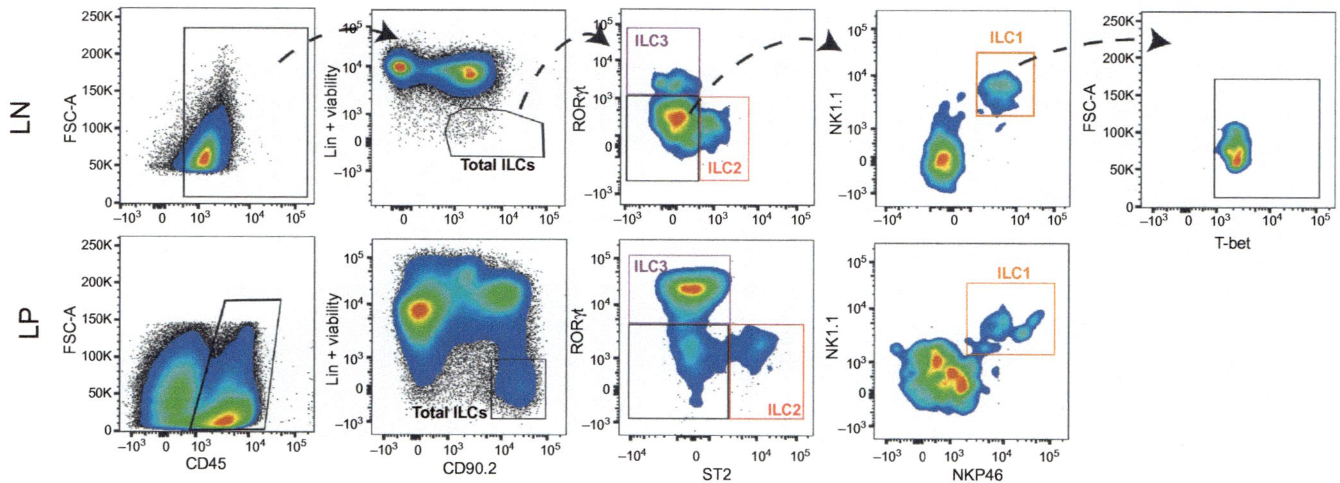

Fig. 2 ILC subset identification by multicolor flow cytometry on representative cell suspensions from mesenteric LNs and lamina propia (LP). From CD45 positive cells, lineage positive cells must be excluded and only CD90 positive cells can be considered as total ILCs. From the total ILC gate, ILC3 can be identified by RORγt expression in comparison to ST2 or GATA3 positive cells that represent the ILC2 population. From the double negative cells, the expression of NK1.1 and NKp46 is assessed and only the double positive cells, being homogenously Tbet positive, are considered as the ILC1 population.

2.4.1 Innate lymphoid cells flow-cytometry activated sorting: Protocol

- Wash and resuspend your cell suspension in cold Sorting Buffer.
- Enrich your cell suspension by depleting CD4, CD8 and CD20 positive cells using magnetic cell sorting following the manufacturer instructions (*Note 9*).
- Stain for surface markers in Sorting buffer during 30 min. Select the markers according to the population you are interested in. For example, to isolate the different ILC2 populations, you can use the following panel: CD45, CD90.2, CD127, ST2, KLRG1 (*Notes 10, 11, 12*).
- When sorting, the gating strategy shown in Fig. 3 can be used.
- Recover your sorted cells in Eppendorf tubes containing 200 μL of pure FCS and keep them always on ice.
- After sorting, count your cells or use the number given during the sorting for further experiments.

2.5 Innate lymphoid cell culture and cytokine stimulation: General considerations

The functional capacity of ILCs depends on the microenvironment and the local cytokines. In general terms, ILCs can be activated by stress signals, microbial compounds, dietary products or cytokines and they respond in an antigen-independent manner. This mode of activation makes them highly reactive and early effectors of the immune response (Vivier et al., 2018). Each ILC subset can be stimulated with a specific combination of cytokines and in turn ILCs will secrete defined soluble products that can be detected either by intracellular staining or by their quantification in the cell-free supernatant, as summarized in the following table (Table 2).

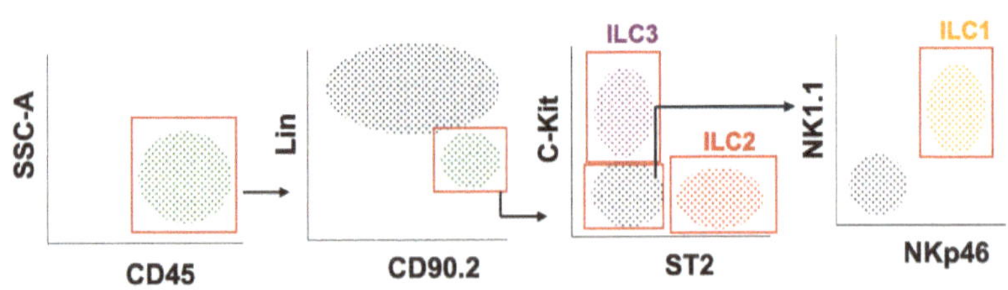

Fig. 3 Gating strategy to sort ILC subsets: The total ILC population corresponds to the Lin-CD90.2^{+} cells within the CD45^{+} cells. Then, from the total ILC population ILC3 can be identified as the cKit^{+}ST2^{-} cells and ILC2 as the cKit^{-}ST2^{+} ones. Finally, from the cKit^{-}ST2^{-} double negative population, ILC1 can be sorted as NK1.1^{+}NKp46^{+} cells.

Table 2 Activation of ILC subsets and their cytokines profiles.

Subset	Stimulation	Product
ILC1	rm-IL-12	IFNγ
	rh-IL-15	TNFα
	rm-IL-18	
ILC2		IL-4
	rm-IL-25	IL-5
	rm-IL-33	IL-13
		IL-9
		Amphiregulin
ILC3		IL-17
	rm-IL-1β	IL-22
	rh-IL-23	LTα
		GM-CSF

rm, Recombinant mouse cytokine; rh, Recombinant human cytokine.

2.5.1 Innate lymphoid cell culture and cytokine stimulation: Protocol

- The culture conditions will vary depending on the ILC subset of interest, and different stimulation media are prepared for the different ILC subsets.
- Distribute the cells into the different conditions and plate them in half of the final volume of ILC culture medium plus 1000 U/mL rh-IL-2 and rh-10 ng/mL IL-7 for all the ILC subsets.
- Prepare a master mix with cytokines and other stimulating components at the double of the final concentration. Then add half of the final volume to each well.
- Stimulation media:
  - o ILC1: rm-IL-12, rh-IL-15 and rm-IL-18 at 10 ng/mL final concentration
  - o ILC2: rm-IL-25 and rm-IL-33 at 50 ng/mL final concentration
  - o ILC3: rm-IL-1β and rh-IL-23 at 10 ng/mL final concentration
- If intracellular staining to detect cytokine production is chosen, add a protein secretion inhibitor during the overnight stimulation (*Note 13*).
- After overnight stimulation, re-stimulate the cells with PMA (50 ng/mL) plus Ionomycin (500 ng/mL) for 1 h at 37 °C before collecting them and start the staining.

- After PMA plus Ionomycin stimulation you can follow the flow cytometry staining protocol mentioned earlier in the chapter (Section 2.3.1).
- If the goal is to recover cell-free supernatant to quantify cytokine secretion, follow the alternative protocol described below:
  - Plate 200 cells/20 μL of ILC culture medium. The initial media should contain 1000 IU/mL rh-IL-2 and rh-10 ng/mL IL-7.
  - Add the other half of the volume containing the stimulating cytokines at the double of the concentration as mentioned before.
  - Incubate your cells over 48 h and recover 30 μL of supernatant to perform cytokine quantification with the method of your choice (*Note 14*).

2.6 Concluding remarks

The rapid improvement in ILC identification and functional evaluation has allowed recent progress in our understanding of their biology and their implications during immune system development, in homeostatic conditions and in disease settings. Through these studies, a higher degree of plasticity than expected and a tissue-dependent diversity of the main ILC subsets can be appreciated. Further functional, transcriptional and epigenetic studies at the single cell level combined with the use of dedicated genetically modified animals will enable to fully decipher the complexity of the ILC family.

2.7 Notes

1. When harvesting LN carefully remove the fat tissue to improve cell purity and viability.
2. Always keep the organs on ice or at 4 °C between the moment of harvesting and processing.
3. After digestion immediately put the tubes on ice while processing in order to stop the digestion and preserve the cells.
4. The fluorochrome used for the lineage antibodies is generally FITC; however, according to the different needs this fluorochrome can either be replaced or the lineage antibodies can be split in different channels.
5. CD127 is a classical surface marker to define the total ILC subset in lymphocyte mixtures. However, upon cell activation it can be downregulated and therefore total ILC frequencies underestimated. In this case CD90.2 (for the C56BL/6 genetic background) should be used to replace, or in addition to, CD127.

6. If only extracellular marker staining is performed, a minimum of 1.5×10^6 cells/well can be used, but if TF will be included in the staining, then a minimum of 3×10^6 cells/well is recommended.
7. If other markers, such as chemokine receptors, are being assessed during the extracellular staining step, incubation should be done at RT instead at 4 °C. Check first the recommendation for the antibodies used in the panel.
8. When adding intracellular markers, cells must be fixed and permeabilized either with a commercially available kit or using homemade solutions. If TF are used for the identification of your target cells, a commercial kit specifically developed for TF staining such as FOXP3 Fix/Perm buffer is highly recommended.
9. One of the recommended methods to enrich your cell suspension is by bead-based magnetic negative enrichment. You can use a cocktail of anti-mouse CD4, CD8 and CD20 magnetic MicroBeads or stain your cells using anti-CD4, anti-CD8 and anti-CD20 antibodies, all labeled with the same fluorochrome (e.g., FITC), followed by incubation with anti-fluorochrome magnetic MicroBeads (e.g., anti-FITC). This process should enrich your ILC population around 10 times; however, a negative impact on cell yield (±30%) should also be taken into account.
10. CD90.1 or CD90.2 is used depending on the genetic background of the mouse strain.
11. CD127 is downregulated from the cell surface in activated cells. Use CD90 if your mice were treated with cytokines, if they are tumor bearing mice or if you suspect your cells might be in an activated state.
12. Different staining panels can be used based on several reports available in the specific literature. As mentioned above, if you want to isolate ILC3 one option is to use the cKit extracellular marker. However, take into account that this marker is also downregulated after activation.
13. Combining protein secretion inhibitors containing both Brefeldin A and Monensin will help to detect the accumulation of intracellular cytokines. These reagents are often toxic if they are used at high doses for prolonged periods of time. Titrating the optimal concentration for the overnight stimulation is highly recommended.
14. If not all wells of the 96-well plate are used, add sterile PBS or buffer in the surrounding empty wells to reduce evaporation during incubation time of the small volume assays.

References

Bando, J. K., Liang, H. E., & Locksley, R. M. (2015). Identification and distribution of developing innate lymphoid cells in the fetal mouse intestine. *Nature Immunology*, *16*(2), 153–160. https://doi.org/10.1038/ni.3057.

Bar-Ephraim, Y. E., & Mebius, R. E. (2016). Innate lymphoid cells in secondary lymphoid organs. *Immunological Reviews*, *271*(1), 185–199. https://doi.org/10.1111/imr.12407.

Dutton, E. E., Camelo, A., Sleeman, M., Herbst, R., Carlesso, G., Belz, G. T., et al. (2017). Characterisation of innate lymphoid cell populations at different sites in mice with defective T cell immunity. *Wellcome Open Research*, *2*, 117. https://doi.org/10.12688/wellcomeopenres.13199.3.

Gasteiger, G., Fan, X., Dikiy, S., Lee, S. Y., & Rudensky, A. Y. (2015). Tissue residency of innate lymphoid cells in lymphoid and nonlymphoid organs. *Science*, *350*(6263), 981–985. https://doi.org/10.1126/science.aac9593.

Gigley, J. P. (2016). The diverse role of NK cells in immunity to toxoplasma gondii infection. *PLoS Pathogens*, *12*(2), e1005396. https://doi.org/10.1371/journal.ppat.1005396.

Gronke, K., Kofoed-Nielsen, M., & Diefenbach, A. (2017). Isolation and flow cytometry analysis of innate lymphoid cells from the intestinal Lamina propria. *Methods in Molecular Biology*, *1559*, 255–265. https://doi.org/10.1007/978-1-4939-6786-5_17.

Liu, M., & Zhang, C. (2017). The role of innate lymphoid cells in immune-mediated liver diseases. *Frontiers in Immunology*, *8*, 695. https://doi.org/10.3389/fimmu.2017.00695.

Monticelli, L. A., Osborne, L. C., Noti, M., Tran, S. V., Zaiss, D. M., & Artis, D. (2015). IL-33 promotes an innate immune pathway of intestinal tissue protection dependent on amphiregulin-EGFR interactions. *Proceedings of the National Academy of Sciences of the United States of America*, *112*(34), 10762–10767. https://doi.org/10.1073/pnas.1509070112.

Sonnenberg, G. F., Monticelli, L. A., Alenghat, T., Fung, T. C., Hutnick, N. A., Kunisawa, J., et al. (2012). Innate lymphoid cells promote anatomical containment of lymphoid-resident commensal bacteria. *Science*, *336*(6086), 1321–1325. https://doi.org/10.1126/science.1222551.

Stier, M. T., Zhang, J., Goleniewska, K., Cephus, J. Y., Rusznak, M., Wu, L., et al. (2018). IL-33 promotes the egress of group 2 innate lymphoid cells from the bone marrow. *The Journal of Experimental Medicine*, *215*(1), 263–281. https://doi.org/10.1084/jem.20170449.

Tait Wojno, E. D., & Beamer, C. A. (2018). Isolation and identification of innate lymphoid cells (ILCs) for Immunotoxicity testing. *Methods in Molecular Biology*, *1803*, 353–370. https://doi.org/10.1007/978-1-4939-8549-4_21.

Vivier, E., Artis, D., Colonna, M., Diefenbach, A., Di Santo, J. P., Eberl, G., et al. (2018). Innate lymphoid cells: 10 years on. *Cell*, *174*(5), 1054–1066. https://doi.org/10.1016/j.cell.2018.07.017.

Vonarbourg, C., Mortha, A., Bui, V. L., Hernandez, P. P., Kiss, E. A., Hoyler, T., et al. (2010). Regulated expression of nuclear receptor RORgammat confers distinct functional fates to NK cell receptor-expressing RORgammat(+) innate lymphocytes. *Immunity*, *33*(5), 736–751. https://doi.org/10.1016/j.immuni.2010.10.017.

Walzer, T., Blery, M., Chaix, J., Fuseri, N., Chasson, L., Robbins, S. H., et al. (2007). Identification, activation, and selective in vivo ablation of mouse NK cells via NKp46. *Proceedings of the National Academy of Sciences of the United States of America*, *104*(9), 3384–3389. https://doi.org/10.1073/pnas.0609692104.

Weizman, O. E., Adams, N. M., Schuster, I. S., Krishna, C., Pritykin, Y., Lau, C., et al. (2017). ILC1 confer early host protection at initial sites of viral infection. *Cell*, *171*(4), 795–808, e712. https://doi.org/10.1016/j.cell.2017.09.052.

CHAPTER EIGHTEEN

A cytofluorimetric assay to evaluate intracellular cytokine production by NK cells

Christian Sordo-Bahamonde[a,b,c], Seila Lorenzo-Herrero[a,b,c], Segundo González[a,b,c], Alejandro López-Soto[a,b,c,]*

[a]Departamento de Biología Funcional, Área de Inmunología, Universidad de Oviedo, Oviedo, Spain
[b]Instituto Universitario de Oncología del Principado de Asturias (IUOPA), Oviedo, Spain
[c]Instituto de Investigación Sanitaria del Principado de Asturias (ISPA), Oviedo, Spain
*Corresponding author: e-mail address: lopezsalejandro@uniovi.es

Contents

Abstract

Natural killer (NK) cells are innate lymphoid cells that, together with $CD8^+$ T lymphocytes, are tightly correlated with immunesurveillance and the elimination of transforming and malignant cells both during early steps of tumorigenesis and metastasis. This capacity is due, but not limited to, their great cytotoxic capacity upon recognition of stress-related activating ligands of NK cells on the surface of tumor cells. Due to the emerging role of NK cells in the response to treatment with immune checkpoint blockers (ICBs) and other immunotherapies, it has become essential to deeply study this immune subset. As mentioned above, NK cell antitumor responses not only rely on their high cytotoxic capacity, cytokine production by this immune subset, such as interferon gamma (IFN-γ) or tumor necrosis factor (TNF-α), also plays a key role on NK cell function. In this chapter, we provide a detailed protocol to measure the intracellular expression of cytokines produced by NK cells upon stimulation. A step-by-step guide to isolate peripheral blood mononuclear cells, purify NK cells from whole blood, co-culture with tumor cell lines and evaluate their cytokine production capacity by flow cytometry is here provided.

Methods in Enzymology, Volume 631
ISSN 0076-6879
https://doi.org/10.1016/bs.mie.2019.05.049

1. Introduction

Natural killer (NK) cells are innate lymphoid cells with a strong cytotoxic activity that allows the immune eradication of virus-infected cells as well as stressed cells during malignant transformation. Unlike T cells, NK cell activity does not involve antigen recognition and, hence, they are able to target cancer cells lacking expression of neoantigens owing to the loss of Major Histocompatibility Complex class I (MHC-I) molecules (Chiossone, Dumas, Vienne, & Vivier, 2018). Instead, NK cells detect the increased expression of stress-regulated molecules on the surface of infected and tumor cells, such as the ligands of the activating receptor killer cell lectin like receptor K1 (KLRK1; best known as NKG2D), which, in humans, include MHC class I polypeptide-related sequence A and B (MICA and MICB) and UL16 binding proteins (ULBPs) (Lopez-Soto, Huergo-Zapico, Acebes-Huerta, Villa-Alvarez, & Gonzalez, 2015). The engagement of NK cell activation receptors triggers a rapid exocytosis of the cytotoxic cargo of these cells as granules containing perforin 1 (PRF1) and granzymes that lead to the killing of the target cell. Likewise, upon activation, NK cells produce diverse cytokines that enhances the recruitment and activation into the tumor microenvironment of members of the adaptive immunity (Deauvieau et al., 2015; Nouri-Shirazi et al., 2000).

A compelling body of data demonstrate that NK cells are key players of the immunosurveillance of cancer cells during the early steps of tumorigenesis and metastatic malignant dissemination of carcinomas (Lopez-Soto, Gonzalez, Smyth, & Galluzzi, 2017). Indeed, NK cell numbers and activity have been correlated with a lower incidence of cancer and a better prognosis in patients with different types of cancer (Lorenzo-Herrero et al., 2018). However, many mechanisms evolved by malignant cells that limit NK cell antitumor activity have been described to date, including epigenetic mechanisms and shedding of NK cell activating ligands (Ferrari de Andrade et al., 2018; Lopez-Soto, Folgueras, Seto, & Gonzalez, 2009). Consequently, therapy-mediated reversion of these tumor immunosubversive mechanisms may represent novel strategies for the management of patients with cancer, which currently constitutes a field of intense research. Classically, antitumor response to therapy with immune checkpoint blockers (ICBs) has been associated almost exclusively with $CD8^+$ T lymphocytes and, indeed, tumor mutational load has been shown to predict clinical response to ICB-based therapies across cancer in patients with advanced disease (Galluzzi, Chan, Kroemer, Wolchok, & Lopez-Soto, 2018; Samstein et al., 2019).

Nevertheless, an emerging key role of NK cells is being enlightened by recent preclinical and clinical findings (Andre et al., 2018; Hsu et al., 2018; Oyer, Gitto, Altomare, & Copik, 2018; Villa-Alvarez et al., 2018) therefore, calling for the development of experimental approaches to deeply evaluate NK cell antitumor function.

As mentioned above, NK cells also participate in the development of immune responses by releasing cytokines and chemokines such as interferon gamma (IFN-γ) or tumor necrosis factor (TNF-α). IFN-γ, a type II interferon, is mainly produced by activated $CD4^+$ Th_1 cells, $CD8^+$ cytotoxic T lymphocytes and NK cells in response to a wide variety of immune or inflammatory stimuli. This cytokine possesses tumor cytostatic and cytotoxic capacity by diverse mechanisms including the induction NK cell activating receptor ligand expression, the arrest of tumor cell proliferation and angiogenesis, as well as the promotion of adaptive immune responses via macrophage activation (Hayakawa et al., 2002; Su et al., 2015; Wang, Jaw, Stutzman, Zou, & Sun, 2012). Conversely, recent studies show that this cytokine may play a role on immune evasion by, for example, increasing PD-L1 expression (Benci et al., 2016; Pai et al., 2019).

In this protocol, we describe a detailed and highly sensitive method to measure the cytokine production by NK cells employing intracellular staining and flow cytometry analysis. Alternatively, an enzyme-linked immunosorbent assay (ELISA) can be performed to measure cytokine production but, nonetheless, intracellular staining show advantages over this method as ELISA assay requires the isolation of NK cells in order to ensure the cytokine producing immune cell subset. Specifically, in this protocol we describe a method to evaluate the intracellular expression of IFN-γ in healthy donor's PBMC-derived NK cells (buffy coats bags) co-cultured with a tumor cell line (K-562).

2. Materials and methods

2.1 Cell culture

2.1.1 Peripheral blood mononuclear cell isolation and culture

1. Peripheral blood mononuclear cells (PBMCs) are isolated from buffy coats from healthy donors by Ficoll density gradient medium (see *Note 1*). Concentrated peripheral blood must be previously diluted in 1 × phosphate-buffered saline (PBS) (137 mM NaCl, 2.7 mM KCl, 4.3 mM Na_2HPO_4, and 1.4 mM KH_2PO_4 in dH_2O) at room temperature (RT) at a 1:1 ratio.

2. Add 4 mL of pre-warmed (RT) Ficoll (Histopaque®-1077, Sigma-Aldrich) in a 15-mL conical tube (see *Note 2*). Then, gently pour 11 mL of previously diluted blood sample on the top of the Ficoll layer avoiding mixing both phases.
3. Immediately after, centrifuge the tubes at 900 × *g* for 22 min at RT. Turn off the break of the centrifuge to avoid disruption of the resulting phases during the stopping process.
4. Add 8 mL of 1 × PBS at RT into a new sterile 15-mL conical tube. Transfer the layer corresponding to PBMCs to this tube with a 1.5-mL sterile Pasteur pipette and kindly mix the content by pipetting up and down (see chapter "Evaluation of NK cell cytotoxic activity against malignant cells by the calcein assay" by Lorenzo-Herrero et al.). Afterward, centrifuge the tube at 250 × *g* for 5 min to wash the isolated PBMCs. Eliminate the supernatant and gently resuspend the cell pellet. Repeat this washing step to completely remove all traces of Ficoll (see *Note 3*).
5. Resuspend the cells in pre-warmed (37 °C) culture medium. Total PBMCs are routinely cultured in Roswell Park Memorial Institute (RPMI) medium (Lonza), supplemented with 10% of heat-inactivated (30 min, 56 °C) fetal bovine serum (FBS) (Sigma-Aldrich, St. Louis, MO), 2 mM glutamine, 1 mM sodium pyruvate, 100 U/mL penicillin and 10 μg/mL streptomycin at 37 °C and 5% CO_2.

2.1.2 NK cell isolation from whole blood and cell culture

The following steps are performed according to the manufacturer's protocol of MACSxpress Whole Blood NK Cell Isolation Kit (Miltenyi Biotec), which allows purification of NK cells by negative selection (see *Note 4*). 1. Reconstitute the MACSxpress beads by adding 7.5 mL of Buffer A (see *Note 5*). Then, prepare the isolation mix for a 1:4 (isolation mix:blood sample) ratio. Thus, for 8 mL of whole blood add 2 mL of reconstituted beads and 2 mL of Buffer B at RT (see *Note 6*). Mix gently by pipetting up and down.

Add 4 mL of isolation mix to 8 mL of whole blood in a sterile 15-mL conical tube. Carefully mix by inversion and incubate for 5 min at RT in a tube rotator (see *Note 7*).

Place the tube into the magnet (MACSxpress Separator) without the lid for 15 min so that the magnetic field induces the labeled cells to adhere to the wall of the tube (see *Note 8*). Erythrocytes will sediment to the bottom.

Collect the supernatant containing purified NK cells and place it in a new 15-mL conical tube by pipetting carefully (see *Note 9*). Finally, centrifuge the supernatant and resuspend NK cells in complete RPMI medium.

2.1.3 Cell line culture

1. The human lymphoblastoid cell line K-562, which was obtained from American Type Culture Collection (ATCC, Manassas, VA), has been widely used as a highly sensitive target for functional assays to evaluate NK cell antitumor activity (see *Note 10*).
2. K-562 cell line is grown as cell suspension culture in complete RPMI medium at 37 °C and 5% CO_2 (see Section 2.1.1) in culture flasks. Cultures should be maintained at 1×10^6 cells/mL by adding fresh pre-warmed medium (37 °C) every 2–3 days. Once every 7–8 days, centrifuge the cell culture at $250 \times g$ for 5 min at RT and completely renew the medium at a density of 1×10^6 cells/mL.
3. A549 human lung carcinoma cells are typically grown in F-12K Medium supplemented with 10% FBS, 2 mM glutamine, 1 mM sodium pyruvate, 100 U/mL penicillin and 10 μg/mL streptomycin at 37 °C and 5% CO_2.

3. Assessment of cytokine production in NK cells by intracellular staining

Stimulation of cytokine production by NK cells can be induced by co-culture with K-562 cell line and/or by adding phorbol 12-myristate 13-acetate (PMA) and ionomycin to the culture medium (see *Note 11*) (see chapter "Cytokine profiling of tumor-infiltrating T lymphocytes by flow cytometry" by Foote et al.). In this protocol, total PBMCs and purified NK cells are co-cultured with K-562 cell line alone or in combination with PMA/ionomycin at a 50:1 or 10:1 (effector:target) ratio, respectively (Fig. 1) (see *Note 12*).

3.1 NK cell-tumor cell co-culture and intracellular cytokine staining assay

1. Count freshly isolated PBMCs or purified NK cells (effector cells) with trypan blue solution (Gibco) (see *Note 13*).
2. Centrifuge an appropriate number of PBMCs or isolated NK cells at $250 \times g$ for 5 min and resuspend the cells in pre-warmed (37 °C) complete RPMI medium. Adjust the volume to 10^7 cells/mL for total PBMCs or 2×10^6 cells/mL for isolated NK cells. Then, transfer

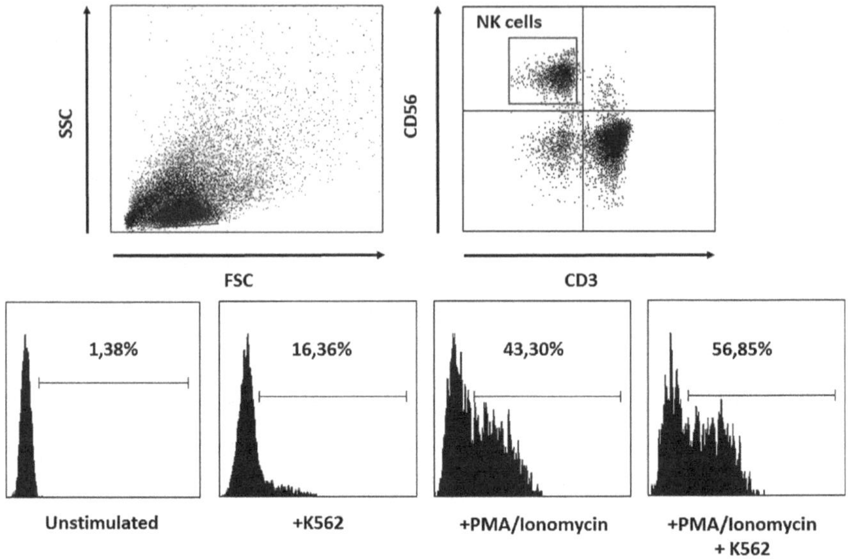

Fig. 1 Assessment of cytokine production in NK cells by intracellular staining. Gating strategy is defined as follows: create a new FSC/SSC-based dot plot to localize the population of PBMCs. Afterward, set a second dot plot and identify the NK cell subset based on the expression of CD56 and CD3 (human NK cells are defined as $CD56^+CD3^-$) to quantify the percentage of IFN-γ-producing NK cells. Finally, a histogram including the NK cell subset showing the PE-positive fraction will allow determination of the percentage of NK cells positive for IFN-γ production. Unstimulated NK cells produce no detectable levels of IFN-γ, whereas exposure to different stimuli results in an increased production of this cytokine.

100 μL of each cell suspension to a 96-well round bottom plate, duplicating each condition (Table 1) (see *Note 14*) for co-culturing with K-562 cells (target cells).

3. Kindly resuspend K-562 cells to separate cell aggregates and transfer them to a 15-mL sterile conical tube. Count target cells by trypan blue exclusion method (see Section 3.1, step 1). Resuspend an appropriate number of target cells in fresh pre-warmed complete RPMI medium at 2×10^5 cells/mL. Finally, add 100 μL of cell suspension to the corresponding wells (Table 1). Additionally, maintain a duplicate of PBMCs/Isolated NK cells alone as a negative control (see *Note 15*). The final volume of each well is 200 μL.
4. As a positive control, a duplicate of total PBMCs or purified NK cells supplemented with 50 nM PMA and 1 μg/mL ionomycin must be included. Unstimulated PBMCs or isolated NK cells are used as negative control.

Table 1 Experimental conditions for co-culturing the K-562 cell line and NK cells.

Condition	PBMCs/ isolated NKs cells (μL/well)	K-562	PMA and ionomycin	Complete RPMI 1640 medium
Negative control	100	–	–	100 μL/well
PMA/ionomycin control	100	–	+	100 μL/well supplemented with PMA/ionomycin
Co-culture	100	100 μL/well	–	–
Positive control	100	100 μL/well	+	5 μL/well supplemented with PMA/ionomycin premix

5. Incubate the plate for 1 h at 37 °C and 5% CO_2. Thereafter, a protein transport inhibitor solution containing 3 μM monensin (BD GolgiStop™) is added to each well in a 5 μL/well premix with RPMI complete medium (see *Note 16*).
6. Incubate the plate for additional 4 h at 37 °C and 5% CO_2.
7. Afterward, transfer the volume of each well to a 5 mL polystyrene cytometry tube (FACs tube) and wash by adding 1 × PBS (250 × *g*, 5 min). Aspirate supernatants and resuspend in a 100 μL/tube premix of specific surface markers of NK cells with fluorochrome-conjugated antibodies.
8. Specific cell lineage antibodies to gate out NK cells must be used including antibodies to positively recognize NK cells, defined as $CD56^+CD3^-$. For that purpose, prepare a 100 μL solution containing 5 μL of allophycocyanin (APC)-conjugated mouse anti-human CD56 (clone 5.9, IgG_{2b}) (Cytognos), 5 μL of fluorescein isothiocyanate (FITC)-conjugated mouse anti-human CD3 (clone UCHT1, IgG_1) (Cytognos) and 90 μL of 1 × PBS per tube (see *Notes 17*). Incubate for 20 min at RT in the dark. Subsequently, wash each tube with 1 × PBS (250 × *g*, 5 min) and discard supernatants.
9. For intracellular staining of cytokines, we use BD Cytofix/Cytoperm Fixation/Permeabilization Kit (BD Biosciences) according to the manufacturer's instructions (see *Note 18*). Briefly, add 200 μL of fixation and permeabilization buffer to each tube and incubate for 20 min on ice in the dark.

10. Then, wash the tubes twice by adding 1 mL of 1 × Perm/Wash buffer and centrifuge at 250 × *g*, 5 min (see *Note 19*). After the first washing step, separate 300 μL of each condition to a new FACs tube to be employed as isotype control (see *Note 20*).
11. Prepare a master mix of intracellular cell staining-compatible antibodies against human cytokines produced by NK cells. Herein, we evaluated the expression of IFN-γ. Hence, add 5 μL of Phycoeryhtrin (PE)-conjugated mouse anti-human IFN-γ (clone 4S·B3, IgG_1; Biolegend) in 45 μL of Perm/Wash buffer per tube and incubate the cells for 30 min on ice (see *Note 21*). Afterward, wash twice with 1 mL 1 × Perm/Wash buffer (250 × *g*, 5 min/see Section 3.1, step 10).
12. Finally, resuspend the cells in 300 μL 1 × PBS and store FACs tubes at 4 °C protected from light until flow cytometry analysis (see *Note 22*).

3.2 Flow cytometry analysis

1. A *CytoFLEX S* cytometer (Beckman Coulter) and *CytExpert 2.1* software are employed in this protocol. Ensure that your equipment detects at least the three-color flow cytometry panel outlined in this chapter (488 and 638 nm channels).
2. First, locate and gate the population of PBMCs based on its forward scatter (FSC) and side scatter (SSC) parameters (see *Note 23*) (Fig. 1).
3. In a new dot plot, select the NK cell population, defined as $CD56^+CD3^-$ (see Section 3.1, step 8) (see *Note 24*).
4. Finally, determine the percentage of IFN-γ-producing NK cells, by setting the NK cell population on a histogram displaying anti-IFN-γ-PE staining and gate the positive fraction (Fig. 1).

4. Notes

1. Vials containing an anticoagulant such as EDTA or sodium heparin are required for the collection of plasma blood samples.
2. The density of Ficoll is affected by temperature, showing optimal results at RT (~20 °C). If cold Ficoll is used, this could result in inappropriate gradient density separation. Also, in order to maximize cell viability, samples must be processed as soon as possible within 24 h upon collection.
3. Extended contact with Ficoll can result into toxicity to PMBCs. Hence recover the mononuclear cell layer and wash the cells by centrifugation

immediately. Repeating this step is important to completely eliminate all Ficoll traces.

4. Negative, rather than positive, selection is preferred for NK cell purification, as this isolation strategy implies no use of antibody or reagent that may alter NK cells. The kit employed in this protocol induces the aggregation and sedimentation of erythrocytes and removes undesired cells by immunomagnetic depletion, maintaining NK cells untouched.
5. Once lyophilized MACSxpress beads are resuspended, they can be stored at 4 °C up to 1 week.
6. The isolation mix must be prepared freshly for every cell separation procedure, so prepare the volume necessary for your experiment following a 1:2 ratio (isolation mix:whole blood).
7. The manufacturer's protocol recommends using a MACSmix™Tube Rotator at 12 rpm.
8. In order to avoid resuspending the erythrocytes or the magnetic-labeled cell subsets, do not move the tube during the separation step.
9. Supernatant contents the cell target fraction; thus, it must be carefully transferred into a new tube while the tube with the beads is still in the magnet. Kindly move the tip up to down while supporting it in the front wall of the tube. Purity of the isolated population can be determined in a small sample of the supernatant by flow cytometry (see Section 3.1, step 8). It is also recommendable to check the viability of isolated NK cells using 7-AAD (7-aminoactinomycin D) or propidium iodide (PI) staining.
10. The K-562 cell line is commonly used as target cells in NK cell assays due to their lack of Human Leukocyte Antigen (HLA) class I and II expression. Although K-562 cell line is the gold standard, assessment of cytokine production can also be evaluated by using alternative suspension or adherent tumor cell lines (Fig. 2).
11. NK cells from total peripheral blood mononuclear cells or isolated NK cells from healthy donors show a poor capability of cytokine production in response to co-culture with tumor cell lines in vitro. In this context, additional stimulus are required in order to induce their activation, like cytokines (IL-2 or IL-15 are commonly employed) or Phorbol 12-myristate 13-acetate (PMA) and ionomycin. It is highly recommended to test different conditions of stimulation when using a new cell line or treatment, since an excessive stimulus by PMA/ionomycin could mask their effect on cytokine production by NK cells.

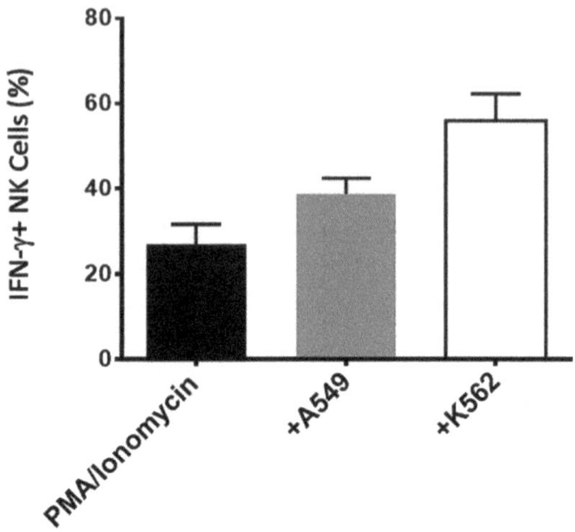

Fig. 2 Comparison of IFN-γ production by NK cells co-cultured with the K-562 gold standard cell line versus the lung carcinoma A549 cell line.

Additionally, it should be noted that PMA/ionomycin can trigger cytokine production by other immune subsets, such as T cells.

12. In this protocol, two different ratios are used; 50:1 for total PBMCs (considering that NK cells commonly represent about 20% of PBMCs in healthy donor samples) and 10:1 (effector:target) ratio when purified NK cells are used. Including appropriate control conditions is crucial. Culturing PBMCs or isolated NK cells alone provides a measurement of the basal cytokine production. The combination of both stimuli (PMA/ionomycin and co-culture with K-562 cell line) leads to an increased production of IFN-γ, as shown in Fig. 1.
13. Use the trypan blue exclusion method and a Neubauer counting chamber to determine the numbers and viability of the samples. Prepare a 1:1 dilution (cell suspension:trypan blue) and incubate for 1–2 min at RT to allow the staining of dead cells. Trypan blue is excluded in viable (white) cells, while dead cells will be stained (blue) owing to incorporation of the dye.
14. Due to the great differences in NK cell responses among donors (Fig. 3), at least two different buffy coats must be employed to ensure reproducibility of the results obtained.
15. Likewise, a duplicate for each condition must be included to minimize variability and increase reliability of the results.
16. Vesicle transport to the Golgi apparatus must be disrupted to prevent cytokine release to the medium. For this purpose, monensin or

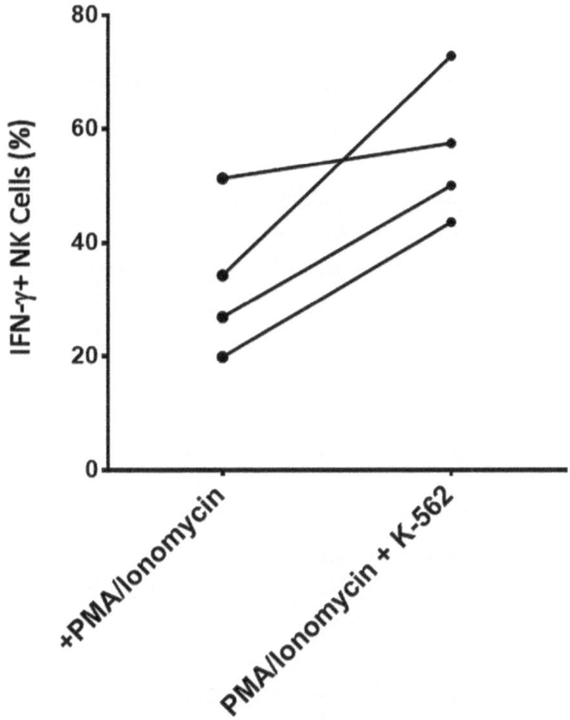

Fig. 3 Variability of stimulation of NK cell-mediated IFN-γ production among samples.

brefeldin A (BFA) must be added after to the co-culture. While BFA (BD GolgiPlug™) indirectly inhibits intracellular protein transport from the Golgi apparatus to the endoplasmic reticulum, monensin (BD GolgiStop™) is a polyether ionophore that blocks acidification of endocytic vesicles (Golgi transmembrane Na^{++}/H^{+} transport), thereby crippling protein secretion. The protein transport inhibitor should be carefully chosen and varies depending on the cytokine and the species employed. Both, monensin and brefeldin must be stored undiluted at 4 °C.

17. When preparing a master mix, it is highly recommended to add 10–15% additional volume of each reagent in order to avoid running out of volume.
18. It is important to avoid cell aggregation by vortexing before adding the fixation/permeabilization solution. This kit includes a formaldehyde- and saponin-based fixation and permeabilization buffer as well as a 10 × Perm/Wash buffer containing saponin to maintain permeabilization.

19. The 10× Perm/Wash solution must be diluted in milli-Q water and maintained on ice during the assay.
20. The isotype control tube allows evaluation of non-specific background signal.
21. Use 1× Perm/Wash buffer to prepare the master mix so that the permeabilization is maintained during this step. Prepare 10–15% more volume than required to avoid running out of volume.
22. Despite cells are fixed and data collection can be postponed, it is highly recommended to acquire samples within 24 h of the assay to get better and more robust results.
23. K-562 cell line has similar FSC and SSC values than PBMCs, so they could remain indistinguishable. Checking target cell autofluorescence and specific effector cell labeling beforehand should avoid any trouble during the assay. Acquire at least 1×10^5 NK cells to get more reliable results and minimize the effect of outliers. The percentage of IFN-γ producing NK cells is calculated as the average result of the duplicates.
24. There are two main subsets of NK cells in peripheral blood classified as $CD56^{bright}$ $CD16^-$ and $CD56^{dim}$ $CD16^+$, since the first one are the main cytokine producers and Herein, we analyzed the cytokine production by total NK cells, but an appropriate gating strategy could be performed in order to determine differences in cytokine production. $CD56^{bright}$ $CD16^-$ are the main cytokine producers and undergo cell activation rapidly. On the other hand, $CD56^{dim}$ $CD16^+$ represent the cytotoxic subset and require longer time intervals for optimal cell activation.

References

Andre, P., Denis, C., Soulas, C., Bourbon-Caillet, C., Lopez, J., Arnoux, T., et al. (2018). Anti-NKG2A mAb is a checkpoint inhibitor that promotes anti-tumor immunity by unleashing both T and NK cells. *Cell*, *175*(7), 1731-1743.e13. https://doi.org/10.1016/j.cell.2018.10.014.

Benci, J. L., Xu, B., Qiu, Y., Wu, T. J., Dada, H., Twyman-Saint Victor, C., et al. (2016). Tumor interferon signaling regulates a multigenic resistance program to immune checkpoint blockade. *Cell*, *167*(6), 1540-1554.e12. https://doi.org/10.1016/j.cell.2016.11.022.

Chiossone, L., Dumas, P. Y., Vienne, M., & Vivier, E. (2018). Natural killer cells and other innate lymphoid cells in cancer. *Nature Reviews. Immunology*, *18*(11), 671–688. https://doi.org/10.1038/s41577-018-0061-z.

Deauvieau, F., Ollion, V., Doffin, A. C., Achard, C., Fonteneau, J. F., Verronese, E., et al. (2015). Human natural killer cells promote cross-presentation of tumor cell-derived antigens by dendritic cells. *International Journal of Cancer*, *136*(5), 1085–1094. https://doi.org/10.1002/ijc.29087.

Ferrari de Andrade, L., Tay, R. E., Pan, D., Luoma, A. M., Ito, Y., Badrinath, S., et al. (2018). Antibody-mediated inhibition of MICA and MICB shedding promotes NK cell-driven tumor immunity. *Science*, *359*(6383), 1537–1542. https://doi.org/10.1126/science.aao0505.

Galluzzi, L., Chan, T. A., Kroemer, G., Wolchok, J. D., & Lopez-Soto, A. (2018). The hallmarks of successful anticancer immunotherapy. *Science Translational Medicine*, *10*(459). https://doi.org/10.1126/scitranslmed.aat7807.

Hayakawa, Y., Takeda, K., Yagita, H., Smyth, M. J., Van Kaer, L., Okumura, K., et al. (2002). IFN-gamma-mediated inhibition of tumor angiogenesis by natural killer T-cell ligand, alpha-galactosylceramide. *Blood*, *100*(5), 1728–1733.

Hsu, J., Hodgins, J. J., Marathe, M., Nicolai, C. J., Bourgeois-Daigneault, M. C., Trevino, T. N., et al. (2018). Contribution of NK cells to immunotherapy mediated by PD-1/PD-L1 blockade. *The Journal of Clinical Investigation*, *128*(10), 4654–4668. https://doi.org/10.1172/JCI99317.

Lopez-Soto, A., Folgueras, A. R., Seto, E., & Gonzalez, S. (2009). HDAC3 represses the expression of NKG2D ligands ULBPs in epithelial tumour cells: Potential implications for the immunosurveillance of cancer. *Oncogene*, *28*(25), 2370–2382. https://doi.org/10.1038/onc.2009.117.

Lopez-Soto, A., Gonzalez, S., Smyth, M. J., & Galluzzi, L. (2017). Control of metastasis by NK cells. *Cancer Cell*, *32*(2), 135–154. https://doi.org/10.1016/j.ccell.2017.06.009.

Lopez-Soto, A., Huergo-Zapico, L., Acebes-Huerta, A., Villa-Alvarez, M., & Gonzalez, S. (2015). NKG2D signaling in cancer immunosurveillance. *International Journal of Cancer*, *136*(8), 1741–1750. https://doi.org/10.1002/ijc.28775.

Lorenzo-Herrero, S., Lopez-Soto, A., Sordo-Bahamonde, C., Gonzalez-Rodriguez, A. P., Vitale, M., & Gonzalez, S. (2018). NK cell-based immunotherapy in cancer metastasis. *Cancers (Basel)*, *11*(1), 29. https://doi.org/10.3390/cancers11010029.

Nouri-Shirazi, M., Banchereau, J., Bell, D., Burkeholder, S., Kraus, E. T., Davoust, J., et al. (2000). Dendritic cells capture killed tumor cells and present their antigens to elicit tumor-specific immune responses. *Journal of Immunology*, *165*(7), 3797–3803.

Oyer, J. L., Gitto, S. B., Altomare, D. A., & Copik, A. J. (2018). PD-L1 blockade enhances anti-tumor efficacy of NK cells. *Oncoimmunology*, 7(11). e1509819https://doi.org/10.1080/2162402X.2018.1509819.

Pai, C. S., Huang, J. T., Lu, X., Simons, D. M., Park, C., Chang, A., et al. (2019). Clonal deletion of tumor-specific T cells by interferon-gamma confers therapeutic resistance to combination immune checkpoint blockade. *Immunity*, *50*(2), 477-492.e8. https://doi.org/10.1016/j.immuni.2019.01.006.

Samstein, R. M., Lee, C. H., Shoushtari, A. N., Hellmann, M. D., Shen, R., Janjigian, Y. Y., et al. (2019). Tumor mutational load predicts survival after immunotherapy across multiple cancer types. *Nature Genetics*, *51*(2), 202–206. https://doi.org/10.1038/s41588-018-0312-8.

Su, X., Yu, Y., Zhong, Y., Giannopoulou, E. G., Hu, X., Liu, H., et al. (2015). Interferon-gamma regulates cellular metabolism and mRNA translation to potentiate macrophage activation. *Nature Immunology*, *16*(8), 838–849. https://doi.org/10.1038/ni.3205.

Villa-Alvarez, M., Sordo-Bahamonde, C., Lorenzo-Herrero, S., Gonzalez-Rodriguez, A. P., Payer, A. R., Gonzalez-Garcia, E., et al. (2018). Ig-like transcript 2 (ILT2) blockade and lenalidomide restore NK cell function in chronic lymphocytic leukemia. *Frontiers in Immunology*, *9*, 2917. https://doi.org/10.3389/fimmu.2018.02917.

Wang, R., Jaw, J. J., Stutzman, N. C., Zou, Z., & Sun, P. D. (2012). Natural killer cell-produced IFN-gamma and TNF-alpha induce target cell cytolysis through up-regulation of ICAM-1. *Journal of Leukocyte Biology*, *91*(2), 299–309. https://doi.org/10.1189/jlb.0611308.

CHAPTER NINETEEN

Applications of microfluidic devices in advancing NK-cell migration studies

Xiaoou Ren[a,b], Abdulaziz Alamri[c], Jolly Hipolito[a,c], Francis Lin[a,b,c,*], Sam K.P. Kung[c,*]

[a]Department of Physics and Astronomy, University of Manitoba, Winnipeg, MB, Canada
[b]Department of Biosystems Engineering, University of Manitoba, Winnipeg, MB, Canada
[c]Department of Immunology, University of Manitoba, Winnipeg, MB, Canada
*Corresponding authors: e-mail address: francis.lin@umanitoba.ca; sam.kung@umanitoba.ca

Contents

Methods in Enzymology, Volume 631
ISSN 0076-6879
https://doi.org/10.1016/bs.mie.2019.05.052

Abstract

Understanding how NK cells interact with tumor cells under specific microenvironment will be informative in development of NK-cell based immunotherapy. Applications of microfluidic devices in in vitro studies of NK-cell migrations offer unique opportunities to examine NK-cell migrations at single-cell under controlled cellular microenvironments. Novel devices can be created and engineered to present precise configuration that mimics cellular microenvironments for cell migration studies. We established previously the first application of a simple Y-shaped device for imaging and analysis of the abilities of the immature and mature DC to regulate murine IL-2 activated NK cell migrations. Here we reported the application of our recent technical development of a novel microfluidic device, which is also called the triple docking device (i.e., D^3-Chip), for the studies of NK-cell migrations in NK-4T1 breast cancer cell interactions in vitro. Key features of this microfluidic device are its pump-free gradient generation, and the three-parallel units design that supports easy setup and parallel comparison of multiple experimental conditions. The cell docking structure enables the prealignment of all NK cells at the same "start" position before their exposures to the test conditions. As a result, quantification of cell displacement toward a chemical gradient can be quantified by enumeration of the number of cells migrated out of the docking structure and their displacements. Such microfluidic devices can be further modified in future to mimic the complex in vivo microenvironments to support more advanced investigations of NK-cell migratory responses in vitro.

1. Introduction

Classical natural killer (NK) cells are bone marrow derived cells that constitute 10–15% of blood lymphocytes (Campbell & Hasegawa, 2013). Unlike T cell, individual NK cell expresses a repertoire of surface receptors that can either be activating or inhibitory upon engagement of their cognate ligands. Functional activation of a NK cell thus depends on the balance of the activation and inhibitory signals it received (Campbell & Hasegawa, 2013; Mahmood, Kanwar, Tran, Zhang, & Kung, 2012). Activated NK cells can induce direct cytotoxic killing of target cells, and/or production of soluble chemokines and cytokines (such as IFN-γ, TNF, and GM-CSF). These effector functions are critical in inducing innate and adaptive immunity in infections and cancer surveillance (Campbell & Hasegawa, 2013).

Unlike the NK-cell mediated cellular cytotoxicity and cytokine-production effector activities, migratory properties of NK cells in different experimental conditions are, in general, relatively less explored. NK cells indeed take clues from various microenvironments to regulate their migratory responses to traffic to peripheral tissues or inflamed lymph nodes to exert their immune-surveillance functions (Campbell & Hasegawa, 2013; Cella, Miller, & Song, 2014). They acquire specific chemokine receptors and nonchemokine family proteins to regulate their preferential migrations to lymph nodes, liver, spleen or inflammatory sites (Bernardini, Sciume, & Santoni, 2013; Pachynski et al., 2012). In NK-dendritic cell (DC) crosstalk (Ferlazzo & Morandi, 2014), we reported soluble factors produced by the immature or LPS-activated mature DCs would differentially regulate chemotactic abilities of the activated NK cells (Alamri et al., 2018; Mahmood, Nandagopal, Sow, Lin, & Kung, 2014). In tumor microenvironments, high level of CX3CL1 expression by breast cancer cells supported recruitment of tumor-infiltrating CD8+ T cells, NK and DC cells (Park, Lee, & Yoon, 2012). Higher level of recruitment of these immune cell types in the tumor was correlated with a good prognosis for disease-free and overall survival. Forced expression of CKb-11 chemokine in breast cancer cell line recruited NK cells to the tumor site and promoted tumor rejections (Braun et al., 2000). The underlying mechanism that determines high vs low infiltration of NK cells into tumor tissues, however, remains to be elucidated in details. It is possible that cancer cells produce factor(s) that prevent the infiltration of NK cells into the tumor environment. Studying NK cell migrations in conditions that mimics tumor microenvironments will therefore provide important insights into our understanding of NK cell immunosurveillance of cancer and in the development of NK cell based immunotherapy.

Microfluidic devices offer unique features that support studies of NK-cell migrations in vitro. Novel devices can be created and engineered to present precise configuration that mimics cellular microenvironments for cell migration studies (Li & Lin, 2011; Lin & Butcher, 2006; Wu, Wu, & Lin, 2013). Live-cell imaging supports examination of these migrating cells at single-cell level under controlled cellular microenvironments (Ren, Levin, & Lin, 2017). Migratory properties (such as speed, migration displacement, and chemotactic index) at the single cell level can be visualized and quantified. We have previously established the first application of a simple Y-shaped device for imaging and analysis of the abilities of the immature and mature DC to regulate murine IL-2 activated NK cell migrations in vitro (Mahmood et al., 2014). Such simple "Y"-type design has two fluidic inlets and a main gradient channel (Mahmood et al., 2014).

Two major limitations of the Y-shaped microfluidic devices are: relatively low throughput and the inability to support cell-cell interaction studies in vitro. Here we reported the application of our recent technical development of a novel microfluidic device (D^3-Chip) in the studies of NK-cell migrations in NK-cancer cell interactions in vitro. The D3-Chip, comprised of three independent microfluidic units (each contains its own set of chemical inlets to create a main gradient channel, cell docking inlet and structures, and waste outlet), is an improved version of the well-established Y-shaped device (Fig. 1A).

2. Preparation of microfluidic devices

2.1 Design of microfluidic devices

A previously designed D^3-Chip (Fig. 1A) (Yang et al., 2017) was applied in this study. This chip was designed based on a previous single-unit device (Wu et al., 2016) with the improved experimental throughput by coordinating three independent test units to fit into a combined view zone. All the parameters of this device were optimized to allow real-time imaging of all test units without the need of changing the stage. When needed, a programmable motorized microscope stage is used to support imaging at different positions. Each independent unit has its own corresponding chemical inlets (2, for the creation of a chemical gradient), cell inlet and outlet. The device therefore supports parallel analyses of three independent cell migrations experiments under different chemical conditions. The D^3-Chip contains a special cell docking area ensures the initial alignment of cells adjacent to a thin barrier region (Fig. 1B). As the size of the loaded cells is larger than the barrier, the cells are "trapped" in the dock before a chemical gradient of interest is applied. Upon stimulation, the cells polarize and squeeze beneath the thin barrier region (3 μm high; 20 μm wide), and migrate toward the gradient direction in the main gradient channel (50 μm high; 220 μm wide). Unique features of this improved microfluidic device include: First, the cell alignment feature better controls the initial positions of all cells relative to the gradient at the beginning, thus allowing precise measurement of relevant parameters that involved in the cell migration and chemotaxis studies. Second, cell migrations through the barrier region stimulated by the chemical gradient partially mimic the process of transmigrated cells to the inflammatory tissues via blood vessels. Third, cell loading and gradient generation applied with all three independent units are fast and consistent, and the time interval between the setup of each unit is negligible. All the advanced features of D^3-Chip are integrated to ensure reliable results extracted from cell migration experiments.

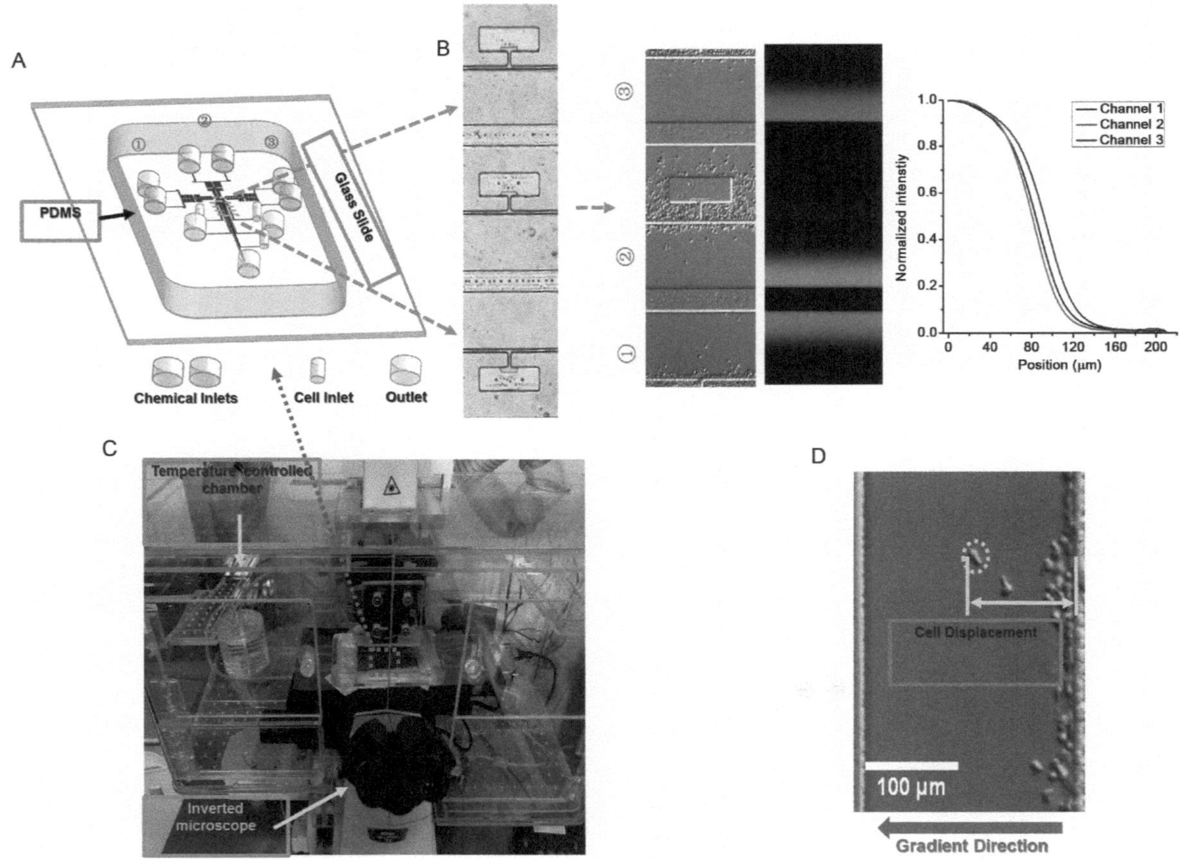

Fig. 1 See figure legend on next page.

2.2 Fabrication of microfluidic devices

The D^3-Chip was achieved by multiheight channels fabricated by two layers using the previously published standard photolithography and soft lithography methods (Yamada et al., 2016; Yang et al., 2017). Detailed fabrication process was conducted in the Nano Systems Fabrication Laboratory (NSFL) at the University of Manitoba. The device feature was designed by SolidWorks software (Dassault Systemes S.A.), and printed to a transparency film at a high resolution (i.e., 24,000 dpi, Fineline Imaging) served as the photomask. The microfeature was negatively patterned onto an oxygen plasma pretreated silicon wafer (Silicon, Inc., ID) by employing UV exposure through the photomask on top of the precoated SU-8 photoresist (MicroChem) layer. The wafer with patterned features was used as the mold to reproduce polydimethylsiloxane (PDMS) (Sylgard 184, Dow Corning) replicas. The PDMS replica was cut off after 2 h of baking in an oven at 80 °C, all the inlets and outlets were punched with proper size. The surface of PDMS replica was cleaned by tape and bonded onto a new glass slide by using an air plasma cleaner (Harrick Plasma). The PDMS channels filled with DI water inside a 4 °C fridge can be kept up to a few days before use.

2.3 Substrate preparation

Rat tail collagen type I (Coring) is used for device coating, to provide a substrate for cell migration. Due to the microscale of the channel, a high concentration of the coating solution is employed to ensure the proper substrate surface density. In our experiments, 20 μg/mL of collagen solution

Fig. 1 Illustration of the D^3-Chip microfluidic system used in the migration study. (A) Schematic illustration of the D^3-Chip. It is comprised of three independent microfluidic units. Each contains its own set of chemical inlets (two inlets to generate a stable chemical gradient), a cell docking inlet and structures, and a waste outlet. (B, left) Actual image of part of the observation zone to show the alignment of the three parallel and independent microfluidic channels (channel 1, 2, 3) created. Each one has its cell docking site and a test channel of defined chemical gradient for analysis of NK-cell migration. (B, middle) An image to show the "docking" of NK cells at a cell docking structure, and the fluorescent images of the gradient profile of each test channel. Fluorescent intensity at each position of the test channel was quantified. (B, right) The profile plots of the gradient in each test channel of the D^3-Chip were consistent, demonstrating the reproducibility of the gradient formation in all three test channels under the D^3-Chip; (C) An image of the microscopic system used in the imaging in the microfluidic cell migration assay; (D) A representative image obtained under the D^3-Chip to demonstrate the imaging of a migrating NK cell at the end of the assay and an illustration of cell migration displacement analysis.

is used for coating the D^3-Chip for 1 h in an incubator at 37 °C. The collagen-coated device is then blocked with completed culture medium for another 30 min in the incubator before cell migration experiment.

3. Generation of chemoattractant gradients

3.1 Principle of microfluidic gradient generation

The D^3-Chip generates a chemoattractant gradient by mixing continuous laminar flows of target chemicals in the microchannels (Lin & Butcher, 2006). The chemical gradient generation of each test unit is rapid, stable, independent, controlled, and pump-free. Depending on the pressure difference of the volumes between the inlet and outlet solutions, the flows are driven from chemical inlets through upstream serpentine channels and downstream main gradient channels to the corresponding outlets. The upstream serpentine channels balance the flows and decrease the flow speed, which ensure a stable gradient generated within a few minutes inside the main gradient channels (Fig. 1B). The D^3-Chip is capable of generating single or overlapped gradients, and rapid gradient switching.

3.2 Measurement of chemo-attractant gradients in microfluidic devices

Fluorescent additives are added into the target chemo-attractant solutions for visualizing and measuring the profile of chemo-attractant gradient in the microchannel using a standard fluorescence microscope. Here, we used FITC-dextran (10 kDa) based on the similar molecular weight of the most chemokine molecules. Typically, <2% of the dye to the total volume is added to the test chemo-attractant preparation, to minimize its effect on the migratory behaviors of cells during experiment. We showed that gradient generation of all three test units in the D^3-Chip are consistent and the gradient profile is stable for at least 1 h, which was sufficient for NK-cell migration experiments (Fig. 1B).

4. Preparations of cells

4.1 Primary NK cells

We isolated primary NK cells from murine splenocytes (either C56BL/6 or BALB/c) using the EasySep mouse NK negative selection kit (StemCell Technologies, Vancouver, BC), according to the manufacturer's protocol. To obtain activated NK cells, we cultured these isolated NK cells at 37 °C and 5% CO_2 in mouse medium containing 10% fetal bovine serum (FBS from Hyclone), 1% PSG (Invitrogen), 1.6mmol/L 2-ME, in the presence of either IL-2 (1000U/mL) or IL-15 (20ng/mL) for 4 days. The purity of the activated NK cells was routinely >90% by flow cytometry (Mahmood et al., 2014).

4.2 Preparation of dendritic cells conditioned medium

Murine bone marrow cells were collected from the femur and tibia, incubated with ACK buffer for 2-min to lyse red blood cells (Alamri et al., 2018). They were seeded ($0.5–1 \times 10^6$ in 1mL per well) in a 24-well plate containing RPMI 1640 (Hyclone) medium supplemented with 1% PSG, 10% FGS, 1.6mmol/L 2-ME and 20ng/mL GM-CSF (Peprotech). On day 3, one-third of the culture medium was replaced by fresh GM-CSF containing complete culture medium to remove nonadherent cells. On day 5, the DC cultures were again replenished with fresh GM-CSF containing complete culture medium to generate immature DC. On day 7, lipopolysaccharide (LPS at 1 μg/μL final concentration) was added to these immature DC cultures for 24-h to obtain LPS-activated mature DC. All the immature and LPS-activated mature DC cells were confirmed by the maturation markers (CD40, CD80, CD86) in flow cytometry (Alamri et al., 2018). Cell-free conditioned medium from either immature DC or LPS-activated mature DC were collected and used in the generation of the chemical gradients in the studies of NK-cell migrations.

4.3 Culture of 4T1 breast cancer tumor cells

Mouse 4T1 cell line (ATCC CRL-2539) produces primary and metastatic breast tumors which resemble stage IV human breast cancer, when injected into BALB/c mice (Pulaski & Ostrand-Rosenberg, 2001). The cells were maintained in complete RPMI1640 Medium supplemented with 10% (final, v/v) fetal bovine serum according to the ATCC culture protocol. Cells are maintained at subconfluence conditions. Adherent cells are detached using 0.25% trypsin, 0.53mM EDTA solution.

5. Experimental setup

5.1 Cell loading

The culture flask contains 4T1 cells was treated with Trypsin-EDTA solution (0.25%, Sigma-Aldrich), and then put inside an incubator (37°C; 5% CO_2)

a few minutes for cell detachment. The 4T1 cells were washed and resuspended in the complete RPMI 1640 culture medium before seeding. Ideally, 200,000 cells in 20 μL volume of the complete culture medium were prepared, loaded into the outlets of all three test units, followed by balancing the corresponding chemical inlets with proper volume of the culture medium in order to make enough cells suspended within the main gradient channels. The chip was then incubated inside the incubator for at least 1 h for the attachment of 4T1 cells. Activated NK cells (either IL-2 or IL-15-activated NK cells) were then collected from the multiwell plate and resuspended in the complete RPMI 1640 culture medium before loading. Typically, 100,000 cells in 10 μL volume of the culture medium were injected inside each cell loading inlets.

5.2 Chemo-attractant preparation and injection

In this study, the conditioned medium of the LPS-activated mature dendritic cells (mDCs) was used as the chemo-attractant (Alamri et al., 2018; Mahmood et al., 2014). Plain RPMI 1640 culture medium supplemented with 10% fetal bovine serum (FBS from Hyclone), 1% PSG (Invitrogen), 1.6 mmol/L 2-ME, was used as the medium control. In the chemo-attractant group, conditioned medium of mDCs and culture medium were injected into the two chemical inlets, respectively, to generate the gradient. In the medium control group, both chemical inlets were injected with the medium control. FITC-Dextran 10 kDa was added to these gradients to confirm the gradient profiles as we described previously.

5.3 Environmental control and image system

A live cell image system that consists of an inverted microscope (Nikon Ti-U) and a temperature-control enclosure chamber (InVivo Scientific) is employed for cell migration experiments (Fig. 1C). The microfluidic device is placed on a programmable motorized stage under the microscope for cell migration imaging. A temperature-controlled chamber encloses the entire microscope stage to maintain a temperature at 37 °C during the experiments.

5.4 Time-lapse optical microscopy

After the desirable temperature at 37 °C was reached and the activated NK cells were loaded, a visualization field was chosen in the main gradient channel with a 10× objective. Differential interference contrast (DIC) images are recorded in a time-lapse manner (10 s interval, 40 min long) during cell migration experiment. The gradient was checked both before and after the experiment.

6. Data analysis

6.1 Image processing

A single D^3-Chip supported analyses of three independent cell migration studies ran in parallel and independently. The experimental conditions tested under these three migration studies in a single D^3-Chip could be either identical (as intra-assay repeats) or different. For examples, we could perform a single D^3-Chip to examine NK-cell migrations under control medium, conditioned medium from immature DC culture and conditioned medium from LPS-activated mature DC culture). Under such designs, each experimental condition would be repeated independently at least two more times under separate D^3-Chips. The time-lapse images were extracted from each set of the experiments, then compared and analyzed within different conditions based on the same device (Fig. 2A).

6.2 Characterization of cell displacement

NK cells that migrated >50 μm (~5 times of the cell size) to the gradient direction within the observation zone by the end of each experiment (40 min) were located, and NIH ImageJ was applied for calibrating the total displacement of each targeted cell in the gradient direction. Quantitative parameters including the number and displacement (Fig. 1D) of migrated cells in the images were extracted, which was presented in the Box Chart (Fig. 2A and B) by using OriginPro software (ver. 2017). Higher migration displacement of activated NK cells toward the gradient of the mDC conditioned medium were observed when compared to the medium control group (Fig. 2A and B).

6.3 Observation of cell-cell interactions and cell tracking

Detailed information of the interactions between 4T1 and the activated NK cells can be further extracted from the time-lapse images. The migration paths of the activated NK cells were tracked by the cell tracking plug-in in NIH ImageJ. In one scenario, we observed multiple NK cells interacted with one 4T1 cell (Fig. 2C). Some of them had firm interactions with the 4T1 cell, and their contact can last for up to 30 min; others had transient interactions with the 4T1 cell, and "left" the 4T1 cells after a few minutes of cell-cell interactions. In another scenario, one NK cell interacted with multiple 4T1 cells during its migration (Fig. 2D). Some interactions can last

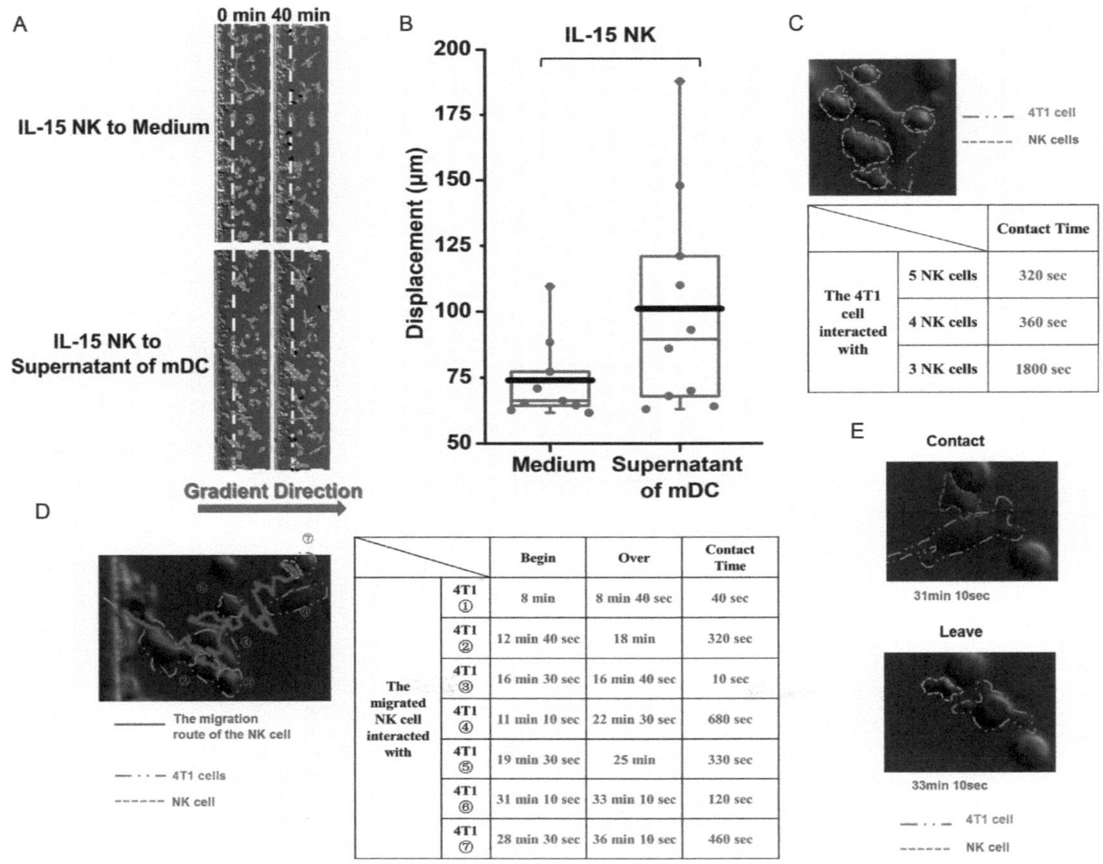

		Contact Time
The 4T1 cell interacted with	5 NK cells	320 sec
	4 NK cells	360 sec
	3 NK cells	1800 sec

		Begin	Over	Contact Time
The migrated NK cell interacted with	4T1 ①	8 min	8 min 40 sec	40 sec
	4T1 ②	12 min 40 sec	18 min	320 sec
	4T1 ③	16 min 30 sec	16 min 40 sec	10 sec
	4T1 ④	11 min 10 sec	22 min 30 sec	680 sec
	4T1 ⑤	19 min 30 sec	25 min	330 sec
	4T1 ⑥	31 min 10 sec	33 min 10 sec	120 sec
	4T1 ⑦	28 min 30 sec	36 min 10 sec	460 sec

Fig. 2 See figure legend on next page.

for a few minutes while others were only a few seconds. In most circumstances, the shape of 4T1 cells continuously changed during the interaction with NK cells, and in some cases significantly "shrunk" after contacting NK cells (Fig. 2E).

6.4 Statistics analysis

Two sample Student's *t*-test available in the OriginPro software (ver. 2017) was used in the statistical analyses of two data groups.

7. Conclusion

Here we reported the application of the D^3-Chip in the live-cell imaging of directed NK-cell migrations and NK-4T1 tumor cell interactions in the controlled gradient environments in vitro. As a proof-of-principle study, we demonstrated that feasibility of studying the number and contact time of NK cells interacting with a cancer cell (Fig. 2C), the tracking of the migration path and contact time of an individual NK cell (Fig. 2D), and the change in cancer cell morphology at the end of the NK-cell interaction (Fig. 2E).

Fig. 2 Representative data to illustrate how measurements of NK-cell migration and cell-cell interaction were extracted from the microfluidic system. (A) IL-15 activated NK cells (black) and 4T1 breast cancer cells (gray) were loaded in the microfluidic device as described. Migrations of the activated NK cells toward the 4T1 cells were compared using gradient generated from either the conditioned medium from LPS-activated mature DC (supernatant of mDC) or the medium control. Representative images of the 4T1-NK cells interactions in the D^3-Chip under different conditions at 0 and 40 min were shown. The white dashed box indicated the section where 4T1 and NK cells were compared; (B) Comparison of the displacement of the IL-15 activated NK cells in either medium control or supernatant of mDC at 40 min. The box chart showed the total displacement of each cell (only cells with displacement >50 μm were included for analysis) in different experimental groups; the bottom and top of the red whisker showed the minimum and the maximum value; the red box shows the migratory cells within the range from 25% to 75% of total cells based on the ranked displacement value; the black bold line indicated the mean displacement; the red line inside the box indicated the median displacement value; (C) The top image showed 5 IL-2 activated NK cells interacting with 1 4T1 cell in a magnified image. The bottom table showed the detailed information about these interactions; (D) The left image showed one migrated IL-2 activated NK cell interacting with multiple 4T1 cells (no. 1–7) in a magnified image. The right table showed the detailed information about these interactions. The dark bold line indicated the migration route of the NK cell. The arrow indicated the start-point; (E) Two images showing the start-point and endpoint of 1 IL-2 activated NK cell interacting with 1 4T1 cell, respectively. The dashed line and dotted line indicated the 4T1 cell and the activated NK cells, respectively, in (C–E).

This microfluidic based approach will therefore support quantitative analyses and comparative studies of different NK cell preparations under a defined chemical gradient (such as conditioned medium from LPS activated mature DC), or conditions that mimic tumor cell microenvironment. This platform will therefore be useful in further our understanding NK-cell migratory properties and tumor cell interactions under different experimental conditions (such as detailed comparisons of different cytokine-activated NK cells, and/or conditioned medium from cancer cells at different time-points). In addition, we observed that conditioned medium from LPS-activated DC promoted activated NK cell migrations toward 4T1 cancer cells (Fig. 2A and B). The latter may suggest the beneficial effect of activating DC at tumor site to promote NK-cell recruitment and its antitumor activity. In future, such microfluidic device could be further modified to mimic the better complex in vivo microenvironment (Zheng et al., 2016) to support more advanced investigations of NK-cell migratory responses under physiologically or pathologically relevant conditions.

Acknowledgments

S.K.P. Kung thanks the support from the Natural Sciences and Engineering Research Council Discovery Grant (RGPIN-2015-04144), and from an Innovation grant from Canadian Cancer Society Research Institute. F. Lin thanks the support from the Natural Sciences and Engineering Research Council of Canada (NSERC) for a Discovery Grant. X. Ren thanks the Graduate Enhancement of Tri-Council Stipends (GETS) program at the University of Manitoba for financial support. A.A. is supported by a scholarship from King Saud University. We thank Ke Yang and the NSFL at the University of Manitoba for their technical support of microfluidic device development.

References

Alamri, A., Rahman, R., Zhang, M., Alamri, A., Gounni, A. S., & Kung, S. K. P. (2018). Semaphorin-3E produced by immature dendritic cells regulates activated natural killer cells migration. *Frontiers in Immunology*, *9*, 1005.

Bernardini, G., Sciume, G., & Santoni, A. (2013). Differential chemotactic receptor requirements for NK cell subset trafficking into bone marrow. *Frontiers in Immunology*, *4*, 12.

Braun, S. E., Chen, K., Foster, R. G., Kim, C. H., Hromas, R., Kaplan, M. H., et al. (2000). The CC chemokine CK beta-11/MIP-3 beta/ELC/Exodus 3 mediates tumor rejection of murine breast cancer cells through NK cells. *Journal of Immunology*, *164*, 4025–4031.

Campbell, K. S., & Hasegawa, J. (2013). Natural killer cell biology: An update and future directions. *The Journal of Allergy and Clinical Immunology*, *132*, 536–544.

Cella, M., Miller, H., & Song, C. (2014). Beyond NK cells: The expanding universe of innate lymphoid cells. *Frontiers in Immunology*, *5*, 282.

Ferlazzo, G., & Morandi, B. (2014). Cross-talks between natural killer cells and distinct subsets of dendritic cells. *Frontiers in Immunology*, *5*, 159.

Li, J., & Lin, F. (2011). Microfluidic devices for studying chemotaxis and electrotaxis. *Trends in Cell Biology*, *21*, 489–497.

Lin, F., & Butcher, E. C. (2006). T cell chemotaxis in a simple microfluidic device. *Lab on a Chip*, *6*, 1462–1469.

Mahmood, S., Kanwar, N., Tran, J., Zhang, M. L., & Kung, S. K. (2012). SHP-1 phosphatase is a critical regulator in preventing natural killer cell self-killing. *PLoS One*, 7, e44244.

Mahmood, S., Nandagopal, S., Sow, I., Lin, F., & Kung, S. K. (2014). Microfluidic-based, live-cell analysis allows assessment of NK-cell migration in response to crosstalk with dendritic cells. *European Journal of Immunology*, *44*, 2737–2748.

Pachynski, R. K., Zabel, B. A., Kohrt, H. E., Tejeda, N. M., Monnier, J., Swanson, C. D., et al. (2012). The chemoattractant chemerin suppresses melanoma by recruiting natural killer cell antitumor defenses. *The Journal of Experimental Medicine*, *209*, 1427–1435.

Park, M. H., Lee, J. S., & Yoon, J. H. (2012). High expression of CX3CL1 by tumor cells correlates with a good prognosis and increased tumor-infiltrating CD8+ T cells, natural killer cells, and dendritic cells in breast carcinoma. *Journal of Surgical Oncology*, *106*, 386–392.

Pulaski, B. A., & Ostrand-Rosenberg, S. (2001). Mouse 4T1 breast tumor model. *Current Protocols in Immunology*, *39*, Chapter 20, Unit, 20.2.1–20.2.16.

Ren, X., Levin, D., & Lin, F. (2017). Cell migration research based on organ-on-chip-related approaches. *Micromachines (Basel)*, *8*, 324.

Wu, J., Hillier, C., Komenda, P., Lobato de, F. R., Santos, S., Levin, D., et al. (2016). An all-on chip method for testing neutrophil chemotaxis induced by fMLP and COPD patient's sputum. *Technology*, *4*, 104–109. S2339547816500035.

Wu, J., Wu, X., & Lin, F. (2013). Recent developments in microfluidics-based chemotaxis studies. *Lab on a Chip*, *13*, 2484–2499.

Yamada, A., Renault, R., Chikina, A., Venzac, B., Pereiro, I., Coscoy, S., et al. (2016). Transient microfluidic compartmentalization using actionable microfilaments for biochemical assays, cell culture and organs-on-chip. *Lab on a Chip*, *16*, 4691–4701.

Yang, K., Peretz-Soroka, H., Wu, J., Zhu, L., Cui, X., Zhang, M., et al. (2017). Fibroblast growth factor 23 weakens chemotaxis of human blood neutrophils in microfluidic devices. *Scientific Reports*, 7, 3100.

Yang, K., Wu, J., Zhu, L., Liu, Y., Zhang, M., & Lin, F. (2017). An all-on-chip method for rapid neutrophil chemotaxis analysis directly from a drop of blood. *Journal of Visualized Experiments: JoVE*, *124*, e55615.

Zheng, F., Fu, F., Cheng, Y., Wang, C., Zhao, Y., & Gu, Z. (2016). Organ-on-a-chip systems: Microengineering to biomimic living systems. *Small*, *12*, 2253–2282.

CHAPTER TWENTY

Complementary approaches to study NKT cells in cancer

Jay A. Berzofsky[a,*], Purevdorj B. Olkhanud[a], Masaki Terabe[b]
[a]Vaccine Branch, Center for Cancer Research, National Cancer Institute, National Institutes of Health, Bethesda, MD, United States
[b]Neuro-Oncology Branch, Center for Cancer Research, National Cancer Institute, National Institutes of Health, Bethesda, MD, United States
*Corresponding author: e-mail address: berzofsj@mail.nih.gov

Contents

Abstract

NKT cells are a small but influential member of the T cell family, recognizing lipids presented by the non-classical MHC-like molecule CD1d rather than peptides presented by classical MHC molecules. They bridge between the innate and adaptive immune systems, serving as rapid responders but also allowing the T cell immune system to recognize lipid antigens, for example derived from tumors or bacteria. They also serve as potent regulatory cells, controlling other immune responses. Type I NKT cells use

Methods in Enzymology, Volume 631
ISSN 0076-6879
https://doi.org/10.1016/bs.mie.2019.08.010

a semi-invariant T cell receptor (TCR) whereas type II use diverse TCRs. Most often, type I NKT cells promote tumor immunity whereas type II tend to suppress it, and the two subtypes crossregulate each other, forming an immunoregulatory axis. Lack of tools to study these important cells has limited the understanding of these, but newer tools have allowed great advances, especially in mouse models. These range from transgenic and knock-out mice to CD1d tetramers carrying ligands for type I or II NKT cells, to antibodies and NKT cell hybridomas. Here we describe these complementary tools and approaches and their use to study NKT cells and their role in the immunology and immunotherapy of cancer.

1. Introduction

NKT cells are a specialized population of T cells that recognize lipid or glycolipid antigens instead of peptides, presented by a non-classical class I MHC-like molecule, CD1d. They bridge between innate and adaptive immunity in that they are early responders like innate cells (Van Kaer, 2007), with constitutive expression of IL-4 and interferon-gamma mRNA that can produce cytokines rapidly, but also contribute to the adaptive immune system by providing the cellular arm with a way to recognize lipid antigens rather than just proteins and their fragments. They also play both effector and regulatory roles. As such, they are of considerable importance in many neoplastic, infectious, and autoimmune diseases.

NKT cells are frequently misunderstood because of their name, which derives from a historical observation of a population of T cells or thymocytes with diminished levels of CD3 expression but with NK markers like NK1.1 (Arase, Arase, Nakagawa, Good, & Onoe, 1993; Arase, Arase, Ogasawara, Good, & Onoe, 1992; Bendelac, Killeen, Littman, & Schwartz, 1994; Levitsky, Golumbek, & Pardoll, 1991). They were subsequently found to use a semi-invariant TCRα chain found in a set of hybridomas, defined as Vα14-Jα18 in mice and Vα24-Jα18 in humans, along with a limited repertoire of Vβ chains, Vβ8, 2, and 7 in mice and Vβ11 in humans (Kronenberg & Gapin, 2002; Lantz & Bendelac, 1994; Taniguchi et al., 1996; Taniguchi, Harada, Kojo, Nakayama, & Wakao, 2003). Also, they were found to recognize glycolipid antigens, as represented by the prototype alpha-galactosylceramide (αGalCer), originally isolated from a marine sponge (Kawano et al., 1997) but now thought to probably derive from a Sphingomonas bacterium symbiotic to the sponge (Kinjo et al., 2005, 2008). Further, it was found that they recognized these antigens presented not by a conventional MHC molecule but by a non-conventional class I-like molecule CD1d (Bendelac et al., 1995), which was non-polymorphic. The lack of CD1d polymorphism explained the conservation of the TCR across mice of many MHC haplotypes.

However, the name Natural Killer T cells, due to the NK markers on true $CD3^+$ T cells, rapidly became problematic. First, NK1.1 (homologous to human CD161) is not expressed in all strains of mice and so was not a universally useful marker, and also is an activation marker upregulated when conventional T cells are activated. Also, some T cells with other NKT characteristics, CD1d restriction and αGalCer recognition, did not express NK1.1. Thus, the definition was changed by a consortium of NKT investigators to be any true (TCR^+) T cell that recognizes lipids presented by CD1d, regardless of NK marker expression (Godfrey, MacDonald, Kronenberg, Smyth, & Van Kaer, 2004). However, the confusing name persisted and has led to misunderstanding among some investigators and continued use of NK markers to identify these cells in some studies, yielding results that cannot be interpreted in terms of the new definition (Berzins, Smyth, & Baxter, 2011). Given their shared specificity for a specific antigen or family of antigens related to αGalCer, this small population (typically about 1% in mouse spleen, up to 15–30% in mouse liver, much more variable but lower in human peripheral blood mononuclear cells, and about <1% in human liver (Exley et al., 2002)) is actually quite large compared to the subset of conventional T cells specific for any single antigen.

Another complication was the discovery by Susanna Cardell and coworkers of a population of T cells recognizing lipids presented by CD1d that did not use the semi-invariant TCRα chain, and generally did not recognize αGalCer (Cardell et al., 1995; Makowska, Kawano, Taniguchi, & Cardell, 2000; Skold, Faizunnessa, Wang, & Cardell, 2000). By the new definition, these met the criteria for NKT cells but were of a different nature, with a much broader TCR repertoire. These were thus considered non-classical NKT cells, and a dichotomy was set up between two major classes of NKT cells, those utilizing the semi-invariant TCR called invariant NKT cells abbreviated iNKT cells, also known as type I NKT cells, and those utilizing other TCRs called non-invariant or type II NKT cells. The type I NKT cells almost all recognize αGalCer-CD1d as a prototype agonist, whereas type II NKT cells generally do not. Because the type II NKT cells have such a broad repertoire of TCRs, almost as broad as that of conventional T cells, it is assumed that they also have a broader repertoire of lipid antigens, but the bounds of that repertoire have not been defined, and it is not yet possible to determine what fraction of total type II NKT cells recognize any given antigen, as the denominator is not known.

Although some publications persist in calling any NK1.1$^+$ (in humans CD161$^+$) CD3$^+$ T cell an NKT cell, this definition can lead to many erroneous results, especially because, as noted earlier, any conventional T cell can upregulate NK1.1 and other NK markers when activated, and because many NKT cells are NK1.1 negative. A number of tools have been developed to more appropriately define and characterize NKT cells and their major divisions, type I and type II, and detect them in cell populations, especially in mouse models. This review will concentrate on those complementary approaches, and their applications in cancer settings.

2. Type I NKT cells

The major approaches to identify type I NKT cells are based on their primary defining characteristics, CD1d restriction, αGalCer recognition, and Vα14 (or Vα24)-Jα18 TCR utilization. Detailed protocols for the study of type I NKT cells in mice have also been published (Kumar, Bezbradica, Stanic, & Joyce, 2017).

2.1 CD1d KO mice

First, mice were made that were deficient in CD1d (CD1d KO) and as expected, these lacked NKT cells but retained all known subsets of conventional CD4 and CD8 T cells (Chen, Chiu, Mandal, Wang, & Wang, 1997; Mendiratta et al., 1997; Smiley, Kaplan, & Grusby, 1997). Thus, NKT cells could be defined as the cells that were uniquely missing in CD1d KO mice. This applied to both type I and type II NKT cells. Of course, defining cells by their absence is problematic, because even though the difference can be used to detect functions of NKT cells by looking for differences in function between wild type (WT) and CD1d KO mice, it cannot be used to identify individual NKT cells. Of note, mice have two CD1d genes on chromosome 3, *cd1d1 and cd1d2*. Although most CD1d KO mice have both genes deleted (Chen et al., 1997; Smiley et al., 1997), there is a strain of CD1d1 KO mice that retain CD1d2 (Mendiratta et al., 1997). A study with KO mice on the C57BL/6 genetic background showed that *cd1d2* has a frame shift mutation that makes *cd1d2* a pseudogene (Park, Roark, & Bendelac, 1998). However, a recent study suggest that *cd1d2* is intact in other strains of mice and contributes to the development of NKT cells (Sundararaj et al., 2018).

2.2 Jα18 KO mice

Second, utilizing the common use of the Jα18 segment of the TCRα chain (originally called Jα281) in type I NKT cells, KO mice were made that lacked this segment (Cui et al., 1997) and thus were selectively deficient in type I NKT cells. Unfortunately, a flaw in the original Jα18 KO mouse construction led to loss of some other Jα segments (Bedel et al., 2012), affecting the conventional TCR repertoire, so these mice have been newly re-made to lack only the Jα18 segment and thus be more uniquely deficient in type I NKT cells. It is recommended to use these newer mice, made by several labs and available from several sources, for all future studies (Chandra et al., 2015; Dashtsoodol et al., 2016; Ren et al., 2017; Zhang et al., 2016). This tool, however, has the same kind of limitation as the CD1d KO mouse, namely that defining cells by their absence in certain mice allows one to look at what functions are absent in those mice that can be attributed to type I NKT cells, but it does not provide a handle to identify individual type I NKT cells. Nevertheless, myriads of valuable studies have used these KO mouse tools in combination to identify functions of NKT cells and to distinguish between ones of type I NKT cells, missing in both CD1d KO and Jα18 KO mice, and those of type II NKT cells, absent in CD1d KO mice, but not Jα18 KO mice.

2.3 TCR Tg mice

The frequency of NKT cells among lymphocytes or leukocytes is very low in many tissues. Transgenic mice expressing the Vα14Jα18 semi-invariant TCR with a T cell deficient genetic background have an increased frequency of type I NKT cells to make it easier to study type I NKT cells (Bendelac, Hunziker, & Lantz, 1996; Kawano et al., 1997). A caveat of using type I NKT cells from those TCR Tg mice is that the promoter used to express the TCR is not the endogenous one. To overcome this issue, lately mice derived from embryonic stem (ES) cells with type I NKT cell nuclei or NKT cell induced pluripotent stem cells (iPS)-derived mice have been developed (Abu-Eid et al., 2014; Wakao et al., 2007) as described below. The low frequency of type I NKT cells in peripheral blood (0.01–2%) of humans also makes it difficult to study them. Even though CD1d and the semi-invariant TCR of type I NKT cells are highly conserved between mice and humans, there are still significant differences. Thus, mouse NKT cells cannot completely recapitulate human NKT cells. Using mice with human CD1d knock-in and Vα24 transgenic TCR may help to gain an understanding of the in vivo behavior of human type I NKT cells (Alvarez et al., 2016), although the TCR transgene can be improved.

2.4 iPS cells and ES cells

Moreover, iPS technology (Takahashi & Yamanaka, 2006) has been utilized to generate functional NKT cells in vitro from either mouse embryonic fibroblasts (MEF) or splenic NKT cells. These iPSC-derived NKT cells produced high levels of the Th1 cytokine IFN-γ without maturation of cell surface phenotype and importantly, suppressed tumor growth in vivo, fulfilling their anti-tumor activity (Watarai, Fujii, et al., 2010). Conversely, using mouse ES cells, a large number of cloned NKT (NKT-ES) cells have been successfully generated in vitro by the transfer of nuclei from mature NKT cells and these NKT-ES cells could mediate adjuvant activity in vivo and cause regression of tumor growth (Watarai, Rybouchkin, et al., 2010). Hence, iPS-derived NKT cells and NKT-ES cells provide new opportunities to study NKT cells in many ways; however, the immunogenicity of these cells should be further examined rigorously.

2.5 CD1d tetramers

The problem of just using gene knockout mice lacking CD1d or Jα18 to identify NKT cells was at least partially overcome when the tetramer technology (Altman et al., 1996) was applied to CD1d. Such MHC tetramers had originally been invented by constructing an enzymatically biotinylatable tail on the C-terminus of MHC class I heavy (alpha) chains, so that they could be biotinylated in only one precise spot. These could then be reacted with fluorescently labeled avidin or streptavidin, that has an almost irreversibly high affinity for biotin (K_D around 10^{-15} M), and that has four binding sites for biotin, forming a tetramer. These MHC tetramers could be loaded with specific peptides and then would bind with high enough affinity to TCRs specific for those peptide-MHC complexes to identify and stain specific T cells for flow cytometry. This revolution in T cell identification was translated to NKT cells by three groups who made CD1d-αGalCer tetramers to identify type I NKT cells (Benlagha, Weiss, Beavis, Teyton, & Bendelac, 2000; Karadimitris et al., 2001; Matsuda et al., 2000). These CD1d tetramers have finally made it possible to identify, study, and even purify by fluorescent-activated cell sorting (FACS), individual type I NKT cells from all strains of mice and even humans. The latter is true because CD1d is monomorphic so that mouse and human CD1d are highly cross-reactive, allowing mouse

CD1d tetramers to be used on human cells and vice versa. Now similar forms of multimers of CD1d that have higher avidity are commercially available.

2.6 NKT hybridomas and lines

Type I NKT cell hybridomas such as DN32.D3 and 24.8.A have also been developed that can be used to examine the specificity of type I NKT TCRs for different glycolipids such as analogues of αGalCer, natural lipids from bacteria and other organisms, and autologous mammalian lipids (Gumperz et al., 2000; Lantz & Bendelac, 1994). However, because the response is simply IL-2 production, as defined by the thymoma parent cell line used to make the hybridoma, they cannot be used to study NKT cell functions. Methods to expand primary type I NKT cells into long-term lines, using either artificial antigen-presenting cells or DCs, have also been published and can be used in this context (Chiba et al., 2009; Li, Tsuji, Schneck, & Webb, 2013; Webb, Bieler, Schneck, & Oelke, 2009).

2.7 Antibodies to TCRs

In addition, because the semi-invariant TCR in humans uses the Vα24 gene segment and the TCR pairs almost exclusively with Vβ11, the combination of antibodies to Vα24 and Vβ11 has been used to identify the type I NKT TCR (Lantz & Bendelac, 1994; Lee, Benlagha, Teyton, & Bendelac, 2002). Of note, using anti-Vα24 is not sufficient to identify type I NKT cells as the Vα24 gene segment is also used by conventional variant TCRα chains. There is even a monoclonal antibody with idiotype-like specificity for the semi-invariant TCR, 6B11 (Exley et al., 2008). In addition, one antibody, NKT14m, specific for the semi-invariant type I NKT TCR of mice, has been published that can have either agonist or antagonist activity, depending on the Fc portion used (Escriba-Garcia et al., 2019; Scheuplein et al., 2015). This antibody also has some limited anti-tumor activity, especially when combined with other agents (Escriba-Garcia et al., 2019).

2.8 Type I NKT subsets

Finally, subsets of type I NKT cells have been defined by their use of different transcription factors, PLZF, T-bet, and RORγt, as NKT1, 2, and 17, with different cytokine profiles (Lee et al., 2015; Lee, Holzapfel, Zhu, Jameson, & Hogquist, 2013). The number of functional subsets has subsequently been expanded even more (Cameron & Godfrey, 2018; Robertson, Berzofsky, & Terabe, 2014).

3. Type II NKT cells

3.1 KO mice

Analogous tools are also available for type II NKT cells, although in a not-completely symmetrical way. For example, whereas type I NKT cells are defined by their absence in both CD1d KO and Jα18 KO mice, type II NKT cells are defined by their absence in CD1d KO mice but presence in Jα18 KO mice. Unfortunately, there are no mice that uniquely lack type II NKT cells, so mice that selectively lack only type II NKT cells cannot be studied for their functionality.

3.2 CD1d-sulfatide tetramers

Type II NKT cells also lack the special feature of a unique semi-invariant TCR to define them (Behar & Cardell, 2000; Cardell et al., 1995; Kato, Berzofsky, & Terabe, 2018). Thus, no single TCR can be used as a marker. The same applies to their lipid specificity. With a broad range of TCRs, they should be able to respond to a broad range of lipids, but only a few such lipid antigens for type II NKT cells have been identified so far. The first and most widely studied is sulfatide, originally isolated from bovine brain myelin sheaths and studied for its role in stimulating type II NKT cells to inhibit experimental allergic encephalitis, a mouse model of multiple sclerosis (Jahng et al., 2004). This group took advantage of this discovery of an antigen for type II NKT cells to make sulfatide-CD1d tetramers to stain T cells with this specificity and characterize them (Arrenberg, Halder, Dai, Maricic, & Kumar, 2010; Halder, Aguilera, Maricic, & Kumar, 2007; Jahng et al., 2004; Weng et al., 2017; Zajonc et al., 2005). However, those tetramers were difficult to make and to export, and have not been widely available. More recently, our lab has made stable CD1d-sulfatide tetramers that appear to be stable and can stain and identify type II NKT cells in mice specific for sulfatide presented by CD1d (Kato et al., 2018). These tetramers have been used to examine their tissue location especially prevalent in liver and lung, their CD4/CD8 usage (more like conventional T cells than type I NKT cells that in mice are mostly CD4 or double negative), their surface markers, and their ability to make different cytokines (Kato et al., 2018) (Kato et al., submitted). However, because we do not have a way to identify all type II NKT cells, we cannot tell what fraction of type II NKT cells are represented by the sulfatide-CD1d-specific subset. We must assume there are other subsets with different specificities that have not yet been identified, which may also have different properties. Thus, type II NKT cells are still incompletely understood and new tools will be needed to address some of these questions.

3.3 Type II NKT hybridomas

There is also a prototype hybridoma XV19 from the Cardell lab (Cardell et al., 1995), with a type II NKT TCR, which happens to respond to sulfatide also, but this cannot be considered representative of the majority of type II NKT cells (Gumperz et al., 2000) as just explained for sulfatide-CD1d tetramers. Nevertheless, this hybridoma has been useful for comparing the specificity of type II vs type I NKT cells for various glycolipids in the range between αGalCer and sulfatide, and various sulfatide analogues (Rhost et al., 2012) (Pasquet manuscript in prep.). Another fairly well characterized type II NKT cell hybridoma is VIII24 described below.

3.4 24αβ-TCR transgenic mice

These mice were made transgenic for a TCR derived from a type II NKT cell hybridoma VIII24 that utilizes Vα3.2 and Vβ9 (Cardell et al., 1995; Skold et al., 2000). Thus, they represent another subset of type II NKT cells that does not recognize sulfatide. These T cells produce large amounts of IFN-γ and IL-4 on activation and have some innate as well as Th1-like properties. Interestingly, the T cells from these mice express Annexin A2, c-kit and Ly6C, which are also expressed in CD8αα thymic cells. They have been found to have immunosuppressive as well as effector functions. Ly6C and c-kit were also found on liver and lung type II NKT cells in our unpublished studies (Kato et al., manuscript submitted).

3.5 4get Jα18 KO mice

Other approaches to take advantage of properties of type II NKT cells other than sulfatide specificity have made use of other transgenic and KO mouse constructs. The 4get mice express green fluorescent protein (GFP) under the IL-4 promoter and so act as a way of identifying IL-4-producing cells. Because type I NKT cells constitutively express IL-4 mRNA, it was reasoned that such GFP-positive T cells in Jα18 KO mice that lack type I NKT cells might be type II NKT cells (Zeissig et al., 2012; Zhao, Weng, Bagchi, & Wang, 2014). These cells responded by IFN-γ production to CD1d-positive bone marrow derived dendritic cells and to some lipids reported to be seen by type II NKT cells, but not to sulfatide (Zhao et al., 2014).

4. NKT cells in cancer

4.1 Type I NKT cells

NKT cells (type I or iNKT) were first observed to play a critical role in cancer immunosurveillance and anti-cancer immunity. This was first observed when αGalCer was shown to have anti-tumor activity and to be an agonist for type I NKT cells (Kobayashi, Motoki, Uchida, Fukushima, & Koezuka, 1995; Morita et al., 1995; Motoki et al., 1995). Chronic administration of αGalCer could even prevent spontaneous tumors in mice treated with the carcinogen methylcholanthrene or in mice transgenic for the HER2 oncogene (Hayakawa, Rovero, Forni, & Smyth, 2003). Even without αGalCer, the efficacy of low-dose IL-12 in treating mouse tumors was found to be dependent on type I NKT cells (Smyth, Taniguchi, & Street, 2000). These studies made use of the KO mice that lacked NKT cells as a tool to identify them. Further, without any intervention, spontaneous rejection of methylcholanthrene-induced sarcomas was found to be dependent on type I NKT cells, making use of Jα18 KO mice that selectively lack them (Smyth et al., 2000). This was confirmed by adoptive transfer of purified liver NKT cells (Crowe, Smyth, & Godfrey, 2002). CD1d-αGalCer tetramers were used to purify type I NKT cells in another model in which Jα18 KO mice were more susceptible to lung nodules produced by i.v. injection of a methylcholanthrene-induced sarcoma, CMS5m, and this susceptibility could be prevented by adoptive transfer of type I NKT cells purified using CD1d-αGalCer tetramers (Nishikawa et al., 2003). Thus, this study combined two different approaches described above for identifying type I NKT cells, the Jα18 KO mice and the CD1d-αGalCer tetramers.

Importantly, although type I NKT cells can make lytic granules and can kill, the protection afforded by these cells did not appear to be mediated by this direct killing of tumor cells, but rather by activation of other effector cells, $CD8^+$ T cells and NK cells (Crowe et al., 2002; Street, Cretney, & Smyth, 2001). Rather, it depended on the ability of the NKT cells to make IFN-γ, but not on their ability to make perforin, as shown using KO mice. Indeed, when NK deficient mice, NKT deficient mice, or RAG KO mice lacking all lymphocytes were reconstituted with different cell populations, it was found that αGalCer-mediated protection depended on sequential production of IFN-γ first by NKT cells and then by NK cells, implying a sequential activation in which NK cells are downstream of NKT cells

(Smyth et al., 2002). These indirect mechanisms may be important because many tumors suppress the expression of CD1d, perhaps to escape direct killing by NKT cells.

Another mechanism by which NKT cells could contribute to anti-cancer immunity was shown to be maturation and activation of dendritic cells (DCs) to make IL-12 (Kitamura et al., 1999; Toura et al., 1999). This DC maturation mechanism improves DC presentation of antigen to and activation of CD8 effector T cells (Fujii, Shimizu, Smith, Bonifaz, & Steinman, 2003), and can be mediated by binding of CD40L on the NKT cells to CD40 on the DCs (Kitamura et al., 1999).

Effects of human tumor growth in cancer patients was observed using antibodies to Vα24 and Vβ11 to identify type I NKT cells in peripheral blood, and it was discovered that these are often decreased in numbers (Crough et al., 2004; Giaccone et al., 2002) or in their function, as measured by IFN-γ production (Dhodapkar et al., 2003; Fujii et al., 2003; Yanagisawa et al., 2002, 2006).

Although most known agonists of type I NKT cells were known to require IFN-γ to protect against cancer, as described above, an exception was found when a new class of type I NKT cell agonists was discovered represented by the prototype lipid beta-mannosylceramide (βManCer) (O'Konek et al., 2011, 2013). This somewhat weaker agonist, that does not induce much cytokine production, was found to also protect against lung metastases, but by a mechanism dependent on TNF-α and nitric oxide synthase (NOS) and not dependent on IFN-γ (O'Konek et al., 2011). Yet, by tetramer staining with either αGalCer-loaded CD1d tetramers or βManCer-loaded CD1d tetramers in different colors, the same type I NKT cells stained with both, so they were not different cells. Moreover, the Vβ usage was the same. When used at suboptimal concentrations, αGalCer and βManCer synergized with each other to protect, further supporting the conclusion that they worked by different mechanisms. Also, unlike αGalCer, βManCer did not induce long-term functional anergy of NKT cells and so could be administered repeatedly (O'Konek et al., 2013). These properties, suggesting that it may work in cases where αGalCer failed, make this molecule attractive for clinical development.

5. Type II NKT cells

In view of all these findings, the field was surprised when we found that NKT cells could also suppress tumor immunity (Terabe et al., 2000). In this case, CD1d KO mice were used to explore the mechanism by which CD4-depletion allowed more effective spontaneous tumor immunosurveillance in a mouse fibrosarcoma model, at a time before the obvious assumption would have been depletion of $CD4^+Foxp3^+$ T regulatory cells (Treg cells). It turned out that CD1d KO mice as well as IL-4Rα KO and STAT6 KO mice were all protected (Kacha, Fallarino, Markiewicz, & Gajewski, 2000; Ostrand-Rosenberg, Grusby, & Clements, 2000; Terabe et al., 2000), and that protection could also be mediated by blocking IL-13 but not IL-4, leading to the conclusion that NKT cell production of IL-13 was responsible for the suppression of tumor immunity (Terabe et al., 2000). The immunity was CD8 T cell dependent, but $CD8^+$ T cells did not have receptors for IL-13, implying that there must be an intermediate cell. Indeed, it was subsequently found that the IL-13 acted on a myeloid cell, probably akin to a myeloid-derived suppressor cell (MDSC) that was $Gr\text{-}1^+$ and $CD11b^+$, and this cell was induced to make TGF-β, which was the downstream molecule that actually suppressed the $CD8^+$ T cells (Terabe et al., 2003). This discovery defined a circuit in which NKT cells were stimulated by tumor growth to make IL-13 which induced a myeloid cell to make TGF-β, which then suppressed the effector cells controlling the tumor (Terabe et al., 2003). A further puzzle in this circuit, why IL-13 was necessary but IL-4 was not even though both signaled through the IL-4Rα chain and STAT6, was resolved when it was discovered that the STAT6 signal was only the first step, but then that synergized with a signal from TNF-α to upregulate a second receptor, the IL-13Rα2, which responds only to IL-13, but not IL-4, and this receptor signaled through AP-1 to induce production of TGF-β (Fichtner-Feigl et al., 2008).

However, the paradox remained that it seemed the same cells that could protect against tumors were also suppressing tumor immunity. This was resolved when it was found in several different mouse tumor models by three different labs that it was type II NKT cells that were sufficient to suppress, whereas type I NKT cells were the protective NKT cells (Terabe et al., 2005). This distinction was made primarily by using a combination of CD1d KO and Jα18 KO mice. The suppressive activity was lost in CD1d KO mice but not in Jα18 KO mice, implying that NKT cells were necessary, but in the absence of type I NKT cells in Jα18 KO mice, type II NKT cells were sufficient to suppress tumor immunity.

These studies, and the observation that Jα18 KO mice seemed to be more suppressed than WT mice in tumor immunity, led to the question of whether type I NKT cells were regulating type II NKT cells

(Ambrosino et al., 2007). Conversely, Ambrosino et al. found that when type II NKT cells were stimulated with sulfatide, they reduced the protection afforded by stimulation of type I NKT cells with αGalCer, in both the subcutaneous fibrosarcoma model and a CT26 lung metastasis model (Ambrosino et al., 2007). Thus, these observations, making use of both KO mice and agonists selective for type I or type II NKT cells, revealed a novel immunoregulatory axis in which type I and type II NKT cells cross-regulated each other, somewhat analogous to the Th1-Th2 axis that has so profoundly affected immunology. Since NKT cells are early responders in many immune responses, the balance along this axis can potentially set the tone for subsequent other adaptive immune responses and thus play a major role in regulating tumor immunity (Ambrosino et al., 2007; Ambrosino, Berzofsky, & Terabe, 2008). This concept was extended when it was found, also using comparisons between WT, Jα18 KO and CD1d KO mice, combined with Treg depletion using anti-CD25 (PC61 monoclonal antibody), that Tregs were the primary suppressive cell in a subcutaneous CT26 tumor model in WT mice where type I NKT cells balanced type II, and in CD1d KO mice where both were absent, but surprisingly, in Jα18 KO mice where type II NKT cells were unopposed, they suppressed equally well so that simple depletion of Tregs was not sufficient to overcome suppression, but rather blockade or elimination of both regulatory cells, Tregs and type II NKT cells, was necessary to eliminate the suppression (Izhak et al., 2013). This conclusion was also confirmed by adoptive transfer of type I NKT cells into Jα18 KO mice, and by stimulation of type II NKT cells using sulfatide, so tetramers and selective agonists were other tools that were used in this study. Thus, one type of T cell, the type I NKT cell, could regulate the balance between two other regulatory cells, Tregs and type II NKT cells, thereby regulating the regulators. These observations were more recently extended to compare the immune regulation when the same tumor, CT26, was growing in the skin or the lung. Using similar tools, it was found that Treg cells were the primary suppressor in the skin in intact WT mice, whereas type II NKT cells were the critical suppressor in the lung, where Treg cells were not necessary (O'Konek et al., 2018). An important take-home message from this finding was that one cannot assume that the same immunotherapy that works for a tumor in one tissue, for example the primary tumor, will necessarily work on the same tumor when it is growing in a different tissue microenvironment, such as a metastasis.

6. Conclusion

Thus, both type I and type II NKT cells play important roles in cancer and cancer immunotherapy. A set of complementary tools, approaches and strategies summarized above have been developed to characterize this small but influential population of T cells and their role in tumor immunity as well as infectious, autoimmune, and other diseases. New tools have opened up a new understanding of these critical effector and regulatory immune cells that may contribute to harnessing their potential for promoting the immunotherapy of cancer as well as for control of other diseases in which the immune system plays a prominent role.

References

Abu-Eid, R., Samara, R. N., Ozbun, L., Abdalla, M. Y., Berzofsky, J. A., Friedman, K. M., et al. (2014). Selective inhibition of regulatory T cells by targeting the PI3K-Akt pathway. *Cancer Immunology Research*, *2*, 1080–1089.

Altman, J. D., Moss, P. A. H., Goulder, P. J. R., Barouch, D. H., McHeyzer-Williams, M. G., Bell, J. I., et al. (1996). Phenotypic analysis of antigen-specific T lymphocytes. *Science*, *274*, 94–96.

Alvarez, R. D., Huh, W. K., Bae, S., Lamb, L. S., Jr., Conner, M. G., Boyer, J., et al. (2016). A pilot study of pNGVL4a-CRT/E7(detox) for the treatment of patients with HPV16+ cervical intraepithelial neoplasia 2/3 (CIN2/3). *Gynecologic Oncology*, *140*, 245–252.

Ambrosino, E., Berzofsky, J. A., & Terabe, M. (2008). Regulation of tumor immunity: The role of NKT cells. *Expert Opinion on Biological Therapy*, *8*, 725–734.

Ambrosino, E., Terabe, M., Halder, R. C., Peng, J., Takaku, S., Miyake, S., et al. (2007). Cross-regulation between type I and type II NKT cells in regulating tumor immunity: A new immunoregulatory axis. *Journal of Immunology*, *179*, 5126–5136.

Arase, H., Arase, N., Nakagawa, K., Good, R. A., & Onoe, K. (1993). NK1.1+ CD4+ CD8- thymocytes with specific lymphokine secretion. *European Journal of Immunology*, *23*, 307–310.

Arase, H., Arase, N., Ogasawara, K., Good, R. A., & Onoe, K. (1992). An NK1.1+ CD4 +8- single-positive thymocyte subpopulation that expresses a highly skewed T-cell antigen receptor V beta family. *Proceedings of the National Academy of Sciences of the United States of America*, *89*, 6506–6510.

Arrenberg, P., Halder, R., Dai, Y., Maricic, I., & Kumar, V. (2010). Oligoclonality and innate-like features in the TCR repertoire of type II NKT cells reactive to a beta-linked self-glycolipid. *Proceedings of the National Academy of Sciences of the United States of America*, *107*, 10984–10989.

Bedel, R., Matsuda, J. L., Brigl, M., White, J., Kappler, J., Marrack, P., et al. (2012). Lower TCR repertoire diversity in Traj18-deficient mice. *Nature Immunology*, *13*, 705–706.

Behar, S. M., & Cardell, S. (2000). Diverse CD1d-restricted T cells: Diverse phenotypes, and diverse functions. *Seminars in Immunology*, *12*, 551–560.

Bendelac, A., Hunziker, R. D., & Lantz, O. (1996). Increased interleukin 4 and immunoglobulin E production in transgenic mice overexpressing NK1 T cells. *The Journal of Experimental Medicine*, *184*, 1285–1293.

Bendelac, A., Killeen, N., Littman, D. R., & Schwartz, R. H. (1994). A subset of CD4+ thymocytes selected by MHC class I molecules. *Science*, *263*, 1774–1778.
Bendelac, A., Lantz, O., Quimby, M. E., Yewdell, J. W., Bennink, J. R., & Brutkiewicz, R. R. (1995). CD1 recognition by mouse NK1+ T lymphocytes. *Science*, *268*, 863–865.
Benlagha, K., Weiss, A., Beavis, A., Teyton, L., & Bendelac, A. (2000). In vivo identification of glycolipid antigen-specific T cells using fluorescent CD1d tetramers. *The Journal of Experimental Medicine*, *191*, 1895–1903.
Berzins, S. P., Smyth, M. J., & Baxter, A. G. (2011). Presumed guilty: Natural killer T cell defects and human disease. *Nature Reviews Immunology*, *11*, 131–142.
Cameron, G., & Godfrey, D. I. (2018). Differential surface phenotype and context-dependent reactivity of functionally diverse NKT cells. *Immunology and Cell Biology*, *96*, 759–771.
Cardell, S., Tangri, S., Chan, S., Kronenberg, M., Benoist, C., & Mathis, D. (1995). CD1-restricted CD4+ T cells in major histocompatibility complex class II-deficient mice. *The Journal of Experimental Medicine*, *182*, 993–1004.
Chandra, S., Zhao, M., Budelsky, A., de Mingo Pulido, A., Day, J., Fu, Z., et al. (2015). A new mouse strain for the analysis of invariant NKT cell function. *Nature Immunology*, *16*, 799–800.
Chen, Y. H., Chiu, N. M., Mandal, M., Wang, N., & Wang, C. R. (1997). Impaired NK1+ T cell development and early IL-4 production in CD1- deficient mice. *Immunity*, *6*, 459–467.
Chiba, A., Cohen, N., Brigl, M., Brennan, P. J., Besra, G. S., & Brenner, M. B. (2009). Rapid and reliable generation of invariant natural killer T-cell lines in vitro. *Immunology*, *128*, 324–333.
Crough, T., Purdie, D. M., Okai, M., Maksoud, A., Nieda, M., & Nicol, A. J. (2004). Modulation of human Valpha24(+)Vbeta11(+) NKT cells by age, malignancy and conventional anticancer therapies. *British Journal of Cancer*, *91*, 1880–1886.
Crowe, N. Y., Smyth, M. J., & Godfrey, D. I. (2002). A critical role for natural killer T cells in immunosurveillance of methylcholanthrene-induced sarcomas. *The Journal of Experimental Medicine*, *196*, 119–127.
Cui, J., Shin, T., Kawano, T., Sato, H., Kondo, E., Toura, I., et al. (1997). Requirement for Valpha14 NKT cells in IL-12-mediated rejection of tumors. *Science*, *278*, 1623–1626.
Dashtsoodol, N., Shigeura, T., Ozawa, R., Harada, M., Kojo, S., Watanabe, T., et al. (2016). Generation of novel Traj18-deficient mice lacking Valpha14 natural killer T cells with an undisturbed T cell receptor alpha-chain repertoire. *PLoS One*, *11* e0153347.
Dhodapkar, M. V., Geller, M. D., Chang, D. H., Shimizu, K., Fujii, S., Dhodapkar, K. M., et al. (2003). A reversible defect in natural killer T cell function characterizes the progression of premalignant to malignant multiple myeloma. *The Journal of Experimental Medicine*, *197*, 1667–1676.
Escriba-Garcia, L., Alvarez-Fernandez, C., Caballero, A. C., Schaub, R., Sierra, J., & Briones, J. (2019). The novel agonistic iNKT-cell antibody NKT14m induces a therapeutic antitumor response against B-cell lymphoma. *Oncoimmunology*, *8*, e1546543.
Exley, M. A., He, Q., Cheng, O., Wang, R. J., Cheney, C. P., Balk, S. P., et al. (2002). Cutting edge: Compartmentalization of Th1-like noninvariant CD1d-reactive T cells in hepatitis C virus-infected liver. *Journal of Immunology*, *168*, 1519–1523.
Exley, M. A., Hou, R., Shaulov, A., Tonti, E., Dellabona, P., Casorati, G., et al. (2008). Selective activation, expansion, and monitoring of human iNKT cells with a monoclonal antibody specific for the TCR alpha-chain CDR3 loop. *European Journal of Immunology*, *38*, 1756–1766.
Fichtner-Feigl, S., Terabe, M., Kitani, A., Young, C. A., Fuss, I., Geissler, E. K., et al. (2008). Restoration of tumor immunosurveillance via targeting of interleukin-13 receptor-alpha 2. *Cancer Research*, *68*, 3467–3475.

Fujii, S., Shimizu, K., Klimek, V., Geller, M. D., Nimer, S. D., & Dhodapkar, M. V. (2003). Severe and selective deficiency of interferon-gamma-producing invariant natural killer T cells in patients with myelodysplastic syndromes. *British Journal of Haematology, 122*, 617–622.

Fujii, S., Shimizu, K., Smith, C., Bonifaz, L., & Steinman, R. M. (2003). Activation of natural killer T cells by alpha-galactosylceramide rapidly induces the full maturation of dendritic cells in vivo and thereby acts as an adjuvant for combined CD4 and CD8 T cell immunity to a coadministered protein. *The Journal of Experimental Medicine, 198*, 267–279.

Giaccone, G., Punt, C. J., Ando, Y., Ruijter, R., Nishi, N., Peters, M., et al. (2002). A phase I study of the natural killer T-cell ligand alpha-galactosylceramide (KRN7000) in patients with solid tumors. *Clinical Cancer Research, 8*, 3702–3709.

Godfrey, D. I., MacDonald, H. R., Kronenberg, M., Smyth, M. J., & Van Kaer, L. (2004). NKT cells: What's in a name? *Nature Reviews. Immunology, 4*, 231–237.

Gumperz, J. E., Roy, C., Makowska, A., Lum, D., Sugita, M., Podrebarac, T., et al. (2000). Murine CD1d-restricted T cell recognition of cellular lipids. *Immunity, 12*, 211–221.

Halder, R. C., Aguilera, C., Maricic, I., & Kumar, V. (2007). Type II NK T cell-mediated anergy induction in type I NK T cells prevents inflammatory liver disease. *The Journal of Clinical Investigation, 117*, 2302–2312.

Hayakawa, Y., Rovero, S., Forni, G., & Smyth, M. J. (2003). Alpha-galactosylceramide (KRN7000) suppression of chemical- and oncogene-dependent carcinogenesis. *Proceedings of the National Academy of Sciences of the United States of America, 100*, 9464–9469.

Izhak, L., Ambrosino, E., Kato, S., Parish, S. T., O'Konek, J. J., Weber, H., et al. (2013). Delicate balance among three types of T cells in concurrent regulation of tumor immunity. *Cancer Research, 73*, 1514–1523.

Jahng, A., Maricic, I., Aguilera, C., Cardell, S., Halder, R. C., & Kumar, V. (2004). Prevention of autoimmunity by targeting a distinct, noninvariant CD1d-reactive T cell population reactive to Sulfatide. *The Journal of Experimental Medicine, 199*, 947–957.

Kacha, A. K., Fallarino, F., Markiewicz, M. A., & Gajewski, T. F. (2000). Spontaneous rejection of poorly immunogenic P1.HTR tumors by Stat6- deficient mice. *Journal of Immunology, 165*, 6024–6028.

Karadimitris, A., Gadola, S., Altamirano, M., Brown, D., Woolfson, A., Klenerman, P., et al. (2001). Human CD1d-glycolipid tetramers generated by in vitro oxidative refolding chromatography. *Proceedings of the National Academy of Sciences of the United States of America, 98*, 3294–3298.

Kato, S., Berzofsky, J. A., & Terabe, M. (2018). Possible therapeutic application of targeting type II natural killer T cell-mediated suppression of tumor immunity. *Frontiers in Immunology, 9*, 314.

Kawano, T., Cui, J., Koezuka, Y., Toura, I., Kaneko, Y., Motoki, K., et al. (1997). CD1d-restricted and TCR-mediated activation of valpha14 NKT cells by glycosylceramides. *Science, 278*, 1626–1629.

Kinjo, Y., Pei, B., Bufali, S., Raju, R., Richardson, S. K., Imamura, M., et al. (2008). Natural Sphingomonas glycolipids vary greatly in their ability to activate natural killer T cells. *Chemistry & Biology, 15*, 654–664.

Kinjo, Y., Wu, D., Kim, G., Xing, G. W., Poles, M. A., Ho, D. D., et al. (2005). Recognition of bacterial glycosphingolipids by natural killer T cells. *Nature, 434*, 520–525.

Kitamura, H., Iwakabe, K., Yahata, T., Nishimura, S., Ohta, A., Ohmi, Y., et al. (1999). The natural killer T (NKT) cell ligand alpha-galactosylceramide demonstrates its immunopotentiating effect by inducing interleukin (IL)- 12 production by dendritic cells and IL-12 receptor expression on NKT cells. *The Journal of Experimental Medicine, 189*, 1121–1128.

Kobayashi, E., Motoki, K., Uchida, T., Fukushima, H., & Koezuka, Y. (1995). KRN7000, a novel immunomodulator, and its antitumor activities. *Oncology Research*, *7*, 529–534.

Kronenberg, M., & Gapin, L. (2002). The unconventional lifestyle of NKT cells. *Nature Reviews. Immunology*, *2*, 557–568.

Kumar, A., Bezbradica, J. S., Stanic, A. K., & Joyce, S. (2017). Characterization and functional analysis of mouse semi-invariant natural T cells. *Current Protocols in Immunology*, *117*, 14.13.11–14.13.55.

Lantz, O., & Bendelac, A. (1994). An invariant T cell receptor alpha chain is used by a unique subset of major histocompatibility complex class I-specific CD4+ and CD4-8- T cells in mice and humans. *The Journal of Experimental Medicine*, *180*, 1097–1106.

Lee, P. T., Benlagha, K., Teyton, L., & Bendelac, A. (2002). Distinct functional lineages of human Va24 natural killer cells. *Journal of Experimental Medicine*, *195*, 637–641.

Lee, Y. J., Holzapfel, K. L., Zhu, J., Jameson, S. C., & Hogquist, K. A. (2013). Steady-state production of IL-4 modulates immunity in mouse strains and is determined by lineage diversity of iNKT cells. *Nature Immunology*, *14*, 1146–1154.

Lee, Y. J., Wang, H., Starrett, G. J., Phuong, V., Jameson, S. C., & Hogquist, K. A. (2015). Tissue-specific distribution of iNKT cells impacts their cytokine response. *Immunity*, *43*, 566–578.

Levitsky, H. I., Golumbek, P. T., & Pardoll, D. M. (1991). The fate of CD4-8- T cell receptor-alpha beta+ thymocytes. *Journal of Immunology*, *146*, 1113–1117.

Li, X., Tsuji, M., Schneck, J., & Webb, T. J. (2013). Generation of mouse iNKT cell lines. *Bio-protocol*, *3*, e419.

Makowska, A., Kawano, T., Taniguchi, M., & Cardell, S. (2000). Differences in the ligand specificity between CD1d-restricted T cells with limited and diverse T-cell receptor repertoire. *Scandinavian Journal of Immunology*, *52*, 71–79.

Matsuda, J. L., Naidenko, O. V., Gapin, L., Nakayama, T., Taniguchi, M., Wang, C. R., et al. (2000). Tracking the response of natural killer T cells to a glycolipid antigen using CD1d tetramers. *The Journal of Experimental Medicine*, *192*, 741–754.

Mendiratta, S. K., Martin, W. D., Hong, S., Boesteanu, A., Joyce, S., & Van Kaer, L. (1997). CD1d1 mutant mice are deficient in natural T cells that promptly produce IL-4. *Immunity*, *6*, 469–477.

Morita, M., Motoki, K., Akimoto, K., Natori, T., Sakai, T., Sawa, E., et al. (1995). Structure-activity relationship of alpha-galactosylceramides against B16-bearing mice. *Journal of Medicinal Chemistry*, *38*, 2176–2187.

Motoki, K., Morita, M., Kobayashi, E., Uchida, T., Akimoto, K., Fukushima, H., et al. (1995). Immunostimulatory and antitumor activities of monoglycosylceramides having various sugar moieties. *Biological & Pharmaceutical Bulletin*, *18*, 1487–1491.

Nishikawa, H., Kato, T., Tanida, K., Hiasa, A., Tawara, I., Ikeda, H., et al. (2003). CD4+ CD25+ T cells responding to serologically defined autoantigens suppress antitumor immune responses. *Proceedings of the National Academy of Sciences of the United States of America*, *100*, 10902–10906.

O'Konek, J. J., Ambrosino, E., Bloom, A., Pasquet, L., Massilamany, C., Xia, Z., et al. (2018). Differential regulation of T-cell mediated anti-tumor memory and cross-protection against the same tumor in lungs versus skin. *Oncoimmunology*, *7*, e1439305.

O'Konek, J. J., Illarionov, P., Khursigara, D. S., Ambrosino, E., Izhak, L., Castillo, B. F., 2nd, et al. (2011). Mouse and human iNKT cell agonist beta-mannosylceramide reveals a distinct mechanism of tumor immunity. *The Journal of Clinical Investigation*, *121*, 683–694.

O'Konek, J. J., Kato, S., Takao, S., Izhak, L., Xia, Z., Illarionov, P., et al. (2013). Beta-mannosylceramide activates type I natural killer T cells to induce tumor immunity without inducing long-term functional anergy. *Clinical Cancer Research*, *19*, 4404–4411.

Ostrand-Rosenberg, S., Grusby, M. J., & Clements, V. K. (2000). Cutting edge: STAT6-deficient mice have enhanced tumor immunity to primary and metastatic mammary carcinoma. *Journal of Immunology*, *165*, 6015–6019.

Park, S. H., Roark, J. H., & Bendelac, A. (1998). Tissue-specific recognition of mouse CD1 molecules. *Journal of Immunology*, *160*, 3128–3134.

Ren, Y., Sekine-Kondo, E., Shibata, R., Kato-Itoh, M., Umino, A., Yanagida, A., et al. (2017). A novel mouse model of iNKT cell-deficiency generated by CRISPR/Cas9 reveals a pathogenic role of iNKT cells in metabolic disease. *Scientific Reports*, *7*, 12765.

Rhost, S., Lofbom, L., Rynmark, B. M., Pei, B., Mansson, J. E., Teneberg, S., et al. (2012). Identification of novel glycolipid ligands activating a sulfatide-reactive, CD1d-restricted, type II natural killer T lymphocyte. *European Journal of Immunology*, *42*, 2851–2860.

Robertson, F. C., Berzofsky, J. A., & Terabe, M. (2014). NKT cell networks in the regulation of tumor immunity. *Frontiers in Immunology*, *5*, 543.

Scheuplein, F., Lamont, D. J., Poynter, M. E., Boyson, J. E., Serreze, D., Lundblad, L. K., et al. (2015). Mouse invariant monoclonal antibody NKT14: A novel tool to manipulate iNKT cell function in vivo. *PLoS One*, *10*, e0140729.

Skold, M., Faizunnessa, N. N., Wang, C. R., & Cardell, S. (2000). CD1d-specific NK1.1+ T cells with a transgenic variant TCR. *Journal of Immunology*, *165*, 168–174.

Smiley, S. T., Kaplan, M. H., & Grusby, M. J. (1997). Immunoglobulin E production in the absence of interleukin-4-secreting CD1-dependent cells. *Science*, *275*, 977–979.

Smyth, M. J., Crowe, N. Y., Pellicci, D. G., Kyparissoudis, K., Kelly, J. M., Takeda, K., et al. (2002). Sequential production of interferon-gamma by NK1.1(+) T cells and natural killer cells is essential for the antimetastatic effect of alpha-galactosylceramide. *Blood*, *99*, 1259–1266.

Smyth, M. J., Taniguchi, M., & Street, S. E. (2000). The anti-tumor activity of IL-12: Mechanisms of innate immunity that are model and dose dependent. *Journal of Immunology*, *165*, 2665–2670.

Smyth, M. J., Thia, K. Y., Street, S. E., Cretney, E., Trapani, J. A., Taniguchi, M., et al. (2000). Differential tumor surveillance by natural killer (NK) and NKT cells. *The Journal of Experimental Medicine*, *191*, 661–668.

Street, S. E., Cretney, E., & Smyth, M. J. (2001). Perforin and interferon-gamma activities independently control tumor initiation, growth, and metastasis. *Blood*, *97*, 192–197.

Sundararaj, S., Zhang, J., Krovi, S. H., Bedel, R., Tuttle, K. D., Veerapen, N., et al. (2018). Differing roles of CD1d2 and CD1d1 proteins in type I natural killer T cell development and function. *Proceedings of the National Academy of Sciences of the United States of America*, *115*, E1204–E1213.

Takahashi, K., & Yamanaka, S. (2006). Induction of pluripotent stem cells from mouse embryonic and adult fibroblast cultures by defined factors. *Cell*, *126*, 663–676.

Taniguchi, M., Harada, M., Kojo, S., Nakayama, T., & Wakao, H. (2003). The regulatory role of Valpha14 NKT cells in innate and acquired immune response. *Annual Review of Immunology*, *21*, 483–513.

Taniguchi, M., Koseki, H., Tokuhisa, T., Masuda, K., Sato, H., Kondo, E., et al. (1996). Essential requirement of an invariant V alpha 14 T cell antigen receptor expression in the development of natural killer T cells. *Proceedings of the National Academy of Sciences of the United States of America*, *93*, 11025–11028.

Terabe, M., Matsui, S., Noben-Trauth, N., Chen, H., Watson, C., Donaldson, D. D., et al. (2000). NKT cell-mediated repression of tumor immunosurveillance by IL-13 and the IL-4R-STAT6 pathway. *Nature Immunology*, *1*, 515–520.

Terabe, M., Matsui, S., Park, J.-M., Mamura, M., Noben-Trauth, N., Donaldson, D. D., et al. (2003). Transforming growth factor-b production and myeloid cells are an effector mechanism through which CD1d-restricted T cells block cytotoxic T lymphocyte-mediated

tumor immunosurveillance: Abrogation prevents tumor recurrence. *The Journal of Experimental Medicine*, *198*, 1741–1752.
Terabe, M., Swann, J., Ambrosino, E., Sinha, P., Takaku, S., Hayakawa, Y., et al. (2005). A nonclassical non-Va14Ja18 CD1d-restricted (type II) NKT cell is sufficient for down-regulation of tumor immunosurveillance. *The Journal of Experimental Medicine*, *202*, 1627–1633.
Toura, I., Kawano, T., Akutsu, Y., Nakayama, T., Ochiai, T., & Taniguchi, M. (1999). Cutting edge: Inhibition of experimental tumor metastasis by dendritic cells pulsed with alpha-galactosylceramide. *Journal of Immunology*, *163*, 2387–2391.
Van Kaer, L. (2007). NKT cells: T lymphocytes with innate effector functions. *Current Opinion in Immunology*, *19*, 354–364.
Wakao, H., Kawamoto, H., Sakata, S., Inoue, K., Ogura, A., Wakao, R., et al. (2007). A novel mouse model for invariant NKT cell study. *Journal of Immunology*, *179*, 3888–3895.
Watarai, H., Fujii, S., Yamada, D., Rybouchkin, A., Sakata, S., Nagata, Y., et al. (2010). Murine induced pluripotent stem cells can be derived from and differentiate into natural killer T cells. *The Journal of Clinical Investigation*, *120*, 2610–2618.
Watarai, H., Rybouchkin, A., Hongo, N., Nagata, Y., Sakata, S., Sekine, E., et al. (2010). Generation of functional NKT cells in vitro from embryonic stem cells bearing rearranged invariant Valpha14-Jalpha18 TCRalpha gene. *Blood*, *115*, 230–237.
Webb, T. J., Bieler, J. G., Schneck, J. P., & Oelke, M. (2009). Ex vivo induction and expansion of natural killer T cells by CD1d1-Ig coated artificial antigen presenting cells. *Journal of Immunological Methods*, *346*, 38–44.
Weng, X., He, Y., Visvabharathy, L., Liao, C. M., Tan, X., Balakumar, A., et al. (2017). Crosstalk between type II NKT cells and T cells leads to spontaneous chronic inflammatory liver disease. *Journal of Hepatology*, 791–800.
Yanagisawa, K., Exley, M. A., Jiang, X., Ohkochi, N., Taniguchi, M., & Seino, K. (2006). Hyporesponsiveness to natural killer T-cell ligand alpha-galactosylceramide in cancer-bearing state mediated by CD11b+ gr-1+ cells producing nitric oxide. *Cancer Research*, *66*, 11441–11446.
Yanagisawa, K., Seino, K., Ishikawa, Y., Nozue, M., Todoroki, T., & Fukao, K. (2002). Impaired proliferative response of V alpha 24 NKT cells from cancer patients against alpha-galactosylceramide. *Journal of Immunology*, *168*, 6494–6499.
Zajonc, D. M., Maricic, I., Wu, D., Halder, R., Roy, K., Wong, C. H., et al. (2005). Structural basis for CD1d presentation of a sulfatide derived from myelin and its implications for autoimmunity. *The Journal of Experimental Medicine*, *202*, 1517–1526.
Zeissig, S., Murata, K., Sweet, L., Publicover, J., Hu, Z., Kaser, A., et al. (2012). Hepatitis B virus-induced lipid alterations contribute to natural killer T cell-dependent protective immunity. *Nature Medicine*, *18*, 1060–1068.
Zhang, J., Bedel, R., Krovi, S. H., Tuttle, K. D., Zhang, B., Gross, J., et al. (2016). Mutation of the Traj18 gene segment using TALENs to generate natural killer T cell deficient mice. *Scientific Reports*, *6*, 27375.
Zhao, J., Weng, X., Bagchi, S., & Wang, C. R. (2014). Polyclonal type II natural killer T cells require PLZF and SAP for their development and contribute to CpG-mediated antitumor response. *Proceedings of the National Academy of Sciences of the United States of America*, *111*, 2674–2679.

CHAPTER TWENTY-ONE

Assessment of IFN-γ and granzyme-B production by *in "sitro"* technology

Claudia Galassi[a,†], Gwenola Manic[b,†], Martina Musella[b,c], Antonella Sistigu[a,b,*,‡], Ilio Vitale[d,e,*,‡]

[a]Istituto di Patologia Generale, Università Cattolica del Sacro Cuore, Rome, Italy
[b]Unit of Tumor Immunology and Immunotherapy, Department of Research, Advanced Diagnostics, and Technological Innovation, IRCCS Regina Elena National Cancer Institute, Rome, Italy
[c]Department of Molecular Medicine, University "La Sapienza", Rome, Italy
[d]Italian Institute for Genomic Medicine (IIGM), Turin, Italy
[e]Candiolo Cancer Institute—FPO, IRCCS, Candiolo, Italy
*Corresponding authors: e-mail address: antonella.sistigu@gmail.com; iliovit@gmail.com

Contents

Abstract

Tumor neantigens (TNAs) and tumor-associated antigens (TAAs) are crucial triggers of anticancer immune responses. Through major histocompatibility complex, such antigens activate T cells, which, by releasing interferon gamma (IFN-γ) and granzyme

† Equally contributed to this article.
‡ Share senior co-authorship.

Methods in Enzymology, Volume 631
ISSN 0076-6879
https://doi.org/10.1016/bs.mie.2019.08.001

B (GRZB), act as crucial effectors against tumor onset and progression. However, in response to immune pressure, cancer cells use different strategies to favor the establishment of an immunosuppressive tumor microenvironment (TME). Elucidating the dynamics of tumor-host co-evolution provides novel opportunities for personalized cancer immunotherapies. The *in sitro* (*in vitro*+*in situ*) technology is an experimental approach involving the preparation of heterocellular cell suspensions from fresh tumors and their use *in vitro*. The *in sitro* experimental setup offers the possibility to (1) analyze immune-related parameters (*e.g.*, quantification of cytokines released in the TME), (2) reveal the mechanism of action of drugs, and (3) unveil crucial cell-intrinsic and cell-extrinsic processes boosting anticancer immune responses. Nonetheless, the *in sitro* technology does not fully recapitulate the complexity of the tissue "*in situ*" nor models the patterns of infiltrating immune cell localization, and hence parallel experimentation should be scheduled. In this chapter we discuss *in sitro* technology to analyze and quantify IFN-γ and GRZB production by T cells either co-cultured with cancer cells in the presence of exogenous adjuvant *stimuli* (*i.e.*, an antibody targeting the immune checkpoint programmed cell death protein 1, and recombinant calreticulin) and boosting with TAAs (*i.e.*, the model SIINFEKL ovalbumin antigen). Specifically, we describe IFN-γ and GRZB quantification by flow cytometry, ELISA and ELISpot technologies.

Abbreviations

Ags antigens
ALP alkaline phosphatase
APC allophycocyanin
DCs dendritic cells
ELISA enzyme-linked immunosorbent assay
ELISpot enzyme-linked immunospot
FBS fetal bovine serum
FITC fluorescein
ISGs IFN-stimulated genes
MDSCs myeloid-derived suppressor cells
MHC major histocompatibility complex
NK natural killer
OVA ovalbumin
PBS phosphate-buffered saline
PE phycoerythrin
PHA phytohemagglutinin
PMA phorbol myristate acetate
recCALR recombinant calreticulin
RPMI Roswell Park Memorial Institute
RT room temperature
TAAs tumor-associated Ags
TILs tumor-infiltrating lymphocytes
TME tumor microenvironment
TNAs tumor neoantigens
Treg regulatory T

1. Introduction

Cancer is a genetic disease commonly initiated and sustained by a sequence of mutations on oncogenic drivers (Pon & Marra, 2015). Such pattern of mutation-driven evolution endows cancer cells with the capability to adapt to the harsh and variable tumor microenvironment (TME) overcoming the nutritional and immunological barriers it constantly imposes, yet it comes at a cost. Indeed, a high mutation rate in cancer cells fosters the emergence of genetic alterations, including non-synonymous point mutations, which are the major potential source for tumor neoantigens (TNAs) (Vitale et al., 2019). This condition is also associated with deregulated expression of tumor-associated antigens (TAAs), which are antigens (Ags) that malignant cells express more, but not uniquely, compared with normal adult and/or embryonic cells (Coulie, Van den Eynde, van der Bruggen, & Boon, 2014). This is relevant for anti-cancer immunity, as the more cancer cells accumulate "non-self" Ags, the higher is their visibility to the immune system (Chen & Mellman, 2013; Matzinger, 2002).

Immune recognition of cancer cells imposes a selection force which shapes the balance between tumor progression and tumor control (Schreiber, Old, & Smyth, 2011; Vitale et al., 2019). Under the continuous pressure imposed by immune effectors, cancer cells take advantage of different mechanisms of immune evasion, including immunoselection of poorly immunogenic variants, activation of immune subversion mechanisms (*e.g.*, the release of anti-inflammatory cytokines or induction of regulatory cells), and expression of immune checkpoint molecules (Schreiber et al., 2011; Vitale et al., 2019). Among the immune cell players exerting constant vigilance within the TME, CD8 T cells are those detecting TNAs and specific for TAAs (Chen & Mellman, 2013). Such ability of CD8 T cells was discovered in mouse tumor models about 40 years ago by Thierry Boon and Hans Schreiber, among the others (De Plaen et al., 1988; Stauss, Van Waes, Fink, Starr, & Schreiber, 1986). As proof of the broadly-recognized relevance of these immune system components, currently tumors are roughly divided into T cell-inflamed, or "hot" tumors, and T cell-non inflamed, or "cold" tumors (Chen & Mellman, 2017; Palucka & Coussens, 2016), being these phenotypes critical factors predicting therapeutic success (Chen & Mellman, 2017).

At the mechanistic level, upon encountering TAAs and TNAs in lymphoid organs, CD8 T cells initiate a multi-phase response impacting on both the immune and tumor systems. Briefly, tumor Ag-specific CD8 T cells sequentially: (1) undergo a multi-log clonal expansion; (2) home to the tumor where interactions with Ags continue, thus driving further on-site expansion and differentiation; (3) exert cytolitic activity; (4) go through an apoptosis-related contraction during which a persistent population of tumor Ag-memory cells is spared so as to confer long-life protection to the host (Klebanoff, Gattinoni, & Restifo, 2006; van der Woude, Gorris, Halilovic, Figdor, & de Vries, 2017; Zhang & Bevan, 2011).

In this context, CD8 T cell-mediated demise of target cells during the cytolitic phase is mainly due to the action of interferon gamma (IFNG, best known as IFN-γ) and granzyme B (GZMB, also known as GRZB) (Davis & Bjorkman, 1988).

IFN-γ is an inflammatory cytokine universally considered a major immune effector (Boehm, Klamp, Groot, & Howard, 1997; Schroder, Hertzog, Ravasi, & Hume, 2004). By modulating the expression of hundreds of IFN-stimulated genes (ISGs), IFN-γ exerts a plethora of biologic functions encompassing inflammation, apoptosis, cell cycle regulation, and gene expression modulation (Schneider, Chevillotte, & Rice, 2014). Specifically, IFN-γ markedly upregulates, on target cells, the major histocompatibility complex class I (MHC-I) and MHC-II (Amaldi, Reith, Berte, & Mach, 1989; Shirayoshi, Burke, Appella, & Ozato, 1988; Steimle, Siegrist, Mottet, Lisowska-Grospierre, & Mach, 1994) as well as the whole Ag processing and presentation machinery, which includes proteasomal molecules, the transporter associated with antigen processing 1/2 (TAP-1/2) and chaperone proteins (Belich, Glynne, Senger, Sheer, & Trowsdale, 1994; Cramer, Nelson, & Klemsz, 2000; Johnson & Pober, 1990; Kern, Steimle, Siegrist, & Mach, 1995; Lah, Hawley, Rock, & Goldberg, 1995; Nandi, Jiang, & Monaco, 1996; Seliger et al., 2001). Through this action, IFN-γ enhances CD8 T cell-mediated cytotoxicity. Moreover, IFN-γ is essential for innate immune cell activation (Pak-Wittel, Yang, Sojka, Rivenbark, & Yokoyama, 2013; Soudja et al., 2014) and for CD8 T cell proliferation and differentiation (Maraskovsky, Chen, & Shortman, 1989). It also displays antiangiogenic (Beatty & Paterson, 2001; Hayakawa et al., 2002; Kammertoens et al., 2017), cytostatic (Bromberg, Horvath, Wen, Schreiber, & Darnell, 1996; Chin et al., 1996; Dighe, Richards, Old, & Schreiber, 1994), proapoptotic (Chin, Kitagawa, Kuida, Flavell, & Fu, 1997; Fulda & Debatin, 2002; Liu et al., 2011; Xu, Fu, Plate, & Chong, 1998), and pronecroptotic (Thapa et al., 2011) properties. Thus, IFN-γ stands out as a gold-standard antitumor cytokine. That said, when hugely and/or chronically present in the TME, IFN-γ is also reported to play a protumorigenic role, exerted by favoring immunosuppression through recruitment of regulatory T (Treg) cells and myeloid-derived suppressor cells (MDSCs) (Brody et al., 2009; Cripps, Wang, Maria, Blumenthal, & Gorham, 2010), and upregulation of metabolic enzymes such as indoleamine 2,3-dioxygenase (IDO) and checkpoint inhibitors such as CD274 (best known as PD-L1) a ligand of programmed cell death 1 (PDCD1; best known as PD-1) (Abiko et al., 2015; Jurgens, Hainz, Fuchs, Felzmann, & Heitger, 2009). Irrespective of these observations, specific IFN-γ signatures and IFN-γ-mediated responses have been positively correlated with patient's survival in different cancer types (Vitale et al., 2019).

GRZB is a serine protease physiologically retained in secretory granules of CD8 T and natural killer (NK) cells (Rousalova & Krepela, 2010). Once cytotoxic cells encounter infected or neoplastic cells, GRZB is vectorially released into immunological synapses and then internalized into target cells where it induces regulated cell death either by direct activation of executioner caspases (*e.g.*, caspase 3) or *via* mitochondrial outer membrane permeabilization (a sign of ongoing regulated cell death) (Anthony, Andrews, Watt, Trapani, & Smyth, 2010; Rousalova & Krepela, 2010; Waterhouse, Sedelies, & Trapani, 2006).

The *in sitro* technology refers to the specific experimental setup developed to help dissect tumor-host crosstalk and co-evolution involving the explant from a fresh tumor of a cell suspension consisting of malignant and non-malignant cells (*i.e.*, stromal and immune cells) and the use of such heterocellular suspension for *in vitro* experimentation (Zitvogel, Pitt, Daillere, Smyth, & Kroemer, 2016). Through such an approach of integrating *in vitro* and *in situ* procedures, the *in sitro* technology offers the possibility to collect immune behavioral data over several days and under variable experimental conditions. Indeed, co-culturing heterocellular suspensions from *ex vivo* explanted tumors allows for recreating and investigating the host–tumor microenvironment. Specific co-culture settings that can be (re)created from fresh tumors for *in sitro* studies include: (1) cancer cells with a pool of $CD45^+$ tumor infiltrating lymphocytes (TILs) like dendritic cells (DCs), macrophages, T cells and NK cells, (2) cancer cells with a selected and isolated subset of TILs (*e.g.*, CD8 T cells), and (3) cancer cells with selected TILs and stromal cells. Of note, challenging this system with chemo-, targeted- and immuno-therapeutics and then analyzing cancer- stromal- and/or immune-related parameters by flow cytometry, enzyme-linked immunosorbent assay (ELISA) and enzyme-linked immunospot (ELISpot), can also lead to the identification and/or validation of crucial mechanisms of the dynamic evolution of tumor-host interplays (Zitvogel et al., 2016).

Here, we describe the use of the *in sitro* technology in the field of oncoimmunology, focusing on the analysis of OT-1 CD8 T cell activation through quantification of IFN-γ and GRZB. We detail protocols of *ex vivo* isolation of MCA.205 ovalbumine (OVA)-expressing tumor cells and $H\text{-}2^b$ histocompatible OT-1 CD8 T cells, and their co-culture and *in vitro* stimulation with exogenous adjuvant *stimuli*. Moreover, we provide methodological details on how to detect and quantify IFN-γ and GRZB by flow cytometry, ELISA and ELISpot. Finally, we discuss the advantages and limitations of the *in sitro* technology, and consider the potential of this methodology for the simplified screening of novel, effective anticancer treatments.

2. Materials

2.1 Disposable

- 75 cm^2 cell culture treated flasks (Corning®, New York, NY)
- 48-well plates (Corning®)
- Elispot MultiScreen®HTS Filter Plates (Millipore, Merck, Darmstadt, Germany)
- 70 μm cell strainer (Falcon, Corning®)
- Cell counting slides
- 15 and 50 mL tubes (Falcon, Corning®)
- 5 mL polystyrene tubes (Falcon, Corning®)
- 1,5 mL eppendorf microtubes (Eppendorf®, Hamburg, Germany)
- Pipet aid and serological pipettes (2 mL, 5 mL, 10 mL, 25 mL, 50 mL)
- Pipettes and sterile tips (1–10 μL, 20–200 μL, 1000 μL)
- Multichannel pipette (20–200 μL)
- Insulin syringes (BD Veo™ insulin syringes with BD Ultra-Fine™ 6 mm x 31G needle, from BD Biosciences, Becton, Dickinson and Company, Franklin Lakes, NJ)
- MACS separation columns (LS, MS Miltenyi Biotec, Bergisch Gladbach, Germany)
- MACS Separators (MiniMACS™, MidiMACS™, Miltenyi Biotec)
- MACS Multistand (Miltenyi Biotec)
- Aluminum foil
- Ice
- Liquid nitrogen
- Timer

- Caliper
- Sterile scalpels, surgical forceps, scissors and pliers
- Plastic wash bottle
- Adsorbent wipes

2.2 Equipment

- Microbiologically controlled animal facility equipped with Class II safety cabinet
- Humidified cell culture incubator at 37°C with 5% CO_2
- Laboratory safety cabinet (Class II)
- Laboratory bench
- Laboratory sink
- Optical microscope
- Refrigerable centrifuge
- Laboratory refrigerator (4°C)
- Ice machine
- Thermostatic bath
- Flow cytometry (such as BD FACSCanto II, BD Biosciences)
- Software for flow cytometry analysis [such as FlowJo software (FlowJo, LLC, Becton, Dickinson and Company)]
- ELISA microplate reader (AID GmbH, Strassberg, Germany)
- ELISpot microplate reader (AID Classic or AID Ispot, AID GmbH).
- Software for ELISpot analysis [such as AID EliSpot software Version 7.0 (AID GmbH)]
- Software for statistical analysis (such as GraphPad Prism, GraphPad Software, San Diego, CA)

2.3 Reagents

- MCA.205-OVA murine fibrosarcoma cells
- Roswell Park Memorial Institute (RPMI) 1640 medium supplemented with 10% (v/v) fetal bovine serum (FBS), 2 mM L-glutamine, 100 IU/mL penicillin G sodium salt, 100 μg/mL streptomycin sulfate, and 50 μg/mL hygromycin (all from Gibco-Thermo Scientific, Thermo Fisher Scientific Inc., Waltham, MA)
- Phosphate-buffered saline (PBS, Gibco-ThermoFisher)
- Trypan blue (Gibco-ThermoFisher)
- Trypsin (Gibco-ThermoFisher)

- MCA.205-OVA hystocompatible OT-1H-2^b C57Bl/6 immunocompetent mice (Charles River, Wilmington, MA)
- Collagenase IV (Sigma-Aldrich, from Millipore-Sigma, Saint-Louis, MO)
- DNase I (Sigma-Aldrich)
- Separation buffer (Miltenyi Biotec)
- CD8a MicroBeads (Miltenyi Biotec)
- Non-tumor cell depletion cocktail (Miltenyi Biotec)
- Anti-mouse PD-1 and rat IgG2a isotype control (BioXCell, West Lebanon, NH)
- Rec-CALR
- SIINFEKL peptide (Sigma-Aldrich)
- Cold PBS-FBS solution (1% FBS in sterile PBS)
- BD Golgi Plug (BD Biosciences)
- Anti-FcγII/III blocking antibody (BD Biosciences)
- Anti-mouse allophycocyanin (APC)-conjugated CD8a (clone 53–6.7, eBioscience™, ThermoFisher) and IgG isotype control (eBioscience™, ThermoFisher)
- Fixation/Permeabilization solution (BD Biosciences)
- BD Perm/Wash buffer (BD Biosciences)
- Anti-mouse fluorescein (FITC)-conjugated IFN-γ (eBioscience™, ThermoFisher) and IgG isotype control (eBioscience™, ThermoFisher)
- Anti-mouse phycoerythrin (PE)-conjugated GRZB (clone NGZB, eBioscience™- ThermoFisher) and IgG isotype control (eBioscience™, ThermoFisher)
- Vitality fixable dead-cell blue dye (L23105, ThermoFisher)
- IFN-γ and GRZB R&D Systems Quantikine ELISA Kits (R&D Systems, Minneapolis, MN) containing:
  - Pre-coated wells
  - Plate sealers
  - Assay Diluent RD1–21
  - IFN-γ or GRZB standards
  - IFN-γ or GRZB positive controls
  - IFN-γ or GRZB Conjugate
  - Wash Buffer
  - Substrate Solution
  - Stop Solution
- Mouse IFN-γ and GRZB ELISpot Kits (Mabtech, Nacka Strand, Sweden) containing:
  - IFN-γ or GRZB Biotinylated Detection Antibodies

- Streptavidin conjugated to Alkaline Phosphatase (ALP)
- Dilution Buffers
- Wash Buffer Concentrate
- ALP-substrate, i.e., 5-bromo-4-chloro-3′-indolyphosphate p′-toluidine salt (BCIP)/nitro blue tetrazolium chloride (NBT) Chromogen
- Mouse IFN-γ or GRZB Positive Control

- 70% ethanol (v/v in sterile distilled water)
- PBS-FBS-Tween solution (0.5% FBS and 0.01% Tween® 20 detergent in sterile PBS)
- Tween 20 (Sigma-Aldrich)

3. Methods

3.1 Cell culture

1. MCA.205 murine fibrosarcoma cells stably transfected to express the OVA Ag (MCA.205-OVA) are routinely cultivated in a Roswell Park Memorial Institute (RPMI) 1640 growth medium supplemented with 10% (v/v) fetal bovine serum (FBS), 2 mM L-glutamine, 100 IU/mL penicillin G sodium salt, 100 μg/mL streptomycin sulfate, and 50 μg/mL hygromycin, under standard culture conditions (*i.e.*, at 37°C and with 5% CO_2) (*see* **Notes 1,2**).
2. When maintenance cultures reach 75–80% confluence (*see* **Note 3**), growth medium is discarded from the cell monolayer, and, upon a wash with pre-warmed phosphate-buffered saline (PBS) to completely remove FBS, cells are detached by incubating them for 1–2 min at 37°C with pre-warmed trypsin (*see* **Notes 4–7**).
3. Growth medium is added to detached cells, which are then centrifuged for 5 min at 1200 rpm at room temperature (RT) and reseeded for maintenance culture and/or experimental procedures into appropriate supports (*see* **Note 8**).

3.2 Tumor injection, explant and processing

1. For *in vivo* injection, MCA.205-OVA murine fibrosarcoma cells are seeded in 75 cm^2 flasks at an approximate density of 5×10^6 cells/flask in 15 mL of growth medium.
2. When cells reach semiconfluence (*i.e.*, at an approximate density of $10–15 \times 10^6$ cells/flask; *see* **Note 9**), growth medium is removed, and

cells are detached with trypsin as reported in subheading 3.1 (steps 2,3) and washed twice for 5 min at 1200 rpm at RT with pre-warmed PBS (*see* **Note 10**).

3. Murine cancer cells are then resuspended at a concentration of 8×10^5 cells in 100 μL of PBS *per* mouse (*see* **Note 11**) and subcutaneously injected in the lower flank of H-2^b histocompatible OT-1 C57Bl/6 immunocompetent mice (females of 6/7-week-old) (*see* **Notes 12,13**) with insulin syringes.
4. Tumor growth is monitored twice *per* week with a common caliper, and when tumors reach a surface (longest dimension × perpendicular dimension) of 40–45 mm^2 (*see* **Note 14**), mice are euthanized (*see* **Note 15**).
5. Tumor draining lymph nodes of euthanized mice are sampled (*see* subheading 3.4) while tumors are excised using sterile scalpels, surgical forceps and pliers under the hood and placed into sterile 1.5 mL eppendorf tubes with 1 mL of growth medium on ice.
6. When all tumors are collected, growth medium is removed and tumors are placed in new 1.5 mL eppendorf tubes with 1 mL of RPMI 1640 medium supplemented with 400 U/mL Collagenase IV and 200 U/mL DNase I (*i.e.*, digestion buffer) (*see* **Note 16**).
7. Tumor samples are cut with sterile surgical scissors until a roughly homogenized solution of 1-to-3 mm tumor pieces is obtained and incubated for 30 min at 37°C.
8. After the incubation with digestion buffer, tumor samples are cut again until tumor pieces are completely dissociated and a dense cell suspension obtained (*see* **Note 17**), transferred into 50 mL tubes through 70 μm cell strainers, and washed twice with growth medium by centrifuging them for 5 min at 1200 rpm at RT before cell count.

3.3 Isolation of OT-1 CD8^{pos} TILs and cancer cells

1. Cell pellet obtained after isolation/processing of tumor-xenografts as described in the previous subheading 3.2 is resuspended in 90 μL of separation buffer with 10 μL of CD8a MicroBeads *per* 1×10^7 cells (*see* **Note 18**).
2. Upon incubation for 10 min at 4°C, 500 μL of separation buffer are added, and samples are transferred into MACS separation columns (*see* **Note 19**).
3. Columns are rinsed with separation buffer, and cell suspension applied onto the column. During this step, unlabeled cells (which also contains

cancer cells) are removed in the flow-through and kept on ice for subsequent purification (*see* description from step 7).

4. Columns are placed on suitable collection tubes, and labeled OT-1 CD8^{pos} T cells are eluted by adding separation buffer and firmly pushing plugers into the columns.
5. The separation step is repeated to reach a purity of approximately 96–98% (*see* **Note 20**).
6. OT-1 CD8^{pos} fraction is then checked for purity by flow-cytometry using a BD FACSCanto II.
7. Meanwhile, unlabeled cancer cell suspension is washed with PBS and centrifuged for 5 min at 1200 rpm at RT.
8. For cancer cell enrichment (performed with non-tumor cell depletion cocktail, Miltenyi Biotec) (*see* **Note 21**), supernatants are completely removed, and cell pellet resuspended in 80 μL of separation buffer plus 20 μL of Non-Tumor Cell Depletion Cocktail *per* 1×10^7 cells.
9. Cells are incubated for 15 min at 4°C, and after the addition of 10 mL of separation buffer followed by centrifugation for 10 min at 1200 rpm at RT, cell pellet is resuspended in 500 μL of separation buffer (up to 1×10^7 cells) and transferred into MACS separation columns (*see* **Notes 18,19**).
10. Columns are rinsed with separation buffer, and cell suspension applied onto the column.
11. Unlabeled cancer cells are collected in the flow-through and columns are washed twice to enrich the cancer cell fraction.
12. Cancer cells are then washed for 5 min at 1200 rpm at RT in growth medium before cell count.

3.4 Isolation of OT-1 CD8^{pos} cells from tumor-draining lymph node

1. Tumor draining lymp nodes are explanted from immunocompetent mice bearing syngeneic tumors implanted as described in subheading 3.2.
2. To isolate effector OT-1 CD8 T cells, organs are enzimatically digested by Collagenase IV (400 U/mL in RPMI 1640) and DNase I (200 U/mL in RPMI 1640) (*see* also subheading 3.2, steps 6–8).
3. OT-1 CD8^{pos} cells purified as described in subheading 3.3, steps 1–6.
4. OT-1 CD8^{pos} cells are then washed for 5 min at 1200 rpm at RT in growth medium before cell count.

3.5 *In sitro* technology

3.5.1 Stimulation of OT-1 CD8^{pos} cells for flow cytometry and ELISA

1. OT-1 CD8^{pos} cells isolated from explanted tumors (subheading 3.3) or draining lymph nodes (subheading 3.4) of xenografted mice are counted, and then 1×10^4 OT-1 CD8^{pos} cells seeded in 48-well plates resuspended in 0.5 mL of growth medium.
2. MCA.205-OVA cancer cells are added to the system at a 1:10 ratio with OT-1 CD8^{pos} cells (1×10^3 MCA205 cancer cells and 1×10^4 OT-1 CD8^{pos} cells) (*see* **Notes 22, 23**).
3. Co-cultured cancer/OT-1 CD8^{pos} cells are incubated for 18-to-24 h at 37°C in the presence of an exogenous adjuvant *stimulus*. For recombinant calreticulin (rec-CALR) testing, MCA.205-OVA cells are pre-incubated with 3 μg of rec-CALR *per* 10^6 cells in 100 μL of PBS on ice for 30 min, followed by three washes. For anti-PD-1 studies, a monoclonal anti-mouse PD-1 or a rat IgG2a isotype control, both at a final concentration of 10 μg/mL, are added. For peptide-based stimulation, OT-1 cells are cultured in the presence of 2 μg/mL OVA SIINFEKL peptide or 3×10^4 MCA.205-OVA cells killed by a 5 min heating at 42°C followed by 1 cycle of freezing/thawing in liquid nitrogen (*see* **Notes 24,25**).
4. For the cytofluorometric assessment of OT-1 CD8^{pos} T cell activation (*see* subheading 3.5.2), 1 μL (*per* mL of cell culture) of BD Golgi Plug containing the protein transport inhibitor brefeldin A is added to the co-culture for additional 6 h so as to promote intracytoplasmic cytokine accumulation (*see* **Notes 26,27**).
5. Alternatively, for ELISA assay (*see* subheading 3.5.3), 24 h after incubation, co-cultured cells are collected and centrifuged for 5 min at 1200 rpm at 4°C (*see* **Note 28**), after which supernatants are transferred in 1.5 mL eppendorf tubes and immediately put on ice for ELISA assay (*see* **Note 29**).

3.5.2 Cytofluorometric assessment of OT-1 CD8^{pos} T cell activation

1. The OT-1 CD8^{pos}-MCA.205-OVA co-cultured cells or OT-1 CD8^{pos} cells stimulated with splenic DCs pulsed with SIINFEKL peptide (refer to Section 3.5.1, step 3) are resuspended in 100 μL of a PBS solution containing 1% FBS and 1 μg of anti-FcγII/III blocking antibody *per* 1×10^6 cells, and incubated for 20 min at 4°C (*see* **Note 30**).
2. For cell surface staining, cells are centrifuged for 5 min at 1200 rpm at 4°C and then resuspended in 100 μL of 1% FBS in cold PBS containing 0.25 μg

of either IgG isotype control or anti-mouse allophycocyanin (APC)-conjugated CD8a (*see* **Note 31**).

3. Cells are incubated for 20 min at 4°C in the dark (*see* **Notes 32,33**) and washed twice in 1% FBS in cold PBS.
4. The vitality fixable dead-cell blue dye is added at a final concentration of 1 μM (*see* **Note 34**).
5. Cells are incubated for 30 min on ice and protected from light and then washed in 1% FBS in cold PBS.
6. Upon centrifugation, cell pellets are resuspended in 100 μL of Fixation/Permeabilization solution and incubated for 20 min at 4°C in the dark (*see* **Note 35**) and then washed twice in 250 μL of cold BD Perm/Wash buffer (*see* **Note 36**).
7. Cell pellets are resuspended with 50 μL of cold BD Perm/Wash buffer containing 0.25 μg of either IgG isotype control or anti-mouse fluorescein (FITC)-conjugated IFN-γ or anti-mouse phycoerythrin (PE)-conjugated GRZB (*see* **Note 37**) and incubated for 20 min at 4°C in the dark.
8. Samples are assessed for CD8a and IFN-γ or GRZB positivity using a BD FACSCanto II and analyzed by means of FlowJo software.
9. During acquisition and analysis, samples are sequentially gated for: morphology, by plotting forward scatter area (FSC-A, x-axis) and side scatter area (SSC-A, y-axis) (gate 1); singlets, by plotting SSC-A (x-axis) and side scatter width (SSC-W, y-axis) (gate 2); viability, by plotting blue laser are (emission 461 nm, x-axis) and SSC-A (y-axis) (*see* **Note 38**) (gate 3); and finally CD8 positivity, by plotting red laser (emission 657 nm, x-axis) and SSC-A (y-axis) (gate 4). On gate 4, cells are analyzed for PE/GRZB and FITC/IFN-γ, by plotting red laser area (emission 573 nm, y-axis) and green laser area (emission 519 nm, y-axis), respectively (*see* **Note 39**) (Fig. 1).

3.5.3 Analysis of CD8^{pos} T cell activation by ELISA assay

1. For IFN-γ and GRZB quantification, 50 μL of Assay Diluent RD1–21 are added to each well of a pre-coated 96 well plate together with either 50 μL of standard, 50 μL of control or 50 μL of supernatants from CD8^{pos} cell cultures obtained as reported in subheading 3.5.1 (step 1–3 and 5) (*see* **Note 40**).
2. After five washes with Wash Buffer, by filling wells with a wash bottle and blotting plate on adsorbent wipes, 100 μL of Mouse IFN-γ or

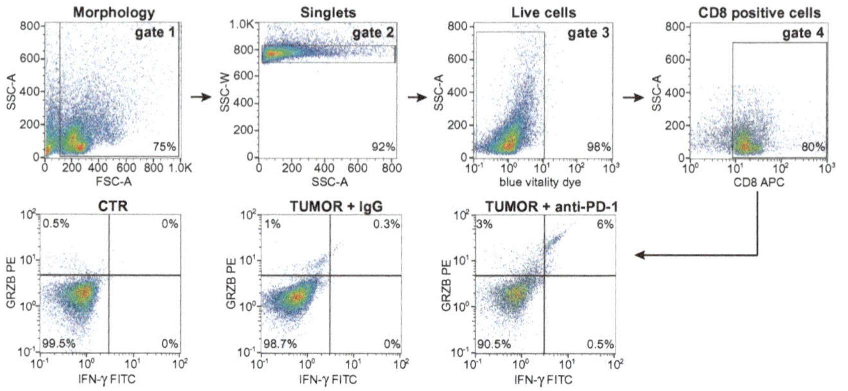

Fig. 1 Cytofluorometric analysis of intracellular IFN-γ and GRZB positivity in purified OT-1 CD8 T cells co-cultured with MCA.205-OVA explanted tumor cells in the presence or not of anti-PD-1 antibody or IgG control. Plots depict the percentages of cells previously gated on morphology, singlets, viability and CD8 positivity.

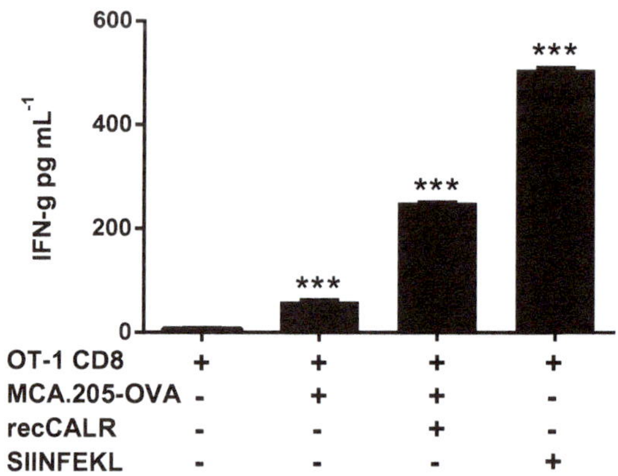

Fig. 2 ELISA quantification of IFN-γ levels in supernatants of *ex vivo* purified *in sitro* tested OT-1 CD8 T cells co-cultured with MCA.205-OVA explanted tumor cells precoated or not with recombinant calreticulin (recCALR) or boosted with the OVA SIINFEKL peptide. Data show mean ± s.d. of triplicate wells. *$P < 0.05$, **$P < 0.01$, ***$P < 0.001$ (one-way ANOVA and Bonferroni post-hoc test) compared to OT-1 CD8 T cells alone.

GRZB Conjugate are added to each well, and samples are covered with plate sealers and incubated for 2 h at RT (*see* **Note 41**).

3. Wells are washed five times with Wash Buffer, after which samples are incubated for 30 min at RT in the dark with 100 μL of Substrate Solution (*see* **Note 42**).
4. One hundred μL of Stop Solution are added and samples analyzed in an ELISA microplate reader set to 450 nm (*see* **Notes 43,44**) (Fig. 2).

3.5.4 Analysis of CD8^{pos} T cell activation by ELISpot assay

1. For quantitative determination of IFN-γ and GRZB by ELISpot assay, wells of an Elispot MultiScreen®HTS Filter Plates are pre-wet for 1 min at RT with 15 μL of 70% ethanol (v/v in sterile distilled water) (*see* **Notes 45,46**).
2. After three washes with 150 μL of sterile PBS (*see* **Note 47**), using a multichannel pipette (20–200 μL), plate is incubated overnight at 4°C with 10 μg/mL of capture anti-IFN-γ or capture anti-GRZB monoclonal antibody resuspended in 100 μL of sterile PBS (*see* **Note 48**).
3. Sixtheen-to-18 h later, coating solution is discarded, and plate wells washed four times with 150 μL of sterile PBS.
4. Membrane blocking is then performed by adding 150 μL of growth medium for at least 2 h at 37°C (*see* **Note 49**).
5. After blocking medium removal, 5×10^3 cancer cells and 5×10^4 CD8^{pos} T cells (isolated as reported in subheadings 3.3, 3.4) are resuspended in 100 μL of growth medium and co-cultured in each plate well (*see* **Notes 50, 51**).
6. Otherwise 5×10^4 OT-1 CD8^{pos} T cells are stimulated with MCA.205-OVA cell lysate or splenic DCs (5×10^3) pulsed with SIINFEKL peptide (refer to subheading 3.5.1, step 3).
7. OT-1 CD8^{pos} T cell - MCA.205-OVA cell co-culture is supplemented with recCALR, or monoclonal anti-mouse PD-1 or rat IgG2a isotype control (refer to subheading 3.5.1, step 3) and incubated for 18-to-48 h (*see* **Note 52**) at 37°C, 5% CO_2 and 95% humidity.
8. Cells are then discarded, and plate wells washed five times with 150 μL of a solution containing 0.5% FBS and 0.01% Tween® 20 detergent in sterile PBS (hereafter referred to as PBS-FBS-Tween solution) (*see* **Notes 53,54**).
9. Plate wells are then incubated for 2 h at 37°C with 100 μL of PBS-FBS-Tween solution containing the detection biotinylated anti-IFN-γ or anti-GRZB antibody at a final concentration of 1 μg/mL (*see* **Note 55**).
10. Upon five washes with 150 μL of PBS-FBS-Tween solution, plate wells are incubated for a time ranging from 45 min-to-1 h at RT with 100 μL of streptavidine-ALP diluted 1:1000 in PBS-FBS-Tween solution (*see* **Note 56**).
11. Plate wells are washed three times with 150 μL of PBS-FBS-Tween solution and three times with 150 μL of sterile PBS (*see* **Note 57**) after which 150 μL of ALP-substrate (BCIP/NBTplus) are added to each well.

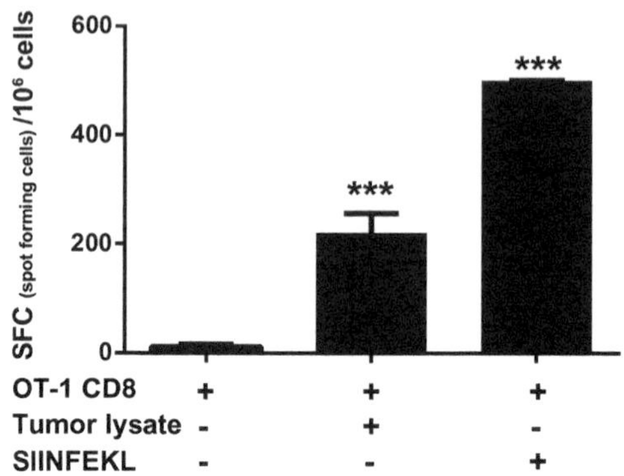

Fig. 3 OVA-specific IFN-γ spot forming cells (SFC) of *in sitro* analyzed OT-1 CD8 T cells stimulated for 48 h with MCA.205-OVA cell lysate (3×10^4 cells) or 2 μg/mL SIINFEKL peptide loaded on splenic DCs. Quantitative data from 3 independent experiments are shown as means ± s.d. *P < 0.05, **P < 0.01, ***P < 0.001 (one-way ANOVA and Bonferroni post-hoc test) compared to OT-1 CD8 T cells alone.

12. Upon an incubation time ranging from 5-to-15 min at RT in the dark (*see* **Note 58**), plate is extensively washed under running tap water to stop reaction.
13. Plate is blotted on absorbent wipes to remove liquid and let dry overnight in the dark (*see* **Note 59**) before analysis using an ELISpot microplate reader for acquisition and the AID EliSpot software Version 7.0 for analysis (Fig. 3).

4. Discussion and concluding remarks

In accordance with the guiding principles for animal experimentation, the scientific community advocates for suitable alternatives to replace or minimize the number of *in vivo* studies. In this context, *in sitro* technology is a valid experimental strategy to investigate the antineoplastic effects of selected compounds using a relatively limited number of animals as it allows for an *ex vivo* reproduction of the TME and draining lymphoid organs. Moreover, the *in sitro* technology has as advantage the possibility to include in the experimental setting either all major immune cell types or only those/that of interest. Specifically, this technology can inform on the efficacy of multiple drugs or drug combinations, administered at different doses, schedules or kinetics at the same time as it makes possible to monitor immune cell activation, proliferation and exhaustion in the presence of the aforementioned *stimuli*. The *in sitro* technology thus offers an affordable and reproducible way to predict the clinical efficacy of tested compounds irrespective of their nature (whether they are small molecules, biologics, cell therapies or combinations thereof), helping unveil their potential mechanism of action in the presence of malignant and non-malignant cells of the TME.

Here, taking advantage of the *in sitro* technology, we demostrated the ability of distinct exogenous *stimuli*, namely anti-PD-1 antibody, recCALR and the OVA peptide SIINFEKL, to (re)activate effector OT-1 $CD8^{pos}$ T cells when co-cultured *in vitro* with MCA.205 OVA-expressing cancer cells. Specifically, by flow cytometry-mediated quantification of intracellular cytokines, we revealed the ability of anti-PD-1 antibody to (re)instate CD8 T cell effector function of producing IFN-γ and GRZB (Fig. 1), which is in line with previous observations (Goldberg et al., 2007). Moreover, through ELISA and ELISpot assays, we showed increased release of IFN-γ from CD8 T cells when these latter are either co-cultured with tumor cell pre-coated with recCALR, which enhances Ag presentation (Hannani et al., 2011) (Fig. 2), or boosted with SIINFEKL peptide and tumor cell lysate (Fig. 3).

That said, it should be taken into account that the *in sitro* technology is not representative of the full complexity of the tissue "*in situ*" and of the immune cell recruitment into the TME. So, correspondence between the *in sitro* models and the *in vivo* environment should be properly ascertained in parallel analyses for instance by tracking immune cell trajectories through *in vivo* reporter systems and *in vitro* microfluidic devices, and by *ex vivo* multiparametric immunohistochemistry. Acknowledging such limitation, the *in sitro* technology is, however, a valid integrated approach for screening different drugs on different microenvironments limiting the number of animals to be used. In conclusion, this technology not only helps explore basic biology, but can also provide critical issues of patient responses to different therapeutic strategies with great impact on translational biology in an era of personalized precision medicine.

Notes

1. Culture conditions may differ for other cell lines.
2. MCA.205-OVA cells optimally grow in 25-, 75-, or 175-cm^2 flasks. As a standard rule, MCA.205-OVA cells take 18-to-24 h for cell doubling and reach 75–80% confluency in 25-, 75-, or 175-cm^2 at 5-to-10 × 10^6, 20-to-30 × 10^6, 40-to-60 × 10^6 cell density, respectively.

3. Cell state may vary if cells are splitted at suboptimal cell densities. Confluency >85% may favor starvation thus inducing cell stress, indeed confluency <30% may select cell clones and change the parental genetic background.
4. Trypsinization may be inhibited by protease inhibitors contained in FBS.
5. Trypsin enzymatic activity is optimal at 37°C, so, as a general guideline, trypsin should be pre-warmed.
6. Trypsin-induced cell detachment widely differs depending on the cell type. However, as a standard rule, long term trypsinization may strip cell surface proteins and trigger cell stress and death.
7. Cell detachment state should be checked by a light microscope.
8. Optimal cell dilutions vary among different cell lines. MCA.205-OVA cells reach 75–80% confluency every 3 days if diluted 1:20.
9. For efficient *in vivo* tumor growth, tumor cells need to be in proliferation, while appropriate cell density is related to cell type. According to our experience, for MCA.205-OVA cells, the optimal monolayer density in 75 cm^2 flasks is of approximately 10 to 15×10^6 cells in 15 mL of growth medium.
10. Rinsing of FBS traces assures unwanted inflammatory reactions that may clear part of the implanted syngeneic tumor before it is able to firmly establish.
11. According to our experience, it is a good practice to prepare single vials for each mouse instead of a unique batch. This guarantees a high reproducibility and accuracy of the number of injected cancer cells.
12. For cancer cells to establish tumors and grow in immunocompetent hosts, histocompatibility needs to be ascertained to avoid immune-mediated rejection.
13. OT-1H-2^b C57Bl/6 immunocompetent mice carry a transgenic T cell receptor responsive to OVA residues 257–264 (SIINFEKL).
14. The time needed to reach a good tumor size (easily explantable and in active proliferation, still within the ethical range of dimension and status) is a function of the cell type. In our hands, subcutaneous injection of 8×10^5 cells ensures the development of 40–45 mm^2 tumors in 8-to-10 days.
15. In accordance with standard directives for animal research.
16. Serum-free medium is recommended, as protease inhibitors in the FBS may limit digestion efficiency (*see* Note 4).
17. To optimize single cell yield, a second mechanical dissociation should be performed before rinsing the digesting enzymes.

18. To increase the number of cells, volumes should be scaled up accordingly.
19. Alternatively, fully automated cell separation (autoMACS Pro Separator, from Miltenyi Biotec) could be performed.
20. At this stage a fraction of CD8a^{neg} cells still lasts. A second separation procedure, by using a new column, is strongly recommended to increase the purity of CD8a^{pos} cells to levels >96%.
21. To improve downstream analyses, depletion of tumor associated cells such as lymphocytes, fibroblasts, and endothelial cells is strongly recommended.
22. Cell number and growth medium volumes could be scaled up or down preserving the 1:10 ratio.
23. The optimal ratio CD8^{pos} cells:cancer cells has been proposed accordingly to previous studies.
24. Adjuvant stimuli doses have been optimized in previous dose-response studies.
25. Co-culture incubation time can be extended by 40-to-48 h.
26. The carboxylic ionophore monensin can be used as alternative protein transport inhibitor. Indeed, monensin promotes intracellular protein accumulation by interacting with the Golgi transmembrane Na^+/H^+ and thus preventing protein secretion from the medial to the trans cisternae of the Golgi complex.
27. As a general rule, kinetic studies should be performed to determine the optimal incubation time. However, exposures to brefeldin A longer than 12 h may be toxic and lead to cell demise.
28. As good practice, supernatants should be mantained at 0-to-4°C before freezing to preserve cytokine stability.
29. Alternatively, supernatants can be immediately stored at −80°C. We recommend to stock supernatants in aliquots of 50 to 100 μL to avoid consecutive freeze and thaw cycles.
30. FcγII/III receptors bind to the Fc portion of antibody-Ag complexes. Blocking of FcγII/III is commonly used in flow cytometry staining experiments to prevent non-specific binding of IgG prior to staining with Ag specific primary antibodies.
31. Optimal antibody concentration has been (and should always be) determined by titration.
32. As a standard rule, cell surface Ag staining should be done with viable, unfixed cells prior to fixation/permeabilization and intracellular cytokine staining.
33. Working in the dark avoids the loss of signal due to photobleaching of fluorochromes.

34. Alternatively, fixable dead-cell dyes (violet, aqua, green) not overlapping with used fluorescences could be employed.
35. At this step, cells can be fixed in 4% paraformaldehyde at 4°C for 20 min, then washed and stored either in PBS/1% FBS at 4°C for up to 72 h or in 90% FCS/10% DMSO at −80°C for >72 h.
36. It is recommended to mantain BD Perm/Wash buffer in washing steps to keep cells permeabilized.
37. BD Perm/Wash buffer should be used to dilute anti-cytokine antibodies to preserve cell permeabilization for intracellular staining.
38. According to the spectral properties of the vital dye.
39. It is of utmost importance to avoid overlapping fluorescences that cannot be distinguished during cytofluorometric acquisition and analysis.
40. Dilution of samples and standards with Assay Diluent ensures an optimally functional sandwich immunoassay preventing non-specific antibody binding, antibody cross-reactivity, or antibody interference.
41. To ensure accurate results, the use of proper adhesion of plate sealers during incubation is recommended.
42. Substrate solution is photosensitive and protection from light is highly recommended. Once added to plate wells, Substrate Solution should change from colorless to gradations of blue depending on IFN-γ concentration.
43. Stop Solution should be added in the same order as Substrate Solution. The color of plate well will turn from gradations of blue to gradations of yellow.
44. To ensure accurate results, the optical density should be determined within 30 min.
45. Once the plate well membranes are pre-wet with ethanol, it is recommended to avoid their drying for the duration of the assay.
46. It is recommended to avoid touching plate well membrane filters with pipette tips when adding reagents and cells in order to preserve membrane integrity.
47. To ensure specific CD8 T cell responses, reagents must be kept sterile.
48. As a standard rule, it is discouraged to perform coating at 37°C and reducing the incubation time, since the efficiency of antibody binding would be significantly reduced.
49. Blocking of non-specific binding sites on the membrane with serum proteins contained in growth medium ensures specificity of anti-IFN-γ or GRZB antibody.

50. The optimal cell concentration is dependent on the intensity of the immune response and should be determined by serial dilution tests.
51. Assay controls such as *i*) culture media without cells; *ii*) tumor cells alone; *iii*) CD8 T cells alone; *iv*) positive control with either non-specific activation using phytohemagglutinin (PHA) or phorbol myristate acetate (PMA); *v*) no primary antibody, are recommended.
52. As for cell concentration, the optimal incubation time is dependent on the intensity of the immune response and should be determined by kinetic tests.
53. The use of a squeeze bottle is suggested as it ensures adequate washing. Alternatively, an automatic plate washer could be used increasing to 1.5 times the number of manual washes.
54. It is recommended not to exceed 0.01% Tween® 20 detergent to prevent the possibility of leakage.
55. As a good practice, secondary antibody should be filtered to avoid non-specific spot formation due to aggregates.
56. Exceeding 1 h incubation is discouraged, since it can result in increased background.
57. Final washes with sterile PBS are performed as Tween® 20 detergent may interfere with spot development.
58. The incubation time of substrate development is a critical step since an exceeding development time will result in increased background.
59. It is highly recommended to protect plate from light in order to preserve spot intensity.

Acknowledgments

IV is supported by the Associazione Italiana per la Ricerca sul Cancro (AIRC IG 2017 grant number 20417) and a startup grant from the Italian Institute for Genomic Medicine (Turin, Italy). AS is supported by the AIRC (Start-Up 2016 #18418) and Ministero Italiano della Salute (RF_GR-2013-02357273).

References

Abiko, K., Matsumura, N., Hamanishi, J., Horikawa, N., Murakami, R., Yamaguchi, K., et al. (2015). IFN-gamma from lymphocytes induces PD-L1 expression and promotes progression of ovarian cancer. *British Journal of Cancer, 112*(9), 1501–1509. https://doi.org/10.1038/bjc.2015.101.

Amaldi, I., Reith, W., Berte, C., & Mach, B. (1989). Induction of HLA class II genes by IFN-gamma is transcriptional and requires a trans-acting protein. *Journal of Immunology, 142*(3), 999–1004.

Anthony, D. A., Andrews, D. M., Watt, S. V., Trapani, J. A., & Smyth, M. J. (2010). Functional dissection of the granzyme family: Cell death and inflammation. *Immunological Reviews, 235*(1), 73–92. https://doi.org/10.1111/j.0105-2896.2010.00907.x.

Beatty, G., & Paterson, Y. (2001). IFN-gamma-dependent inhibition of tumor angiogenesis by tumor-infiltrating CD4+ T cells requires tumor responsiveness to IFN-gamma. *Journal of Immunology, 166*(4), 2276–2282.

Belich, M. P., Glynne, R. J., Senger, G., Sheer, D., & Trowsdale, J. (1994). Proteasome components with reciprocal expression to that of the MHC-encoded LMP proteins. *Current Biology, 4*(9), 769–776.

Boehm, U., Klamp, T., Groot, M., & Howard, J. C. (1997). Cellular responses to interferon-gamma. *Annual Review of Immunology, 15*, 749–795. https://doi.org/10.1146/annurev.immunol.15.1.749.

Brody, J. R., Costantino, C. L., Berger, A. C., Sato, T., Lisanti, M. P., Yeo, C. J., et al. (2009). Expression of indoleamine 2,3-dioxygenase in metastatic malignant melanoma recruits regulatory T cells to avoid immune detection and affects survival. *Cell Cycle, 8*(12), 1930–1934. https://doi.org/10.4161/cc.8.12.8745.

Bromberg, J. F., Horvath, C. M., Wen, Z., Schreiber, R. D., & Darnell, J. E., Jr. (1996). Transcriptionally active Stat1 is required for the antiproliferative effects of both interferon alpha and interferon gamma. *Proceedings of the National Academy of Sciences of the United States of America, 93*(15), 7673–7678.

Chen, D. S., & Mellman, I. (2013). Oncology meets immunology: The cancer-immunity cycle. *Immunity, 39*(1), 1–10. https://doi.org/10.1016/j.immuni.2013.07.012.

Chen, D. S., & Mellman, I. (2017). Elements of cancer immunity and the cancer-immune set point. *Nature, 541*(7637), 321–330. https://doi.org/10.1038/nature21349.

Chin, Y. E., Kitagawa, M., Kuida, K., Flavell, R. A., & Fu, X. Y. (1997). Activation of the STAT signaling pathway can cause expression of caspase 1 and apoptosis. *Molecular and Cellular Biology, 17*(9), 5328–5337.

Chin, Y. E., Kitagawa, M., Su, W. C., You, Z. H., Iwamoto, Y., & Fu, X. Y. (1996). Cell growth arrest and induction of cyclin-dependent kinase inhibitor p21 WAF1/CIP1 mediated by STAT1. *Science, 272*(5262), 719–722.

Coulie, P. G., Van den Eynde, B. J., van der Bruggen, P., & Boon, T. (2014). Tumour antigens recognized by T lymphocytes: at the core of cancer immunotherapy. *Nature Reviews Cancer, 14*(2), 135–146. https://doi.org/10.1038/nrc3670.

Cramer, L. A., Nelson, S. L., & Klemsz, M. J. (2000). Synergistic induction of the Tap-1 gene by IFN-gamma and lipopolysaccharide in macrophages is regulated by STAT1. *Journal of Immunology, 165*(6), 3190–3197.

Cripps, J. G., Wang, J., Maria, A., Blumenthal, I., & Gorham, J. D. (2010). Type 1 T helper cells induce the accumulation of myeloid-derived suppressor cells in the inflamed Tgfb1 knockout mouse liver. *Hepatology, 52*(4), 1350–1359. https://doi.org/10.1002/hep.23841.

Davis, M. M., & Bjorkman, P. J. (1988). T-cell antigen receptor genes and T-cell recognition. *Nature, 334*(6181), 395–402. https://doi.org/10.1038/334395a0.

De Plaen, E., Lurquin, C., Van Pel, A., Mariame, B., Szikora, J. P., Wolfel, T., et al. (1988). Immunogenic (tum-) variants of mouse tumor P815: Cloning of the gene of tum- antigen P91A and identification of the tum- mutation. *Proceedings of the National Academy of Sciences of the United States of America, 85*(7), 2274–2278.

Dighe, A. S., Richards, E., Old, L. J., & Schreiber, R. D. (1994). Enhanced in vivo growth and resistance to rejection of tumor cells expressing dominant negative IFN gamma receptors. *Immunity, 1*(6), 447–456.

Fulda, S., & Debatin, K. M. (2002). IFNgamma sensitizes for apoptosis by upregulating caspase-8 expression through the Stat1 pathway. *Oncogene, 21*(15), 2295–2308. https://doi.org/10.1038/sj.onc.1205255.

Goldberg, M. V., Maris, C. H., Hipkiss, E. L., Flies, A. S., Zhen, L., Tuder, R. M., et al. (2007). Role of PD-1 and its ligand, B7-H1, in early fate decisions of CD8 T cells. *Blood, 110*(1), 186–192. https://doi.org/10.1182/blood-2006-12-062422.

Hannani, D., Sistigu, A., Kepp, O., Galluzzi, L., Kroemer, G., & Zitvogel, L. (2011). Prerequisites for the antitumor vaccine-like effect of chemotherapy and radiotherapy. *Cancer J, 17*(5), 351–358. https://doi.org/10.1097/PPO.0b013e3182325d4d.

Hayakawa, Y., Takeda, K., Yagita, H., Smyth, M. J., Van Kaer, L., Okumura, K., et al. (2002). IFN-gamma-mediated inhibition of tumor angiogenesis by natural killer T-cell ligand, alpha-galactosylceramide. *Blood, 100*(5), 1728–1733.

Johnson, D. R., & Pober, J. S. (1990). Tumor necrosis factor and immune interferon synergistically increase transcription of HLA class I heavy- and light-chain genes in vascular endothelium. *Proceedings of the National Academy of Sciences of the United States of America, 87*(13), 5183–5187.

Jurgens, B., Hainz, U., Fuchs, D., Felzmann, T., & Heitger, A. (2009). Interferon-gamma-triggered indoleamine 2,3-dioxygenase competence in human monocyte-derived dendritic cells induces regulatory activity in allogeneic T cells. *Blood, 114*(15), 3235–3243. https://doi.org/10.1182/blood-2008-12-195073.

Kammertoens, T., Friese, C., Arina, A., Idel, C., Briesemeister, D., Rothe, M., et al. (2017). Tumour ischaemia by interferon-gamma resembles physiological blood vessel regression. *Nature, 545*(7652), 98–102. https://doi.org/10.1038/nature22311.

Kern, I., Steimle, V., Siegrist, C. A., & Mach, B. (1995). The two novel MHC class II transactivators RFX5 and CIITA both control expression of HLA-DM genes. *International Immunology*, 7(8), 1295–1299.

Klebanoff, C. A., Gattinoni, L., & Restifo, N. P. (2006). CD8+ T-cell memory in tumor immunology and immunotherapy. *Immunological Reviews, 211*, 214–224. https://doi.org/10.1111/j.0105-2896.2006.00391.x.

Lah, T. T., Hawley, M., Rock, K. L., & Goldberg, A. L. (1995). Gamma-interferon causes a selective induction of the lysosomal proteases, cathepsins B and L, in macrophages. *FEBS Letters, 363*(1–2), 85–89.

Liu, F., Hu, X., Zimmerman, M., Waller, J. L., Wu, P., Hayes-Jordan, A., et al. (2011). TNFalpha cooperates with IFN-gamma to repress Bcl-xL expression to sensitize metastatic colon carcinoma cells to TRAIL-mediated apoptosis. *PLoS One, 6*(1). e16241https://doi.org/10.1371/journal.pone.0016241.

Maraskovsky, E., Chen, W. F., & Shortman, K. (1989). IL-2 and IFN-gamma are two necessary lymphokines in the development of cytolytic T cells. *Journal of Immunology, 143*(4), 1210–1214.

Matzinger, P. (2002). The danger model: A renewed sense of self. *Science, 296*(5566), 301–305. https://doi.org/10.1126/science.1071059.

Nandi, D., Jiang, H., & Monaco, J. J. (1996). Identification of MECL-1 (LMP-10) as the third IFN-gamma-inducible proteasome subunit. *Journal of Immunology, 156*(7), 2361–2364.

Pak-Wittel, M. A., Yang, L., Sojka, D. K., Rivenbark, J. G., & Yokoyama, W. M. (2013). Interferon-gamma mediates chemokine-dependent recruitment of natural killer cells during viral infection. *Proceedings of the National Academy of Sciences of the United States of America, 110*(1), E50–E59. https://doi.org/10.1073/pnas.1220456110.

Palucka, A. K., & Coussens, L. M. (2016). The basis of Oncoimmunology. *Cell, 164*(6), 1233–1247. https://doi.org/10.1016/j.cell.2016.01.049.

Pon, J. R., & Marra, M. A. (2015). Driver and passenger mutations in cancer. *Annual Review of Pathology, 10*, 25–50. https://doi.org/10.1146/annurev-pathol-012414-040312.

Rousalova, I., & Krepela, E. (2010). Granzyme B-induced apoptosis in cancer cells and its regulation (review). *International Journal of Oncology, 37*(6), 1361–1378.

Schneider, W. M., Chevillotte, M. D., & Rice, C. M. (2014). Interferon-stimulated genes: A complex web of host defenses. *Annual Review of Immunology, 32*, 513–545. https://doi.org/10.1146/annurev-immunol-032713-120231.

Schreiber, R. D., Old, L. J., & Smyth, M. J. (2011). Cancer immunoediting: Integrating immunity's roles in cancer suppression and promotion. *Science*, *331*(6024), 1565–1570. https://doi.org/10.1126/science.1203486.
Schroder, K., Hertzog, P. J., Ravasi, T., & Hume, D. A. (2004). Interferon-gamma: An overview of signals, mechanisms and functions. *Journal of Leukocyte Biology*, *75*(2), 163–189. https://doi.org/10.1189/jlb.0603252.
Seliger, B., Schreiber, K., Delp, K., Meissner, M., Hammers, S., Reichert, T., et al. (2001). Downregulation of the constitutive tapasin expression in human tumor cells of distinct origin and its transcriptional upregulation by cytokines. *Tissue Antigens*, *57*(1), 39–45.
Shirayoshi, Y., Burke, P. A., Appella, E., & Ozato, K. (1988). Interferon-induced transcription of a major histocompatibility class I gene accompanies binding of inducible nuclear factors to the interferon consensus sequence. *Proceedings of the National Academy of Sciences of the United States of America*, *85*(16), 5884–5888.
Soudja, S. M., Chandrabos, C., Yakob, E., Veenstra, M., Palliser, D., & Lauvau, G. (2014). Memory-T-cell-derived interferon-gamma instructs potent innate cell activation for protective immunity. *Immunity*, *40*(6), 974–988. https://doi.org/10.1016/j.immuni.2014.05.005.
Stauss, H. J., Van Waes, C., Fink, M. A., Starr, B., & Schreiber, H. (1986). Identification of a unique tumor antigen as rejection antigen by molecular cloning and gene transfer. *The Journal of Experimental Medicine*, *164*(5), 1516–1530.
Steimlé, V., Siegrist, C. A., Mottet, A., Lisowska-Grospierre, B., & Mach, B. (1994). Regulation of MHC class II expression by interferon-gamma mediated by the transactivator gene CIITA. *Science*, *265*(5168), 106–109.
Thapa, R. J., Basagoudanavar, S. H., Nogusa, S., Irrinki, K., Mallilankaraman, K., Slifker, M. J., et al. (2011). NF-kappaB protects cells from gamma interferon-induced RIP1-dependent necroptosis. *Molecular and Cellular Biology*, *31*(14), 2934–2946. https://doi.org/10.1128/MCB.05445-11.
van der Woude, L. L., Gorris, M. A. J., Halilovic, A., Figdor, C. G., & de Vries, I. J. M. (2017). Migrating into the tumor: A roadmap for T cells. *Trends in Cancer*, *3*(11), 797–808. https://doi.org/10.1016/j.trecan.2017.09.006.
Vitale, I., Sistigu, A., Manic, G., Rudqvist, N. P., Trajanoski, Z., & Galluzzi, L. (2019). Mutational and antigenic landscape in tumor progression and cancer immunotherapy. *Trends in Cell Biology*, *29*(5), 396–416. https://doi.org/10.1016/j.tcb.2019.01.003.
Waterhouse, N. J., Sedelies, K. A., & Trapani, J. A. (2006). Role of bid-induced mitochondrial outer membrane permeabilization in granzyme B-induced apoptosis. *Immunology and Cell Biology*, *84*(1), 72–78. https://doi.org/10.1111/j.1440-1711.2005.01416.x.
Xu, X., Fu, X. Y., Plate, J., & Chong, A. S. (1998). IFN-gamma induces cell growth inhibition by Fas-mediated apoptosis: Requirement of STAT1 protein for up-regulation of Fas and FasL expression. *Cancer Research*, *58*(13), 2832–2837.
Zhang, N., & Bevan, M. J. (2011). CD8(+) T cells: Foot soldiers of the immune system. *Immunity*, *35*(2), 161–168. https://doi.org/10.1016/j.immuni.2011.07.010.
Zitvogel, L., Pitt, J. M., Daillere, R., Smyth, M. J., & Kroemer, G. (2016). Mouse models in oncoimmunology. *Nature Reviews Cancer*, *16*(12), 759–773. https://doi.org/10.1038/nrc.2016.91.

CHAPTER TWENTY-TWO

Assessment of IFNγ responsiveness in patient-derived xenografts

Jordan J. Cardenas[a], Camila Robles-Oteiza[a], Katerina Politi[b,c,d,*]

[a]Department of Immunobiology, Yale School of Medicine, New Haven, CT, United States
[b]Department of Pathology, Yale School of Medicine, New Haven, CT, United States
[c]Department of Medicine (Section of Medical Oncology), Yale School of Medicine, New Haven, CT, United States
[d]Yale Cancer Center, Yale University School of Medicine, New Haven, CT, United States
*Corresponding author: e-mail address: katerina.politi@yale.edu

Contents

Abstract

Patient-derived xenografts are a useful tool in cancer immunology, as they allow researchers to study human cancers in vivo when starting with a relatively small amount of human tumor tissue. These models make it possible to study tumor cell-intrinsic changes that occur in response to external stimuli including cytokines like interferon gamma (IFNγ) that are important for effective anti-tumor immune responses. IFNγ responsiveness can be measured by assessing surface expression of MHC class I on

Methods in Enzymology, Volume 631
ISSN 0076-6879
https://doi.org/10.1016/bs.mie.2019.10.027

tumor cells, the molecule on which tumor antigens are presented to cytotoxic T cells in the tumor microenvironment. Low levels of MHC class I and lack of responsiveness have been associated with resistance to T-cell directed therapies like immune checkpoint inhibitors. In this chapter, we present a protocol for an assay to screen patient-derived xenografts for their responsiveness to IFNγ. The results of this assay can be used as a starting point for uncovering cancer cell-intrinsic mechanisms of resistance to immunotherapies in patient tumors.

1. Introduction

Preclinical models of human cancer are essential for gaining insight into this collection of diseases. Efforts to collect and propagate primary tumor samples from patients have led to the generation of patient-derived models of cancer, mostly engrafted as xenografts (patient-derived xenografts, PDXs) into immunodeficient mice. PDX models offer certain advantages over other mouse models of cancer, including the ability to expand and study a human tumor tissue in depth in vivo without the need for a large amount of the primary patient tissue. PDX models also recapitulate the variability that exists within human tumors, even among those of the same histological subtype. In contrast to human cancer cell lines cultured in vitro, by growing in the mouse, PDXs develop in the presence of other cell types (e.g., stromal cells and vasculature) which can influence tumor biology. As a result of these features, PDXs are valuable tools for identifying tumor-intrinsic factors relating to tumor progression and can be used as preclinical models for cancer therapies, especially those that target tumor cells (Uthamanthil, Tinkey, & de Stanchina, 2017).

Cancer immunotherapies, such as immune checkpoint inhibitors (ICIs) that leverage a patient's own immune system to attack tumor cells, are transforming the cancer treatment landscape and their success has spurred research efforts in the field of cancer immunology. The immune system has already evolved many anti-tumor mechanisms, though transformed cells that develop into invasive tumors are able to circumvent these mechanisms in a process known as immune escape (Choi et al., 2018). The principle behind ICIs is to block inhibitory interactions on immune cells that dampen their ability to exert anti-tumor activity. Among these therapies, blockade of CTLA-4 and the PD-1/PD-L1 axis are the most broadly used, including as first- and second-line therapies in several cancers including melanoma and lung cancer (Webb et al., 2018). Despite the success of these therapies compared to chemotherapy, there is tremendous variability in response rates between different cancer types. Moreover, resistance to immunotherapy, whether it be primary or acquired (occurring after an initial response to therapy), is not well understood (Zaretsky et al., 2016). One roadblock to understanding resistance to ICIs is the challenge of developing preclinical models from human specimens, since human tumor tissue requires in vivo propagation in immunodeficient mice that cannot be used to test immunotherapeutics directly (Choi et al., 2018). However, PDXs can be useful models to study tumor-intrinsic mechanisms of resistance to therapy. For example, the ability of a tumor to respond to interferon gamma (IFNγ) stimulation is critical for response to immune checkpoint inhibitors (Sharma, Hu-Lieskovan, Wargo, & Ribas, 2017). Responsiveness to IFNγ can be examined by measuring upregulation of cell surface MHC-I by flow cytometry following stimulation with the cytokine (Gettinger et al., 2017). PDXs (and patient-derived tumor cells) that do not express surface MHC-I and do not upregulate MHC-I after stimulation with IFNγ are likely able to evade CD8+ T-cell recognition and thus be resistant to immunotherapy. Therefore, this approach allows the identification of PDXs unresponsive to IFNγ stimulation irrespective of the mechanism (genetic, epigenetic, or signaling) that underlies this defect.

In this chapter we present a protocol for the generation and maintenance of PDX lines that then can be used to study tumor-intrinsic mechanisms of resistance to immunotherapy. Additionally, we describe a protocol to measure the IFNγ responsiveness of PDXs. The use of PDX models for this purpose has the potential to foster a greater understanding of mechanisms of resistance to ICIs and ultimately determine strategies to overcome this resistance.

2. Establishment and maintenance of PDX lines

2.1 Primary tumor cell injection

There are numerous methods for establishing and maintaining PDX lines to study human cancers. The establishment of a PDX requires tumor cells directly from patients, most commonly pieces of human tumors obtained via biopsy or surgery or from fluid obtained from pleural effusions or ascites. The patient cells are then engrafted into a mouse, commonly subcutaneously in the flank of the mouse. These are referred to as subcutaneous PDX models. Orthotopic PDXs are generated when the tumor cells are implanted into the same tissue as the tissue of origin of the patient tissue (Hidalgo et al., 2014). Here we describe studies in subcutaneous PDXs.

Methods for suspending human tumor pieces or cells for implantation vary greatly among labs that work with these models, and it is important to record the suspension and implantation method used for the establishment of each PDX. Fresh tumor pieces can be directly injected subcutaneously with a large-gauge needle or resuspended in Matrigel before injection. Alternatively, the pieces can be digested into a single-cell suspension using various methods, including the methods outlined later in this chapter. These single-cell suspensions can again be injected directly or resuspended in Matrigel. The use of any one method may impact the probability of successful engraftment; therefore, it is important to establish the optimal approach for the tumor type under investigation (Hidalgo et al., 2014).

A final important consideration when establishing a PDX line is the strain of mouse used. In order to prevent rejection of the human tumor tissue, an immunodeficient strain of mouse must be used. There are a number of immunodeficient strains that have been optimized throughout the years for the generation of PDXs (Shultz et al., 2014). The first strain used for engraftment of human tumor cells was the nude mouse which has a mutation in the *Foxn1* gene that causes the mice to be athymic and unable to develop T cells. However, these mice still have functional innate immune systems, including potent NK cells. Another strain that is still commonly used is the NOD-*scid* mouse, which has defects in the activity of innate cells such as NK cells, macrophages, and dendritic cells, as well as the absence of T and B adaptive immune cells. Further optimization of this strain by mutating the IL2 receptor-gamma, which plays a major role in cytokine signaling for both the innate and adaptive immune systems, yielded the NOD-scid-gamma (NSG) mouse. NSG mice are an extremely immunodeficient strain and have been valuable for increasing engraftment rates of PDXs.

2.2 Passaging and maintaining a PDX line

A successfully engrafted PDX will begin to grow as a subcutaneous tumor in the mouse. In order to maintain the line, the PDX must be passaged regularly. While institutional animal care policies may vary, a subcutaneous tumor of a volume of over 1 cm^3 is generally deemed to be the upper size limit for these tumors. Depending on the downstream analyses to be performed on the PDX and the amount of tumor tissue needed for each assay and passage, PDXs can be harvested at smaller tumor volumes. Indeed, PDXs can be passaged when the tumor volume is as small as 150–200 mm^3 by re-injecting tumor pieces or a single-cell suspension into a new mouse. Section 4.2

outlines a protocol for harvesting tumor tissue for passaging. Tumor pieces or single-cell suspensions can also be suspended in freezing media, allowed to slow-freeze then preserved in liquid nitrogen for injection at a later date. For each subsequent passage, it is important to record the passage number.

2.3 Critical data for fidelity of records associated with PDX lines

We have alluded to the vast amount of variation that exists in techniques dealing with the generation and study of PDX models of human cancer. To ensure that certain standard criteria are followed when generating and characterizing PDXs and that specific data are collected about the models, the PDX models minimal information standard (PDX-MI) was developed (Meehan et al., 2017). The PDX-MI describes four different categories of information that should be collected and recorded when a PDX model is developed. These categories include: (1) clinical details (information such as patient demographics and tumor classification), (2) information on model creation (including mouse strain, engraftment rates, etc.), (3) model quality assurance data (e.g., assessment of the human origin of the PDX tumors), and (4) model study (e.g., assays performed, treatments administered, etc.). This information should be diligently kept for each PDX during analysis to ensure that the most appropriate PDXs are used for experiments and to avoid errors.

3. In vivo stimulation of PDX lines with interferon gamma

Interferon gamma (IFNγ) is the sole type II interferon and has diverse functions in the immune microenvironment. It is secreted and recognized by several different immune cells and has a wide array of immunostimulatory functions that include but are not limited to inducing T cell activation, antigen presentation and activating macrophages (Ni & Lu, 2018). While PDX models are not useful to study patient immune systems, these models can be used to study tumor-intrinsic factors that contribute to tumor immunogenicity and response to therapy, such as the responsiveness of a tumor to IFNγ stimulation. On non-immune cells IFNγ induces expression of antigen presentation machinery, leading to an increase in MHC class I on the cell surface, which is essential for recognition of cancer cells by cytotoxic T cells (Cheon, Borden, & Stark, 2014). Therefore, it is possible to assess the responsiveness of a PDX tumor to IFNγ (gamma-responsiveness) using fluorescence activated cell sorting (FACS) analysis to measure the levels of MHC class I on the surface of a cancer cell as a readout (Gettinger et al., 2017; Legrier et al., 2016).

4. Protocol

4.1 Equipment

1. 1-mL syringes and 18 and 27-G needles
2. Calipers
3. Dissecting tools (scissors and forceps)
4. Razor blades
5. 70-μm filters
6. 15-mL conical tubes
7. 50-mL tubes
8. Laboratory centrifuge
9. 96-well, U-bottom plate
10. Titer tubes
11. Flow cytometer

4.2 Reagents and buffers

1. Phosphate-buffered saline (PBS)
2. Recombinant human interferon gamma (R&D, 285-IF)
3. Wash buffer (RPMI-1640 + 2% FBS)
4. Digest buffer (wash buffer + 0.5 mg/mL collagenase + 1 μg/mL DNase)
5. ACK lysis buffer
6. FACS buffer (PBS + 2% FBS)
7. Fluorophore-conjugated antibodies for FACS staining:

Antibody/stain	Dilution	Vendor
Brilliant Violet 711 anti-mouse CD45	1:400	BioLegend #103147
APC/Cyanine7 anti-mouse CD45	1:400	BioLegend #103116
Alexa Fluor 488 anti-human EpCAM	1:400	BioLegend #324209
Brilliant Violet 605 anti-human MHC-I	1:100	BioLegend #311432
LIVE/DEAD fixable red dead cell stain	1:2000	Invitrogen #L23102

4.3 Intratumoral injection of interferon-γ (Fig. 1)

3.1. Begin with at least two subcutaneous tumors of the same PDX growing in separate NSG mice. Analysis can be performed once the tumor has grown large enough to have tissue to re-passage into a new mouse, to flash freeze, and to run FACS analysis, about 150–200 mm^3.

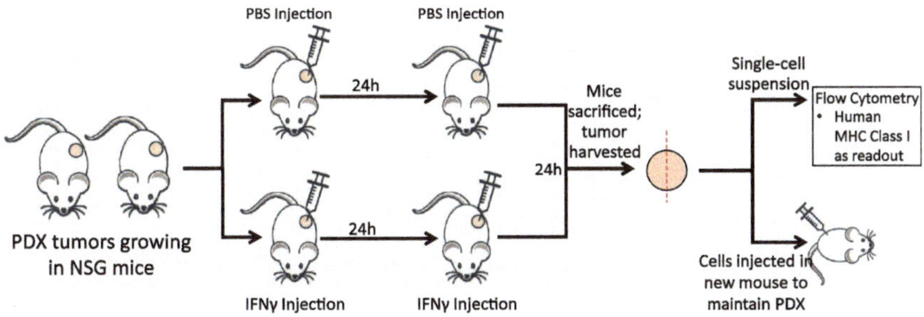

Fig. 1 Schematic representation of the IFNγ stimulation protocol. Begin with at least two tumors of the same PDX growing in different NSG mice. Once tumors reach a sufficient volume, mice are injected intratumorally with either IFNγ or PBS every 24h. After 48h (two injections), mice are sacrificed and tumors are harvested. A single-cell suspension is prepared and stained for flow cytometric analysis. A portion of PDX tissue should be saved to re-passage into a new mouse or frozen down to maintain the line, either as tumor pieces or a single-cell suspension.

3.2. Measure and record the length and width of the tumor growing in the mouse using calipers. Calculate the tumor volume using the equation Tumor Volume = Length $\times$ Width^2 $\times$ 0.52, using the shorter of the two measurements as the width measurement (Euhus, Hudd, LaRegina, & Johnson, 1986; Tomayko & Reynolds, 1989).

3.3. Dilute stock recombinant human IFNγ in PBS to a concentration of 125,000 units per 100-μL injection.

3.4. Using a 27-G needle, inject 100 μL of diluted IFNγ directly into the tumor of one of the two PDX tumors. Inject 100 μL of PBS into the other PDX tumor as a negative control for IFNγ stimulation.

3.5. Administer a second dose of IFNγ/PBS after 24 h.

3.6. 48 h after the initial IFNγ injection, the PDX tumor is ready to be extracted for analysis.

4.4 Tumor digestion and single-cell suspension

4.1 Record tumor measurements in mm^3.

4.2 Euthanize the mouse according to the appropriate institution-described protocol.

4.3 Pin the animal to a cork board, such that the tumor faces upward.

4.4 Spray the tumor and surrounding fur thoroughly with 70% ethanol.

4.5 Using forceps, lift the skin close to the tumor and make a small incision along one edge of the tumor. Insert scissors between the skin and the tumor, and carefully separate the tumor from the skin. Cut away separated skin, then cut along the bottom edge of the tumor to separate it from the mouse abdomen. Transfer tumor to PBS until ready to digest.

4.6 At this point, the tumor can be divided for desired analyses. A portion of the tumor should be finely chopped using a razor blade. If the PDX needs to be passaged this can be done using an 18-G needle to inject the minced tumor into a new NSG mouse. A portion of the tumor can be flash frozen in liquid nitrogen for further analysis. A piece of about 25 mm^3 or greater is necessary for FACS analysis.

4.7 Transfer the portion of the tumor devoted to FACS to a dry Petri dish. Using a razor blade, finely chop the tumor until no large pieces remain. Wash the razor blade with digest buffer and top the dish up to 3–5 mL of digest buffer, depending on the size of the tumor.

4.8 Incubate the tumor for 45 min at 37 °C.

4.9 Transfer the contents of the Petri dish into a 70-μm filter in a 50-mL tube that has been pre-wet with wash buffer. Rinse the Petri dish with wash buffer to ensure that no tumor remains. Alternate pressing the remaining tumor sample through the filter using the plunger of a syringe and rinsing the filter with wash buffer. Do not exceed a total volume of 15 mL.

4.10 Transfer the filtered sample to a 15-mL conical tube. Spin at 1500 RPM for 5 min at 4 °C. For large tumor samples (greater than 500 mm^3), centrifugation at 2500 RPM for 8 min can be performed instead.

4.11 Optional: If there are visible red blood cells in the cell pellet, lyse the sample with ACK buffer. Thoroughly resuspend the pellet in 1 mL of ACK lysis buffer and incubate at room temperature for 2 min. Top the volume up to 15 mL with PBS, and spin at 1500 RPM for 5 min at 4 °C.

4.12 Resuspend the cell pellet in an appropriate volume of FACS buffer and pass the sample through a 70-μm filter.

4.5 Flow cytometry staining

5.1 Transfer approximately 1×10^6 cells (a minimum of 200,000 cells should be used to ensure that a sufficient number of cancer cells are stained since stromal cells are also present in the PDX digest) for each desired flow cytometry panel, as well an additional 1×10^6 cells for an unstained sample, into a 96-well U-bottom plate.

5.2 Wash the samples with FACS buffer: top up the volume in each well to 200 μL with FACS buffer and spin the plate at 1650 RPM for 2 min at 4 °C.

5.3 Protecting samples from direct light exposure, prepare a stain mixture of all desired surface markers (see Section 4.1) for appropriate dilutions; each well of 1×10^6 cells will be stained with 100 μL of stain mixture.

5.4 Protecting samples from direct light exposure, add 100 μL of the prepared stain mixture to each sample and pipette up and down to mix. Stain for 30 min on ice in the dark.

5.5 Wash the samples with FACS buffer: top up the volume in each well to 200 μL with FACS buffer and spin the plate at 1650 RPM for 2 min at 4 °C. Repeat this step one time.

5.6 Resuspend each sample in 200 μL FACS buffer and transfer the samples to flow cytometry tubes.

5.7 Analyze the samples on a flow cytometer.

5. Analysis

5.1 Markers (Fig. 2)

This IFNγ stimulation assay is a useful tool to analyze PDX samples. When investigating PDX lines established from immunotherapy-treated patients, the gamma-responsiveness of the tumor cells may be one indicator of whether the tumor is permissive to immune stimulation. Responsiveness to IFNγ is an indicator of a tumor-intrinsic factor that can affect sensitivity to immune checkpoint inhibitors. To perform this assay, it is essential that specific antibody stains are used to ensure that the correct population of cells within the PDX is captured for analysis.

The first consideration when performing FACS analysis of any sample is to ensure that the relative size of the cells, as seen by forward- and side scatter, is appropriate for the cell type being analyzed. In the case of PDX tumors, cell size tends to be less uniform and more variable. Therefore, a generous cell size gate is necessary to capture all of the tumor cells. This variation in tumor cell size makes it difficult to distinguish the tumor cell population based on size alone, so more markers are necessary to exclude unwanted cell types. In any flow cytometry analysis, a live/dead stain should be used to gate on the live cells from the tumor sample, as dead cells have the potential to auto-fluoresce in the analysis channels.

Despite the immunocompromised state of NSG mice, mouse CD45 positive cells (e.g., myeloid cells) may be present in PDX tumors. Therefore, staining for mouse CD45 will enable elimination of these cells from the population being analyzed. Similarly, since the tumors are taken directly from human tissue, it is possible for a small population of human

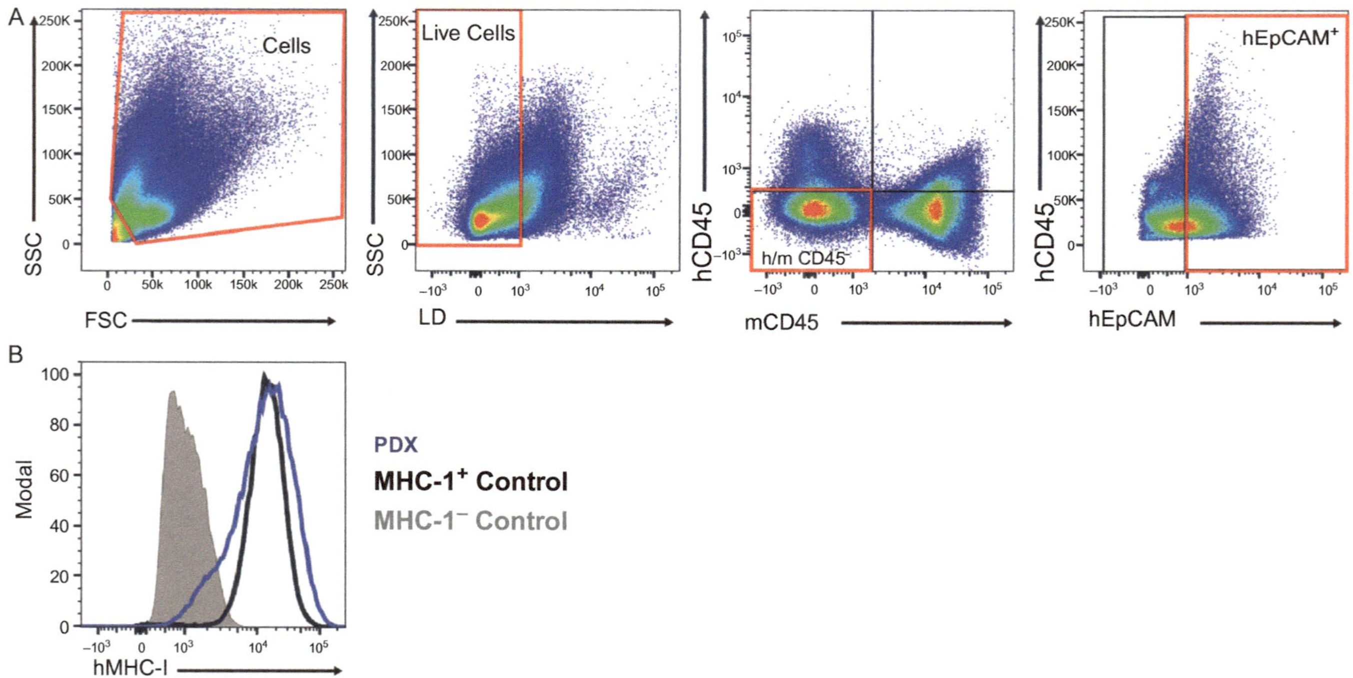

Fig. 2 Representative flow cytometry gating of PDX samples. (A) First plot shows the forward scatter (FSC) on the X axis by the side scatter (SSC) on the Y axis (After gating out of doublets). Debris in the bottom-left corner of the plot is excluded, but cells are gated generously by FSC and SSC as tumor cells are not uniform in size. Generous gating here allows for all possible tumor cells to be captured. Within this initial gate, cells are plotted using the Live/Dead (LD) stain on the X axis while the SSC remains on the Y axis. Cells in the population that are positive in the Live/Dead channel are excluded from analysis, as these are dead cells. The live cell population is then plotted with both human and mouse CD45. Since the tumor cells are negative for both human and mouse CD45, the double-negative population is used from this plot. Within this population, additional tumor cell markers can be used, depending on the tumor type being analyzed. In this case, human EpCAM is used as a tumor marker, and the population of positive cells is analyzed for human MHC class I expression. (B) Example histogram of human MHC class I expression readout. The PDX sample is shown in blue. Human MHC-I positive (black) and negative (filled, gray) are shown on the same plot to compare expected MHC-I expression of known MHC-positive and -negative cell lines. In this example, H1975 cells are used as an MHC-I^{+} control, while Daudi cells are used as a MHC-I^{-} control. Examples of PDXs that upregulate MHC Class I as a result of stimulation with IFNγ can be found in Gettinger et al. (2017).

immune cells to remain in the tumor, especially during early PDX passages. The likelihood of this occurring in subsequent passages is very low. Nonetheless, the samples should be stained with human CD45 in order to exclude these human immune cells. Another molecule that can be helpful for ensuring that only human tumor cells are being analyzed is to stain for mouse MHC class I. Any mouse stroma or other cells that have made it into the tumor before processing would be positive for mouse MHC class I, so exclusion of these cells from the population is useful. Furthermore, it is helpful to stain for human tumor markers in order to positively identify the population of human tumor cells. However, it is important to be aware of the heterogeneity in expression of any marker used for a given cancer type so that no desired human tumor cells are excluded. For example, human EpCAM can be used to identify epithelial-derived tumor cells. However, some tumors can downregulate EpCAM expression so this marker is not helpful in these cases. Western blotting or immunohistochemistry for the marker of interest on the PDX can be performed prior to FACS analysis to establish whether the marker is expressed abundantly and uniformly in tumor cells.

The main readout for IFNγ responsiveness of a PDX is FACS analysis for cell surface human MHC class I. Under normal conditions, MHC class I surface expression is upregulated upon stimulation with IFNγ. When analyzing PDX samples, an IFNγ-responsive PDX will exhibit an increase in the fluorescence intensity of the fluorophore conjugated to the human MHC-I antibody in the IFNγ-stimulated PDX sample compared to the PBS-treated PDX. A PDX may be non-responsive to IFNγ for a variety of reasons. If the PDX is completely negative for MHC class I with and without IFNγ stimulation, this may indicate a defect in the IFNγ pathway. For example, loss-of-function mutations in JAK1 or JAK2 could prevent the interferon gamma receptor from signaling downstream in cells. Additionally, mutations in antigen presentation machinery genes, like β2-microglobulin (*B2M*), could also prevent induction of MHC-I in a cell. Moreover, if the PDX expresses some level of MHC-I without stimulation and is non-responsive to IFNγ, this may indicate other defects in IFNγ signaling in the tumor that can then be further investigated.

5.2 Control cell lines

Proper control cell lines should be stained in tandem with the PDX samples as positive and negative controls for normal human MHC-I expression and IFNγ responsiveness (Fig. 2B). As a negative control for human MHC-I, the

Daudi line can be used (ATCC® CCL-213™). This is a B cell lymphoma line that is negative for MHC-I. There are a variety of human cancer cell lines that are positive for MHC-I, such as H1975 (ATCC® CRL-5908™), a lung cancer cell line. It may prove useful to test human cancer cell lines that match the cancer type of the PDX samples being tested for their expression of human MHC-I, as well as their responsiveness to in vitro stimulation with IFNγ.

6. Summary

Patient-derived xenografts are a powerful tool for modeling human cancer. While the maintenance of PDX lines can be costly and time consuming, these models offer the advantage of being useful for a variety of investigative assays that generally cannot be conducted on limited amounts of primary patient tumor tissue. One such assay described in this chapter is an approach to analyze the responsiveness of patient tumors to IFNγ, a critical immunomodulatory molecule found in the tumor microenvironment. This assay provides a tool to functionally assess IFNγ signaling in a tumor and can be used to determine the frequency of defects in this pathway in a given tumor type. Moreover, wider application of this assay on an array of immunotherapy-resistant PDX samples has the potential to reveal the role of IFNγ responsiveness in the response of tumors to immunotherapies and the mechanisms that underlie defects in the pathway.

Acknowledgments

This work was supported by NIH/NCI R01 CA230275 (K.P.) and F99 CA245819 (C.R.O.).

References

Cheon, H., Borden, E. C., & Stark, G. R. (2014). Interferons and their stimulated genes in the tumor microenvironment. *Seminars in Oncology*, *41*(2), 156–173.

Choi, Y., Lee, S., Kim, K., Kim, S., Chung, Y., & Lee, C. (2018). Studying cancer immunotherapy using patient-derived xenografts (PDXs) in humanized mice. *Experimental & Molecular Medicine*, *50*(8), 99.

Euhus, D. M., Hudd, C., LaRegina, M. C., & Johnson, F. E. (1986). Tumor measurement in the nude mouse. *Journal of Surgical Oncology*, *31*, 229–234. https://doi.org/10.1002/jso.2930310402.

Gettinger, S., Choi, J., Hastings, K., Truini, A., Datar, I., Sowell, R., et al. (2017). Impaired HLA class I antigen processing and presentation as a mechanism of acquired resistance to immune checkpoint inhibitors in lung cancer. *Cancer Discovery*, 7(12), 1420–1435.

Hidalgo, M., Amant, F., Biankin, A. V., Budinská, E., Byrne, A. T., Caldas, C., et al. (2014). Patient-derived xenograft models: An emerging platform for translational cancer research. *Cancer Discovery*, *4*, 998–1013.

Legrier, M. E., Bièche, I., Gaston, J., Beurdeley, A., Yvonnet, V., Déas, O., et al. (2016). Activation of IFN/STAT1 signaling predicts response to chemotherapy in oestrogen receptor-negative breast cancer. *British Journal of Cancer, 114*(2), 177–187.

Meehan, T., Conte, N., Goldstein, T., Inghirami, G., Murakami, M. A., Brabetz, S., et al. (2017). PDX-MI: Minimal information for patient-derived tumor xenograft models. *Cancer Research*, 77(21), e62–e66.

Ni, L., & Lu, J. (2018). Interferon gamma in cancer immunotherapy. *Cancer Medicine*, 7(9), 4509–4516.

Sharma, P., Hu-Lieskovan, S., Wargo, J. A., & Ribas, A. (2017). Primary, adaptive, and acquired resistance to cancer immunotherapy. *Cell, 168*(4), 707–723.

Shultz, L. D., Goodwin, N., Ishikawa, F., Hosur, V., Lyons, B. L., & Greiner, D. L. (2014). Human cancer growth and therapy in immunodeficient mouse models. *Cold Spring Harbor Protocols*, 7, 694–708. https://doi.org/10.1101/pdb.top073585.

Tomayko, M. M., & Reynolds, C. P. (1989). Determination of subcutaneous tumor size in athymic (nude) mice. *Cancer Chemotherapy and Pharmacology, 24*, 148–154.

Uthamanthil, R., Tinkey, P., & de Stanchina, E. (2017). *Patient derived tumor xenograft models: Promise, potential and practice*. London: Academic Press.

Webb, E. S., Liu, P., Baleeiro, R., Lemoine, N. R., Yuan, M., & Wang, Y. (2018). Immune checkpoint inhibitors in cancer therapy. *Journal of Biomedical Research, 32*(5), 317–326.

Zaretsky, J. M., Garcia-Diaz, A., Shin, D. S., Escuin-Ordinas, H., Hugo, W., Hu-Lieskovan, S., et al. (2016). Mutations associated with acquired resistance to PD-1 blockade in melanoma. *The New England Journal of Medicine, 375*, 819–825.

CHAPTER TWENTY-THREE

Real-time cell analysis (RTCA) to measure killer cell activity against adherent tumor cells *in vitro*

Hans-Heinrich Oberg, Christian Peters, Dieter Kabelitz*, Daniela Wesch

Institute of Immunology, University Hospital Schleswig-Holstein and Christian-Albrechts University Kiel, Kiel, Germany

*Corresponding author: e-mail address: dietrich.kabelitz@uksh.de

Contents

Abstract

The enhancement of immune responses against tumor cells is a main focus of cancer immunotherapy. Immunotherapeutic approaches comprise a broad range of clinical applications including adjuvant therapies, check point inhibitors, cellular therapies, oncolytic viruses or targeted biologics such as bispecific antibodies. The usage of bispecific antibodies is one promising approach to enhance cytotoxicity and to selectively target effector cells to tumor-associated antigens. Here, we discuss the real-time cell analysis system as a suitable *in vitro* method to determine the interaction of tumor cell with effector cells alone or within a heterogeneous mixture of immune cells in

Methods in Enzymology, Volume 631
ISSN 0076-6879
https://doi.org/10.1016/bs.mie.2019.07.020

peripheral blood or within tumor-infiltrating cells. The determination of cytotoxic effector cell activity using the real-time cell analyzer is highly useful to monitor the dynamic cellular interplay over extended periods of time.

1. Introduction

Several methods can provide information about cytotoxicity of effector cells against tumor cells. For decades, the gold standard to measure cytotoxicity against tumor cells was the chromium ^{51}(Cr) release assay. While cytotoxicity against tumor cells is usually measured after 4h in the ^{51}Cr release assay, it can be monitored with the real-time cell analysis (RTCA; ACEA Biosciences, San Diego, CA, USA) in real time up to several days. This RTCA assay, available as single plate (96-well plate) or multi-plate (6 × 96-well plate) or high-throughput plate (384-well plate), can be used as an alternative method to measure cytotoxicity against adherent tumor cells without the incorporation of any labels (Davenport et al., 2015; Messaoudene et al., 2014; Oberg, Peipp, et al., 2014). Besides cytotoxicity, the measurement of proliferation or cell death to screen quality of cells or effects of anti-viral-, anti-bactericidal- or other drugs can also be performed with the RTCA assay. An application overview is described in detail under https://www.aceabio.com/products/rtca-sp.

To measure killing of suspension cell targets rather than adherent ones, the *E*-plate wells can be incubated with appropriate tethering reagents as described elsewhere (Cerignoli et al., 2018). The measurement of the impedance of the (adherent) cells in arbitrary cell index units reflects changes in cellular parameters such as cell integrity, morphology or cell number. The impedance is analyzed with electronic sensors located at the bottom of the applied *E*-plates. Tumor cells are allowed to adhere over a time course of 24h to ensure that they are in a linear growth phase. At this time, graded numbers of effector cells, different concentrations of reagents to be tested, or the combination of both is added. If the addition of effector cells or reagents under study induces a loss of impedance, this corresponds to killing of tumor cells. A positive as well as a negative control is applied in this system to precisely evaluate and quantify the tumor cell lysis. The loss of impedance correlates with the effector-cell-mediated lysis of tumor cells. Although the assay is quite expensive, a potent advantage of measuring impedance over an extended time course is that it enables us to observe whether small numbers of effector cells (as likely to be present at the tumor site) can completely lyse tumor cells or if tumor cells can regenerate when lysis is incomplete (Oberg et al., 2016, 2018). To save costs, plates can be washed with cell lysing detergents once and re-used for preliminary experiments including the titration of tumor cells to analyze their adherence capacity. For the majority of the experiments, however, the application of unused plates is recommended. In this chapter, we illustrate the application of RTCA to measure effector cell cytotoxicity in the interaction with tumor cells in the absence or presence of other immune cells.

2. Tumor cell lines

In this study, we applied pancreatic ductal adenocarcinoma cell lines (PDAC) as target cells for the RTCA assay. Alternatively, any other adherent tumor cell line or primary tumor material as described elsewhere can be used (Oberg et al., 2018; Oberg, Peipp, et al., 2014).

PDAC is the fourth most lethal tumor in western countries with a poor prognosis. The overall 5-year survival rate is only ~6% and therapeutic options are limited due to resistance of PDAC to radio- or chemotherapy. Only 10–20% of the patients are accessible for surgical resection, while 80% of the patients have an advanced disease with distant metastases at the time of diagnosis (Leung et al., 2012; Siegel, Miller, & Jemal, 2015). High-grade differentiated PDAC tumors are often associated with the expression of human epidermal growth factor receptor 2 (HER2, ErbB2) which stimulates proliferation of tumor cells (Harder et al., 2012). Since anti-HER2 therapies are successful for the treatment of HER2-expressing tumors, HER2 is often selected as a tumor target antigen (Loibl & Gianni, 2017; Sadeghi, Olevsky, & Hurvitz, 2014). A bispecific antibody which targets effector cells to HER2-expressing PDAC cells was applied in several assays of this study. Besides redirecting distinct effector cell populations including T lymphocytes to the tumor site, the bispecific antibody enhanced the cytotoxic activity of T lymphocytes (Brinkmann & Kontermann, 2017). The development of bispecific antibodies has been successfully introduced into clinical application, e.g. blinatumomab for treatment of relapsed or refractory B-cell precursor acute lymphoblastic leukemia (Topp et al., 2014). Targeting solid tumors with bsAb is more complex and is under intensive investigation.

PDAC cell lines of different origin and grade of differentiation including BxPC3-, Panc1-, Capan-1- and -2 cells were kindly provided by Dr. Christian Röder, Institute for Experimental Cancer Research UKSH/CAU, Kiel.

BxPC3-, Panc1-, Capan-2 cells were initially established from primary tumors (stages G1–G3) while Capan-1 cells were generated out of liver metastases (stage G1) (Sipos et al., 2003). All PDAC cell lines were cultured in RPMI 1640 supplemented with 2 mM L-glutamine, 25 mM Hepes, 100 U/mL penicillin, 100 μg/mL streptomycin, 10% FCS (complete culture medium) under regular conditions (5% CO_2, humidified, 37 °C). The genotype of tumor cell lines was recently confirmed by short tandem repeats analysis and the absence of mycoplasma was routinely determined once per month by RT-PCR.

3. Effector cells

γδ T cells are attractive effector cells for T-cell based immunotherapy, due to their high plasticity and their potent killing of a broad range of tumor cells in a HLA non-restricted manner. γδ T cells comprise a small subset of CD3-positive T cells in the peripheral blood but are a major T-cell population in the mucosal tissue (Chitadze, Oberg, Wesch, & Kabelitz, 2017). Although the T-cell receptor repertoire of blood- and intestinal γδ T cells is different, both subsets can infiltrate into tumors. γδ T cells infiltrate in higher numbers than αβ T cells in the malignant ductal epithelium as well as in the neighboring stroma of PDAC patients (Helm et al., 2014; Oberg et al., 2016). In this study, we discuss the RTCA system using γδ T cells as effector cells. Alternatively, however, any other effector cell can be applied in the RTCA.

To isolate resting γδ T cells out of peripheral blood or to analyze γδ T cells within peripheral blood mononuclear cells (PBMC), the latter were isolated from leukocyte concentrates obtained from adult healthy donors through the Department of Transfusion Medicine (UKSH, Campus Kiel). In accordance with the Declaration of Helsinki, written informed consent was provided by all blood donors, and the research was approved by the relevant institutional review board of Kiel University Medical Faculty (D445/18). PBMC were isolated from heparinized blood by Biocoll density gradient centrifugation. To isolate γδ T cells by negative selection, the T-cell receptor $\gamma\delta^+$ T Cell Isolation Kit (Miltenyi Biotec, Bergisch Gladbach, Germany) was used according to the manufacturer's instructions. The purity of the negatively isolated γδ T cells was >96%. Resting γδ T cells or PBMC were co-cultured with PDAC cells (24 h after their adherence) in medium with 50 U/mL rIL-2 (Novartis, Basel, Switzerland) in the absence or presence of 300 nM Bromohydrin-pyrophosphate (BrHPP; Innate Pharma, Marseille, France) or 1 μg/mL bispecific antibody [$(HER2)_2 \times V\gamma9$], both of which selectively activate γδ T cells. The generation and binding capacity of the bispecific antibody [$(HER2)_2 \times V\gamma9$] is described elsewhere (Oberg et al., 2015, 2018; Oberg, Peipp, et al., 2014). Vγ9Vδ2 γδ T cells in the blood as well as within tumor-infiltrating T cells recognize phosphorylated antigens such as isopentenyl pyrophosphate (IPP) which are intermediates of the mevalonate pathway of isoprenoid synthesis and which are overproduced by tumor cells. IPP and synthetic analogs like BrHPP induce a selective 400–1000-fold expansion of γδ T cells in the presence of 50 U/mL rIL-2 (Novartis, Basel, Switzerland) (Espinosa et al., 2001; see chapter "*In vitro* expansion of Vγ9Vδ2 T cells for immunotherapy" by Peters et al., this volume). Alternatively, tumor-infiltrating immune cells as well as *in vitro* expanded γδ T-cell lines (Oberg et al., 2018; see chapter "*In vitro* expansion of Vγ9Vδ2 T cells for immunotherapy" by Peters et al., this volume) instead or PBMC can be used for RTCA. The isolation of tumor-infiltrating immune cells is described elsewhere (Helm et al., 2014; Oberg et al., 2018).

3.1 Equipment

1. 5% CO_2-incubator (Heracell 240, Thermo Fisher Scientific, Langenselbold, Germany)
2. xCELLigence RTCA Analyzer with a station in a CO_2-incubator and Control Unit with RTCA software outside the CO_2-incubator (ACEA Biosciences, San Diego, CA, USA)
3. Microscopes (Axioskop, Axiovert 10, both from Carl Zeiss, Göttingen, Germany)
4. Pipettor (Brand, Wertheim, Germany)
5. Pipettes, 0.1–1000 μL and Multichannel pipette (Eppendorf, Hamburg, Germany)

3.2 Consumables

1. 96-well *E*-plates (ACEA Biosciences)
2. Serological pipettes, 1–50 mL (Sarstedt, Nümbrecht, Germany)
3. Pipette tips, 10–1000 μL (Sarstedt)
4. Reaction tubes, 15- and 50 mL (Greiner-Bio-One GmbH, Frickenhausen Germany)
5. Cell culture flasks, 50-, 250-, or 650 mL (Greiner-Bio-One GmbH)
6. 24-, 12-, or 6-well plates (Greiner-Bio-One GmbH)

3.3 Reagents

1. PBS, phosphate-buffered saline (Biochrom, Berlin, Germany)
2. Complete culture medium (e.g. RPMI 1640 with L-glutamine, Hepes, antibiotics and human or fetal serum)
3. Biocoll, density 1.077 g/mL, isotonic (Biochrom)
4. 0.05% (w/v) Trypsin/0.02% (w/v) EDTA (both from Biochrom) in *aqua ad injectabilia*
5. Triton-X-100 in *aqua ad injectabilia* (Merck, Darmstadt, Germany)
6. 0.1% Trypan blue (w/v) in PBS
7. Phosphorylated antigens (PAg): e.g. 300 nM BrHPP; Innate Pharma, Marseille, France)
8. Bispecific antibody: 0.1–1 μg/mL tribody $[(HER2)_2 \times V\gamma9]$

4. Protocol

4.1 Determination of the optimal cell number of adherent tumor cells

1. Viability of the cells is analyzed microscopically.
2. Remove adherent tumor cells from flasks by treatment with PBS and thereafter with trypsin/EDTA in the 5% CO_2-incubator (the period of time depends on the adherence of the tumor cell).
3. Wash the tumor cells twice, count the viable cell number (e.g. microscopically with a Neubauer chamber and Trypan blue dye exclusion) and prepare different tumor cell dilutions.
4. Add 50 μL of (complete) culture medium to the 96-well *E*-plates and measure the background impedance in the RTCA for 15–30 min.
5. Ten to 15 min after the background has balanced, adherent tumor cells are seeded in different cell numbers (5000–15,000 per well) in triplicates in a volume of 50 μL into the 96-well plates. They will passively adhere at the bottom of the *E*-plates covered with gold electrodes.
6. After 30 min at room temperature, the *E*-plate is placed into the RTCA Unit located in the humidified 5% CO_2 incubator.
7. The impedance of the cells expressed as an arbitrary unit called cell index (CI) is measured every 5 min for up to 24 h or longer (dependent on growth of the tumor cells).

4.2 Addition of effector cells and stimuli

1. After tumor cells have reached the linear growth phase, *E*-plates are taken out of the incubator and placed on pre-warmed pads inside a

laminar flow to avoid a sudden drop of temperature. The data acquisition is paused during this time.

2. The desired stimuli such as pAg BrHPP or bispecific antibodies or medium (as control) are added in a volume of 50 µL to the *E*-plates. If other stimuli are used which require a preincubation of target cells, this stimuli can be incubated in the RTCA Unit for an additional time period.
3. Effector cells are added at different effector to target (E/T) ratio in triplicates in a volume of 50 µL directly or after incubation with the substances to the *E*-plates which are positioned on pre-warmed pads.
4. As a control, target cells or effector cells alone with or without substances are applied as well as the treatment of target cells with a final concentration of 1% Trition-X-100 as a positive control for maximal lysis.
5. For the precise analysis of cytotoxicity, the cellular impedance measurement is reduced from every 5 min to every minute for the indicated time points. The user has the possibility to define each time interval individually.
6. Effector cell-induced tumor cell lysis is indicated as the loss of impedance of tumor cells, i.e. decrease of the CI.

5. RTCA-system

5.1 Monitoring the impedance of target cells

The impedance is measured by electronic sensors which measure ionic environment in the absence of cells. The addition of cells alters the local ionic environment of the *E*-plate wells and induces changes of cellular impedance which reflects changes in cellular parameters such as cell proliferation and morphological changes (e.g. spreading, adherence) of the applied PDAC cells. As shown in Fig. 1, an increased cell number of the different PDAC cells (from 5×10^3 to 12.5×10^3) resulted in an enhanced cellular impedance expressed as CI. By analyzing four different PDAC cells, the RTCA revealed an obvious difference between the various tumor cell lines regarding the time course kinetics of changes in impedance. While BxPC3 cells with a cell number $>7.5 \times 10^3$ resulted in a CI of 2 after nearly 24 h, Panc1 cells with the same cell number acquired a CI of 1 after a similar time period. In contrast, the Capan-2 and Capan-1 cell lines took almost 60–70 h to obtain a CI of 1, and only with the highest cell number of 12.5×10^3 cells seeded per well. The measured impedance can be affected by the adherence properties of tumor cells, the quality of the gold electrodes, the expression intensity of adhesion molecules on tumor cells, and the cell viability.

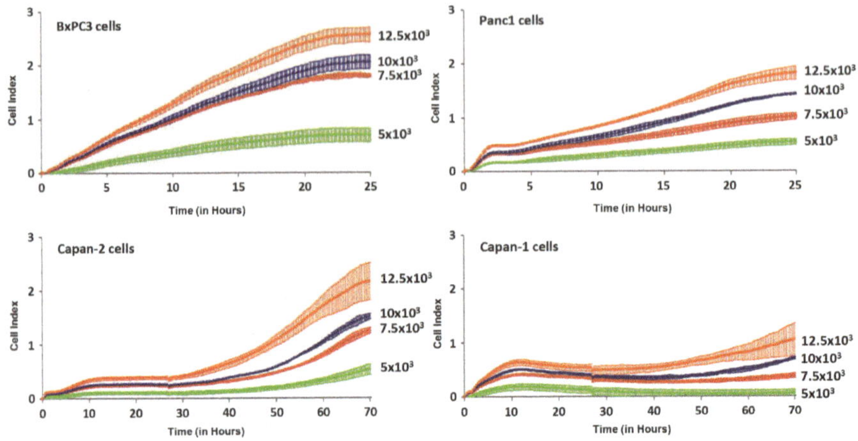

Fig. 1 Titration of PDAC cell numbers in a RTCA assay. (A–D) Different cell numbers of the indicated PDAC cell lines were cultured in complete medium for 25–70h on an *E*-plate covered at the bottom with electronic sensors that measured the impedance of the adherent tumor cells expressed as an arbitrary unit called cell index (CI) every 5min. The variance in the widths of the tracing is due to the standard deviation. The average of triplicates and standard deviation were calculated for each cell number.

The tumor cells pass through different phases as follows: (i) attachment, (ii) lag, (iii) linear growth, and (iv) steady state/confluence. An optimal CI to proceed with the experiment by adding effector cells is reached when the tumor cells are in the linear growth phase.

Taken together, the application of different target cell numbers reveals informative insights into the impedance of target cell numbers and their time course dependency.

5.2 Analysis of the effector-mediated lysis of target cells

7.5×10^3 Panc1 cells or BxPC3/wells were chosen as an optimal cell number to co-culture them with 375,000–45,000 effector cells for further 50h and to avoid an overgrowth in the wells where no lysis takes place. The CI of the tumor cell was 1 after 24h of adherence and before the addition of effector cells. The addition of effector cells such as resting γδ T cells in medium at an E/T ratio of 50:1 only very slightly changed the CI of Panc1 cells, which were initially seeded with 7.5×10^3 Panc1 cells/wells for 22h (Fig. 2A and B). Similar results were obtained with lower E/T ratios of 25:1 to 6:1 (data not shown). These data reveal that γδ T cell cytotoxicity against Panc1 cells was not induced in the absence of a stimulus. A weak effect was due to the

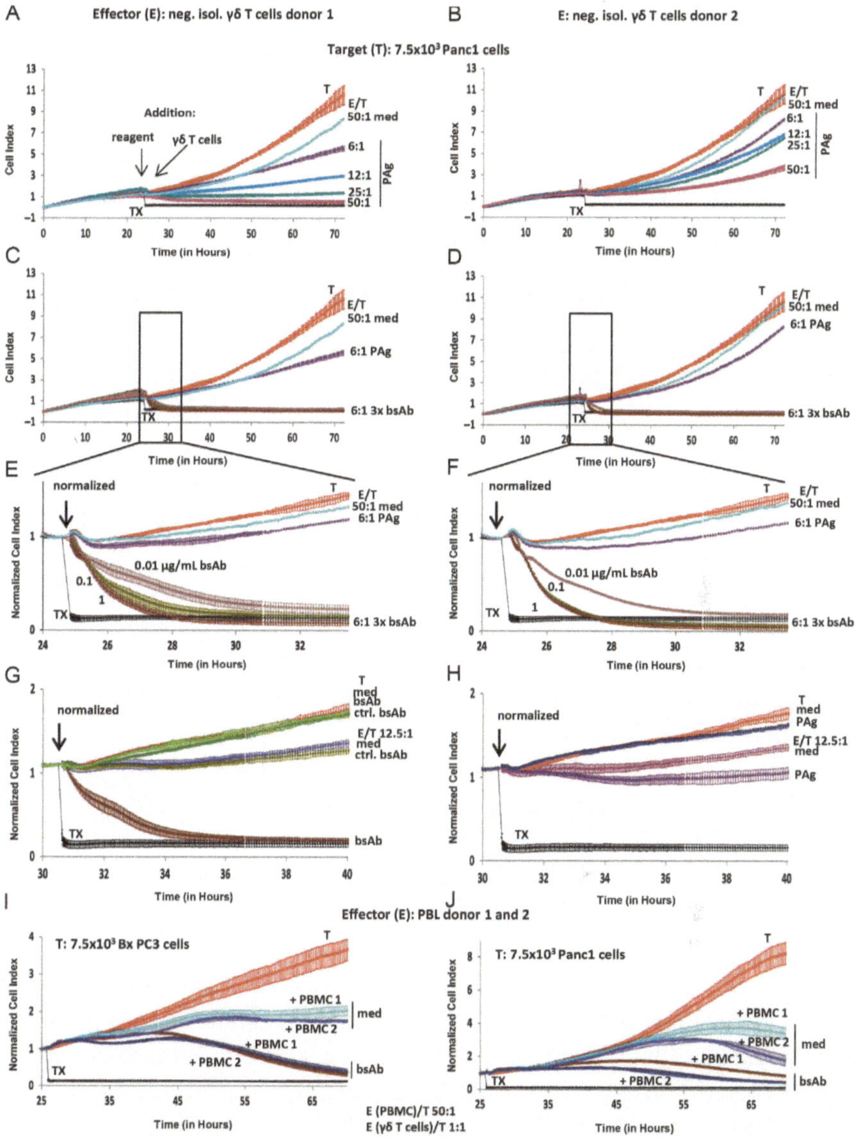

Fig. 2 γδ T cell-mediated cytotoxicity against PDAC cells. Indicated tumor cells [T] were seeded 24 h before addition of effector cells [E]. (A–H) Cytotoxicity mediated by negatively isolated, resting γδ T cells or (I–J) peripheral blood mononuclear cells (PBMC) in co-cultures with (A–H and J) Panc1 cells or BxPC3 cells (I) at the indicated effector/target (E/T) ratios was analyzed by RTCA over the indicated time period. (I–J) The E/T ratio of 1:1 refers to γδ T cells (within PBMC) to tumor cells. (A–J) Tumor cells (T) alone or co-cultures were set up in 50 U/mL rIL-2 medium only or were additionally stimulated with 300 nM phosphorylated antigen BrHPP (pAg) or with the indicated concentrations of bispecific Ab (bsAb) [$(HER2)_2 \times V\gamma9$] or 1 µg/mL control construct. Triton-X-100 (TX) was applied as a positive control to induce maximal lysis in tumor cells. (E–H) The CI was normalized to 1 shortly before the addition of effector cells or medium, pAg or bsAb. The bold arrows indicate the time of normalization. CI was then measured every minute for additional indicated hours. The loss of tumor cell impedance and thus a decrease of CI correlated with lysis of tumor cells. The average of triplicates and standard deviation were calculated; one representative experiment out of three is shown.

application of IL-2 which is necessary for activation of γδ T cells. Non-adherent effector cells such as T cells or NK cells displayed a negligible CI as described elsewhere (Cerignoli et al., 2018; Oberg et al., 2015, 2018; Oberg, Peipp, et al., 2014). After stimulation of γδ T cells with phosphorylated antigens (pAg), γδ T cell cytotoxicity is induced as measured by the reduction of CI for Panc1 cells which was dependent on the number of added γδ T cells (Fig. 2A). Similar results with resting γδ T cells of an additional donor are shown in Fig. 2B. The loss of CI correlates with the γδ T cell-mediated tumor cell lysis. To ensure that a decrease in the CI is not caused by changes in the adherence of viable tumor cells, the lysis of tumor cells can be confirmed in parallel by another cytotoxic assay such as a ^{51}Cr-release assay as shown elsewhere (Oberg, Peipp, et al., 2014). After initial activation, resting γδ T cells require 10–20 h before they exert measurable cytotoxicity, whereas pre-activated γδ T cells cultured with IL-2 exert cytotoxic activity more rapidly (Oberg et al., 2016; Oberg, Kellner, et al., 2014; Oberg, Peipp, et al., 2014). The RTCA system enables for the analysis of CI over an extended time course and therefore holds an important advantage over endpoint assays (such as the ^{51}Cr-release assay), because it allows for the determination of cytotoxicity kinetics as well as for the observation of tumor cell regeneration due to an incomplete lysis (Oberg et al., 2016, 2018).

Regarding the low frequency of γδ T cells at the tumor site, the favorable characteristics of bispecific antibodies in enhancing cytotoxicity of effector cells was characterized and compared with the pAg-stimulation at a low E/T ratio of 6:1. The single-chain bispecific antibody construct used here has a specificity for Vγ9 of the γδ T cell receptor and HER2 expressed by Panc1 cells (Oberg, Peipp, et al., 2014). The addition of different concentrations of bispecific antibody [$(HER2)_2 \times V\gamma9$] enhanced the cytotoxicity also at low E/T ratio in two different donor samples in comparison to pAg (Fig. 2C and D).

Since the initial adherence in different wells of the *E*-plate can differ slightly, the CI can be normalized to 1 after having reached the linear growth phase and before the addition of antibody constructs, other substances, or effector cell suspensions as shown in Fig. 2E and F. The normalization together with the view on a time interval of 10 h instead of 50 h revealed that a concentration of 0.1–1 μg/mL of bispecific antibody was more effective than a lower concentration with 0.01 μg/mL (Fig. 2E and F). In addition, the study revealed that the bispecific antibody added to tumor cells in the absence of effector cells did not influence the CI, similar to the application of a control construct which had no specificity for HER2 (Fig. 2G). The addition of effector cells in medium with an E/T ratio of 12.5:1 slightly decreased the CI comparable to the addition of the control bispecific antibody, whereas the bispecific antibody [$(HER2)_2 \times V\gamma9$] drastically decreased the CI similar to the effect induced by Triton-X-100 (for maximal lysis) (Fig. 2E–G). The application of pAg to the tumor cells alone did not influence their impedance while a weak decrease was observed in combination with effector cells in comparison to the medium control (Fig. 2H). Again, the loss of impedance induced by bispecific antibody [$(HER2)_2 \times V\gamma9$] was more pronounced than the pAg-induced activation of effector cells.

Similar results of a superior decrease of impedance by bispecific antibody [$(HER2)_2 \times V\gamma9$]-activated effector cells were also observed when using BxPC3 cells as targets instead of Panc1 cells (Fig. 2I), and with activation of γδ T cells within the PBMC at an E/T ratio of 1:1 (γδ T cells to target cells) (Fig. 2I and J).

5.3 Calculations and statistic

The raw data files were exported from the RTCA software (version 2.0.0.1301, Copyright© 2004–2012, ACEA) to Microsoft Excel [version 14.0.7128.5000, (32-bit)] for further analysis. The mean of Triton-X-100 samples was calculated and defined as 100% lysis over several hours after addition of effector cells. The percentage of lysis of each sample was calculated compared to the appropriate control samples and was normalized to maximal lysis with Triton-X-100. To summarize different experiments, the fold change in CI of the different determinations at a distinct time points can be calculated by using the formula as follows: $CI - \text{Fold change} = 1 - \left(\left(1 - \left(\frac{S}{C}\right) \times \left(\frac{1}{1 - \frac{M}{C}}\right)\right)\right)$; S = CI value of the sample; C = value of the medium control; M = Mean of CI value of Triton-X-100 sample. Time point zero was defined as first measurement after addition of T cells.

The current RTCA Software Pro is an integrated software package to analyze the data with a standardized computer hard drive. By using the xIMT software (ACEA Biosciences), the calculation of percentage cytolysis and KT_{50} (the required time to lyse 50% of the target cells) can be determined as described elsewhere (Cerignoli et al., 2018).

6. Summary

An accumulation of γδ T cells in the ductal epithelium of PDAC tissue suggests an important role of γδ T cells in the immune response against PDAC which is known for its strongly immunosuppressive microenvironment. The RTCA provides a tool to determine the functional capacity of γδ T cells or other cytotoxic effector cells alone or within a heterogenous population (e.g. PBMC or tumor-infiltrating immune cells isolated from tumor tissue of cancer patients) in a time-dependent manner. Low effector/target ratios can be applied, and further, RTCA allows determination of a possible regeneration of tumor cells if their lysis is incomplete. In addition, the efficiency of substances like bispecific antibodies that enhance cytotoxicity or overcome possible immunosuppressive effects—as we postulated—can be analyzed.

Taken together, RTCA provides a sensitive and reproducible assay to determine cytotoxicity against tumor cells of interest and may provide an excellent assessment of whether a particular personalized tumor treatment strategy will be effective. It is also a potent platform to test new strategies that aim to optimize the killing of PDAC (and other tumor) cells.

Acknowledgments

We gratefully acknowledge Dr. Christian Röder for providing PDAC cell lines and his help with the short tandem repeat analysis. Bispecific antibodies were produced in cooperation with Prof. Matthias Peipp (Department of Stem Cell Transplantation and Immunotherapy, UKSH, Kiel) and Dr. Christian Kellner (Department of Transfusion Medicine, Cell Therapeutics and Haemostaseology, LMU Munich).

References

Brinkmann, U., & Kontermann, R. E. (2017). The making of bispecific antibodies. *mAbs, 9*, 182–212.

Cerignoli, F., Abassi, Y. A., Lamarche, B. J., Guenther, G., Santa, A. D., Guimet, D., et al. (2018). In vitro immunotherapy potency assays using real-time cell analysis. *PLoS One, 13*, e0193498.

Chitadze, G., Oberg, H. H., Wesch, D., & Kabelitz, D. (2017). The ambiguous role of gammadelta T lymphocytes in antitumor immunity. *Trends in Immunology, 38*, 668–678.

Davenport, A. J., Jenkins, M. R., Cross, R. S., Yong, C. S., Prince, H. M., Ritchie, D. S., et al. (2015). CAR-T cells inflict sequential killing of multiple tumor target cells. *Cancer Immunology Research, 3*, 483–494.

Espinosa, E., Belmant, C., Pont, F., Luciani, B., Poupot, R., Romagne, F., et al. (2001). Chemical synthesis and biological activity of bromohydrin pyrophosphate, a potent stimulator of human gamma delta T cells. *The Journal of Biological Chemistry, 276*, 18337–18344.

Harder, J., Ihorst, G., Heinemann, V., Hofheinz, R., Moehler, M., Buechler, P., et al. (2012). Multicentre phase II trial of trastuzumab and capecitabine in patients with HER2 overexpressing metastatic pancreatic cancer. *British Journal of Cancer, 106,* 1033–1038.

Helm, O., Mennrich, R., Petrick, D., Goebel, L., Freitag-Wolf, S., Roder, C., et al. (2014). Comparative characterization of stroma cells and ductal epithelium in chronic pancreatitis and pancreatic ductal adenocarcinoma. *PLoS One, 9,* e94357.

Leung, L., Radulovich, N., Zhu, C. Q., Organ, S., Bandarchi, B., Pintilie, M., et al. (2012). Lipocalin2 promotes invasion, tumorigenicity and gemcitabine resistance in pancreatic ductal adenocarcinoma. *PLoS One, 7,* e46677.

Loibl, S., & Gianni, L. (2017). HER2-positive breast cancer. *Lancet, 389,* 2415–2429.

Messaoudene, M., Fregni, G., Fourmentraux-Neves, E., Chanal, J., Maubec, E., Mazouz-Dorval, S., et al. (2014). Mature cytotoxic CD56(bright)/CD16(+) natural killer cells can infiltrate lymph nodes adjacent to metastatic melanoma. *Cancer Research, 74,* 81–92.

Oberg, H. H., Grage-Griebenow, E., Adam-Klages, S., Jerg, E., Peipp, M., Kellner, C., et al. (2016). Monitoring and functional characterization of the lymphocytic compartment in pancreatic ductal adenocarcinoma patients. *Pancreatology, 16,* 1069–1079.

Oberg, H. H., Kellner, C., Gonnermann, D., Peipp, M., Peters, C., Sebens, S., et al. (2015). Gammadelta T cell activation by bispecific antibodies. *Cellular Immunology, 296,* 41–49.

Oberg, H. H., Kellner, C., Gonnermann, D., Sebens, S., Bauerschlag, D., Gramatzki, M., et al. (2018). Tribody [(HER2)2xCD16] is more effective than trastuzumab in enhancing gammadelta T cell and natural killer cell cytotoxicity against HER2-expressing cancer cells. *Frontiers in Immunology, 9,* 814.

Oberg, H. H., Kellner, C., Peipp, M., Sebens, S., Adam-Klages, S., Gramatzki, M., et al. (2014). Monitoring circulating gammadelta T cells in cancer patients to optimize gammadelta T cell-based immunotherapy. *Frontiers in Immunology, 5,* 643.

Oberg, H. H., Peipp, M., Kellner, C., Sebens, S., Krause, S., Petrick, D., et al. (2014). Novel bispecific antibodies increase gammadelta T-cell cytotoxicity against pancreatic cancer cells. *Cancer Research, 74,* 1349–1360.

Sadeghi, S., Olevsky, O., & Hurvitz, S. A. (2014). Profiling and targeting HER2-positive breast cancer using trastuzumab emtansine. *Pharmacogenomics and Personalized Medicine, 7,* 329–338.

Siegel, R. L., Miller, K. D., & Jemal, A. (2015). Cancer statistics, 2015. *CA: A Cancer Journal for Clinicians, 65,* 5–29.

Sipos, B., Moser, S., Kalthoff, H., Torok, V., Lohr, M., & Kloppel, G. (2003). A comprehensive characterization of pancreatic ductal carcinoma cell lines: Towards the establishment of an in vitro research platform. *Virchows Archiv, 442,* 444–452.

Topp, M. S., Gokbuget, N., Zugmaier, G., Klappers, P., Stelljes, M., Neumann, S., et al. (2014). Phase II trial of the anti-CD19 bispecific T cell-engager Blinatumomab shows hematologic and molecular remissions in patients with relapsed or refractory B-precursor acute lymphoblastic leukemia. *Journal of Clinical Oncology, 32,* 4134–4140.

CHAPTER TWENTY-FOUR

Analysis of cancer cell-intrinsic immune regulation in response to CD8+ T cell attack

Natalie J. Neubert[†], Laure Tillé[†], Christophe Martignier[†], Silvia A. Fuertes Marraco[†], Daniel E. Speiser[*,†]
Department of Oncology, University of Lausanne, Lausanne, Switzerland
*Corresponding author: e-mail address: doc@dspeiser.ch

Contents

Abstract

Immunotherapies against cancer continue to improve, but many cancers show primary or secondary resistance. Novel research strategies are necessary to reach a comprehensive understanding of the underlying mechanisms. There is increasing evidence that

[†] Conceived and designed the experiments: N.J.N., L.T., S.A.F.M., D.E.S. Performed the experiments: N.J.N., L.T., S.A.F.M. Analyzed the data: N.J.N., L.T. Wrote the manuscript: N.J.N., C.M., S.A.F.M., D.E.S.

Methods in Enzymology, Volume 631
ISSN 0076-6879
https://doi.org/10.1016/bs.mie.2019.05.035

T cells themselves provoke immune escape of cancer cells. In this chapter we describe a co-culture system to analyze the dynamic interplay between T cells and cancer cells. Using human melanoma cell lines and T cell clones, we obtained reproducible and comparable results despite the high heterogeneity of tumor cells. We show the feasibility of differential protein and gene expression analysis of melanoma cells isolated from our culture system. Thus, the system allows quantifying broadly the differential gene expression in melanoma cells upon interaction with T cells, revealing immune-related reactions in cancer cells. Many parts of this chapter were previously published in an original paper and are reproduced here for this book.

1. Introduction

Immunotherapy against cancer is based on the principle of therapeutic activation of cytotoxic CD8+ T lymphocytes (CTLs), resulting in increased cytotoxicity and cytokine production by tumor antigen-specific T cells. A growing number of solid malignancy types may respond to immunotherapy such as adoptive T cell transfer with autologous tumor infiltrating T cells, and inhibitory receptor blockade (checkpoint blockade) (Khalil, Smith, Brentjens, & Wolchok, 2016; Ribas & Wolchok, 2018; Rosenberg & Restifo, 2015; Sharma & Allison, 2015). However, the majority of patients have tumors with primary or secondary resistance to immunotherapy, calling for intense research efforts to identify the underlying mechanisms and to develop therapeutic strategies to overcome resistance. Several lines of research showed that CTLs themselves contribute to provoke escape from immunotherapy. This was evidenced by the finding of long-term escape variants of tumor cells resistant to immune attack (Abouzahr et al., 2006; Speiser et al., 2006), in both mouse models (Landsberg et al., 2012; McGray et al., 2013) and in data from human bulk tumors (Schramm, Campain, Scolyer, Yang, & Mann, 2011; Shin et al., 2017; Zaretsky et al., 2016). However, immune alterations in cancer cells are not necessarily the result of the selection of escape mutants due to long-term immune pressure; they are also observed short-term. In particular, the modulation of expression of immune genes in presence of tumor infiltrating T cells has been called "adaptive resistance" (Taube et al., 2012) and is often involved in immune escape mechanisms. This is best illustrated by the frequent upregulation of PD-L1 in tumors infiltrated by T cells (Tumeh et al., 2014), a mechanism that can be overcome by therapeutic blockade of the PD-1/PD-L1 axis representing the broadly most successful approach for immunotherapy of human solid cancers (Ribas & Wolchok, 2018; Sharma & Allison, 2015).

Regulation of immune genes, however, is highly complex and involves many functions and triggers that have yet to be investigated, with the aim to identify the ones that are (co-)responsible for therapy escape. Therefore, we developed a co-culture system between human cancer cells and CD8+ T cells enabling the characterization of immune alterations in cancer cells in response to T cell attack (Fig. 1). Here we present materials, controls, culture parameters and readouts of this system. This book chapter is based on the report we published previously in Frontiers in Immunology (2016; 7:326), entitled "A Well-Controlled Experimental System to Study Interactions of Cytotoxic T Lymphocytes with Tumor Cells", with permission of the publisher. For more details we refer to this paper (Neubert et al., 2016) and the cited references.

2. Methods

2.1 Cells and cell cultures

CD8+ T cell clones and melanoma cell lines were established at the University Hospital of Lausanne. The study was approved by the local ethics committee. Patients consented to participation. Melanoma cell lines were established from metastatic surgery specimens. They were cultured in RPMI 1640—GlutaMAX™-I, complemented with 10% heat-inactivated FCS (PAA), 1.1 μM Arginine (Sigma Aldrich), 0.48 μM Asparagine (Sigma Aldrich), 11.25 μM Glutamine (Gibco), 10 mM Hepes (Gibco), and 100 U/mL of penicillin/streptomycin (Gibco). Where indicated, the medium was supplemented with IFNγ (222 U/mL; Peprotech), TNFα (50 ng/mL; Peprotech) or both TNFα and IFNγ (10 ng/mL and 222 U/mL). Peripheral blood mononuclear cells (PBMCs) of melanoma patients or healthy donors were used to establish CD8+ T cell clones. The clones were maintained as previously described (Lövgren et al., 2012) and were cultured in RPMI 1640—GlutaMAX™-I, supplemented with 100 U/mL of penicillin/streptomycin, 2 mM L-glutamine, 1 mM non-essential amino acids, 1% Na pyruvate, 0.1 mg/mL Kanamycin (all Gibco), 5×10^{-5} 2β-mercaptoethanol (Sigma) and 8% human serum. Human serum was produced in house from serum of 30 male A+ donors collected from the CRS Interregional Transfusion Center Bern. Serum quality was evaluated based on proliferation, and using mixed lymphocyte culture and mycoplasma PCR. All melanoma cell lines and T cell clones were regularly controlled and confirmed mycoplasma-free by PCR.

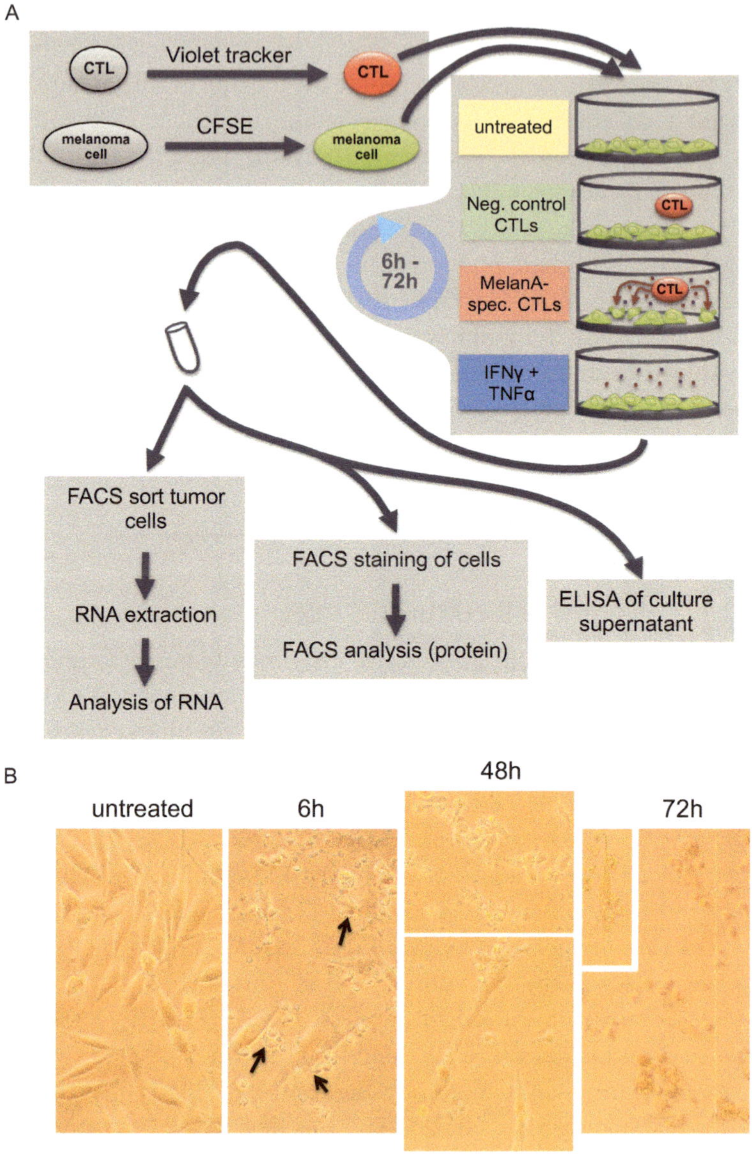

Fig. 1 CTL-melanoma cell co-culture (A) Experimental setup. Depending on downstream analysis, co-cultures were harvested and analyzed after 6, 24, 48 or 72 h. (B) Co-culture of a melanoma-specific CTL clone and a melanoma cell line at the indicated time points. At 6 h, arrows point out visible direct CTL-melanoma cell contacts. In the untreated condition, melanoma cells were cultured alone in absence of CTLs or cytokines. Two photos are shown at the time points 48 h and 72 h, with the purpose to show more cells than what fitted in a single photo.

2.2 Fluorescence-activated cell sorting (FACS)

1. To distinguish the two cell types, cells were fluorescently labeled before the co-culture (Fig. 1). Melanoma cells were labeled with 1 μM CFSE (Cell trace CFSE cell proliferation kit, Molecular Probes) and CD8+ T cell clones were labeled with 1 μM Violet tracker (Cell trace Violet Cell Proliferation kit, Molecular Probes).
2. For analysis of intracellular protein expression, BrefeldinA (final concentration: 10 μg/mL, Sigma Aldrich) was added to the culture medium for the last 4 h of culture.
3. To distinguish live and dead cells, the harvested cells were stained with LIVE/DEAD® Fixable Aqua Dead Cell Stain Kit (Life Technologies) for 30 min. Then cells were fixed at 4 °C overnight in PBS supplemented with 1% Formaldehyde, 2% Glucose and 5 mM NaAzide.
4. Afterward, to stain cell surface proteins, the cells were washed once with FACS buffer (PBS containing 5 mM EDTA, 0.2% BSA and 0.2% NaAzide) and stained with antibodies for 20 min.
5. To stain for intracellular proteins, cells were permeabilized with 0.1% saponin-containing FACS buffer and stained with antibodies in the same buffer for 30 min. Stainings were done with the following anti-human antibodies: Melan-A (clone A103, produced and labeled in house), HLA-DR (clone LN3, eBioscience), HLA Class I (clone W6/32, Biolegend), CD8 (clone SK1, BD), IFNγ (clone B27, BD) and TNFα (clone MAb11, BD). Following matched isotype control antibodies were used: mouse IgG1 (produced in house, batch 24.11.14), mouse IgG2b kappa (clone eBMG2b, eBioscience), mouse IgG2a kappa (clone MOPC173, Biolegend), mouse IgG1 kappa (clone MPC-11, BD), IgG1 kappa (clone MOPC-21, BD). PE-labeled A2-ELA-specific tetramers (No. 404, TC Metrix) were used to stain Melan-A-specific CD8+ T cell clones.
6. Cells were acquired with the Gallios flow cytometer (Beckman Coulter, 3-laser configuration) and FlowJo (Tree Star Inc., v9.7.7) was used for data analysis. FACS sorting was done with an Astrios cytometer (Beckman Coulter).

For FACS sorting, cells were stained with 3 μg/mL DAPI.

2.3 RNA extraction for gene expression analysis

Melanoma cells were directly sorted into suspension media (PBS containing 10% BSA and 0.5 M EDTA). RNA extraction was performed with the Pico

Pure™ RNA isolation kit (Arcturus) according to the manufacturer's Protocol B "RNA Extraction from Cell Pellets" and including the optional freezing at −80 °C directly after cell lysis as well as the optional on-column DNA digest. RNA was quantified with the Qubit fluorometer (Life Technologies) and the quality of the samples was determined with the 2100 Bioanalyzer (Agilent).

2.4 NanoString analysis

To quantify the RNA recovered from cells, a custom-made set of probes by the iGE3 Genomics platform (University of Geneva), the Prep Station and nCounter devices (NanoString Technologies) were used as previously described (Fernandez et al., 2012). The data were analyzed with the R software (version 3.2.2) and the NanoStringNorm package (version 1.1.19). Average counts of the negative controls were subtracted from the raw counts and the data were log2-transformed. Then undefined values (the ones resulting from log transformation of zero count values) were replaced with "NA." A shift based on the normalization genes was defined using simple regression with the mean as offset. When log fold-changes were computed, a value of 10 was added to all normalized counts, the counts were averaged per condition, fold-changes were computed (treated/untreated) and eventually log2 transformed. To reduce the bias toward extremely high fold-change values when a gene had a very low expression level in the untreated condition we added 10 to all counts before fold-change calculations.

3. Selection of melanoma cell lines and CD8+ cytotoxic T lymphocyte (CTL) clones

The model is based on *in vivo* expanded cell lines. We selected four human melanoma cell lines from the numerous cell lines that we had generated from melanoma metastases or tumor infiltrated lymph node specimens in the context of our clinical investigations (Neubert et al., 2017). The selected cell lines were Me275 (from patient LAU50), Me290 (LAU203), T1015A (LAU1015) and T1185B (LAU1185). We only used cell lines that had been passaged for less than 6 months at the time of the experiments. Since we aimed at studying antigen-specific CTL/melanoma cell interactions, we chose melanoma cell lines expressing a specific antigen (Melan-A) and HLA Class I as determined by flow cytometry (Fig. 2A). All melanoma cell lines were derived from HLA-A2 positive patients.

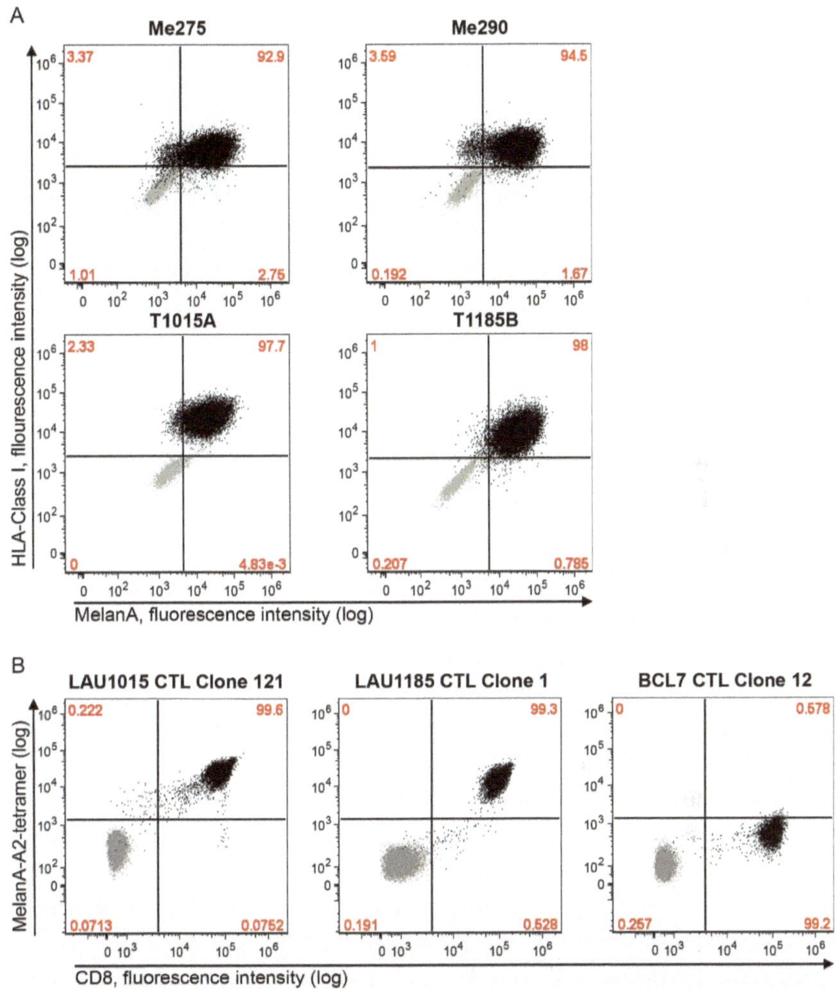

Fig. 2 Expression of typical markers of the two cell types used for our experiments. Melanoma cell lines express the target antigen Melan-A, and CTLs express Melan-A-specific T cell receptors (TCRs). (A) The melanoma cell lines were Melan-A and HLA Class I positive as revealed by flow cytometry analysis using the pan-HLA Class I specific antibody W6/32. Specifically stained melanoma cell lines are shown in black and isotype-matched controls in gray. Red numbers indicate gate frequency of stained samples. (B) Melan-A-specific CTLs (Clone 121 and Clone 1), but not Epstein-Barr-virus-specific CTLs (Clone 12) used as negative control stained positively with Melan-A/HLA-A2 tetramers, indicating that they expressed Melan-A-specific TCRs on their surface. All CTLs were CD8 positive. Stained CTLs are displayed in black and unstained CTLs in gray. This and following figures have been previously published in Neubert, N.J., Soneson, C., Barras, D., Baumgaertner, P., Rimoldi, D., Delorenzi, M., et al. (2016). A well-controlled experimental system to study interactions of cytotoxic T lymphocytes with tumor cells. *Frontiers in Immunology, 7*, 326. https://doi.org/10.3389/fimmu.2016.00326.

In parallel we selected melanoma antigen (Melan-A) specific CTL clones from the large panel of clones that we had established previously (Speiser et al., 2008). We selected two CTL clones specific for HLA-A2 and the Melan-A_{26-35} decapeptide EAAGIGILTV, namely clone 1 from patient LAU1185 and clone 121 from patient LAU1015 (Fig. 2B). Both clones were capable of killing at least 50% of the tumor cells in a standard 4-h killing assay (Lövgren et al., 2012), at a CTL:melanoma cell ratio of 10:1 (Fig. 3). These potent CTLs were chosen expecting that they would exert substantial immune pressure and consequently provoke changes in the melanoma cells in co-cultures.

4. Establishment of the melanoma cell/T cell co-culture

Melanoma cell lines differ substantially in their growth rate. To establish comparable growth conditions, we pre-determined for each cell line the seeding density at which cells would not be confluent before 6 days in any of the culture conditions. Depending on the cell line, between 35,000 and 70,000 melanoma cells were seeded per 24-well in 2 mL. These seeding densities were chosen to allow only one medium change during the experiment (on the day that CTLs were added), such that we prevented medium changes during co-culture and prevented the removal of secreted factors that might play roles in the co-culture. In parallel, we also treated melanoma cells alone with the two cytokines IFNγ and TNFα at doses that reproduce the presence of CTLs (Neubert et al., 2017).

The schedule of the co-culture procedure is shown in Fig. 3. We seeded the melanoma cell lines 3 days prior to the start of the co-culture, to allow them to adhere, in 24-wells and 2 mL volume. At the start of the co-culture (day 0), 1 mL of medium was discarded and replaced with 1 mL of fresh medium containing the respective "treatment," i.e., CTL clones, cytokines (IFNγ and TNFα) or no treatment. On day 0 we added 30 U/mL of IL2 to all cultures. In setup experiments, 30 U/mL of IL2 was determined as the lowest concentration of IL2 that assured persistence of unstimulated CTLs (data not shown). We found similar protein expression with or without IL2 (data not shown), thus the presence of IL2 did not influence the melanoma cells. Depending on the condition of the melanoma cell lines, growth speed differed slightly between experiments. To calculate the appropriate number of CTLs for the chosen CTL:melanoma ratio the number of melanoma cells was counted in two additional control wells on day 0.

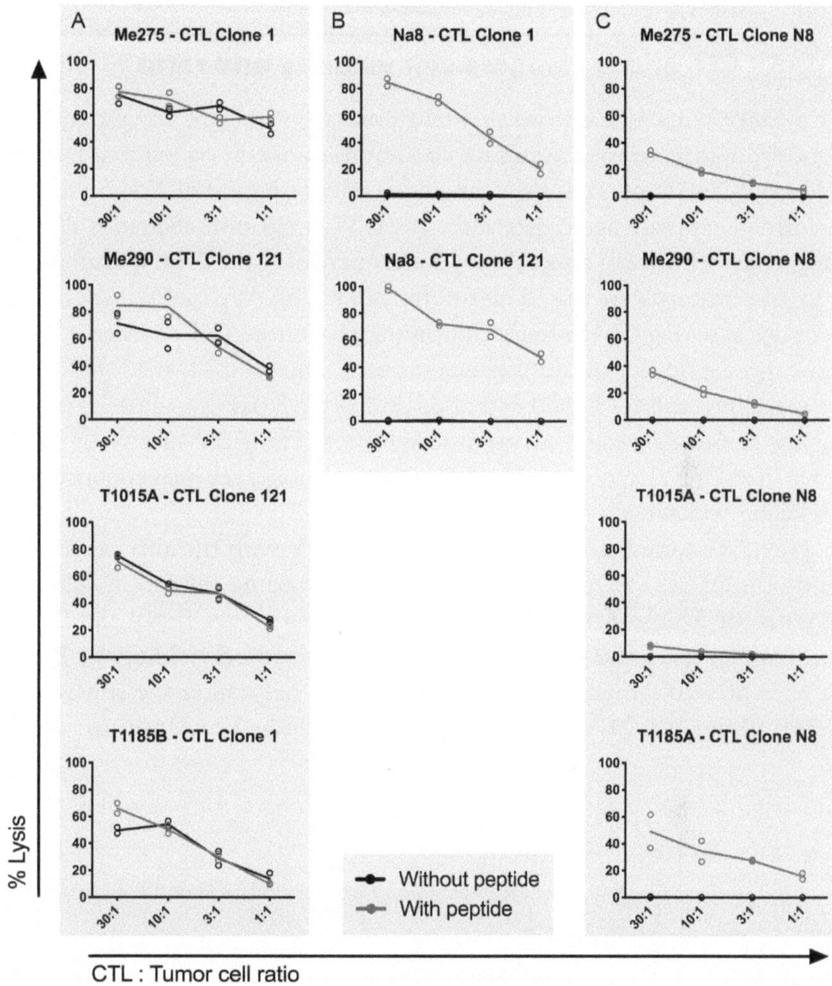

Fig. 3 Killing capacity of Melan-A-specific CTLs measured in a 4h 51Cr release assay. (A) Melan-A-positive melanoma cell lines were co-cultured with Melan-A-specific CTLs at the indicated ratios. The Melan-A peptide ELAGIGILTV was used as positive control. (B) As control, a Melan-A-negative melanoma cell line (Na8) was co-cultured with two different Melan-A-specific CTL clones. The CTLs did not kill the melanoma cells in absence of synthetic Melan-A peptide ELAGIGILTV. (C) As control, Melan-A-positive melanoma cell lines were cultured with the YFV-specific CTL clone N8. Killing was only seen in presence but not in absence of synthetic YFV peptide LLWNGPMAV. $N = 2$, error bars indicate standard deviation.

5. CTL and melanoma cell viability and ratio

Since our aim was to isolate melanoma cells viable for downstream analyses, we optimized the co-culture conditions to obtain co-existence of both CTLs and melanoma cells over the whole culture period of 3 days. First, we performed experiments to determine the CTL:melanoma cell ratios allowing persistence of both cell types. As expected, increasing the CTL:melanoma cell ratio proportionally increased melanoma cell killing. We found three moderate ratios showing co-existence during the co-culture (3:1, 1:1 and 1:3) and chose those ratios for our further experiments. Due to CTL-mediated killing, the melanoma cell:CTL ratio decreased during the 3 day co-culture. The finding that Melan-A-specific but not non-specific CTLs (EBV-specific or Yellow Fever Virus (YFV)-specific CTLs) killed the melanoma cell lines confirmed that the killing was antigen-specific (Fig. 4).

Next, we tested different harvesting methods with the aim to recover a maximum of cells from co-cultures. Since melanoma cells are fragilized in presence of Melan-A-specific CTLs or their cytokines IFNγ and TNFα, quick and gentle harvesting was required. The method yielding the highest number of viable melanoma cells consisted of harvesting the supernatant, then washing the 24-well once with PBS, followed by 2min incubation

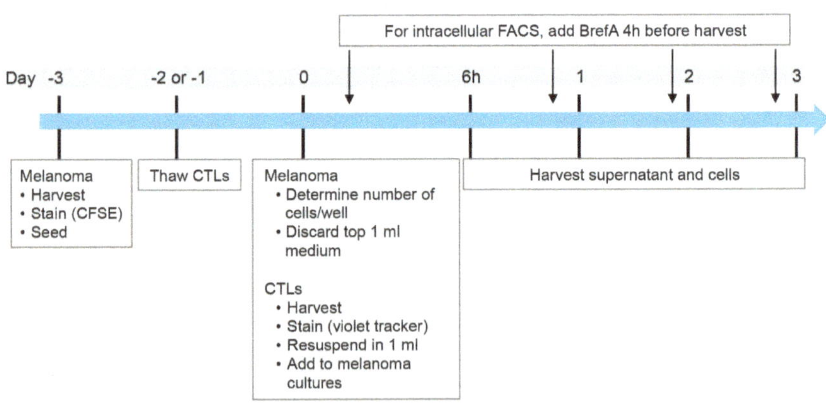

Fig. 4 Schedule of co-cultures. Melanoma cell lines are thawed at least 1 week prior to the experiment, to assure stable cell behavior in the co-cultures. CTLs are thawed at day −2 and the viability is assessed at day −1, which leaves time to thaw new CTLs in case of insufficient viability. Melanoma cells are stained with CFSE and re-seeded to start the co-culture experiment. To stain for intracellular cytokines in CTLs, we added BrefeldinA at the indicated time points. Supernatant was taken before addition of BrefeldinA or from replicate cultures without BrefeldinA.

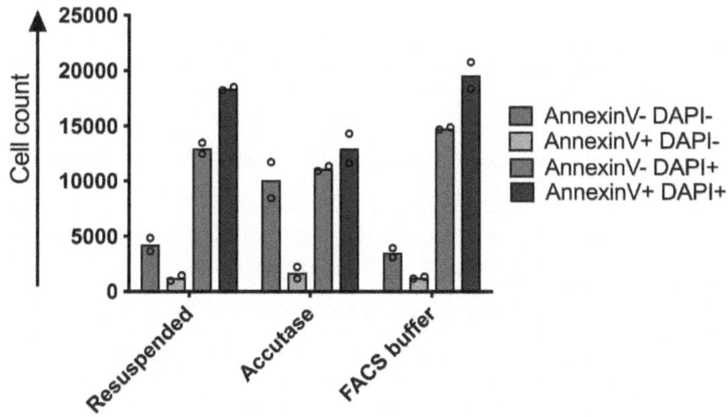

Fig. 5 Melanoma cell viability depends on the method of harvesting. For the first condition ("Resuspended"), cells were detached by repeated pipetting of the medium. For the second condition ("Accutase"), supernatant was removed, the 24-wells were washed once with PBS, cells were briefly incubated with 100 μL accutase at 37 °C, and detached cells were collected with 500 μL fresh medium. And for the third condition ("FACS buffer"), supernatant was removed, 1 mL FACS buffer was added (PBS supplemented with 5 mM EDTA, 0.2% BSA and 0.2% NaAzide), cells were left 5–10 min on ice and then detached cells were collected. Shown is the mean of two replicates.

with 100 μL accutase at 37 °C, addition of 0.5 mL fresh culture medium to stop the reaction and finally collecting the detached cells (Fig. 5). We collected and pooled all liquids including supernatant and washes to avoid loss of cells. To quantify apoptotic cells, harvested samples were stained with fresh AnnexinV-binding buffer (14 mM NaCl, 2.5 mM CaCl2, 10 mM Hepes) containing AnnexinV-FITC (BD Pharmingen) and DAPI nucleic acid stain (Molecular Probes) at 150 ng/mL.

Rapid processing was crucial for the quality of RNA that we wanted to use for expression analyses. We further minimized interventions on the cells after harvest, to optimize viability of cells and achieve high quality data. For instance, to achieve lysis of sorted cells for RNA extraction within 2 h from start of harvesting, no staining was performed on harvested cells except for addition of DAPI for dead cell exclusion. This was possible because we had labeled melanoma cell lines and CTLs with CFSE and violet tracker prior to co-culture. Thus, co-cultures were harvested, centrifuged once, and directly resuspended for sorting (DAPI was added on the cell suspension just prior acquisition). Finally, since sorting was performed sequentially, we also harvested the samples sequentially, with the goal to avoid waiting times of the harvested cells before sorting and to organize comparable timings among samples before and during sorting.

6. CTL function

Besides killing, CTLs secrete IFNγ and TNFα. Both killing and cytokine production depend on antigen-specific interaction. IFNγ and TNFα protein expression was quantified as a measure for T cell activation. For the intracellular staining of IFNγ and TNFα we added BrefeldinA (blocking vesicle transport) during 4 h before harvesting the co-cultures. FACS analysis showed that antigen-specific but not control CTLs expressed IFNγ and TNFα within 6 h after start of the co-culture (Fig. 6). In addition, IFNγ was more strongly expressed than TNFα. Not surprisingly, the percentages of cytokine-producing CTLs were highest in the co-cultures with the lowest CTL to melanoma cell ratio (1:3). This was expected because highest effector function in T cells is found when stimulator (target) cells outnumber T cells (Mahnke et al., 2012).

7. Kinetic of protein expression

Next, we wanted to know whether the co-culture is a valid system to investigate CTL-tumor cell interactions. Therefore, we analyzed two markers on the melanoma cell lines known to upregulate upon exposure to IFNγ or CTLs: HLA-DR and HLA Class I (Basham & Merigan, 1983; Lanza et al., 1995). While all cell lines expressed HLA Class I when untreated, its expression further increased in all cell lines upon co-culture (Fig. 7). HLA-DR expression also increased in three melanoma cell lines that had no or very low HLA-DR expression when left untreated, whereas in a 4th cell line, T1015A, that expressed HLA-DR constitutively, no increase was observed. Taken together, HLA Class I and HLA-DR behaved as expected with increased expression upon exposure to CTLs or their cytokines, supporting that the co-culture system is a valid approach to investigate tumor cell reactions to CTLs. We chose the 1:1 CTL:melanoma cell ratio at the 48 h time-point for future protein expression analysis, because strong protein expression changes were measured and, furthermore, sufficient numbers of living melanoma cells could be harvested for reliable analysis.

8. mRNA screen of co-cultured melanoma cells

In parallel to protein expression analysis by FACS, we investigated mRNA expression. We compared melanoma cell lines cultured alone for

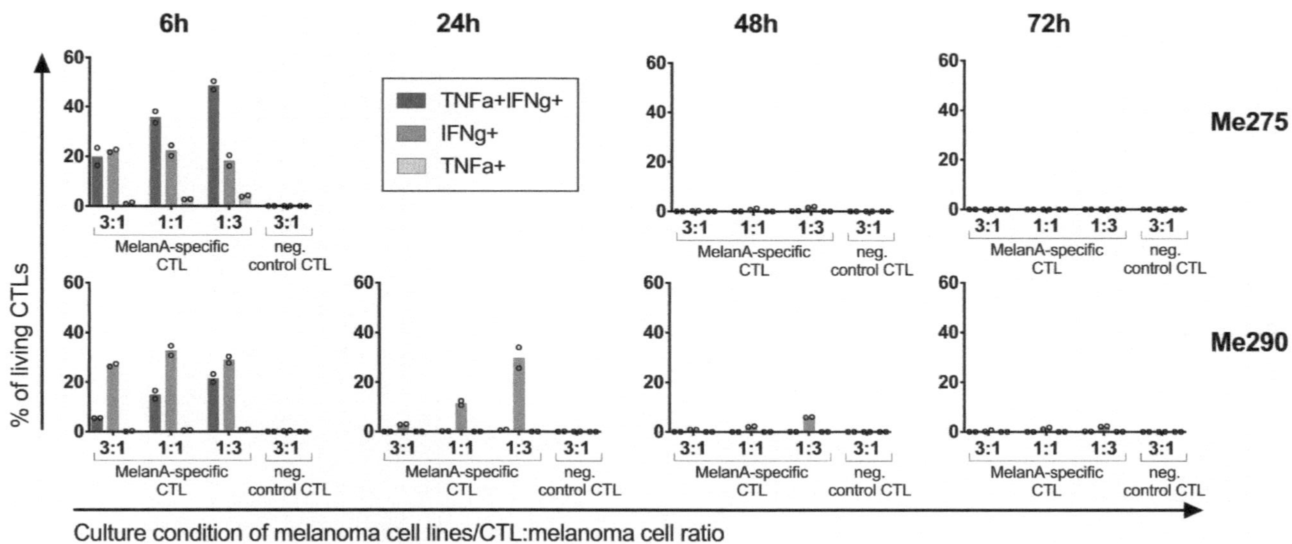

Fig. 6 Cytokine production by co-cultured antigen-specific CTLs. Intracellular accumulation of IFNγ and TNFα in Melan-A-specific CTLs isolated from two co-cultures with the melanoma cell lines Me275 (top row, data not determined for 24 h) and Me290 (bottom row). Negative control (EBV-specific) CTLs did not accumulate intracellular cytokines. Cytokine expression was measured by flow cytometry after intracellular staining of the co-cultured cells.

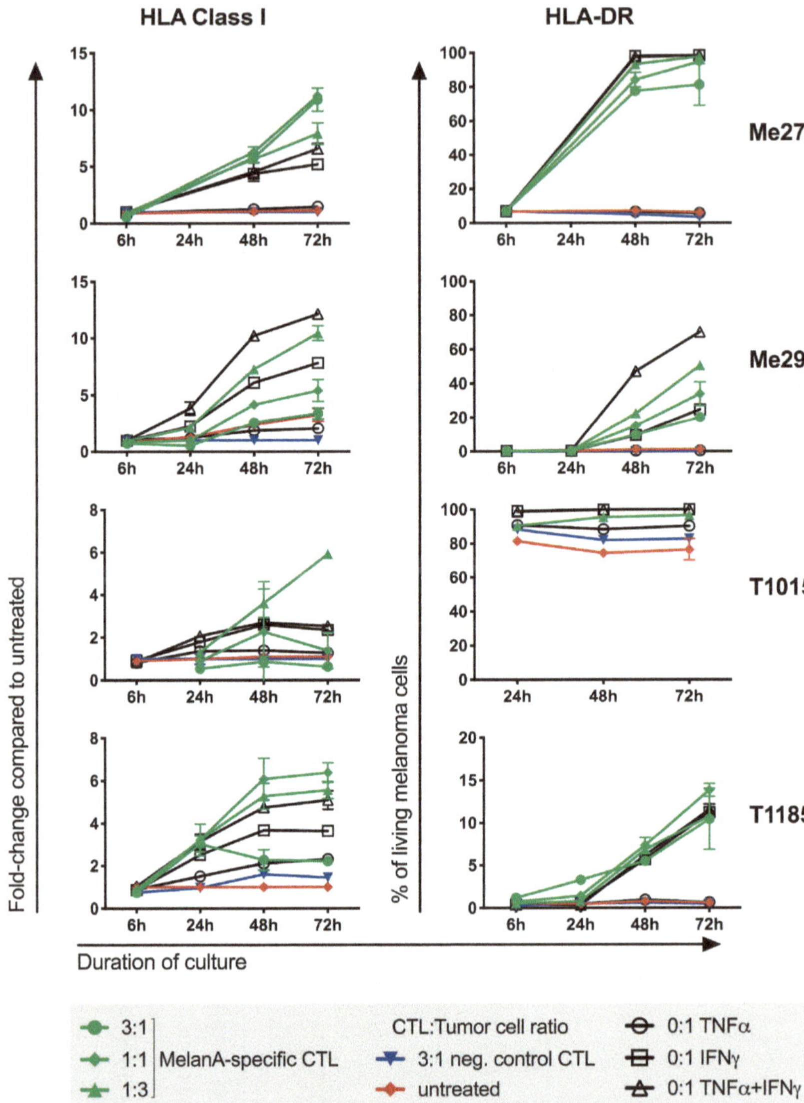

Fig. 7 Increased HLA Class I and HLA-DR expression by melanoma cells upon exposure to CTLs or the CTL-derived cytokines IFNγ and TNFα. Exposure of different melanoma cell lines to Melan-A-specific CTLs (green) or CTL-derived cytokines (black) but not to non-specific CTLs (blue) provoked changes of protein expression on the melanoma cells. Untreated cells are shown in red. HLA Class I was detected using the pan-HLA Class I specific antibody W6/32 and calculated as fold-change compared to untreated cells. HLA-DR is presented as % positive of living cells. Measurements were performed by flow cytometry.

24 h to melanoma cell lines cultured with Melan-A-specific CTLs. Following culture, we aimed to isolate at least 35,000 living melanoma cells from the co-culture by FACS based on staining and gating as shown in Fig. 8 in order to obtain sufficient mRNA for analysis. Depending on the culture condition (i.e., considering melanoma cell killing by CTLs), we predetermined that pooling of three to twelve 24-wells with identical cultures was necessary to obtain sufficient viable melanoma cells after 24 h culture. As mentioned previously, we labeled melanoma cells with CFSE and CTLs with violet tracker for separate identification of melanoma cells and CTLs.

For RNA analysis we used the NanoString technology for several reasons: it is highly quantitative and sensitive, allowing the detection of several hundred individual mRNAs with a high dynamic range, in a single reaction, with a small quantity of RNA (100 ng RNA); furthermore it is a hybridization-based approach that does not require additional amplification or enzymatic steps (introducing less unwanted bias (Geiss et al., 2008)) and has relatively low cost.

For the NanoString panel, we assembled a set of 185 genes (described in (Neubert et al., 2017)) designed for monitoring immune gene regulation in melanoma cells and including control genes (Fig. 9A). Out of the eight chosen control genes, three (*DIRC1, SLC26A3* and *PHKG1*) were not used for normalization because they showed very low expression levels. Therefore, we based normalization on the remaining control genes (*ALG12, GUSB, RPLP0, KRBA2* and *ADAT2*), which showed sufficiently high gene expression (Fig. 9B).

We checked cell purity after cells sorting, but also in the gene expression (NanoString) data of the melanoma cells on behalf of a series of CTL markers (*CD3D, CD3E, CD3G, CD8A* and *CD8B*). The normalized expression values of all CTL markers were below the threshold for background in all samples, except for *CD3D*, for which two culture conditions (Me275 and T1185B with Melan-A-specific Clone 1) were slightly above the threshold but still very low. Thus, potential CTL contamination was negligible.

9. Assessment of HLA upregulation in a larger melanoma cell panel

Since we wanted to know whether melanoma cells generally show the type of gene expression observed in the four cell lines studied so far, we determined protein expression in a larger panel of melanoma cells

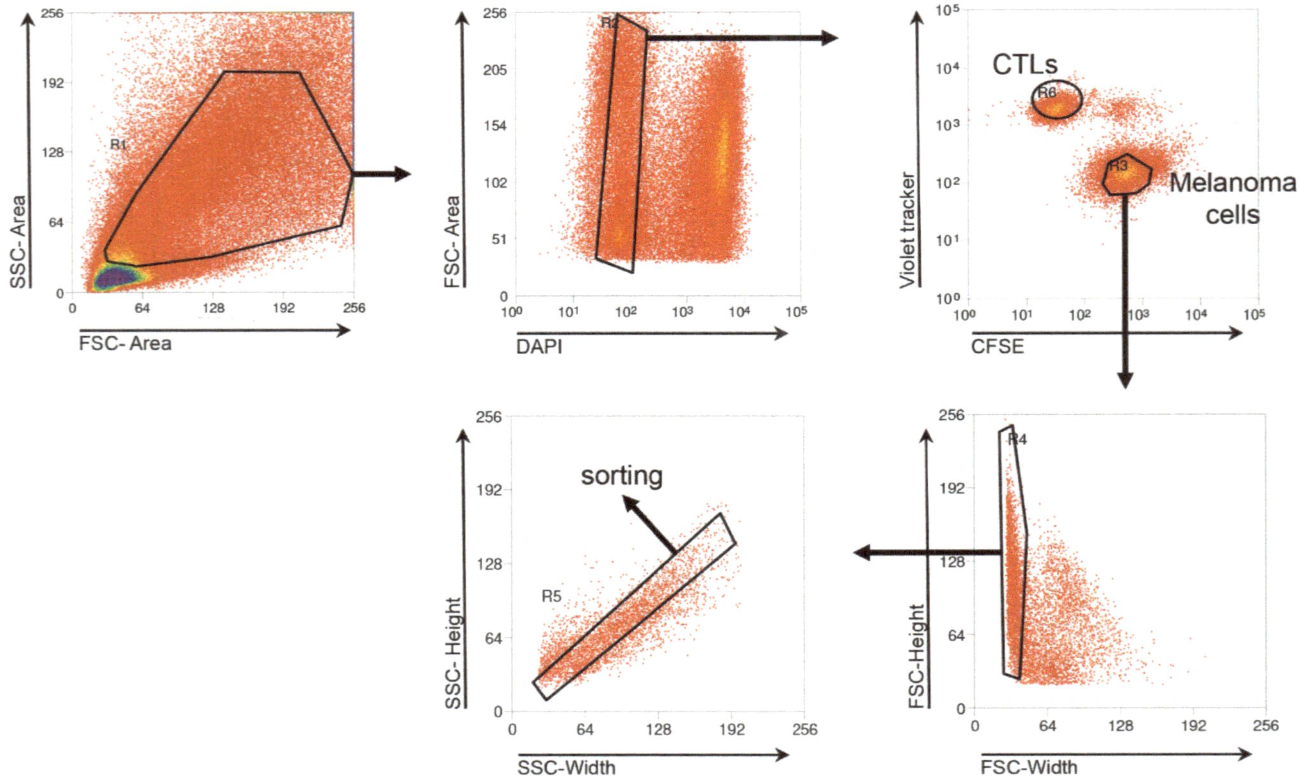

Fig. 8 Strategy of cell sorting for gene expression analysis of melanoma cells. Melanoma cells were stained and gated as illustrated. DAPI negative CFSE positive cells were chosen, cell doublets were excluded.

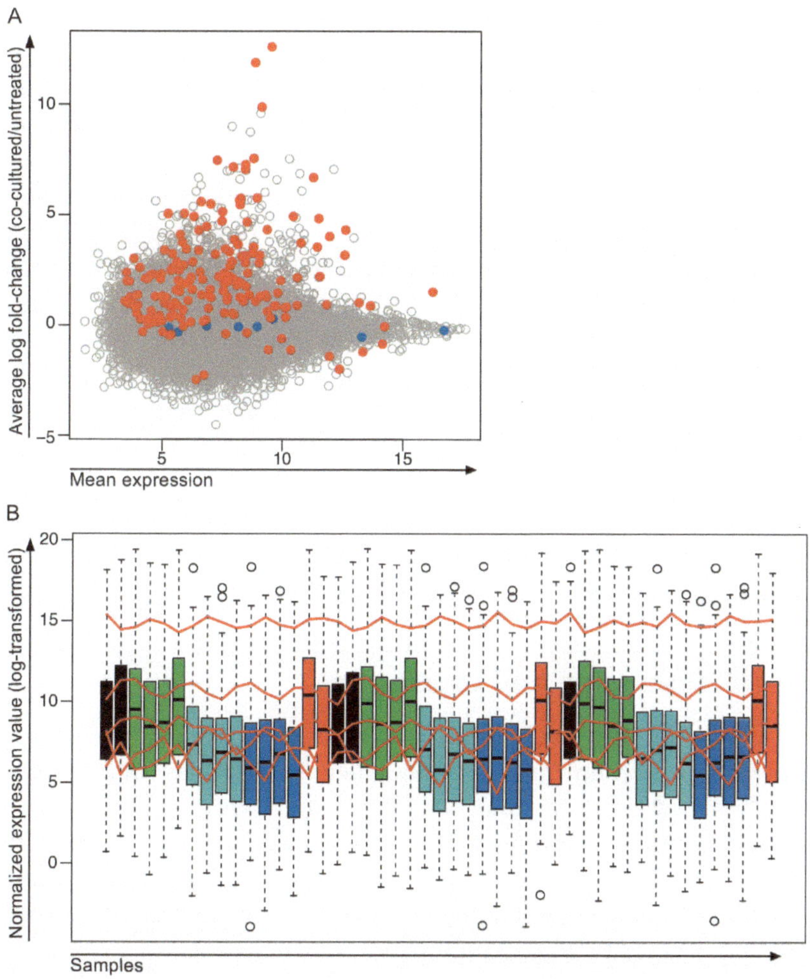

Fig. 9 Stable expression of normalization genes. (A) Quantification of the185 genes that were later chosen for quantification by NanoString. The data show the average expression measured by microarray (as described in (Neubert et al., 2016)) from all samples on the x-axis and the average log-fold change between the co-cultured and the non-co-cultured samples across the 3 melanoma cell lines (Me290, Me275 and T1185B) on the y-axis. For the NanoString, the normalization genes (blue) were chosen to span a wide range of expression values but without changes between the treated and untreated cell lines. Red dots, genes selected for quantification by NanoString; blue dots, normalization genes for NanoString; gray dots, genes that were only measured by microarray. (B) Normalized gene expression values of all samples analyzed by NanoString. Red lines indicate expression levels of genes used for normalization (order from highest to lowest expression: *RPLP0, GUSB, ALG12, ADAT2* and *KRBA2*). Treatment color code: Black, Melan-A-specific clone 1; green, cytokines; cyan, negative control CTLs; blue, untreated; red, Melan-A-specific Clone 121. Samples are shown in the same order as they were run on the NanoString cartridge.

(Neubert et al., 2017). Based on our finding that the two cytokines IFNγ plus TNFα triggered highly similar gene and protein expression in melanoma cells as co-cultures with antigen-specific CTLs, we incubated the melanoma cells with IFNγ plus TNFα and performed FACS analyses after 48 h.

Data obtained from this system are published in our recent papers (Neubert et al., 2018, 2017). Here we show the results for expression of the two human MHC Classes, HLA Class I and HLA Class II in our panel of 16 melanoma cells. Two representative histograms are shown to visualize the expression pattern of surface HLA-DR (Fig. 10A). Over half of the cell lines expressed constitutively HLA-DR at variable levels, often with a bimodal distribution. Nevertheless, the cytokine treatment induced HLA-DR upregulation ($P < 0.0001$) in 15 out of 16 cell lines (Fig. 10B). Two representative histograms are also shown to visualize the expression pattern of HLA Class I (Fig. 11A). The samples presented very few HLA Class I negative cells and the results were therefore calculated in fold changes. All of the 14 tested cell lines upregulated HLA Class I after IFNγ plus TNFα treatment (Fig. 11B).

10. Discussion and conclusion

CTLs not only can kill cancer cells but also provoke changes of the tumor microenvironment with consequences that may impact on the patient's clinical outcome. Our *in vitro* co-culture system of melanoma cell lines and melanoma-specific CTLs allows studying cellular interactions using human specimens, with well-defined and reproducible culture conditions. This allows obtaining viable cells for downstream analyses generating broad and detailed biological information.

As expected, our data validated that melanoma cells respond by rapidly upregulating their HLA Class I and II expression. Importantly, as we published previously (Neubert et al., 2017), this rapid upregulation does not only apply to HLA gene expression, but to the expression of most other immune genes that we studied. In fact, the majority of melanoma cells show broad and conserved immune gene regulation despite their extraordinary genomic heterogeneity.

Our *in vitro* system allows to study functional interactions between the cancer cell and the CTL, conversely it does not allow direct conclusions on *in vivo* functions and their consequences for tumor growth and patient survival. This requires much more extended studies with multiple techniques and approaches, both pre-clinically and clinically. Nevertheless,

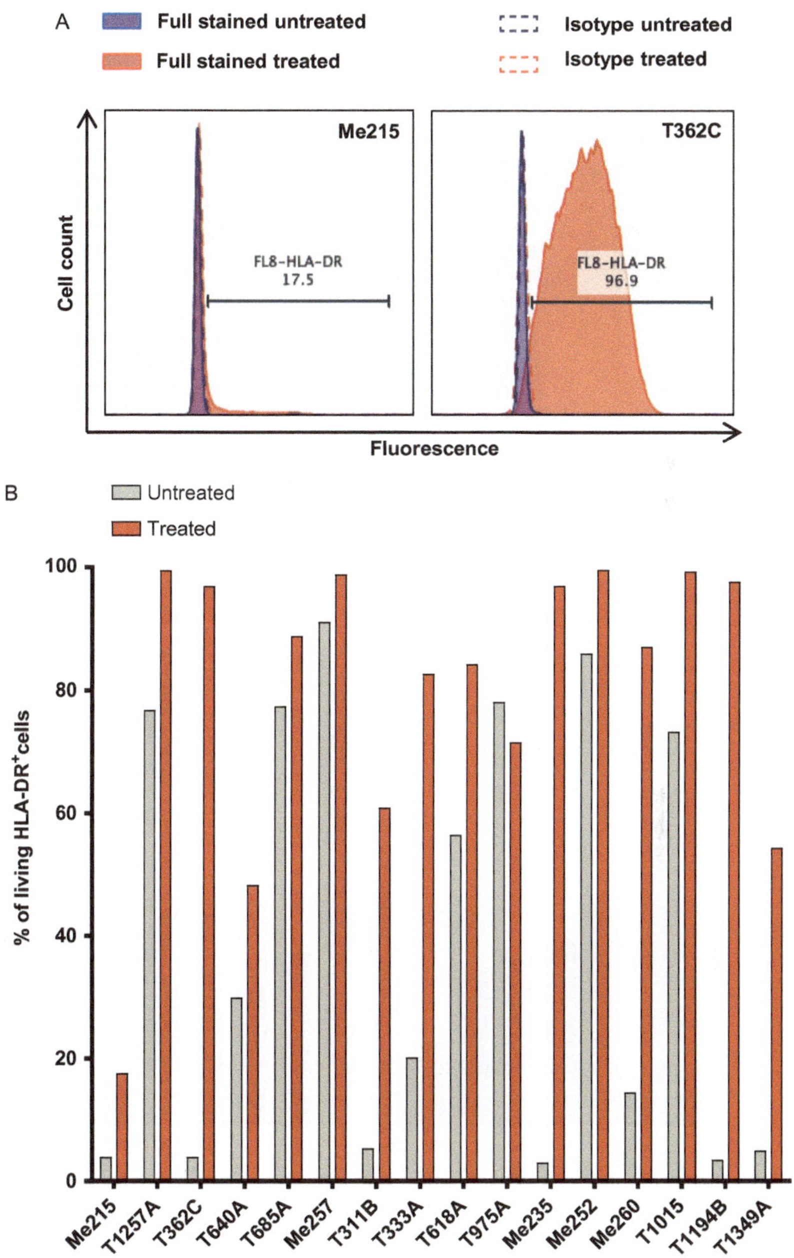

Fig. 10 HLA-DR expression by melanoma cell lines before and after 48 h treatment with IFNγ plus TNFα. (A) Histograms of two representative melanoma cell lines are shown to visualize the expression pattern of HLA-DR. (B) Six out of 16 of the untreated melanoma cell lines did not express HLA-DR at baseline, while the treatment induced the upregulation of HLA-DR in 15 out of 16 cell lines.

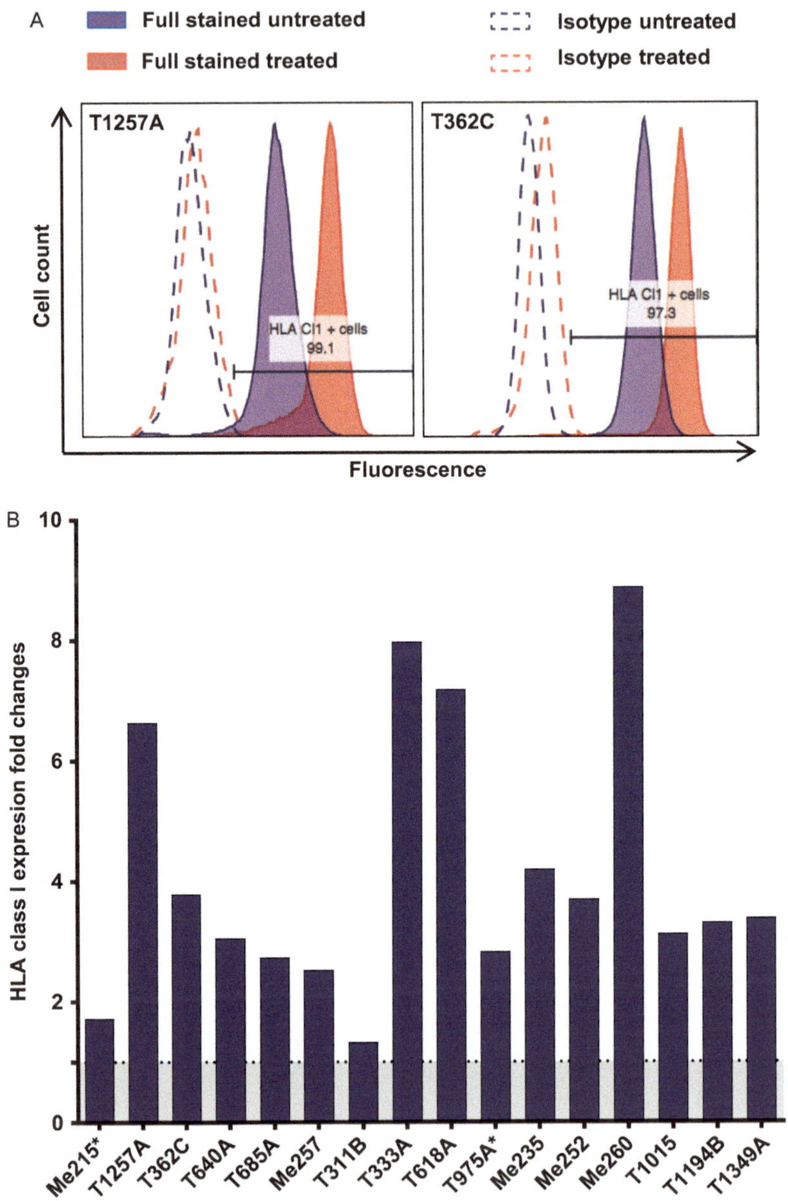

Fig. 11 HLA Class I expression by melanoma cell lines before and after 48 h IFNγ plus TNFα treatment. (A) Two representative histograms are shown to visualize HLA Class I expression. (B) All the tested melanoma cell lines upregulated HLA Class I after IFNγ plus TNFα treatment. HLA Class I was detected using the pan-HLA Class I specific antibody W6/32. The two cell lines marked with * (Me215, T975A) were tested in a separate experiment.

our system enables pinpointing potentially important immune regulatory features of cancer cells in reaction to CD8+ T cells. We found that many changes of immune functions in cancer cells occur rapidly within hours and a few days of T cell attack.

The observed rapid functions may take place despite possible long-term genomic alterations that may determine outgrowth of selected cancer cells with more aggressive behavior, often also implied in therapy resistance. This is well known for oncogene functions. For immune functions, research is still in a relatively early phase. Therefore, it can be expected that pathogenic immune alterations will increasingly be found, following the prototype observation that Jak loss-of-function mutations confer immunotherapy resistance in human melanoma through defective interferon signaling (Shin et al., 2017; Zaretsky et al., 2016).

In vitro, MHC loss can be observed in selected tumor cells from long-term co-cultures with tumor antigen-specific T cells, as e.g. described by Abouzahr et al. (2006). In addition, tumor cell variants may be resistant to cytotoxicity through acquisition of mechanisms inhibiting apoptosis (reviewed in Fulda, 2007). Again, the outgrowth of tumor cell variants requires long-term cultures; the short-term system presented here is not suited for this purpose. Interestingly, we did not see major upregulation of anti-cell death genes in the melanoma cells, suggesting that the melanoma cells we studied were not specialized or selected for cell death resistance.

We think that similar co-cultures can also be performed with other tumor antigens, since there is no obvious reason why HLA-A2/Melan-A should mediate a different quality of interaction as opposed to other tumor T cell epitopes. We have chosen this antigen because HLA-A2 is the most frequent HLA class I allele in the Caucasian population, and Melan-A is expressed by most melanomas. In addition, Melan-A-specific T cell clones are relatively easy to generate and handle, as compared to other clones, likely because of the high precursor frequency of Melan-A-specific CD8 T cells. Thus, on average they are "younger" as compared to T cells specific for other tumor antigens which are much more rare and thus often rapidly terminally differentiated, not well growing *in vitro*.

Finally, this co-culture system can also be used to investigate the consequences of T cell-tumor cell interactions on the CTLs. We have performed preliminary experiments by sorting CTLs based on staining with Violet tracker and DAPI, but analyzed no further on the CTL counterpart of the co-culture. Control conditions would need to be evaluated and validated, for example by using tumor cells transfected or not with the targeted antigen, or by using CTLs engineered with two different TCRs, one specific and another one not specific for antigens presented by the tumor cells.

In conclusion, our short-term co-culture system is well suited to characterize the early and dynamic interplay of melanoma cells and CTLs derived from human specimens. It allows the isolation of sufficient viable cancer cells for high quality protein and mRNA expression data. Our system has the potential to address questions of resistance mechanisms that will likely gain importance as T cell-based immunotherapies are increasingly applied in modern Oncology.

Acknowledgments

We thank the patients and their families for their dedicated collaboration. For their numerous contributions, we would also like to thank all our lab members and all our scientific and medical colleagues who have provided all together outstanding support for this work. Funding was kindly provided by the Swiss Cancer League (3507-09-2014, 3971-08-2016), the Swiss National Science Foundation (310030_179459, 320030_152856), the ISREC Foundation (on "Cancer and Immunology"), the Medic Foundation, the Emma Muschamp Foundation, Alfred and Annemarie von Sick (Switzerland), the Wilhelm Sander Foundation (Germany), the Cancer Research Institute (USA), Ludwig Cancer Research (USA), and the Campbell Family Institute for Breast Cancer Research, Princess Margaret Cancer Centre (Canada).

Conflict of interest

The authors declare no potential conflict of interest.

References

Abouzahr, S., Bismuth, G., Gaudin, C., Caroll, O., van Endert, P., Jalil, A., et al. (2006). Identification of target actin content and polymerization status as a mechanism of tumor resistance after cytolytic T lymphocyte pressure. *Proceedings of the National Academy of Sciences of the United States of America*, *103*(5), 1428–1433. https://doi.org/10.1073/pnas.0510454103.

Basham, T. Y., & Merigan, T. C. (1983). Recombinant interferon-gamma increases HLA-DR synthesis and expression. *Journal of Immunology (Baltimore, Md: 1950)*, *130*(4), 1492–1494.

Fernandez, P., Solenthaler, M., Spertini, O., Quarroz, S., Rovo, A., Lovey, P.-Y., et al. (2012). Using digital RNA counting and flow cytometry to compare mRNA with protein expression in acute leukemias. *PLoS One*, 7(11), e49010. https://doi.org/10.1371/journal.pone.0049010.

Fulda, S. (2007). Inhibitor of apoptosis proteins as targets for anticancer therapy. *Expert Review of Anticancer Therapy*, 7(9), 1255–1264. https://doi.org/10.1586/14737140.7.9.1255.

Geiss, G. K., Bumgarner, R. E., Birditt, B., Dahl, T., Dowidar, N., Dunaway, D. L., et al. (2008). Direct multiplexed measurement of gene expression with color-coded probe pairs. *Nature Biotechnology*, *26*(3), 317–325. https://doi.org/10.1038/nbt1385.

Khalil, D. N., Smith, E. L., Brentjens, R. J., & Wolchok, J. D. (2016). The future of cancer treatment: Immunomodulation, CARs and combination immunotherapy. *Nature Reviews Clinical Oncology*, *13*(5), 273–290. https://doi.org/10.1038/nrclinonc.2016.25.

Landsberg, J., Kohlmeyer, J., Renn, M., Bald, T., Rogava, M., Cron, M., et al. (2012). Melanomas resist T-cell therapy through inflammation-induced reversible dedifferentiation. *Nature*, *490*, 412–416. https://doi.org/10.1038/nature11538.

Lanza, L., Peirano, L., Bosco, O., Contini, P., Filaci, G., Setti, M., et al. (1995). Interferons up-regulate with different potency HLA class I antigen expression in M14 human melanoma cell line. Possible interaction with glucocorticoid hormones. *Cancer Immunology, Immunotherapy: CII*, *41*(1), 23–28.

Lövgren, T., Baumgaertner, P., Wieckowski, S., Devêvre, E., Guillaume, P., Luescher, I., et al. (2012). Enhanced cytotoxicity and decreased CD8 dependence of human cancer-specific cytotoxic T lymphocytes after vaccination with low peptide dose. *Cancer Immunology, Immunotherapy: CII*, *61*(6), 817–826. https://doi.org/10.1007/s00262-011-1140-1.

Mahnke, Y. D., Devêvre, E., Baumgaertner, P., Matter, M., Rufer, N., Romero, P., et al. (2012). Human melanoma-specific CD8(+) T-cells from metastases are capable of antigen-specific degranulation and cytolysis directly ex vivo. *Oncoimmunology*, *1*(4), 467–530. https://doi.org/10.4161/onci.19856.

McGray, A. J. R., Hallett, R., Bernard, D., Swift, S. L., Zhu, Z., Teoderascu, F., et al. (2013). Immunotherapy-induced CD8+ T cells instigate immune suppression in the tumor. *Molecular Therapy: the Journal of the American Society of Gene Therapy*, *22*(1), 206–218. https://doi.org/10.1038/mt.2013.255.

Neubert, N. J., Schmittnaegel, M., Bordry, N., Nassiri, S., Wald, N., Martignier, C., et al. (2018). T cell-induced CSF1 promotes melanoma resistance to PD1 blockade. *Science Translational Medicine*, *10*(436). https://doi.org/10.1126/scitranslmed.aan3311.

Neubert, N. J., Soneson, C., Barras, D., Baumgaertner, P., Rimoldi, D., Delorenzi, M., et al. (2016). A well-controlled experimental system to study interactions of cytotoxic T lymphocytes with tumor cells. *Frontiers in Immunology*, *7*, 326. https://doi.org/10.3389/fimmu.2016.00326.

Neubert, N. J., Tillé, L., Barras, D., Soneson, C., Baumgaertner, P., Rimoldi, D., et al. (2017). Broad and conserved immune regulation by genetically heterogeneous melanoma cells. *Cancer Research*, *77*(7), 1623–1636. https://doi.org/10.1158/0008-5472.CAN-16-2680.

Ribas, A., & Wolchok, J. D. (2018). Cancer immunotherapy using checkpoint blockade. *Science (New York, N.Y.)*, *359*(6382), 1350–1355. https://doi.org/10.1126/science.aar4060.

Rosenberg, S. A., & Restifo, N. P. (2015). Adoptive cell transfer as personalized immunotherapy for human cancer. *Science (New York, N.Y.)*, *348*(6230), 62–68. https://doi.org/10.1126/science.aaa4967.

Schramm, S.-J., Campain, A. E., Scolyer, R. A., Yang, Y. H., & Mann, G. J. (2011). Review and cross-validation of gene expression signatures and melanoma prognosis. *Journal of Investigative Dermatology*, *132*(2), 274–283. https://doi.org/10.1038/jid.2011.305.

Sharma, P., & Allison, J. P. (2015). The future of immune checkpoint therapy. *Science (New York, N.Y.)*, *348*(6230), 56–61. https://doi.org/10.1126/science.aaa8172.

Shin, D. S., Zaretsky, J. M., Escuin-Ordinas, H., Garcia-Diaz, A., Hu-Lieskovan, S., Kalbasi, A., et al. (2017). Primary resistance to PD-1 blockade mediated by JAK1/2 mutations. *Cancer Discovery*, *7*(2), 188–201. https://doi.org/10.1158/2159-8290.CD-16-1223.

Speiser, D. E., Baumgaertner, P., Barbey, C., Rubio-Godoy, V., Moulin, A., Corthesy, P., et al. (2006). A novel approach to characterize clonality and differentiation of human melanoma-specific T cell responses: Spontaneous priming and efficient boosting by vaccination. *Journal of Immunology (Baltimore, Md: 1950)*, *177*(2), 1338–1348. https://doi.org/10.4049/jimmunol.177.2.1338.

Speiser, D. E., Baumgaertner, P., Voelter, V., Devêvre, E., Barbey, C., Rufer, N., et al. (2008). Unmodified self antigen triggers human CD8 T cells with stronger tumor reactivity than altered antigen. *Proceedings of the National Academy of Sciences of the United States of America*, *105*(10), 3849–3854. https://doi.org/10.1073/pnas.0800080105.

Taube, J. M., Anders, R. A., Young, G. D., Xu, H., Sharma, R., McMiller, T. L., et al. (2012). Colocalization of inflammatory response with B7-h1 expression in human melanocytic lesions supports an adaptive resistance mechanism of immune escape. *Science Translational Medicine*, *4*(127), 127ra37. https://doi.org/10.1126/scitranslmed.3003689.

Tumeh, P. C., Harview, C. L., Yearley, J. H., Shintaku, I. P., Taylor, E. J. M., Robert, L., et al. (2014). PD-1 blockade induces responses by inhibiting adaptive immune resistance. *Nature*, *515*(7528), 568–571. https://doi.org/10.1038/nature13954.

Zaretsky, J. M., Garcia-Diaz, A., Shin, D. S., Escuin-Ordinas, H., Hugo, W., Hu-Lieskovan, S., et al. (2016). Mutations associated with acquired resistance to PD-1 blockade in melanoma. *The New England Journal of Medicine*, *375*(13), 2835–2840. https://doi.org/10.1056/NEJMoa1604958.

CHAPTER TWENTY-FIVE

A flow cytometry-based method to screen for modulators of tumor-specific T cell cytotoxicity

Javier Santos, Jesús Ogando, Rosa Ana Lacalle*, Santos Mañes*
Department of Immunology and Oncology, Centro Nacional de Biotecnología (CNB/CSIC), Madrid, Spain
*Corresponding authors: e-mail address: rlacalle@cnb.csic.es; smanes@cnb.csic.es

Contents

Abstract

Cancer immunotherapy relies on the ability of immune cells to kill malignant cells. The cytotoxic T lymphocyte (CTL) response is perhaps the functional measure that best reflects cell-mediated immunity against cancer. Straightforward methods that facilitate quantitative evaluation of the potency of compounds that can modulate T cell-mediated, tumor antigen-specific immune responses are central in the screening cascade when searching for new immunotherapeutic agents for cancer. Here we describe a simple, sensitive method, based on flow cytometry analyses, to quantitatively measure cytotoxicity *in vitro*. We provide examples that validate its feasibility and specificity using CD8$^+$ T lymphocytes specific for a surrogate tumor antigen, and a blocking antibody for the inhibitory PD-1/PD-L1 axis. This method can nonetheless be applied to the screening of virtually any cytotoxicity modulatory compound, including antibodies and small molecules or T cell-based therapies, and can be scaled up for high-throughput workflow and automation.

Methods in Enzymology, Volume 631
ISSN 0076-6879
https://doi.org/10.1016/bs.mie.2019.02.040

1. Introduction

Cell-mediated cytotoxicity is integral in the immune response to intracellular pathogens and cancer cells in the body. Although cell-induced cytotoxicity can involve innate immune cells such as natural killer (NK) cells or macrophages, we will focus here on that mediated by cytotoxic T lymphocytes (CTL), as they are central components of the adaptive immune system involved in this process. In physiological conditions, the CTL response requires the interaction of at least three types of immune cells, involving both the innate and adaptive immune arms (Castellino & Germain, 2006). These include (i) the professional antigen-presenting cells such as dendritic cells and macrophages, which present the antigens in the context of major histocompatibility complex (MHC) class I and class II molecules, (ii) the $CD4^+$ T lymphocytes that recognize the MHC class II/antigen complex and provide the necessary help for response to T-dependent antigen, and (iii) the $CD8^+$ T lymphocytes that, after interaction of their T cell receptor (TCR) with the MHC class I/antigen complex, differentiate into antigen-specific CTL with effector function.

Cytotoxicity needs physical contact of CTL with infected or malignant cells to trigger their specific death. The contact area between the effector and target cells is termed the cytotoxic synapse (De la Roche, Asano, & Griffiths, 2016). CTL can kill their recognized cellular targets via two pathways. One is a Ca^{2+}-dependent process that involves secretion or exocytosis of cytotoxic granules to the synaptic cleft, causing necrosis of the target cell (Menasche et al., 2008). These cytolytic granules are comprised mainly of perforins, which form pores in the membrane, and granzymes, a group of serine proteases that act in the cytosol of the targeted cells. The second pathway is Ca^{2+}-independent and involves the induction of apoptosis in target cells via interaction of the Fas ligand (FasL), expressed on the CTL surface, with the Fas receptor in the target cells (Halle, Halle, & Frster, 2017). Cytokine production by immune cells enhances the cytotoxic effect by increasing apoptotic signals and recruiting more immune cells with effector capacity.

In this chapter we describe a flexible, flow cytometry-based method for quantitative determination of the CTL response, which might be applicable for high-throughput screening of small compounds, antibodies or T cell-based therapeutics designed to increase CTL-mediated cytotoxicity.

2. From $CD8^+$ T lymphocytes to armed CTL

Naïve T lymphocytes are quiescent cells that circulate between the secondary lymphoid organs and the bloodstream, in search of their specific antigen. Antigen recognition takes place in the peripheral lymphoid organs and involves interaction between the TCR and the antigenic peptide presented by MHC molecules. In the case of $CD8^+$ T lymphocytes, the antigen must be presented by the MHC class I molecules. This interaction between the TCR and the MHC-peptide complex leads to the formation of a specialized structure termed immune synapse (Monks, Freiberg, Kupfer, Sciaky, & Kupfer, 1998; Potter, Grebe, Freiberg, & Kupfer, 2001). The structure of this synapse comprises a series of concentric supramolecular activation clusters (SMAC), each containing specific protein complexes and having specific activities. The central (c)SMAC concentrates the TCR, the co-receptors (CD4 or CD8), and associated signaling molecules such as the Src leukocyte-specific tyrosine kinase (Lck) and the zeta-chain-associated protein (ZAP-70). The cSMAC is surrounded by the peripheral (p)SMAC, which concentrates adhesion molecules such as the $\alpha_L\beta_2$ integrin (LFA-1). The most external ring, termed distal (d)SMAC, is formed by the grouping of filamentous actin (F-actin) involved in the retrograde flow and clustering of receptors at the cSMAC (Le Floc'h & Huse, 2015; O'Keefe, Blaine, Alegre, & Gajewski, 2004).

Formation of the immune synapse leads to T cell activation, expansion, and differentiation, transforming "unarmed" naïve $CD8^+$ T lymphocytes into heavily "armed" effector CTL loaded with cytolytic granules that contain granzymes and perforins. Once these armed cells recognize their targets in the periphery, a new immunological synapse is formed. The purpose of this synapse is not to prime the CTL, but rather to facilitate polarized secretion of cytolytic granules toward the point of contact between the CTL and the target cell (Stinchcombe, Majorovits, Bossi, Fuller, & Griffiths, 2006). Formation of an immunological synapse is not an absolute requirement for target lysis (Purbhoo, Irvine, Huppa, & Davis, 2004); nonetheless, the cytotoxic synapse with an ordered pSMAC and polarized granule delivery offers a clear advantage for efficient cell killing.

The immune and cytotoxic synapses are very similar from a structural point of view, including centrosome polarization and F-actin reorganization at the dSMAC. There are nevertheless differences between priming and cytotoxic synapses. Immune synapses involved in priming prolong naïve $CD8^+$ T cell interactions with antigen-presenting cells for several hours (Bousso & Robey, 2003). In contrast, cytotoxic synapses are more dynamic; they cause firm adhesion of the CTL to the target, reorganization of the centrosome, release of granules to the synapse cleft, and detachment from the target; all these processes take place within minutes. Drawing on Cajal's definition of neural synapses as "protoplasmic kisses," Berke defined the cytotoxic synapse as the "kiss of death" (Berke, 1995).

3. Suppression of the CTL response by the PD-1/PD-L1 axis

It is estimated that CTL can kill 2.4 specific B-cell targets per hour; the tumor microenvironment (TME) decreases CTL killing rate to <0.2 per hour (Breart, Lemaitre, Celli, & Bousso, 2008; Dustin & Long, 2010). This is an effect of the prevailing immunosuppression in the TME, which may be caused by immunoregulatory cell subtypes (regulatory T cells, myeloid-derived suppressor cells, etc.), and/or by tumor cell-intrinsic mechanisms that restrict CTL activity. One of these mechanisms involves receptor programmed cell death-1 (PD-1, CD279) and its ligand, PD-L1 (B7-H1, CD274).

The inhibitory receptor PD-1 acts physiologically as an immune checkpoint that maintains peripheral self-tolerance. It is expressed in immune-privileged organs, and PD-1-deficient mice spontaneously develop autoimmune diseases (Nishimura et al., 2001; Nishimura, Nose, Hiai, Minato, & Honjo, 1999). Pathogens and tumors hijack PD-1 inhibitory signals as an evasive immunosurveillance mechanism. PD-L1 is upregulated in many human cancers, whereas PD-1 is highly expressed on tumor-infiltrating lymphocytes (Topalian, Drake, & Pardoll, 2015). The interaction of PD-L1 on the malignant cell surface with PD-1 on the T cell acts as a brake for CTL cytotoxicity (Iwai et al., 2002). This led to the hypothesis that PD-L1 in tumor cells serve as a molecular shield that hampers tumor cell lysis mediated by PD-1-expressing CTL (Juneja et al., 2017; Zou, Wolchok, & Chen, 2016). This view is supported by studies in animal models and clinical trials, which show that PD-1 or PD-L1 blockade increases the polyfunctionality of tumor-infiltrating CTL, thus reinvigorating immune responses in the TME (Gubin et al., 2014). To date, five therapeutic antibodies that target the PD-1/PD-L1 pathway have been approved by the US Food and Drug Administration (FDA) for several cancers, given their superior clinical benefit compared to classical chemotherapy-based treatments in a subset of patients (Ribas, 2012). Although a chapter in this series centers on assays for PD-1/PD-L1 inhibitor development, we have used the PD-1/PD-L1 axis to exemplify the feasibility of our assay for screening purposes.

4. Overview of the CTL cytotoxicity assay

Cell-mediated cytotoxicity is defined as the lysis of target cells by CTL effector cells. CTL activity is triggered by TCR recognition of an antigen in the MHC-I context expressed by the target cell. As a source of effector cells, our assay uses $CD8^+$ T cells from the OT-I transgenic mouse, a model in which most $CD8^+$ T cells express a TCR specific for the ovalbumin $(OVA)_{257-264}$ peptide. Effector cells are generated by activation of OT-I cells with $(OVA)_{257-264}$-loaded splenocytes, which provide B lymphocytes to present the antigen correctly. Nevertheless, cytotoxic T cells can be also generated from human T cells using well-established protocols (Ho, Nguyen, Wolfl, Kuball, & Greenberg, 2006). Human T cells must be used when the cytotoxic agents to be tested are suspected to be species-specific (for instance, a humanized antibody for an inhibitory human T cell receptor). In these situations, target cells should be also of human origin.

The target cells are E.G7-OVA, an OVA-expressing cell clone derived from the EL4 mouse lymphoma (established from a lymphoma induced in a C57BL/6 mouse by 9,10-dimethyl-1,2-benzanthracene), and E.G7-OVA-PD-L1 cells, an E.G7-OVA clone that overexpresses PD-L1. In this system, the cytotoxic activity of CTL cells is partially blocked by the inhibitory PD-1/PD-L1 pathway; this scenario permits analysis of compounds designed to bypass the immunosuppressive pathway induced by PD-1 or that collaborates with PD-1-blocking agents to enhance CTL killing activity.

CellTrace 5-(6)-carboxyfluorescein succinimidyl ester (CFSE)-labeled E.G7-OVA or E.G7-OVA-PD-L1 cells are co-cultured with OT-I blasts, and dead cells are identified by propidium iodide (PI) staining. The double CFSE/PI labeling defines four groups of cells: living target cells (green); dead target cells (green and red); dead effector cells (red), and live effector cells (unstained). Cell population analysis by flow cytometry allows evaluation of target cell death and quantitative measurement of $CD8^+$ T cell cytotoxic activity (Fig. 1). Treatment of effector OT-I blasts with compounds or antibodies that might affect their CTL activity converts this test into a screening method to seek new modulators of CTL-mediated cytotoxicity. The assay is

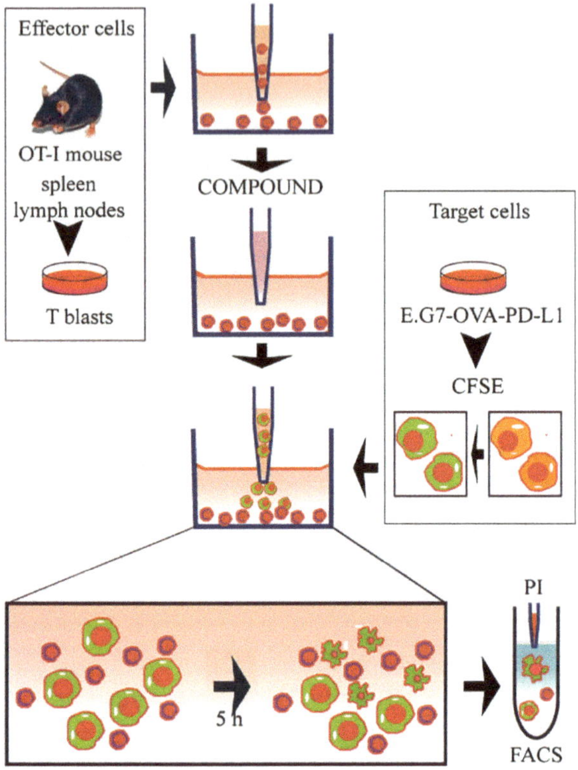

Fig. 1 General scheme of the cytotoxicity assay. Target cells are labeled with CFSE and co-cultured with OT-1 effector cells for 5 h. For screening, vehicle or compounds are added when necessary. Dead cells are then PI-stained and analyzed by flow cytometry to determine the percentage of cytotoxicity. *PI*, propidium iodide; *FACS*, fluorescence-activated cell sorter.

also flexible, and can determine the cytotoxic capacity of modified effector T cells (CAR-T or TCR-transgenic cells), making it suitable for screening of T cell-based therapies.

5. Generation of stable E.G7-OVA-PD-L1-expressing cells

1. Maintain E.G7-OVA cells (American Type Culture Collection [ATCC], Manassas, VA, USA) between 0.1 and 1.0×10^6 cells/mL in complete medium (RPMI 1640 with 10% fetal bovine serum (FBS), 2 mM L-glutamine, 1.0 mM sodium pyruvate, 0.05 mM 2-mercaptoethanol, 100 U/mL penicillin, 100 µg/mL streptomycin and 0.2 mg/mL G418). Refresh cell culture medium every 2–3 days (see Note 1).

2. Seed E.G7-OVA cells (15.0×10^6) into a new flask at a concentration of 0.3×10^6 cells/mL the day before nucleofection (see Note 2).
3. One hour before transfection, fill a 12-well plate with 1 mL per well complete medium and pre-incubate the plate in a humidified 37 °C/5% CO_2 incubator.
4. Pellet E.G7-OVA cells (5.0×10^6 per condition) at $350 \times g$ for 5 min at room temperature (RT). Remove supernatant completely.
5. Resuspend the cell pellets in 100 μL Nucleofector Solution L, avoiding air bubbles. Cells should be in contact with this solution no longer than 15 min.
6. Mix 2 μg of plasmid DNAs, pCMV3-mPD-L1-HA (mouse PD-L1 ORF mammalian expression plasmid, C-HA tag; Sino Biological Inc., Beijing, China), pCMV3-hyg empty vector or pMax-GFP plasmid (positive control) with the cell suspensions (see Note 3).
7. Transfer DNA/cell suspensions carefully into nucleofector cuvettes, avoiding bubbles.
8. Introduce the cuvettes into an Amaxa Nucleofector device (Lonza Group, Basel, Switzerland). Select Program C-009 (see Note 4).
9. Once the program is finished, as rapidly as possible, add 500 μL pre-warmed complete medium into the cuvettes, avoiding bubbles (see Note 5). Transfer cell suspensions quickly into the pre-warmed 12-well plate and place them in a humidified 37 °C/5% CO_2 incubator.
10. Check transfection efficiency after 24 h, using the pMax-GFP transfected cells for fluorescence microscopy (see Note 6). Renew medium for all conditions. At this point, add the selective antibiotic hygromycin B (0.6 mg/mL; Sigma-Aldrich, St. Louis, MO, USA) in every condition.
11. After 7 days, clone the stable E.G7-OVA-PD-L1-transfected cells by limiting dilution (see Note 7). Seed cell suspensions ranging from 0.5 to 4 cells/100 μL into 96-well flat-bottom microplates with complete medium supplemented with 0.6 mg/mL hygromycin B. Use two plates for each cell dilution.
12. After 6, 12 and 24 h, inspect plates carefully using an optical microscope to select those wells that contain a single cell.
13. Five days later, check for cell clones that show good cell proliferation and confluence. Transfer cells from the selected wells into a 48-well flat-bottom microplate. Maintain cell clones at optimum confluence by diluting with complete medium and 0.6 mg/mL hygromycin B.

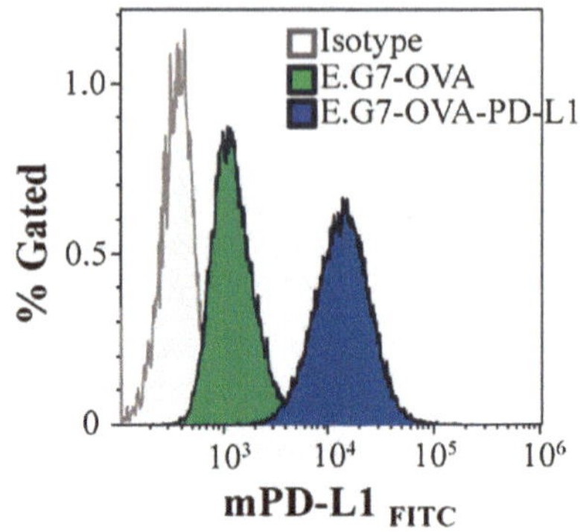

Fig. 2 Characterization of E.G7-OVA-PD-L1 cell clones. Flow cytometry histograms showing PD-L1 labeling in E.G7-OVA-PD-L1 and parental E.G7-OVA cells. Control isotype staining is shown.

14. Check PD-L1 expression in the E.G7-OVA-PD-L1 cell clones selected by flow cytometry (anti-CD274-FITC; Clone MIH6, OriGene Technologies GmbH, Herford, Germany), using parental E.G7-OVA as reference (Fig. 2).
15. Select the E.G7-OVA-PD-L1 cell clone that expresses the highest PD-L1 level and maintain it in complete medium supplemented with 0.2 mg/mL hygromycin B. Check regularly for PD-L1 by flow cytometry.

6. Generation of *ex vivo*-activated OT-I cells

1. Carefully remove the spleen and lymph nodes from an OT-I mouse (C57BL/6-Tg (TcraTcrb) 1100Mjb/J; Jackson Laboratory, Bar Harbor, ME, USA) and maintain at 4 °C in phosphate-buffered saline (PBS) pH 7.4 supplemented with 3% FBS.
2. Place spleen and lymph nodes in a 40-μm cell strainer on top of a 50 mL Falcon-type tube and mash the lymphoid organs with a plunger end of a 5 mL syringe. Rinse with 5 mL PBS plus 3% FBS to obtain a cell suspension. Discard the strainer with the organ membranes.
3. Centrifuge the cell suspension at 350 × *g* for 5 min at RT; remove the supernatant.
4. Resuspend the cell pellet in 2 mL erythrocyte lysis buffer 1 × (see Note 8) and incubate in a 37 °C water bath for 1 min. Terminate the lysis reaction by rapid addition of 40 mL PBS plus 3% FBS.

5. Spin cells (350 × *g*, 5 min, RT) and resuspend the pellet in 2 mL complete RPMI medium (see Note 9). Count live cells by trypan blue exclusion in a Neubauer chamber at a 1/200 cell dilution. Usually, $0.5–1 \times 10^8$ splenocytes are obtained from a single mouse spleen.
6. Seed spleen cells at 2×10^6/mL in a T75 flask, in complete RPMI-1640 medium supplemented with 10 pM $OVA_{257–264}$ (OVA; InvivoGen, San Diego, CA, USA) (see Notes 10 and 11).
7. After 2 days, count cells, and centrifuge at 350 × *g* for 5 min at RT. Remove as much supernatant as possible to eliminate the OVA peptide.
8. Seed the activated OT-I cells (0.2×10^6 cells/mL) in complete RPMI medium supplemented with 50 U/mL recombinant murine IL-2 (PeproTech, Rocky Hill, NJ, USA) (see Note 12).
9. After 24 h, check OT-I cell expansion and activation. Stain 10^5 OT-1 cells with anti-CD8-FITC (clone 53-6.7, Beckman Coulter, Indianapolis, IN, USA), anti-CD25-PE (clone PC61, BD-Pharmingen, San Jose, CA, USA), anti-CD69-PE-Cy7 (clone H1.2F3, eBioscience, San Diego, CA, USA) and anti-PD-1-APC-eFluor780 (clone J43, eBioscience) antibodies. Typical flow cytometry analysis results are shown in Fig. 3. OT-I cells can then be used for the cytotoxicity assay (see below). Good results can be obtained in the cytotoxicity assay using a minimum of 50% double-positive $CD25^+CD69^+$ OT-I cells.

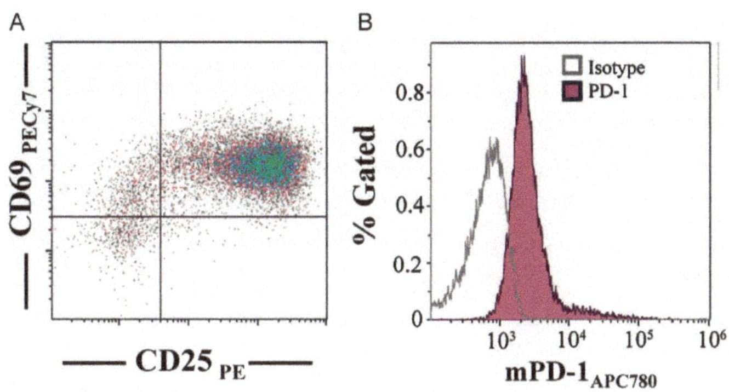

Fig. 3 Characterization of effector OT-I cells. (A) Dot plot obtained by flow cytometry, showing expression of the activation markers CD69 and CD25 in OT-I cells at day 3 post-activation. (B) Single parametric histogram showing PD-1 expression in activated OT-I cells. Isotypic control is shown.

7. Cytotoxicity assay

The cytotoxicity assay was performed in a 96-well flat-bottom plate. The components, added to each well sequentially, were 50 μL effector cells, 10 μL compound (in this case, an anti-PD-L1 antibody) or vehicle (here, an isotype-matched IgG antibody), and 50 μL target cells. A constant number of target cells and a variable number of effectors were added per well, depending on the ratio being studied (see Note 13); for example, for a 5:1 ratio, 3.75×10^5 effector and 7.5×10^4 target cells were seeded per well. Each ratio was assayed in triplicate.

1. Add the necessary number of E.G7-OVA and E.G7-OVA-PD-L1 target cells at a concentration of 1×10^6 cells/mL in complete RPM1-1640 medium, to a 15 mL Falcon-type tube containing 0.5 μM CFSE (Invitrogen, Carlsbad, CA, USA) (see Note 14). Incubate tubes for 30 min at 37 °C with caps loosened.
2. Pellet the labeled cells by centrifugation and wash twice with complete RPMI medium. Resuspend cells at 7.5×10^4 cells/50 μL in complete RPMI medium.
3. Calculate the number of OT-1 effector cells (determined as above) required for the highest E:T ratio. Then prepare serial dilutions in complete RPMI medium to prepare the subsequent ratios. For instance, for a 5:1 E:T ratio, 3.75×10^5 OT-I cells/50 μL cells would be needed.
4. Add the test compounds (or vehicles) in a final volume of 10 μL to the wells containing OT-I cells. In this example, anti-PD-L1 blocking antibody (Clone MIH6) or isotype control IgG2a (Thermo Fisher Scientific, Waltham, MA, USA) were added at 100 μg/mL (see Note 15). Incubate the OT-I effector cells and the antibodies for 15 min at RT.
5. Add 7.5×10^4 CFSE-labeled target cells to the wells in 50 μL. Prepare wells containing only effector cells or only target cells as controls of non-cytotoxic dead cells for each cell type. Add RPMI-1640 complete medium and vehicle to control wells to bring all samples to the same final volume (110 μL).
6. Incubate the effector and target cells for 5 h at 37 °C/5% CO_2 (see Note 16).
7. After incubation, place the 96 well microplate on ice (4 °C). Transfer the whole volume of each well to flow cytometry tubes containing 100 μL PBS, and add 10 μL propidium iodide (PI; Thermo Fisher Scientific) immediately before analysis in the Cytomics FC-500 Flow Cytometer (Beckman Coulter, Brea, CA, USA) at a high flow rate (see Notes 17 and 18). Acquire at least 10,000 events in the target cell gate.

For gating purposes, first analyze a tube containing a mixture of unlabeled target and effector cells; set the negative peak between zero and 10^1. This will allow you to set up the population by size and complexity in the forward scatter (FSC)/side scatter (SSC) histogram. In our example, gate *a* was applied to exclude cell debris (Fig. 4A). The population gated in "*a*" was analyzed for CFSE and PI expression (Fig. 4B). The double parametric dot plot identifies four populations: living target cells (Q3), dead target cells (double positive; Q2), dead effector cells (Q1), and live effector cells (Q4) (Fig. 4B). The cytotoxicity values of the effector cells were defined as the percentage of dead target cells in total target cells. First, target cells were identified and gated by CFSE expression (Fig. 4C). Dead target cells were then further identified by gating in region *b* and by PI staining (Fig. 4D). The number of events in region *c* corresponds to the percentage of dead target cells. The difference between this percentage and that of dead target cells in the absence of effector OT-I cells indicates the percentage of OT-I-induced cytotoxicity.

8. Concluding remarks

The method presented here provides a rapid, reliable approach with which to quantify the cytotoxicity induced by tumor-specific $CD8^+$ T cells in culture conditions, responding to chemical or biological cues, in a format compatible with high-throughput workflow and automation (Joslin et al., 2018). Flow cytometry allows the quantification of cell numbers as well as of the viability of both target and effector $CD8^+$ T cells. When the CFSE-labeled E.G7-OVA cells are targeted by OT-I cells, the effector cells use perforin to create holes in the target cell membrane, which enables PI entry. Double labeling by CFSE and PI readily distinguishes dead target cells. The compounds can nonetheless induce target cell cytotoxicity independently of the cytotoxic activity mediated by the effector OT-I cells. It is therefore important to determine target cell cytotoxicity in a specific set of control conditions, including the absence of OT-I cells. Only such an experimental setup allows discrimination between OT-I-mediated cytotoxicity and other non-immune-mediated tumor cell death.

To illustrate this assay, we show representative cytotoxicity curves obtained in two different experimental setups. In the first, we compared OT-I-mediated cytotoxicity of E.G7-OVA and E.G7-OVA-PD-L1 cells, and observed that PD-L1 overexpression reduced the OT-I cytotoxic response (Fig. 4E), which indicates that the system is dependent on the

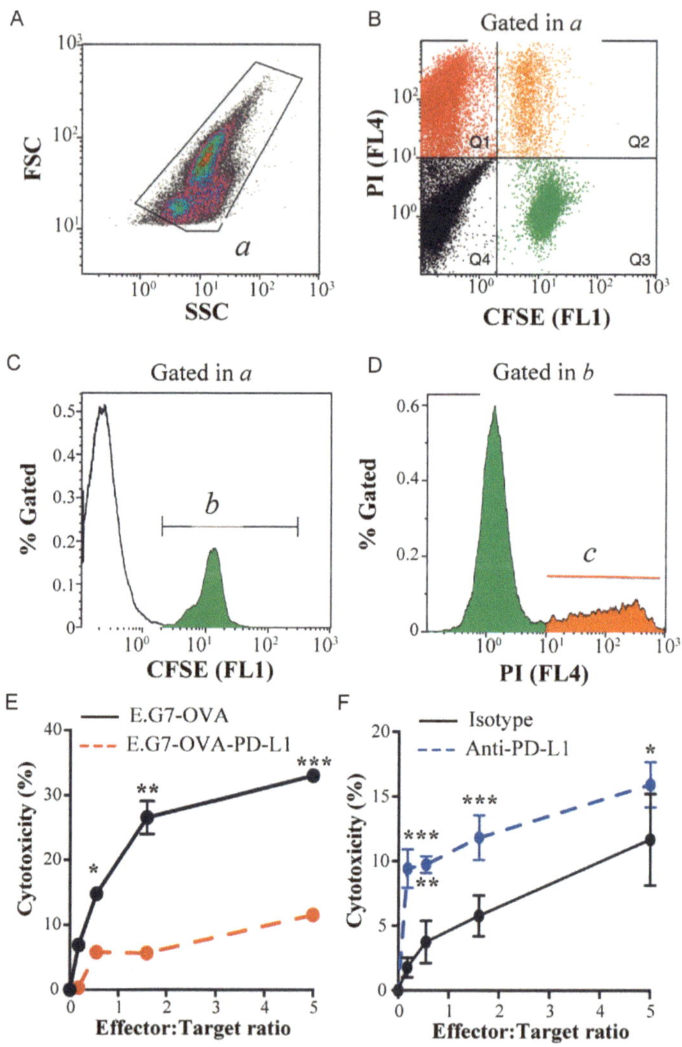

Fig. 4 Gating strategy used in flow cytometry-based cytotoxicity assay and examples. (A) FCS vs SSC density plot. Gate *a* was drawn to select target and effector cells. (B) Dual color fluorescence density plot. Cells were first gated by FSC/SSC and then by double CFSE/PI staining. Quadrants Q1 to Q4 are marked. (C) Single parametric histogram showing CFSE expression in gated target cells; region *b* identifies green cells. (D) Monoparametric histogram showing PI staining in the gated green-labeled target cells; region *c* defines the green and red double-positive cells in the population. (E) Curves showing OT-I-mediated cytotoxicity of E.G7-OVA (*solid line*) and E.G7-OVA-PD-L1 cells (*dotted*) at different E:T ratios. (F) Curves showing the percentage of cytotoxicity of E.G7-OVA-PD-L1 targeted cells incubated with OT-I cells in the presence of a blocking anti-PD-L1 antibody (*dotted line*) or an isotype control (*solid line*). For *E* and *F*, data are mean ± SEM of triplicates at the indicated E:T ratios. $^{*}P < 0.05$; $^{**}P < 0.01$; $^{***}P < 0.001$, two-way ANOVA with Bonferroni post-hoc test.

PD-1/PD-L1 axis. In the second system, an anti-PD-L1 blocking antibody was added to prevent the PD-1 inhibitory effect on the OT-I cytotoxic response. In this case, treatment of effector cells with the blocking antibody increased the capacity of OT-1 to kill cells at different E:T ratios (Fig. 4F). This result clearly demonstrates the utility of the assay described in this chapter as a screening method to identify agents able to modulate the cytotoxicity response of tumor-specific $CD8^+$ T cells.

8.1 Notes

1. Optimal culture conditions might differ for different cell lines.
2. Cells must be in exponential growth phase before nucleofection; fix the seeding conditions for the cell line of interest.
3. For GFP-encoding plasmids, nucleofection efficiency can be determined directly in the transfected cells.
4. The nucleofection program might vary for distinct cell lines or could be different in other nucleofector devices.
5. Avoid leaving the cell suspension in the cuvette in contact with Nucleofection Solution L for extended periods of time after nucleofection, as this reduces cell viability and gene transfer efficiency.
6. Transfection efficiency can be checked by other methods, such as flow cytometry.
7. Other techniques to generate stable cell clones can be used depending on conditions. For example, if cells express GFP, cytofluorometry-assisted cell sorting can be used to obtain a cell clone in one step from the transfected cells. Some cytometers, such as the BD FACSAria series, allow direct sorting of single cells into 96-well plates, thus obviating tedious, time-consuming work.
8. The erythrocyte lysis buffer 10 × stock solution is composed of 8.26 g NH_4Cl, 1.19 g $NaHCO_3$, 200 μL 0.5 M EDTA pH 8, deionized water to 100 mL, and adjusted to pH 7.3. The solution is passed through a 0.22 μm filter. The 1 × working solution should be diluted in deionized water at the time of the assay.
9. Splenocyte growth medium is RPMI 1640 containing 10% FBS, 2 mM L-glutamine, 1.0 mM sodium pyruvate, 0.05 mM 2-mercaptoethanol, 100 U/mL penicillin, 100 μg/mL streptomycin, and 0.1 mM non-essential amino acids.
10. The $OVA_{257–264}$ peptide is a class I (Kb)-restricted peptide epitope of OVA, presented by the class I MHC molecule H-2Kb (sequence: SIINFEKL).

11. A 10 pM $OVA_{257-264}$ peptide concentration is sufficient to promote OT-I activation and expansion. The initial number of splenocytes incubated with $OVA_{257-264}$ peptide depends on the final number of $CD8^+$ T cells required for the cytotoxicity assay. An initial number of 5×10^6 splenocytes (day 0) will usually produce approximately 20×10^6 activated OT-I cells at day 3 post-activation.
12. Recombinant murine IL-2 is used to promote T cell growth and differentiation.
13. The E:T ratio can be in a wide range from 0.1:1 to 100:1, depending on target cells used.
14. CFSE diffuses into all target cells and the probe acetate group is cleaved by intracellular esterases to produce carboxyfluorescein succinimidyl ester. Other fluorescent probes can be used to label target cells; optimal incubation time must be determined experimentally.
15. Select a suitable isotype-matched control antibody as control for the assay.
16. Shorter incubation times are recommended for higher E:T ratios.
17. PI penetrates and labels DNA in dead or seriously damaged cells, excluding live cells. Other red fluorescent dyes, such as 7-AAD, can be also used.
18. Incubation with PI should be carefully controlled, as cell labeling increases with time. This can have a negative effect on the results.

Acknowledgments

We thank lab members for helpful discussion, R.M. Peregil for technical assistance, M.C. Moreno and S. Escudero for help with flow cytometry, and C. Mark for excellent editorial assistance. This work was funded by grants from the Spanish Ministerio de Economía y Competitividad (MINECO) (SAF2017-83732-R; AEI/FEDER, EU), the Comunidad de Madrid (B2017/BMD-3733; Immunothercan-CM), and the Merck-Salud Foundation. J.S. is supported by a predoctoral fellowship from the MINECO and the EU European Social Fund.

References

Berke, G. (1995). The CTL's kiss of death. *Cell*, *81*, 9–12.

Bousso, P., & Robey, E. (2003). Dynamics of CD8+ T cell priming by dendritic cells in intact lymph nodes. *Nature Immunology*, *4*, 579–585.

Breart, B., Lemaitre, F., Celli, S., & Bousso, P. (2008). Two-photon imaging of intratumoral CD8+ T cell cytotoxic activity during adoptive T cell therapy in mice. *The Journal of Clinical Investigation*, *118*, 1390–1397.

Castellino, F., & Germain, R. N. (2006). Cooperation between CD4+ and CD8+ T cells: When, where, and how. *Annual Review of Immunology*, *24*, 519–540.

De la Roche, M., Asano, Y., & Griffiths, G. M. (2016). Origins of the cytolytic synapse. *Nature Reviews. Immunology*, *16*, 421–432.

Dustin, M. L., & Long, E. O. (2010). Cytotoxic immunological synapses. *Immunological Reviews*, *235*, 24–34.

Gubin, M. M., Zhang, X., Schuster, H., Caron, E., Ward, J. P., Noguchi, T., et al. (2014). Checkpoint blockade cancer immunotherapy targets tumour-specific mutant antigens. *Nature*, *515*, 577–581.

Halle, S., Halle, O., & Frster, R. (2017). Mechanisms and dynamics of T cell-mediated cytotoxicity in vivo. *Trends in Immunology*, *38*, 432–443.

Ho, W. Y., Nguyen, H. N., Wolfl, M., Kuball, J., & Greenberg, P. D. (2006). In vitro methods for generating CD8+ T-cell clones for immunotherapy from the naive repertoire. *Journal of Immunological Methods*, *310*, 40–52.

Iwai, Y., Ishida, M., Tanaka, Y., Okazaki, T., Honjo, T., & Minato, N. (2002). Involvement of PD-L1 on tumor cells in the escape from host immune system and tumor immunotherapy by PD-L1 blockade. *Proceedings of the National Academy of Sciences of the United States of America*, *99*, 12293–12297.

Joslin, J., Gilligan, J., Anderson, P., Garcia, C., Sharif, O., Hampton, J., et al. (2018). A fully automated high-throughput flow cytometry screening system enabling phenotypic drug discovery. *SLAS Discovery: Advancing Life Sciences R & D*, *23*, 697–707.

Juneja, V. R., McGuire, K. A., Manguso, R. T., Lafleur, M. W., Collins, N., Haining, W. N., et al. (2017). PD-L1 on tumor cells is sufficient for immune evasion in immunogenic tumors and inhibits CD8 T cell cytotoxicity. *The Journal of Experimental Medicine*, *214*, 895–904.

Le Floc'h, A., & Huse, M. (2015). Molecular mechanisms and functional implications of polarized actin remodeling at the T cell immunological synapse. *Cellular and Molecular Life Sciences*, *72*, 537–556.

Menasche, G., Menager, M. M., Lefebvre, J. M., Deutsch, E., Athman, R., Lambert, N., et al. (2008). A newly identified isoform of Slp2a associates with Rab27a in cytotoxic T cells and participates to cytotoxic granule secretion. *Blood*, *112*, 5052–5062.

Monks, C. R., Freiberg, B. A., Kupfer, H., Sciaky, N., & Kupfer, A. (1998). Three-dimensional segregation of supramolecular activation clusters in T cells. *Nature*, *395*, 82–86.

Nishimura, H., Nose, M., Hiai, H., Minato, N., & Honjo, T. (1999). Development of lupus-like autoimmune diseases by disruption of the PD-1 gene encoding an ITIM motif-carrying immunoreceptor. *Immunity*, *11*, 141–151.

Nishimura, H., Okazaki, T., Tanaka, Y., Nakatani, K., Hara, M., Matsumori, A., et al. (2001). Autoimmune dilated cardiomyopathy in PD-1 receptor-deficient mice. *Science*, *291*, 319–322.

O'Keefe, J. P., Blaine, K., Alegre, M. L., & Gajewski, T. F. (2004). Formation of a central supramolecular activation cluster is not required for activation of naive CD8+ T cells. *Proceedings of the National Academy of Sciences of the United States of America*, *101*, 9351–9356.

Potter, T. A., Grebe, K., Freiberg, B., & Kupfer, A. (2001). Formation of supramolecular activation clusters on fresh ex vivo CD8+ T cells after engagement of the T cell antigen receptor and CD8 by antigen-presenting cells. *Proceedings of the National Academy of Sciences of the United States of America*, *98*, 12624–12629.

Purbhoo, M. A., Irvine, D. J., Huppa, J. B., & Davis, M. M. (2004). T cell killing does not require the formation of a stable mature immunological synapse. *Nature Immunology*, *5*, 524–530.

Ribas, A. (2012). Tumor immunotherapy directed at PD-1. *The New England Journal of Medicine*, *366*, 2517–2519.

Stinchcombe, J. C., Majorovits, E., Bossi, G., Fuller, S., & Griffiths, G. M. (2006). Centrosome polarization delivers secretory granules to the immunological synapse. *Nature*, *443*, 462–465.

Topalian, S. L., Drake, C. G., & Pardoll, D. M. (2015). Immune checkpoint blockade: A common denominator approach to cancer therapy. *Cancer Cell*, *27*, 450–461.

Zou, W., Wolchok, J. D., & Chen, L. (2016). PD-L1 (B7-H1) and PD-1 pathway blockade for cancer therapy: Mechanisms, response biomarkers, and combinations. *Science Translational Medicine*, *8*, 328rv4.

CHAPTER TWENTY-SIX

Evaluation of NK cell cytotoxic activity against malignant cells by the calcein assay

Seila Lorenzo-Herrero[a,b,c], Christian Sordo-Bahamonde[a,b,c], Segundo González[a,b,c], Alejandro López-Soto[a,b,c,*]
[a]Departamento de Biología Funcional, Área de Inmunología, Universidad de Oviedo, Oviedo, Spain
[b]Instituto Universitario de Oncología del Principado de Asturias (IUOPA), Oviedo, Spain
[c]Instituto de Investigación Sanitaria del Principado de Asturias (ISPA), Oviedo, Spain
*Corresponding author: e-mail address: lopezsalejandro@uniovi.es

Contents

Abstract

The immune system exerts a tight cancer surveillance to eradicate nascent and developing tumors. Natural killer (NK) cells rely on their cytotoxic activity to eliminate abnormal or stressed cells, including those undergoing malignant transformation. Thereupon, NK cells, along with $CD8^+$ T cells, are effector immune cells with a pivotal role in cancer immunosurveillance. Due to the outstanding effectivity of immunotherapy in cancer management, strategies aimed at boosting NK cell-mediated tumor recognition and elimination are being thoroughly studied, such as approaches to induce tumor surface expression of NK cell activating ligands. Additionally, the development of methods to evaluate NK cell cytotoxic responses towards malignant cells with distinct immunogenic profiles is of paramount relevance. In this chapter, we detail a simple but highly sensitive and reproducible methodological approach to quantify in vitro NK cell-mediated lysis of a given target cell, e.g. a tumor cell, based on calcein-AM staining and fluorescence detection. Further, this method can be employed to determine the

Methods in Enzymology, Volume 631
ISSN 0076-6879
https://doi.org/10.1016/bs.mie.2019.05.036

impact of chemical modulators or antibodies blocking cell surface proteins on the antitumor capacity of NK cells, with the aim to improve anticancer immune responses and elucidate the underlying mechanisms regulating NK cell activity.

1. Introduction

The immune system exerts a constant surveillance to protect the organism against threats such as infections, but it is also in charge of eliminating nascent tumors. Among the distinct immune populations, natural killer (NK) cells stand out as a key component of the antitumor arsenal, together with cytotoxic $CD8^+$ T lymphocytes (CTLs), due to their ability to recognize and eliminate abnormal or stressed cells, including those undergoing malignant transformation (Chiossone, Dumas, Vienne, & Vivier, 2018; Morvan & Lanier, 2016). NK cells are able to modulate tumor immunogenicity, sometimes leading to tumor elimination, and have been shown to exert a relevant control of metastasis, limiting the dissemination of tumor cells (Lopez-Soto, Gonzalez, Smyth, & Galluzzi, 2017).

NK cells are characterized by their dynamic cytotoxic activity. Upon recognition of a target cell, a strong activating signal triggers the release of preformed secretory granules in the cytoplasm of the NK cell into the immunological synaptic cleft established. These granules contain cytolytic proteins, mainly perforin and granzymes, which are responsible for the lysis of the target cell. More specifically, perforin can open large transmembrane pores that allow the diffusion of granzymes into the target cell, hence initiating the apoptotic death of the diseased cell (Voskoboinik, Whisstock, & Trapani, 2015).

Unlike T cells, NK cell activation and function are not triggered by engagement of specific peptide-MHC complexes, since they rely on the detection of cell surface ligands by an array of NK cell activating and inhibitory receptors. The effector activity of these immune cells is regulated by the resulting balance of signals transduced through these receptors. Natural killer group 2, member D (KLRK1; best known as NKG2D) is a central activating receptor in cancer immunosurveillance mainly expressed by NK cells $CD8^+$ T cells, $\gamma\delta$ T cells and NKT-like cells (Lopez-Soto, Huergo-Zapico, Acebes-Huerta, Villa-Alvarez, & Gonzalez, 2015). NKG2D ligands, MHC class I polypeptide-related sequence A/B (MICA/B) and UL16 binding protein (ULBP) molecules (ULBP1–6) in humans, are stress-regulated molecules highly expressed on the surface of transformed and infected cells (Gonzalez, Lopez-Soto, Suarez-Alvarez, Lopez-Vazquez, & Lopez-Larrea, 2008; Guillerey, Huntington, & Smyth, 2016). However, tumor cells evolve a variety of immunosubversive strategies to avoid NK cell-mediated responses including, but not limited to, the repression of activating self-molecules such as NKG2D ligands (Kaiser et al., 2007; Lopez-Soto, Folgueras, Seto, & Gonzalez, 2009). Therefore, strategies to boost tumor cell surface expression of NK cell activating ligands are being extensively studied to increment tumor immunogenicity and, hence, promote NK cell-mediated tumor clearance (Chiossone, Vienne, Kerdiles, & Vivier, 2017; Ferrari de Andrade et al., 2018; Galluzzi, Chan, Kroemer, Wolchok, & Lopez-Soto, 2018).

Calcein-AM based NK cell cytotoxicity assay is a reliable and highly sensitive method widely employed to evaluate the killing activity of NK cells against a given target cell. This protocol allows the identification of differences in NK cell antitumor responses in vitro linked to modifications of the immunogenic profile of tumor cells (e.g. surface upregulation of NK cell activating ligands (Acebes-Huerta et al., 2016; Lopez-Soto et al., 2013)) or the NK cell itself (e.g. dysregulation of activating or inhibitory receptors (Andre et al., 2018; Zhang et al., 2018)). Alternatively, NK cell cytotoxicity can be assayed by methods that gauge the release of ^{51}Cr, as detailed in chapter 30 of this book. Nonetheless, these methods only detect the final lysis of the target, which results from the balance between the cytotoxic capacity of NK cells and the resistance of the target cells. Therefore, it is recommended to complement NK cell cytotoxicity evaluation to degranulation assays and/or cytokine production (detailed in chapter 19 of this book) that measure the degree of activation of NK cells in the experimental settings studied.

Herein, we describe in detail an in vitro cytotoxicity assay based on the quantification of the lysis of a target cell line through calcein-AM staining and fluorescence detection. We evaluate the NK cell response to K-562 tumor cells, a reference cell line widely employed in the study of NK cell antitumor activity. With minor variations, this protocol can be adapted to other experimental conditions.

2. Cell culture

1. Human chronic myelogenous leukemia K-562 cells are routinely cultured at 37 °C in 5% CO_2, in RPMI 1640 medium (Lonza) supplemented with 10% heat-inactivated fetal bovine serum (FBS) (Sigma-Aldrich,

St. Louis, MO), 1 mM sodium pyruvate, 2 mM L-glutamine, 100 U/mL penicillin and 10 μg/mL streptomycin (see Notes 1–3).

2. Every 6–7 days, transfer the cell culture to an appropriate conical tube and centrifuge cells at 250 ×*g* for 5 min at room temperature (RT). Eliminate supernatant and resuspend the cell pellet in fresh pre-warmed (37 °C) growth medium at 10^6 cells/mL approximately (see Note 4).

3. Isolation of peripheral blood mononuclear cells (PBMCs)

1. Collect buffy-coats from healthy donors one day prior to the assay (see Notes 5 and 6). Under sterile conditions, dilute the concentrated blood with 1× PBS (2.7 mM KCl, 4.3 mM Na_2HPO_4, 1.4 mM KH_2PO_4 and 137 mM NaCl in dH_2O) (1:1) in a 50 mL conical tube (one per healthy donor) and mix lightly by inversion (see Note 7).
2. PBMCs are isolated by Ficoll density gradient. For that purpose, place 15 mL of Histopaque-1077® (Sigma Aldrich, St. Louis, MO) pre-warmed at RT in a 50 mL conical tube (see Note 8). Carefully pour 35 mL of diluted blood on top of the Histopaque, avoiding mixing and establishing two separated phases.
3. Right afterwards, centrifuge the tubes at 900 ×*g* for 22 min at RT, setting the brake off to avoid disruption of the gradient layers. Once the centrifugation is over, handle the tubes carefully and transfer the layer corresponding to the mononuclear cells to a 15 mL conical tube containing 10 mL 1× PBS with a plastic Pasteur pipette or similar (see Note 9). Invert the tubes gently for mixing.
4. Centrifuge the recovered PBMCs at 250 ×*g* for 5 min at RT and eliminate the supernatant. Wash the cell pellet with 1× PBS once more and, if the supernatant remains cloudy, carry out a third washing step (see Note 10). Resuspend the cell pellet in fresh pre-warmed (37 °C) growth medium avoiding the formation of clumps (see Note 11).
5. To determine cell number, prepare a dilution 1:1 with 0.4% Trypan Blue solution (Gibco® - Thermo Fisher Scientific, Carlsbad, CA) and incubate for 2–3 min. Transfer the cells to a Neubauer chamber and count the amount of alive (non-dyed) cells.
6. PBMCs are maintained at 2×10^6 cells/mL at 37 °C in 5% CO_2, in complete RPMI medium supplemented as described in Section 2.1.

If necessary, supplement the cell culture with 50 U/ml interleukin 2 (IL-2; ORF Genetics) (see Note 12).

4. Target cell labeling

1. Prepare the Assay Medium supplementing 1% FBS (Sigma-Aldrich, St. Louis, MO), 1 mM sodium pyruvate, 2 mM L-glutamine, 100 U/mL penicillin and 10 μg/mL streptomycin to RPMI medium (Lonza).
2. Prepare a staining solution by mixing 1 μL of 10 mM calcein-AM (calcein-acetoxymethyl; Thermo Fisher Scientific, Carlsbad, CA) per 1 mL of 1 × PBS and protect it from direct light until labeling the target cells (see Note 13).
3. Determine cell density of K-562 culture as described before (see Section 3.5) (see Notes 14).
4. Centrifuge an appropriate number of cells with 1 × PBS in a 5 mL tube at 250 ×*g* for 5 min at RT (see Note 15). Wash the cell pellet once more to eliminate residual growth medium (see Note 16). After discarding the supernatant, vortex the cell pellet briefly to make sure no clumps remain.
5. Resuspend the cell pellet in staining solution at a density of 10^6 cells/mL and mix by pipetting. Incubate the cells for 30 min at 37 °C in the dark, mixing every 10 min by inversion.
6. Right afterwards, add 1 volume of pre-warmed (37 °C) growth medium to the cell suspension and centrifuge at 250 ×*g* for 5 min at RT. Wash the pellet with growth medium once more. If considered necessary, add a third wash to remove the excess of dye (see Note 17).
7. Resuspend the cell pellet in Assay Medium and determine cell density as described (see Section 3.5). Adjust cell concentration to 2×10^5 cells/mL (see Note 18).

5. Effector/target cell co-culture

1. Stain an appropriate number of target cells as detailed above and resuspend them at an adequate cell concentration (see Section 4).
2. Determine cell concentration of PBMC culture with a Neubauer chamber. Prepare a dilution 1:1 with Trypan Blue solution to discriminate dead cells (see Note 19).

Table 1 Experimental conditions for a conventional calcein-AM-based cytotoxicity assay.

	Effector cells	K-562 cells	Assay medium
Background Signal	–	–	200 μL/well
Spontaneous Target Cell Lysis	–	100 μL/well	100 μL/well
Maximal Target Cell Lysis	–	100 μL/well	100 μL/well
Co-Culture	100 μL/well	100 μL/well	-

3. Centrifuge a suitable number of PBMCs at 250 ×g for 5 min at RT (see Note 20). Aspirate supernatant and resuspend the cell pellet in Assay Medium.
4. To plate the co-culture, add 100 μL of each cell suspension in a U-bottom 96-well culture plate (Corning Inc., Corning, NY) to set the following conditions in triplicates (Table 1). To seed the effector cells, a sequential dilution method is used in this protocol. For this purpose, adjust PBMC concentration to 10^7 cells/mL to plate the E:T ratio at 50:1. Afterwards, modify cell concentration to 5×10^6 and 2×10^6 cells/mL to set the E:T ratios at 25:1 and 10:1 respectively (see Note 21). The final volume per well is 200 μL.
5. Right after, incubate for 4 h at 37 °C and 5% CO_2 in the dark.

6. Fluorometric analysis

1. 30 min before the end of the co-incubation, add 10 μL of lysis buffer (1% Triton X-100 in dH_2O) to each of the maximal lysis condition triplicates and mix by pipetting up and down. Then, incubate for 30 min at 37 °C.
2. Centrifuge the 96-well plate for 250 ×*g* for 5 min at RT. Carefully transfer 100 μL of supernatant from each well into a new 96-well plate.
3. Measure the fluorescence of the assay at 485 nm excitation and 530 nm emission wavelengths in a suitable plate reader (Varioskan™ Lux Multimode plate reader).

4. To calculate the percentage of specific NK cell-mediated lysis, first subtract the background signal corresponding to the Assay Medium to each condition (see Note 22). Then, apply the following formula:

$$\text{Specific cytotoxicity }(\%) = \frac{(\text{mean fluorescence in coculture} - \text{mean fluorescence in spontaneous lysis})}{(\text{mean fluorescence in maximal lysis} - \text{mean fluorescence in spontaneous lysis})} \times 100$$

5. As shown in Fig. 1A, NK cell-mediated lysis of K-562 cells increases with the E:T ratio, along with the number of effector NK cells present in the co-culture. NK cell responses are variable among the different donors, which correlates with the distinct percentage of NK cells within each PBMC population (Fig. 1B) (see Note 23).

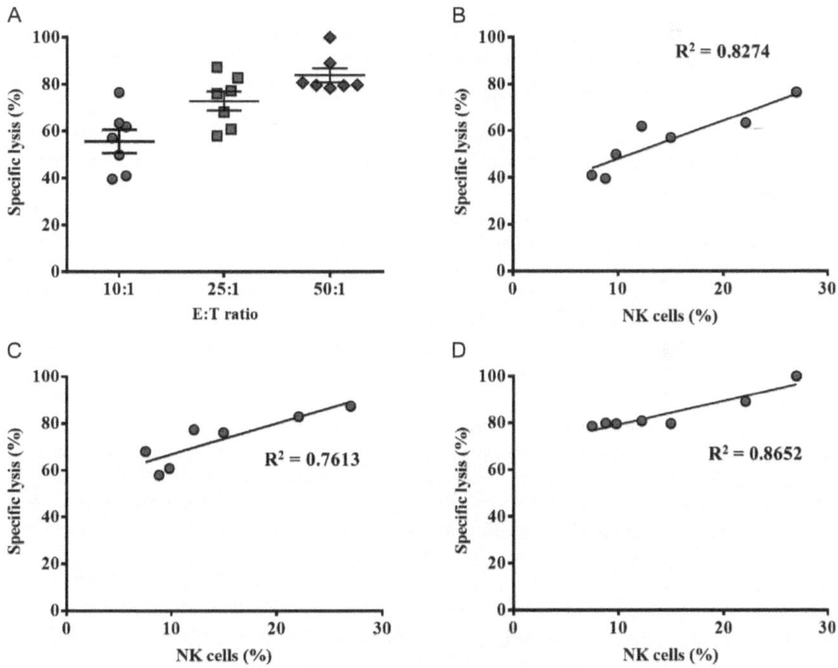

Fig. 1 NK-cell mediated lysis of K-562 cells. K-562 cells were stained with 10 μM calcein-AM and co-incubated with PBMCs from healthy donors at three different E:T ratios (10,1, 25:1, 50:1). After 4 h of co-culture, fluorescence was measured in the supernatant and specific lysis was calculated as described in Section 6. (A) Each dot represents the specific cell lysis obtained for each individual experiment and dark lines correspond to mean ± standard error of the mean (SEM). Correlation between K-562 cell lysis and percentage of NK cells within the PMBC population for each E:T ratio: 10:1 (B), 25:1 (C) and 50:1 (D) is calculated. The relative percentage of NK cells was determined by flow cytometry.

7. Concluding remarks

This protocol provides a reliable method to quantify the lysis of a target cell triggered by NK cell cytotoxic activity. This approach is highly sensitive compared to other methods, such as flow cytometry evaluation, as it relies on the release of cleaved calcein into the supernatant of the co-culture upon cell death, and the fluorometric detection allows identification of slight variations on cell lysis. Further, unlike radioactivity-based assays that gauge NK cell cytotoxic activity (e.g. Chromium-51 release assays) this method represents no hazardous risk for the researcher.

With minor modifications, this method can be adapted to different target (tumor) cells and further allows evaluation of the cytotoxic capacity of NK cells in different malignancies that affect this immune population, as well as determine NK cell-mediated lysis after exposure to chemical modulators or blocking antibodies against different cell surface proteins.

8. Notes

1. K-562 cells lack human leukocyte antigen (HLA) class-I and -II expression, and, consequently, are a gold-standard cell line for evaluating NK cell effector activity.
2. K-562 cells grow as a suspension and can be maintained in 25 cm^2 or 75 cm^2 flasks, as well as in 60 mm- or 100 mm-diameter Petri dishes for cell culture depending on the laboratory requirements.
3. Fresh pre-warmed (37 °C) growth medium is normally added every 2–3 days to maintain the cell culture, based on the American Type Culture Collection (ATCC, Manassas, VA) recommendations.
4. Optimal culture conditions may substantially vary depending on the cell line.
5. Usually, cytotoxicity assays are performed with at least two different sets of PMBCs to ensure the reproducibility of the results.
6. Alternatively, unprocessed whole blood from healthy donors can be employed as well, although the amount of isolated PBMCs per mL of blood is considerably lower than that of buffy-coats.
7. Buffy-coats have a high cell density and erythrocytes can aggregate in large clumps, trapping mononuclear cells and, hence, hampering the lymphocyte recovery. Diluting blood with 1 × PBS palliates this problem.

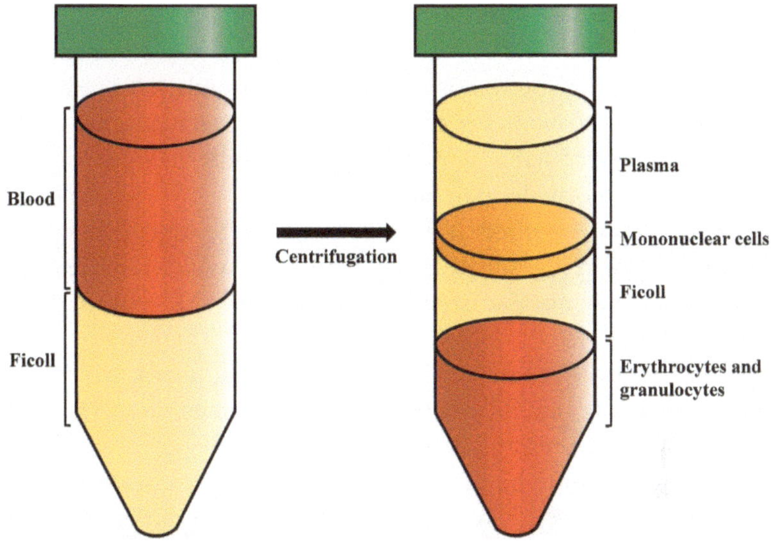

Fig. 2 Isolation of PBMCs by Ficoll density gradient. Mononuclear cells appear in a thin cloudy interphase between the plasma and Ficoll layers.

8. The optimal density of Histopaque is achieved at temperatures of 18 °C to 20 °C, according to the manufacturer.
9. The four differentiated layers that result from gradient formation are the following: a top plasma layer; a white cloudy layer, where the mononuclear cells remain; the Histopaque layer and, at the bottom, a cell pellet with erythrocytes and granulocytes (Fig. 2).
10. Histopaque solution can be toxic to cells, so it is recommended to avoid prolonged contact and wash the cell pellet immediately after recovery of the mononuclear cell layer to eliminate remaining traces of the solution as much as possible.
11. In some cases, erythrocytes remain in the mononuclear cell layer, contaminating the sample recovered. Pre-warming Histopaque favors erythrocyte sedimentation, as low temperatures prevent erythrocytes to sediment through the high-density solution. As an alternative, erythrocytes can be partially removed based on their higher adherence to the plastic tube walls compared to PBMCs. For that purpose, carefully resuspend the cell pellet after the last washing step and transfer the supernatant containing the majority of PBMCs to a new 15 mL tube, leaving the red blood cell layer attached to the tube. This method can reduce the number of mononuclear cells recovered.

12. NK cell recognition of distinct target cells is highly variable and, therefore, cytotoxicity rates differ among cell lines, being especially low for poorly immunogenic tumor cell lines. IL-2 stimulation of PBMCs can improve NK cell cytotoxic activity against a given target and, additionally, more reproducible and comparative results between experiment replicates are normally obtained.
13. Calcein-AM is supplied as a powder. Once resuspended, store it in single-use working aliquots (5–10 μL) to avoid repeated freeze-thawing. This reagent should be kept at -20 °C and protected from light.
14. Adherent cells can be stained attached to the culture plate. Nonetheless, this alternative can render heterogeneous staining among the cell population. Optimal results are achieved using single cell suspensions.
15. Cell staining should be performed with a number of cells higher than that required for the cytotoxicity assay, since cell numbers are usually reduced owing to consecutive washing steps.
16. Calcein-AM is a non-fluorescent analog of calcein that bears an acetoxymethyl (AM) group, a modification that allows the transport of the compound into live cells, while the original compound only penetrates dead cells. Certain serum components may have esterase activity, which can result in cleavage of the AM group of the calcein, thereby reducing the available dye to stain live cells and increasing the basal cell lysis fluorescence of the cytotoxicity assay due to staining of non-viable cells. Consequently, it is advisable to thoroughly wash the cell pellet to avoid the presence of serum in the cell suspension during the labeling.
17. Occasionally, spontaneous target cell lysis fluorescence can be higher than expected, decreasing the interval between spontaneous and maximal cell lysis and making the results difficult to interpretation. To improve this difference, ensure dye excess removal by increasing the number of washing steps after staining.
18. Dye and cell concentrations can vary among different cell lines. Therefore, determine the optimal staining settings and cell concentration for your target cell line by evaluating fluorescence intensity at basal and maximum conditions and cell viability after dye incubation (Fig. 3).
19. Cytotoxicity assays can be carried out using IL-2-activated NK cells purified from the peripheral blood of donors. To this end, a wide variety of human NK cell isolation kits, based on magnetic cell labeling and

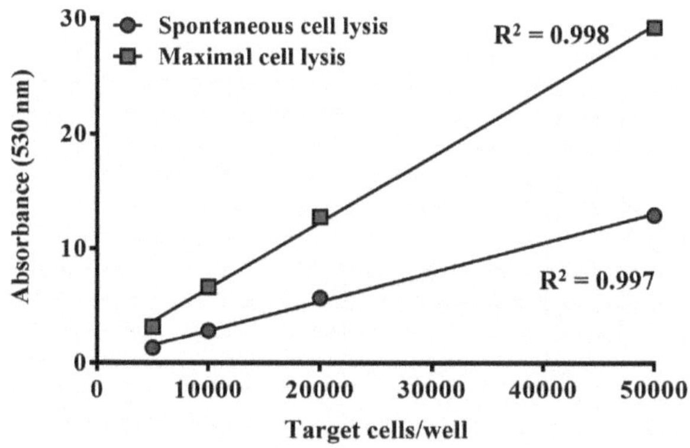

Fig. 3 Determination of optimal target cell concentration. K-562 cells were stained with 10 μM calcein-AM and plated into six wells at the densities detailed (5000–50,000 cells/well). Maximal and spontaneous cell lysis were determined as described in Section 6. The assay shows a linear relationship in both conditions. The optimal cell concentration should maximize the differences between maximal and spontaneous lysis, thus minimizing the cell density to limit the number of PBMCs required.

separation, are commercially available. Alternatively, human NK cell lymphoma cell lines, such as NKL or NK-92, can also be employed as effector cells.

20. Herein, the cytotoxicity assay is performed with three different effector:target (E:T) ratios: 50:1, 25:1 and 10:1. Final target cell number stays the same for each condition (20,000 cells per well) while effector cell numbers vary depending on the ratio. Taking this into consideration, calculate an appropriate number of PBMCs to plate the co-incubation.
21. As an alternative to sequential dilutions, prepare three different PBMC suspensions corresponding to each E:T ratio. This method requires higher numbers of PBMCs as each suspension should have an extra 10% of cells to account for minor pipetting errors.
22. Fluorescence values for maximal lysis should, at least, double those corresponding to spontaneous lysis to ensure the sensitivity of the assay.
23. This method can also be employed to measure antibody-dependent cell-mediated cytotoxicity (ADCC), by pre-incubating target cells with a suitable antibody. Fig. 4 depicts an example analyzing the response of NK cells against Raji target cells in the presence or absence of rituximab (10 μg/mL), an anti-CD20 antibody.

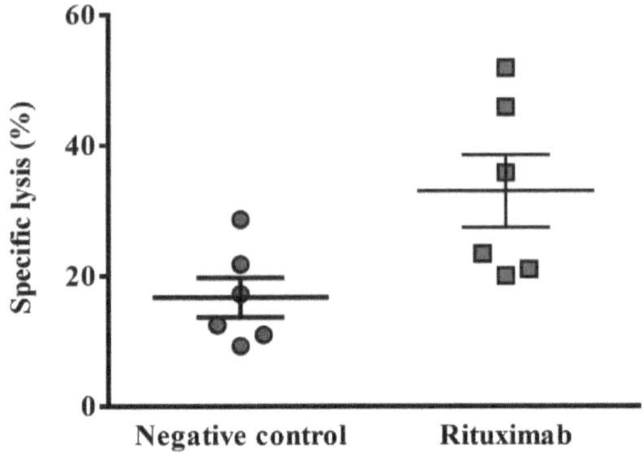

Fig. 4 Evaluation of ADCC by calcein-AM assay. Raji cells were stained with 10 µM calcein-AM and preincubated with the anti-CD20 monoclonal antibody rituximab (10 µg/mL). Right after, target cells were co-incubated with PBMCs from healthy donors for 4 h and NK-cell mediated cytotoxicity was evaluated as described in Section 6.

Acknowledgments

This work was supported by a Spanish grant of Instituto de Salud Carlos III (PI16/01485). S.L.H. holds a predoctoral Severo Ochoa Fellowship (BP14–150) from FICYT of Principado de Asturias, Spain.

References

Acebes-Huerta, A., Lorenzo-Herrero, S., Folgueras, A. R., Huergo-Zapico, L., Lopez-Larrea, C., Lopez-Soto, A., et al. (2016). Drug-induced hyperploidy stimulates an anti-tumor NK cell response mediated by NKG2D and DNAM-1 receptors. *Oncoimmunology*, *5*(2), e1074378. https://doi.org/10.1080/2162402X.2015.1074378.

Andre, P., Denis, C., Soulas, C., Bourbon-Caillet, C., Lopez, J., Arnoux, T., et al. (2018). Anti-NKG2A mAb is a checkpoint inhibitor that promotes anti-tumor immunity by unleashing both T and NK cells. *Cell*, *175*(7), 1731–1743 e1713. https://doi.org/10.1016/j.cell.2018.10.014.

Chiossone, L., Dumas, P. Y., Vienne, M., & Vivier, E. (2018). Natural killer cells and other innate lymphoid cells in cancer. *Nature Reviews. Immunology*, *18*(11), 671–688. https://doi.org/10.1038/s41577-018-0061-z.

Chiossone, L., Vienne, M., Kerdiles, Y. M., & Vivier, E. (2017). Natural killer cell immunotherapies against cancer: Checkpoint inhibitors and more. *Seminars in Immunology*, *31*, 55–63. https://doi.org/10.1016/j.smim.2017.08.003.

Ferrari de Andrade, L., Tay, R. E., Pan, D., Luoma, A. M., Ito, Y., Badrinath, S., et al. (2018). Antibody-mediated inhibition of MICA and MICB shedding promotes NK cell-driven tumor immunity. *Science*, *359*(6383), 1537–1542. https://doi.org/10.1126/science.aao0505.

Galluzzi, L., Chan, T. A., Kroemer, G., Wolchok, J. D., & Lopez-Soto, A. (2018). The hallmarks of successful anticancer immunotherapy. *Science Translational Medicine*, *10*(459), 14. https://doi.org/10.1126/scitranslmed.aat7807.

Gonzalez, S., Lopez-Soto, A., Suarez-Alvarez, B., Lopez-Vazquez, A., & Lopez-Larrea, C. (2008). NKG2D ligands: Key targets of the immune response. *Trends in Immunology*, *29*(8), 397–403. https://doi.org/10.1016/j.it.2008.04.007.

Guillerey, C., Huntington, N. D., & Smyth, M. J. (2016). Targeting natural killer cells in cancer immunotherapy. *Nature Immunology*, *17*(9), 1025–1036. https://doi.org/10.1038/ni.3518.

Kaiser, B. K., Yim, D., Chow, I. T., Gonzalez, S., Dai, Z., Mann, H. H., et al. (2007). Disulphide-isomerase-enabled shedding of tumour-associated NKG2D ligands. *Nature*, *447*(7143), 482–486. https://doi.org/10.1038/nature05768.

Lopez-Soto, A., Folgueras, A. R., Seto, E., & Gonzalez, S. (2009). HDAC3 represses the expression of NKG2D ligands ULBPs in epithelial tumour cells: Potential implications for the immunosurveillance of cancer. *Oncogene*, *28*(25), 2370–2382. https://doi.org/10.1038/onc.2009.117.

Lopez-Soto, A., Gonzalez, S., Smyth, M. J., & Galluzzi, L. (2017). Control of metastasis by NK cells. *Cancer Cell*, *32*(2), 135–154. https://doi.org/10.1016/j.ccell.2017.06.009.

Lopez-Soto, A., Huergo-Zapico, L., Acebes-Huerta, A., Villa-Alvarez, M., & Gonzalez, S. (2015). NKG2D signaling in cancer immunosurveillance. *International Journal of Cancer*, *136*(8), 1741–1750. https://doi.org/10.1002/ijc.28775.

Lopez-Soto, A., Huergo-Zapico, L., Galvan, J. A., Rodrigo, L., de Herreros, A. G., Astudillo, A., et al. (2013). Epithelial-mesenchymal transition induces an antitumor immune response mediated by NKG2D receptor. *Journal of Immunology*, *190*(8), 4408–4419. https://doi.org/10.4049/jimmunol.1202950.

Morvan, M. G., & Lanier, L. L. (2016). NK cells and cancer: You can teach innate cells new tricks. *Nature Reviews. Cancer*, *16*(1), 7–19. https://doi.org/10.1038/nrc.2015.5.

Voskoboinik, I., Whisstock, J. C., & Trapani, J. A. (2015). Perforin and granzymes: Function, dysfunction and human pathology. *Nature Reviews. Immunology*, *15*(6), 388–400. https://doi.org/10.1038/nri3839.

Zhang, Q., Bi, J., Zheng, X., Chen, Y., Wang, H., Wu, W., et al. (2018). Blockade of the checkpoint receptor TIGIT prevents NK cell exhaustion and elicits potent anti-tumor immunity. *Nature Immunology*, *19*(7), 723–732. https://doi.org/10.1038/s41590-018-0132-0.

CHAPTER TWENTY-SEVEN

^{51}Cr-release to monitor NK cell cytotoxicity

Leslie Elsner[a], Ralf Dressel[a,b,*]
[a]Institute of Cellular and Molecular Immunology, University Medical Center Göttingen, Göttingen, Germany
[b]DZHK (German Center for Cardiovascular Research), Partner Site Göttingen, Göttingen, Germany
*Corresponding author: e-mail address: rdresse@gwdg.de

Contents

Abstract

Natural killer (NK) cells are cytotoxic lymphocytes that belong to the innate lymphoid cells. They have the ability to attack tumor cells that lack ligands of inhibitory NK receptors and/or express ligands of activating NK receptors. The most important ligands of inhibitory NK receptors are major histocompatibility complex (MHC) class I molecules, which are down regulated on many tumor cells. Several ligands of activating NK receptors, in contrast, are up regulated on tumor cells due to cellular and genotoxic stress. Due to these features, NK cells constitute an important element of the tumor immunosurveillance and are explored as a cellular tool for tumor immunotherapy. Therefore, it is important to be able to monitor the cytotoxic activity of NK cells reliably in preclinical animal models as well as in patients. It is also of interest to determine the susceptibility of tumor cells toward NK cell-mediated cellular cytotoxicity. The ^{51}Cr-release assay monitors the lysis of target cells by NK cells and is suitable to address both questions. We describe here the experimental set-up of this classic cytotoxicity assay in detail.

Methods in Enzymology, Volume 631
ISSN 0076-6879
https://doi.org/10.1016/bs.mie.2019.05.037

1. Introduction

Natural killer (NK) cells have been discovered due to their ability to kill certain tumor cell lines without needing a prior antigen-dependent activation (Kiessling, Klein, Pross, & Wigzell, 1975; Kiessling, Klein, & Wigzell, 1975), which naïve cytotoxic T lymphocytes (CTL) require to subsequently execute their cytolytic activity. In later years, it turned out that the activity of NK cells is controlled by a set of germline-encoded inhibitory and activating NK receptors that interact with ligands expressed on target cells (Caligiuri, 2008). Moreover, it was discovered that NK cells do not only kill tumor cell lines but also cells infected by certain viruses (Bastidas-Legarda & Khakoo, 2018). Today, NK cells are considered as part of the group of innate lymphoid cells (ILC) together with the more recently discovered ILC1, ILC2, and ILC3 populations that have T helper cell-like functions in the innate branch of lymphocytes (Freud, Mundy-Bosse, Yu, & Caligiuri, 2017; Vivier et al., 2018).

A frequent feature of cells that are attacked by NK cells is a missing or low expression of major histocompatibility complex (MHC) class I molecules. Therefore, the missing-self hypothesis was developed to explain the regulation of NK cell killing (Ljunggren & Kärre, 1990). It explained that NK cells are normally inhibited by self-MHC class I molecules, which function as ligands for inhibitory NK receptors (Kärre, 2008). Tumor cells that loose MHC class I molecules, thereby escaping tumor antigen-specific CTL, can become targets of NK cells. Similarly, virus-infected cells, in which MHC class I molecules are down regulated by viral mechanisms, can become targets of NK cells. Later activating NK receptors were discovered that bind to ligands, which are absent on normal cells, but become induced uponcellular or genotoxic stress (Gasser & Raulet, 2006). Such ligands were designated as "stress-induced self" antigens (Medzhitov & Janeway, 2002). The best example of such an activating NK receptor is the natural killer group 2D (NKG2D) receptor, which binds to MHC class I polypeptide-related sequence A (MICA) and B (MICB) as well as UL16 binding protein (ULBP) 1–6 as stress-inducible ligands (Carapito & Bahram, 2015). Thus, the balance between signals from inhibitory and activating NK receptors determines the cytotoxic activity of NK cells (Vivier, Ugolini, Blaise,Chabannon, & Brossay, 2012). NK cells are also mediators of the antibody-dependent cell-mediated cytotoxicity (ADCC). They can kill target cells, which are marked by antibodies, via engagement of the Fc gamma receptor CD16. ADCC is an important mechanism for the therapeutic effect of several antibodies used in cancer therapy such as rituximab (Okroj, Osterborg, & Blom, 2013).

The fact that freshly isolated or resting NK cells kill certain tumor cell lines without any prior stimulation does not mean that the cytotoxic activity of resting NK cells cannot be further stimulated (Dressel, 2017). Numerous protocols have been described to stimulate NK cells *in vitro* (Granzin et al., 2017). Many of them involve the culture of NK cells in presence of cytokines including interleukin (IL)-2, IL-12, IL-15, and IL-18. Thereby, NK cells can become able to attack target cells which are resistant to unstimulated NK cells (Gröschel et al., 2017). Notably, it is currently explored in clinical studies whether activated NK cells can be used for immunotherapy of cancer (Hofer & Koehl, 2017; Suen, Lee, Leung, Pan, & Li, 2018).

NK cells kill target cells in a cell contact-dependent manner. They use the granule-exocytosis pathway that involves the secretion of perforin or granzymes as well as death receptors such as the FasL-Fas pathway to induce cell death (Martinez-Lostao, Anel, & Pardo, 2015). The NK cell-mediated cytotoxicity can be measured at the site of the NK cell or at the site of the target cell. During granule exocytosis, LAMP-1 (CD107a) is transported to the plasma membrane of NK cells. Therefore, staining of the degranulation marker CD107a at the cell surface of NK cells by flow cytometry can be used as a parameter to determine the activity of NK cells (Schönberg, Hejazi, & Uhrberg, 2012). Labeling of target cells with the radioactive isotope Chromium-51 (^{51}Cr) before co-culturing them with NK cells and measuring the NK cell-mediated release of ^{51}Cr is a classical assay to determine NK cell cytotoxicity at the site of the target cell (Pross & Jondal, 1975). The ^{51}Cr-release assay has been used earlier to determine cytotoxic effects of other lymphocytes. Initially, it has been developed to measure cytotoxic effects of spleen cells, which were presumably alloreactive CTL (Brunner, Mauel, Cerottini, & Chapuis, 1968). Notably, the assay is useful to measure both the activity of NK cells from different sources, e.g., different patients, against a standard cell line and the susceptibility of different target cell lines, such as different tumor cell lines, toward normal NK cells derived, e.g., from blood donors. K-562 is a cell line that is highly susceptible to killing by human NK cells that was obtained from a patient with chronic myelogenous leukemia in terminal blast crises and is of the erythroleukemia type (Lozzio & Lozzio, 1975). YAC-1 is a mouse T cell lymphoma cell line that is readily killed by mouse and rat NK cells (Kiessling, Klein, & Wigzell, 1975). Therefore, these cell lines are used frequently as positive controls or standard target cells in experiments assaying NK cell cytotoxicity.

^{51}Cr is a radioactive isotope of chromium having a half-life of 27.7 days and decaying with emission of gamma rays. Radioactive sodium chromate ($Na_2^{51}CrO_4$) is taken-up by eukaryotic cells by a still unresolved mechanism and the chromium is bound in an unspecific manner to cellular proteins so that it is spontaneously released only at a low rate. However, it is readily released to the cell culture medium, if NK cells kill labeled target cells. More target cells are expected to be killed and more ^{51}Cr is expected to be released to the cell culture supernatant, if the ratio of NK cells to NK cell-sensitive target cells becomes higher. Therefore, the cytotoxicity of NK cells is usually determined in ^{51}Cr release assays at several effector to target (E:T) ratios.

Here we provide a detailed description of the ^{51}Cr-release assay and its application to monitor NK cell cytotoxicity as it is used in our laboratory. Notably, the assay is also useful to measure the activity of other cytotoxic cells such as CTL or γδT cells (Dressel, Elsner, Novota, Kanwar, & Fischer von Mollard, 2010; Dressel, Lübbers, Walter, Herr, & Günther, 1999; Steinle, Groh, & Spies, 1998).

2. Cell culture of target cells

1. Human K-562 and mouse YAC-1 (Kiessling, Klein, & Wigzell, 1975) target cells grow readily in suspension culture in $NaHCO_3$-buffered Dulbecco's modified Eagle medium (DMEM) supplemented with 10% fetal calf serum (FCS), 2 mM L-glutamine, 1 mM sodium pyruvate, 50 μM 2-mercaptoethanol, 100 U/mL penicillin, and 100 μg/mL streptomycin (Gröschel et al., 2017; Isernhagen et al., 2015) (see Notes 1–3).
2. As non-adherent cells, both K-562 and YAC-1 cells can be easily maintained by splitting cultures and replacement of fresh medium. Start cultures should have about 2×10^5 cells/mL and 80% of the cell culture volume can be discarded and replaced by fresh medium every second or third day to keep the cells in the exponential growth phase. For the ^{51}Cr-release assay, it is essential that vitality of targets cells is not impaired before the assay by suboptimal culture conditions. Therefore, it is recommended to split the K-562 or YAC-1 target cell cultures at the day before the ^{51}Cr-release assay is performed (see Note 4).
3. If other cells than K-562 or YAC-1 are intended to be used as target cells in ^{51}Cr-release assays, very different culture conditions might be required. It is essential to establish optimal culture conditions for each target cell type to minimize the proportion of spontaneously dying cells because this will critically affect the extent of spontaneous release of ^{51}Cr in the cytotoxicity assay.

3. Isolation of human NK cells

1. Human NK cells can be readily isolated from peripheral blood, but NK cells from other sources, e.g., tumor tissue or spleen might also be used to address specific questions in particular experiments. However, procedures to isolate NK cells need to be adapted to such alternative sources.
2. If NK cells are obtained from blood, usually peripheral mononuclear cells (PBMC) are isolated in the first step by Ficoll-Hypaque gradient centrifugation or a similar method (see Note 5).
3. Usually, 5–15% of PBMC are NK cells. Most commonly, magnetic cell sorting techniques are used to purify NK cells (see Note 6). These techniques are based on positive selection of cells that express NK cell markers such as CD56 or NKp46 or on the depletion of all non-NK cells. Since procedures for positive selection involve binding of antibodies to NK receptors, they can result in the activation of NK cells and the separated cells are conjugated with magnetic beads. Therefore, negative selection protocols that lead to the purification of so-called untouched NK cells might be preferred in experiments that aim to investigate resting NK cells or NK cells that are supposed to be activated *in vitro* by defined protocols. Positive selection procedures, on the other hand, are usually less expensive and are sometimes faster.
4. The purity of the NK cells should be determined after the isolation procedure by flow cytometry. Commonly used markers for human NK cells are CD56 and NKp46 (CD335). Moreover, NK cells do not express the T cell marker CD3. With a minimal set of markers, the proportion of human NK cells can be determined as $CD3^{-}CD56^{+}$ or $CD3^{-}NKp46^{+}$ cells in flow cytometry. Further NK cell markers, such as CD16, CD94, NKG2C, NKG2D, and killer-cell immunoglobulin-like receptors (KIR), can be included in the staining protocol to gain more information on the distribution of NK cell subtypes in the isolated NK cell population.

4. Isolation of murine NK cells

1. Murine NK cells are frequently isolated from spleen, but blood and other tissues are also possible sources.
2. If the spleen is used as source of NK cells, it needs to be removed under sterile conditions and placed into cell culture medium. In a Petri dish, the

organ is cut with sterile forceps and scissors into several pieces. Single cell suspensions can be obtained by gently pressing the organ pieces with the plunger of a syringe through the mesh of a cell strainer or by using a Tenbroeck homogenizer designed to isolate lymphocytes.

3. Next, red blood cells (RBC) are removed by incubation for 5 min in a RBC lysis buffer consisting of 155 mM NH_4Cl, 10 mM $KHCO_3$, and 0.1 mM EDTA (pH 7.4–7.8). Afterward, the cells are washed in cell culture medium or phosphate-buffed saline (PBS) and spun down at 300 g for 10 min before being resolved in medium or PBS as required by the next step.
4. Murine NK cells can be readily purified from splenocytes by magnetic cell separation methods (see Note 7).
5. The proportion of mouse NK cells can be determined after separation by flow cytometry as $CD3^-NKp46^+$ or $CD3^-CD49b^+$ cells (see Note 8). In C57BL/6 mice, the NK cell marker NK1.1 can be used alternatively. However, this marker is not present in most other mouse strains. For detection of rat NK cells, the markers NKR-P1 and NKp46 are commonly used. Depending on the research question, further NK cell markers, such as CD27, CD94, NKG2C, NKG2D, and Ly49, can be included in the staining protocol for a more detailed characterization of the isolated NK cell population.

5. Activation of NK cells

1. While the classical NK cell targets such as K-562 and YAC-1 are readily killed by freshly isolated resting NK cells, other target cells are susceptible to activated NK cells only (Gröschel et al., 2017; Kruse et al., 2015). NK cells can be activated *in vitro* by cytokines, exposure to ligands of activating NK receptors, NK target cells, or combinations of such stimuli (Granzin et al., 2017). In addition, numerous protocols exist to not only activate but also expand NK cells *in vitro* (Koehl et al., 2016).
2. A classical and simple protocol for the activation of NK cells is the stimulation with IL-2 (see Note 9). For this purpose, $2–5 \times 10^6$ purified NK cells/mL are cultured in complete medium supplemented with IL-2 (100 U/mL) for 3–5 days.
3. In other protocols, NK cells are exposed to a combination of cytokines, e.g., IL-12 (5 ng/mL), IL-15 (1 ng/mL), and IL-18 (50 ng/mL) for 3–5 days. A combination of cytokines that activates NK cells within 16 h is IL-2

(100 U/mL) plus IL-12 (5 ng/mL) and IL-18 (50 ng/mL). However, the ability of NK cells to kill specific target cells might vary depending on the protocol that has been used for stimulation (see Note 10).

6. ^{51}Cr-release assay

6.1 Preparation of target cells

1. Harvest and count the target cells (see Notes 11 and 12) and transfer 0.75×10^6 cells (a cell number that is sufficient for one 96-well plate) to a 15-mL tube.
2. Spin the target cells down, discard the supernatant, and resuspend the cells in 100 μL FCS.
3. Move the target cells to a laboratory that is equipped and (if required) licensed for working with radioactive substances (see Note 13).
4. Add 1.85 MBq $Na_2^{51}CrO_4$ to the target cell suspension and incubate them for 60–90 min at 37 °C.
5. Wash the target cells three times: add 10 mL PBS or cell culture medium, centrifuge the tube at 300 g for 10 min, discard the supernatant and repeat this step two times. Suspend the cells finally in 15 mL cell culture medium with 10% FCS so that they have a concentration of 5×10^4 cells/mL (see Note 14).

6.2 Preparation of killer cells

1. Harvest the NK cells, transfer them to a 15-mL tube, spin them down (300 g, 10 min), discard the supernatant, and resuspend the cells in 5 mL complete medium before counting them.
2. Centrifuge the NK cells again (300 g, 10 min), discard the supernatant, and resuspend them in complete medium with 10% FCS to achieve the desired cell concentration (see Note 14). The cell concentration depends on the E:T ratios that are planned for the respective experiment (see Notes 15 and 16).
3. Add 100 μL medium with 10% FCS to each well of a round-bottomed 96-well plate (see Note 14).
4. Add 100 μL of the NK cell suspension to the wells of row A of the plate. The experimenter can choose the number of columns that are used for one experimental condition. Typically, the experiments are performed at least in technical triplicates so that the NK cells are added to the first three columns (1–3) of the plate for one NK cell-target cell combination.

Depending on the research question, another target cell line could be exposed to the same NK cell population in the next three columns (4–6) of the plate or another NK cell population could be used as effector cells for the same target. Up to four NK cell-target cell combinations can be investigated in triplicates on one 96-well plate at seven different E:T ratios (Fig. 1).

5. Next, a serial dilution of the NK cells is performed to allow for the analysis of target cell lysis at different E:T ratios. For this purpose, the cells in the wells of row A are carefully resuspended with a multichannel micropipette and 100 μL of the cell suspension are transferred to the wells in row B, in which 100 μL culture medium are already present (see step 3). The cells are resuspended in the wells of row B before 100 μL of the suspensions are transferred to the wells of row C. This step is repeated for the following rows up to row G. Importantly, the 100 μL cell suspension from the wells of row G are discarded and not transferred to the wells of row H allowing for the determination of spontaneous ^{51}Cr-release from target cells that will be later added to the wells of row H of the plate (see below).

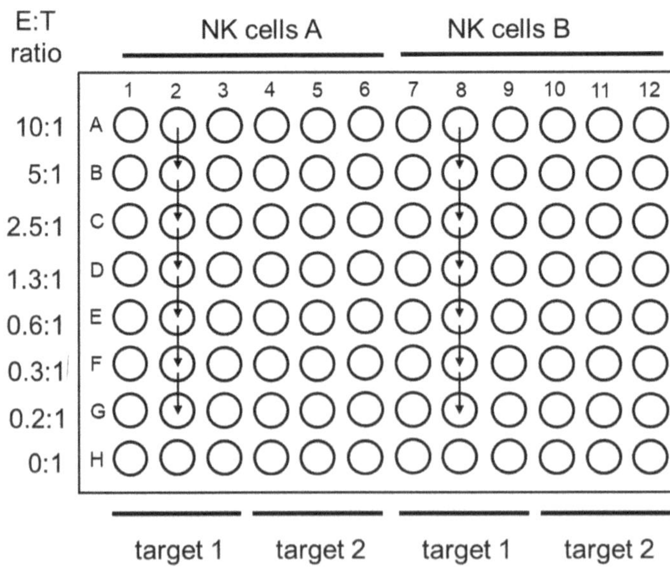

Fig. 1 Schematic representation of a 96-well plate with a set-up of a ^{51}Cr-release assay testing two NK cell populations (NK cells A and NK cells B) against two target cell lines (target 1 and target 2) in triplicates at seven E:T ratios. The set-up of the serial NK cell dilution leading to E:T ratios ranging from 10:1 to 0.2:1 is indicated representatively by arrows in columns 2 and 8. To the wells in row H, no NK cells are added allowing for the determination of the spontaneous ^{51}Cr-release of the target cell lines.

6. In some experimental set-ups, additional conditions need to be tested, e.g., antibodies or inhibitors that might interfere with the NK cell-mediated cytotoxicity. In such experiments, 50 μL medium can be given initially to each well of the plate (see step 3) and the NK cells are added in a volume of 50 μL. The serial dilution is then performed by transferring 50 μL cell suspension to the wells of next rows of the plate (see step 5). This allows now adding any required reagent dissolved in 50 μL medium to the wells.
7. Move the plate with the serial dilution of NK cells to a laboratory that is equipped and (if required) licensed for working with radioactive substances (see Note 13).

6.3 Set-up of the ^{51}Cr-release assay

1. Transfer the suspension of target cells to a Petri dish (see Note 17) and add 100 μL of the cell suspension (i.e., 5×10^3 cells) to those wells of the 96-well plate containing the serial dilution of NK cells (see above), in which the killing of the respective target cells is planned to be assessed. This can be done best with a multi-channel micropipette. At the end of this step, each well of the 96-well plate contains 200 μL cell culture medium.
2. Close the plate with its lid and centrifuge it for 3 min at 50 g in a centrifuge equipped with a rotor for 96-well plates (see Note 18).
3. Incubate the plate for 4 h at 37 °C in an incubator with 5%-CO_2 atmosphere (see Notes 19 and 20).

6.4 Harvesting and measuring of the ^{51}Cr-release assay

1. At the end of the incubation time, centrifuge the plate for 3 min at 50 g (see Note 21).
2. Transfer carefully 50 μL of the supernatant in each well to a new 96-well plate that is suitable for counting radioactivity (see Note 22). This can be done most easily with a 8- or 12-channel micropipette. Make sure that the cell pellet at the bottom of the plates is not affected at this step by introducing the pipette tips in a 45° angel into the wells.
3. Add 5 μL of a 10%-Triton X100 solution to the wells of the 96-well plate, in which the assay had been set up, to lyse the remaining cells (see Note 23).
4. Resuspend and lyse the cell pellets in the wells carefully by passing the medium several times through the tips of a micropipette. This can be

done most easily with a 8- or 12-channel micropipette. Transfer 50 μL of the suspension of the sediment to another 96-well plate that is suitable for counting radioactivity.

5. Add to each well of the plates containing the supernatant and the sediment 200 μL scintillation fluid (see Note 24), close the plates with a foil that is resistant to the scintillation fluid, and shake the plates at least for 1 min on an automatic shaker to ensure that the probes are mixed well with the scintillation fluid.
6. Measure the radioactivity in the wells of the plates in a radioactivity counter (see Note 25) and obtain the results as counts per minute (cpm) for each well.

7. Data analysis

1. Calculate the proportion of ^{51}Cr-release (R) from the target cells in each well using the following formula (see Note 26):

$$R\,[\%] = \frac{\text{cpm supernatant} \times 4 \times 100}{((\text{cpm sediment} \times 3) + \text{cpm supernatant})}$$

2. Calculate the mean of the ^{51}Cr-release (R) in the wells that contained target cells but no NK cells, i.e., the mean of the triplets in row G in the described experiment. This is the spontaneous ^{51}Cr-release (SR) of target cells. It is not caused by NK cells, but it can be due to the release of ^{51}Cr from intact cells and the release of ^{51}Cr from cells that spontaneously died during the time of the assay. In a typically 4-h-^{51}Cr-release assay using K-562 or YAC-1 target cells, the SR is around 10%. It might be higher for other target cell lines. However, if the SR is higher than about 20%, it could indicate that the culture conditions of the target cells are suboptimal and the results of the assay should be interpreted with caution.
3. Calculate the specific lysis of target cells in each well, i.e., the proportion of target cells that were killed by NK cells during the incubation time using the following formula:

$$\text{specific lysis}\,[\%] = R\,[\%] - \text{mean of SR}\,[\%]$$

4. Calculate the mean of technical replicates and quantify the amount of variation, e.g., by calculation of the standard deviation.
5. Display the results in a graph showing the specific lysis in relation to the E:T ratios (Fig. 2).

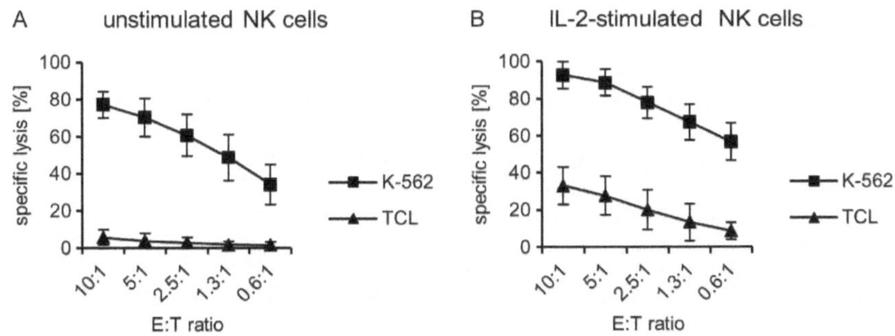

Fig. 2 (A) Means and standard deviations (SD) of the specific lysis of K-562 cells and a tumor cell line (TCL) by freshly purified NK cells is shown at different E:T ratios as determined in triplicates by a ^{51}Cr-release assay. (B) The specific lysis of these target cells by IL-2-activated NK cells (4 days of culture with 100 U/mL IL-2) has been analyzed in parallel. K-562 cells are highly susceptible to resting and IL-2-activated NK cells. Nonetheless, K-562 cells are slightly more efficiently killed by IL-2-activated than resting NK cells. The tumor cell line, in contrast, is resistant to resting NK cells and moderately killed by IL-2-activated NK cells only.

8. Concluding remarks

This chapter provided a detailed description of an application of the classic ^{51}Cr-release assay to monitor NK cell cytotoxicity. Although this method requires working with a radioactive isotope, which is an obvious disadvantage, it is still widely used by researchers in the field of cellular cytotoxicity who have access to a laboratory equipped for radioactive work. The assay has several advantages: It offers the possibility to determine both, the cytotoxic activity of NK cells and the susceptibility of target cells toward NK cell-mediated cytotoxicity. If in addition to the assay, also the counting of radioactivity is done in a 96-well plate format, it can be easily scaled up. Between four and eight plates can be handled in parallel by experienced experimenters and (depending on the instrument) counted automatically overnight. Since the quite robust ^{51}Cr-release assay has been widely used by many researcher, the principle results obtained can be compared to many published data sets, which might help to interpret new results. Nonetheless, the direct quantitative comparison of data obtained in different laboratories is usually not possible due to many parameters, which are specific to individual laboratories.

9. Notes

1. K-562 and YAC-1 cells can be purchased from American Type Culture Collection (ATCC, Manassas, VA, USA) (see: http://www.lgcstandards-atcc.org/products/all/CCL-243.aspx?geo_country=de#

generalinformation and https://www.lgcstandards-atcc.org/products/all/TIB-160.aspx?geo_country=de#generalinformation). ATCC recommends to culture K-562 cells in Iscove's Modified Dulbecco's Medium (IMDM) and YAC-1 cells in Roswell Park Memorial Institute (RPMI)-1640 medium. However, both cell lines can be easily adapted to other culture media such as DMEM.

2. K-562 and YAC-1 cells can be routinely maintained in 10-cm Ø Petri dishes for bacterial cultures or other dishes or flasks for non-adherent cells.
3. All cell lines should routinely be tested to exclude mycoplasma contamination.
4. The viability of target cell lines should be tested before being used in ^{51}Cr-release assays, if it is doubtful.
5. Heparinized blood should be diluted 1:1 with PBS before density gradient centrifugation.
6. Several companies offer kits with detailed instructions for separation of human NK cells by magnetic cell sorting techniques.
7. Several companies offer kits with detailed instructions for separation of mouse and rat NK cells by magnetic cell sorting techniques.
8. CD49b is a NK cell marker that is present in all mouse strains that have been tested. The marker is also known as DX5 since it is recognized by the DX5 monoclonal antibody.
9. PBMC can also be cultured for 3–5 days in the presence of IL-2 (100 U/mL) to increase the cytotoxic activity of NK cells present among the PBMCs. The resulting cells are designated as lymphokine-activated killer (LAK) cells. They can be generated easily and might be useful in experiments, in which it is not important to work with a well-characterized effector cell population.
10. Depending on the purpose and design of the experiment, it might be necessary to establish specific protocols for the activation of NK cells.
11. If target cells are used that are adherent to cell culture dishes, appropriate conditions for detachment need to be chosen. Many cell lines can be detached by an incubation in 0.25% trypsin/2 mM EDTA. In this case, plates are washed with PBS before the pre-warmed trypsin/EDTA solution is added (2 mL for a 10-cm Ø Petri dish). The detachment process should be inspected by light microscopy and kept to the minimal time. After detachment, target cells should be harvested in cell culture medium with FCS to inactivate the trypsin. Cell clumps may be dissolved by passing the target cell suspension several times

through a 5 or 10-mL pipette. However, they should be treated as gentle as possible to avoid later an increased spontaneous release of ^{51}Cr. Moreover, the target cells should be in the exponential growth phase for best labeling results.

12. Reducing the time of exposure of target cells to trypsin to the minimum reduces the proteolysis of cell surface molecules, such as adhesion molecules or ligands of NK receptors that are required for the killing of target cells by NK cells.

13. The regulations for working with radioactive substances vary in different countries. Make sure that all applicable legal requirements are fulfilled before starting the work. The required amount of radioactive sodium chromate needs to be ordered timely, since ^{51}Cr has a rather short half-live time of only 27.7 days. Alternatively, cytotoxicity assays, which do not require radioactive substances, e.g., europium-based assays, can be used to determine the lysis of target cells by NK cells (Blomberg et al., 1996), if no equipment to work with radioactive substances is available.

14. Commonly, NK cells are cultured in RPMI-1640 medium. However, depending on the requirements of target cells other media, such as DMEM, can be used alternatively. In case that no incubator with CO_2-supply is available in the laboratory used for work with radioactive substances, 4-(2-hydroxyethyl)-1-piperazineethanesulfonic acid (HEPES)-buffered medium supplemented with 10% FCS can be used alternatively. Some researchers prefer using V-bottom instead of round-bottom 96-well plates because this shape of the wells force the effector and target cells into even closer contact. However, for the selection of the plates also the requirements of the target cells need to be taken into account.

15. If the experiment employs highly purified but resting NK cells as effector cells and the typical target cell lines (K-562 or YAC-1) as target cells, E:T ratios starting at 10:1 are usually suitable requiring at this point an NK cell concentration of 1×10^6/mL. An E:T ratio of 5:1 requiring a cell concentration of 5×10^5/mL could be sufficient, if pre-activated NK cells are used. LAK cells, which might contain many non-NK cells, could require much higher E:T ratios. e.g., 100:1. Then the cell concentration needs to be set to 1×10^7/mL at this step.

16. For many experimental settings, appropriate E:T ratios need to be individually determined.

17. Use Petri dishes for bacterial cultures at this step to avoid a rapid adherence of target cells.
18. This step brings the NK and target cells rapidly into close contact so that the NK cells can execute their contact-dependent cellular cytotoxicity without delay.
19. If HEPES-buffered DMEM has been used, an incubator without 5%-CO_2 atmosphere can be used.
20. The standard incubation time of 4 h might be extended up to 24 h, if the cytotoxic activity of NK cells against a particular target cell population is very low. However, this will likely lead to an increased spontaneous release of ^{51}Cr. The optimal incubation time depends on the experimental conditions and varies in different experimental settings.
21. This centrifugation step is employed to ensure that the radioactivity that is still present within intact target cells is collected in the cell pellet at the bottom of the wells. It might be dispensable, if the 96-well plate is handled with care avoiding any disturbance of the cell pellet formed during the 4 h incubation period.
22. We are using 96-well plates for a MicroBeta^2 counter from PerkinElmer Life Sciences. If other counters are used to measure radioactivity, plates or tubes that fit to this equipment are needed instead at this step.
23. Lysing the target cells at this step is crucial, if strongly adherent target cells are used that cannot be resuspended easily by pipetting.
24. This step is required, if a liquid scintillation counter is used to measure the radioactivity. Alternatively, ^{51}Cr can also be counted directly in a gamma counter, which does not require a scintillation cocktail.
25. We use a MicroBeta^2 counter from PerkinElmer Life Sciences running a program for detecting ^{51}Cr in a 96-well-plate format. Each well is measured for 1 min and the counts per minute (cpm) are recorded. The output is most convenient in the format of an Excel sheet.
26. The formula takes into consideration the proportions of volume that are taken for measurements in the described experiment, i.e., 50 μL of 200 μL supernatant and 50 μL of 150 μL sediment. The denominator ((cpm sediment × 3) + cpm supernatant) represents the total amount of ^{51}Cr that has been introduced to a well by adding the ^{51}Cr-labeled target cells. The formula needs to be adapted, if other proportions of the probes are taken for counting the radioactivity. We prefer to determine the total amount of ^{51}Cr for every well of the plate to adjust for the variability that can occur during manual pipetting. However, it is

alternatively appropriate and even more commonly done to measure the maximum release of ^{51}Cr only in some representative samples of labeled target cells. This procedure has the advantage to reduce the amount of required and then radioactively contaminated materials. The specific ^{51}Cr-release (SR) from the target cells is then calculated using the following formula:

$$\mathrm{SR}\,[\%] = \frac{(\mathrm{cpm\,sample\,release} - \mathrm{cpm\,spontaneous\,release}) \times 100}{(\mathrm{cpm\,maximum\,release} - \mathrm{cpm\,spontaneous\,release})}$$

Acknowledgments

Our work on NK cells has been supported by grants from the Deutsche Forschungsgemeinschaft (GRK 1034; SFB 1002 TP C05), the European Union (MRTN-CT-2004-512253; FP7-PEOPLE-2012-ITN-315963), and the DZHK (81X2300139).

References

Bastidas-Legarda, L. Y., & Khakoo, S. I. (2018). Conserved and variable natural killer cell receptors: Diverse approaches to viral infections. *Immunology, 156*(4), 319–328.

Blomberg, K., Hautala, R., Lovgren, J., Mukkala, V. M., Lindqvist, C., & Akerman, K. (1996). Time-resolved fluorometric assay for natural killer activity using target cells labelled with a fluorescence enhancing ligand. *Journal of Immunological Methods, 193*, 199–206.

Brunner, K. T., Mauel, J., Cerottini, J. C., & Chapuis, B. (1968). Quantitative assay of the lytic action of immune lymphoid cells on 51-Cr-labelled allogeneic target cells in vitro; inhibition by isoantibody and by drugs. *Immunology, 14*, 181–196.

Caligiuri, M. A. (2008). Human natural killer cells. *Blood, 112*, 461–469.

Carapito, R., & Bahram, S. (2015). Genetics, genomics, and evolutionary biology of NKG2D ligands. *Immunological Reviews, 267*, 88–116.

Dressel, R. (2017). Collaboration of heat shock protein 70 and stress-induced NKG2D ligands in the activation of NK cells against tumors. *Current Immunology Reviews, 13*, 56–63.

Dressel, R., Elsner, L., Novota, P., Kanwar, N., & Fischer von Mollard, G. (2010). The exocytosis of lytic granules is impaired in Vti1b- or Vamp8-deficient CTL leading to a reduced cytotoxic activity following antigen-specific activation. *Journal of Immunology, 185*, 1005–1014.

Dressel, R., Lübbers, M., Walter, L., Herr, W., & Günther, E. (1999). Enhanced susceptibility to cytotoxic T lymphocytes without increase of MHC class I antigen expression after conditional overexpression of heat shock protein 70 in target cells. *European Journal of Immunology, 29*, 3925–3935.

Freud, A. G., Mundy-Bosse, B. L., Yu, J., & Caligiuri, M. A. (2017). The broad spectrum of human natural killer cell diversity. *Immunity, 47*, 820–833.

Gasser, S., & Raulet, D. H. (2006). The DNA damage response arouses the immune system. *Cancer Research, 66*, 3959–3962.

Granzin, M., Wagner, J., Kohl, U., Cerwenka, A., Huppert, V., & Ullrich, E. (2017). Shaping of natural killer cell antitumor activity by ex vivo cultivation. *Frontiers in Immunology, 8*, 458.

Gröschel, C., Hübscher, D., Nolte, J., Monecke, S., Sasse, A., Elsner, L., et al. (2017). Efficient killing of murine pluripotent stem cells by natural killer (NK) cells requires activation by cytokines and partly depends on the activating NK receptor NKG2D. *Frontiers in Immunology*, *8*, 870.

Hofer, E., & Koehl, U. (2017). Natural killer cell-based cancer immunotherapies: From immune evasion to promising targeted cellular therapies. *Frontiers in Immunology*, *8*, 745.

Isernhagen, A., Malzahn, D., Viktorova, E., Elsner, L., Monecke, S., von Bonin, F., et al. (2015). The MICA-129 dimorphism affects NKG2D signaling and outcome of hematopoietic stem cell transplantation. *EMBO Molecular Medicine*, 7, 1480–1502.

Kärre, K. (2008). Natural killer cell recognition of missing self. *Nature Immunology*, *9*, 477–480.

Kiessling, R., Klein, E., Pross, H., & Wigzell, H. (1975). "Natural" killer cells in the mouse. II. Cytotoxic cells with specificity for mouse Moloney leukemia cells. Characteristics of the killer cell. *European Journal of Immunology*, *5*, 117–121.

Kiessling, R., Klein, E., & Wigzell, H. (1975). "Natural" killer cells in the mouse. I. Cytotoxic cells with specificity for mouse Moloney leukemia cells. Specificity and distribution according to genotype. *European Journal of Immunology*, *5*, 112–117.

Koehl, U., Kalberer, C., Spanholtz, J., Lee, D. A., Miller, J. S., Cooley, S., et al. (2016). Advances in clinical NK cell studies: Donor selection, manufacturing and quality control. *Oncoimmunology*, *5*, e1115178.

Kruse, V., Hamann, C., Monecke, S., Cyganek, L., Elsner, L., Hübscher, D., et al. (2015). Human induced pluripotent stem cells are targets for allogeneic and autologous natural killer (NK) cells and killing is partly mediated by the activating NK receptor DNAM-1. *PLoS One*, *10*, e0125544.

Ljunggren, H. G., & Kärre, K. (1990). In search of the 'missing self': MHC molecules and NK cell recognition. *Immunology Today*, *11*, 237–244.

Lozzio, C. B., & Lozzio, B. B. (1975). Human chronic myelogenous leukemia cell-line with positive Philadelphia chromosome. *Blood*, *45*, 321–334.

Martinez-Lostao, L., Anel, A., & Pardo, J. (2015). How do cytotoxic lymphocytes kill cancer cells? *Clinical Cancer Research*, *21*, 5047–5056.

Medzhitov, R., & Janeway, C. A., Jr. (2002). Decoding the patterns of self and nonself by the innate immune system. *Science*, *296*, 298–300.

Okroj, M., Osterborg, A., & Blom, A. M. (2013). Effector mechanisms of anti-CD20 monoclonal antibodies in B cell malignancies. *Cancer Treatment Reviews*, *39*, 632–639.

Pross, H. F., & Jondal, M. (1975). Letter: Spontaneous cytotoxic activity as a test of human lymphocyte function. *Lancet*, *1*, 335–336.

Schönberg, K., Hejazi, M., & Uhrberg, M. (2012). Protocol for the clonal analysis of NK cell effector functions by multi-parameter flow cytometry. *Methods in Molecular Biology*, *903*, 381–392.

Steinle, A., Groh, V., & Spies, T. (1998). Diversification, expression, and gamma delta T cell recognition of evolutionarily distant members of the MIC family of major histocompatibility complex class I-related molecules. *Proceedings of the National Academy of Sciences of the United States of America*, *95*, 12510–12515.

Suen, W. C., Lee, W. Y., Leung, K. T., Pan, X. H., & Li, G. (2018). Natural killer cell-based Cancer immunotherapy: A review on 10 years completed clinical trials. *Cancer Investigation*, *36*, 431–457.

Vivier, E., Artis, D., Colonna, M., Diefenbach, A., Di Santo, J. P., Eberl, G., et al. (2018). Innate lymphoid cells: 10 years on. *Cell*, *174*, 1054–1066.

Vivier, E., Ugolini, S., Blaise, D., Chabannon, C., & Brossay, L. (2012). Targeting natural killer cells and natural killer T cells in cancer. *Nature Reviews. Immunology*, *12*, 239–252.

Kolibri 360 GmbH
Ferdinand-Jühlke-Straße 7
99095 Erfurt, Deutschland
produktsicherheit@kolibri360.de